KU-315-349

THE COLOURING, BRONZING AND PATINATION OF METALS

THE COLOURING, BRONZING AND PATINATION OF METALS

A manual for the fine metalworker and sculptor

Cast bronze

Cast brass

Copper and copper-plate

Gilding metal

Sheet yellow brass

Silver and silver-plate

Richard Hughes and Michael Rowe

*crafts*council

Every effort has been made to ensure that the information in this book, which is based on the authors' experience, is accurate. However, the authors and publisher take no responsibility for any harm which may be caused by the use or misuse of any materials or processes mentioned herein; nor is any condition or warranty implied.

© 1982 Richard Hughes and Michael Rowe

ISBN 0 903798 60 3

All rights reserved.
No part of this book may be reproduced in any form without the written permission of the copyright holders.

Published by the Crafts Council
11/12 Waterloo Place
London SW1Y 4AU

Designed by Philip Miles

Printed in Great Britain
by Burgess & Son (Abingdon) Ltd
Abingdon, Oxfordshire

JACOB KRAMER COLLEGE LIBRARY

CLASS

739.
15

1964

Greenhead £19-00 1.83

Picture credits

All photographs of samples and bowls by Ian Dobbie.

Line drawings by Karen Hughes.

Other photographs: Page 10 British Museum; page 12 Victoria & Albert Museum (Crown Copyright reserved); page 14 National Tourist Board of Greece; page 16 National Gallery of Art, Washington (Samuel H. Kress Collection); page 17, left, reproduced by permission of the Trustees, the Wallace Collection, London; right, British Museum; page 18 Courtauld Institute of Art; page 19 Victoria & Albert Museum (Crown Copyright reserved); page 20 Sotheby & Co., London; page 21 the Henry Moore Foundation.

CONTENTS

PREFACE

The decorative colouring and patination of metals extends back over many hundreds of years and is found in all the great metalworking traditions. Although there has been a renewed interest in metal colouring recently, some of the most interesting techniques used in these traditions, which were predominantly concerned with the colouring of copper and its various alloys, have been largely neglected. This has been partly due to the increasing difficulty in obtaining recipes and detailed information about procedures. In this book we have tried to meet the needs of sculptors and fine metalworkers by presenting recipes and practical techniques that relate to materials that are currently available.

Recipes have been collected from as wide a range of sources as possible, and each has been tested with a selection of cast and sheet materials. Repeat tests have been carried out, both on test samples and larger scale objects, and as far as possible procedures have been refined and new techniques developed. The results of this research are presented in this book. With the exception of a few potentially very hazardous recipes which were not tested all the results have been included, from the excellent to the mediocre, so that as rich a source of information as possible is made available to the metalworker. Metal colouring is not an exact science but depends to some extent on the skill and judgement of the individual. Some of the recipes that have proved intractable to the authors will no doubt be taken up and used to excellent effect by others. In addition to presenting recipes and techniques, we have tried to include relevant information relating to safety; a subject that is almost totally neglected in the older literature. Modern terminology is adopted throughout, to avoid the confusion caused by archaic chemical names and systems of weights and measures. Glossaries dealing with the older terminology are also included for reference, together with a bibliography.

The principle application for metal colouring and patination is, as it has traditionally been, in the field of decorative fine metalwork and sculpture. We hope that this collection of recipes and techniques, and the test results illustrated in the colour plates, will give some idea of the potential of colouring, bronzing and patination for the artist craftsman.

ACKNOWLEDGEMENTS

The authors gratefully acknowledge the financial support of the Research Committee of Camberwell School of Art and Crafts, and the financial assistance of the Crafts Council, whose joint support has enabled the research on which this book is based to be carried out. We are also grateful to Camberwell School of Art for the generous provision of laboratory and workshop facilities in the Department of Silversmithing and Metalwork, the Science Division of the Department of Typographic Design and Printing, and the Department of Sculpture. We would like to extend out particular thanks to Ian Tregarthen Jenkin and Laurence Sparey of Camberwell School of Art, to the Publications and also the Grants and Loans Sections of the Crafts Council, for their support and encouragement throughout the course of the research programme and their valuable assistance in the preparation of this book.

We are indebted to Dr Fred Young of Birmingham University (formerly of the Science Division, Camberwell School of Art) for his invaluable advice and guidance in the chemistry and safety aspects of the research, and to Dr John Tampion of the Polytechnic of Central London for his assistance in clarifying a number of matters relating to the chemistry involved in the later stages of research. The valuable contributions by Glennys Wild, Keeper of the Department of Applied Art, Birmingham City Museum and Art Gallery and Mr F.L. Temple, formerly of A. Edward Jones Ltd, Silversmiths, Birmingham, in matters both technical and historical, are also most gratefully acknowledged.

The research programme has benefitted greatly from the invaluable information provided by Mr Kato, Mr Eiro Niiyama, Mr Hara Kinjiro, and the Tokyo Geijutsu Daigaku, and we are most grateful for the generous assistance of Mr Charles Rudd of Brunel University in the difficult task of translation. The translated material most generously provided by Mr David Walker of the Western Australian Institute of Technology has also proved to be most useful.

In connection with the literature search, our particular thanks are due to Anne Carpenter, formerly of the City University, Miss A.M. Allott of the Sheffield City Libraries, and to Mr Leslie Jones of the Metals Society. In addition, we would like to express our thanks to the staff of the Science Museum Library, the Science Reference Library, the Patent Office Library, the British Library, and the libraries of the Victoria and Albert Museum, Brunel University, the University of Surrey, and the Royal College of Art, for their patient help in tracing and obtaining much of the material that has been essential to us in our work. We are also most grateful for the provision of facilities by the Central Library Resources Service of the Inner London Education Authority, and in particular for the generous assistance provided by Antony Daws, Pamela Munns and Daphne Sage, in carrying out the computer search for more recent material.

We would like to express our thanks to the many individuals and organisations who have responded to our field search and provided us with much useful information and advice. In particular we would like to thank Mr Ian McIntyre and the staff of the Department of Conservation and Technical Services of the British Museum, Mr Gaskin of the Art Bronze foundry, Alan Bradley of the Morris Singer foundry, Meridian Bronze Foundry, Rob Prewitt of Camberwell School of Art foundry, Tissa Ranasinghe of the Royal College of Art foundry, Dr Cameron of the Monumental Brass Society, Mr Peter Gainsbury of the Worshipful

Company of Goldsmiths, Dr M.W. Pascoe of the Science Division of Camberwell School of Art (formerly of the British Museum, Department of Conservation and Technical Services), the Copper Development Association, and the Deutsches Kupfer Institut.

We are also grateful for the additional material provided by Alistair McCallum, Elizabeth Holder, Judy McCaig, Douglas Steakley, Bakri Yehia, Graham Arthur, Nick Stanbury, Cynthia Cousens, and Leonard Smith; and for workshop assistance from Clive Burr, Richard Fairhall, Tim Ayers, Chris Russell, Clive Hardy and Andy Mitchell.

Richard Hughes and Michael Rowe. *May 1982*

HISTORICAL INTRODUCTION

The role of colour in sculpture and the decorative arts has obviously been of prime importance from the first. Colour has been used to create lifelikeness in sculpture, it has been used symbolically in both religious and secular contexts, and more generally to unify and enhance the decorative aspects of work in all cultures. Although in one sense colour can be thought of as a secondary characteristic of sculpture and craft work, in that it is often applied in a process that is distinct from that of making the object, in many cases it is a key factor in the visual coherence and significance of objects and therefore of prime importance.[1] In addition to its role in finished work, colour has also been an important factor in the control of craft processes. It is a directly observable symptom of change and has been used as an indicator of alterations in the state of materials, and in changes of temperature, where no quantitative measure was available. The correct temperature for achieving a particular tempering of steel is still judged, in the workshop, by observing the sequence of colours produced as the metal is progressively heated. Similarly, the annealing and soldering temperatures of metals such as copper and silver are judged by observing changes in the colour of the hot object. Most of the basic processes involved in the preparation and working of metals involves the use of heat, which results in the conversion of the surface metal into oxides. Annealing operations on copper, for example, produce red, brown and black oxides which may under certain circumstances become tenaciously adherent. The surface of the metal will also be altered by the effects of the atmosphere, causing gradual changes which initially tarnish the metal with thin uneven colouring, and which may eventually cause the whole surface to become converted into a basic mineral product which is typically green in colour.[2] These coloured surface products, and the more drastic corrosion products and incrustations produced by burial, were noted by the earliest metalworkers and in many cases formed the basis of a developing chemical technology.[3] The extent to which these coloured surface products were valued in their own right is not known, but it is certainly true that in some cases metals that were most resistant to this type of attack were the most prized, and gold and to a lesser extent silver were reserved for use in the more important artefacts because of their 'nobility'. Even if the early metalworkers regarded these natural corrosion products as a mere nuisance to be combatted, in many instances the same cannot be said of the inheritors of their traditions. On a number of occasions in the history of metalwork, there has been a revival of interest in the metalwork and sculpture carried on at earlier periods, and the ancient artefacts of an earlier period, complete with the natural patination generally produced by long burial, have been assiduously collected and studied. The revivals of interest in classicism, notably in renaissance Italy and the Chin dynasty in 12th century China, carried with them a desire to imitate the natural patination associated with the prized artefacts and eventually led to a concern with artificial patination as a finishing process in its own right for contemporary work.

Early evidence of colouring techniques

Clearly the intimate relationship between the patination of this type that was practised as a craft process and the natural products produced by time and circumstance, means that it is virtually impossible to be certain that a particular ancient artefact has been deliberately patinated. There are however some surface colourings which are extremely unlikely to have been

1. A high proportion of the sculpture of the ancient world was coloured. In many cases statues were painted, including those from Greece, the middle east, far east and pre-colombian South America. Romanesque statues and most Gothic statues were also painted. Visually many of these works must have been more closely related to the brightly coloured totem figures of North America or the painted Buddha images of Nepal, than to the monochrome form in which they have come down to us.

2. The patina products tend to reach limiting values for their basicity, corresponding with the natural minerals—brochantite (basic copper sulphate, $CuSO_4.3Cu(OH)_2$), malachite (basic copper carbonate, $CuCO_3.Cu(OH)_2$) and atacamite (basic copper chloride, $CuCl_2.3Cu(OH)_2$). See Vernon and Whitby, Bib. 378, 379, 380; Gettens, Bib. 149, 151; Matsumo, Bib. 253. Plenderleith, Bib. 287, 285.

3. See brief general mention in Smith, Bib. 337; and occasional references in Smith, Bib. 342. More detailed accounts are given by Levey, Bib. 232; Lucas and Harris, Bib. 239; and Forbes, Bib. 136.

4. See Smith, Bib. 337, for a brief discussion, and Smith, Bib. 336 for a fuller treatment.

5. McKerrell and Tylecote, Bib. 259.

6. Smith, Bib. 337.

7. There are two distinct ways of selectively removing the baser constituent of an alloy from the surface. Cementation is carried out hot using chlorides, for example, and depends on solid state diffusion of the base metal to the surface of the alloy, where it is removed by chemical reaction. The other process, a parting reaction, is carried out cold using mineral acids or their equivalent, which selectively dissolve the base metal at the surface. See Smith, Bib. 337. Similar processes were used by the pre-Columbian metalworkers of South America. See Lechtman, Bib. 228.

8. See Plenderleith, Bib. 286, for a discussion of this artefact.

9. See Lechtman, Bib. 228, for minerals that might have been used prior to the availability of mineral acids.

Chinese bronze ritual vessel (*Ku*) with natural patination induced by long-term burial in the ground. Shang style, 13th–14th century BC.

10. For a general introduction to ancient Chinese bronzes, see Watson, Bib. 387. Further information on casting technique and alloys are given in Barnard, Bib. 56. See also Ferguson, Bib. 123.

11. Discussions of certain kinds of corrosion are given by Collins, Bib. 93. Plenderleith, Bib. 287, provides a more extended account which is broader and includes consideration of incrustation.

produced by natural processes and which appear to have been used on copper and bronze objects, testifying to an early interest in surface coating and patination for decorative or protective purposes. In some Anatolian metalwork dating from the third millenium BC, bronze castings were partially covered with contrasting areas of a silvery finish, which have been identified as an intermetallic arsenic compound. Similar silvery-white finishes have been noted on sheet copper artefacts from Egypt at about the same date, produced by arsenic coating.[4] Arsenic coatings are also found on most of the early bronze age axes and halberds from northern Britain in the second millenium, which are thought to have been produced either by the 'sweating' of the alloy during casting, or alternatively by dipping.[5] These finishes, produced by various methods, are identifiable as having been deliberately produced because of the very special conditions that would be required to produce them, which could not reasonably be expected to have occurred through the agency of natural processes. The majority of chemically induced patinas are not distinguishable in this way, and the evidence for patination of this kind is provided only by a restricted number of objects in which the surface colouring is partial and selective. The Uratian bull heads which are composed of two different alloys provide one such example. The horns are made of a brass-like material and are left uncoloured, while the main part consists of bronze which is coated with a black patina, which was clearly produced artificially.[6]

Another type of surface treatment for which there is early evidence is distinct from patination although related to it, and consists of the superficial enrichment of alloys by the partial removal of one of the constituents.[7] The naturally occurring gold found in the middle east contains a substantial proportion of silver, which could not be removed by smelting, giving it a rather white appearance. A spear head that was unearthed in the Ur excavations had the superficial appearance of pure gold, but was found to be actually composed of an alloy containing as little as thirty percent gold, providing evidence that the surface had been enriched by chemical treatment.[8] Surface enrichment of this kind involves the use of chemicals related to the mineral acids,[9] which were commonly used at later dates in the chemical patination of copper and its alloys. Although this in no way implies that such patination was carried out at an early date, it does give an indication that the range of chemicals involved was known, and used in a metalworking context.

Chinese metalworking and patination

Chinese metalworking in the non-ferrous metals is probably most familiar in the form of the cast bronze ritual vessels, and related objects, which were produced at various periods in history and which are particularly noted for extremely rich and colourful patination.[10] As with the artefacts from other early metalworking traditions, it is not known to what extent artificial patination was carried out at an early date. The art of bronze working in the form associated with the ritual vessels appears in China in the second millenium, surviving examples tending to come from the Shang dynasty (1523–1028 BC) and from the subsequent Chou dynasty (1028–249 BC). They are characterised by the use of linear relief which varies in its degree of emphasis from the fine linear decoration of early Shang, through the extreme high relief associated with late Shang and early Chou, to the low relief treatment used in the late Chou period. These linear relief treatments were produced by modelling or cutting directly into the surface of the mould. The intaglio lines of later bronzes were often filled with black or coloured materials, and eventually with inlays of other metals or semi-precious stones, but it is generally thought that the early bronzes relied solely on the contrast of the linear relief for decorative effect. It is thought that the particularly fine patination that is often seen on these early objects was a natural product, and that the bronzes were originally left with the natural colour of the metal. Most of the surviving early objects only survived because they had been buried, and the alkaline soils of China are known to produce only mildly corrosive rather than destructive effects, resulting in colourful patination.[11] Although this is probably true of the early bronzes, artificial patination was certainly practised at a later date.

One form of artificial patination, that was certainly used during the Chou dynasty from about the fifth century BC, was primarily associated with the production of bronze mirrors. Many of these were decorated, either wholly or partially, with an extremely resistant black patina whose precise nature is still not known. The process is thought to have involved selective etching of the surface, probably with strong acids, subsequent darkening by a process possibly involving a compound of tin, and finally coating with an impervious 'vitreous' layer of unknown composition.[12] Similar selective etching techniques, using hard lacquers which were cut through to provide partial resists, were used in the so called 'pattern-etched' surfaces found on swords and other weapons of about the same period. The etched pattern is finished so that it is co-planar with the metal surface, and provides a colour contrast.[13]

A large proportion of the bronzes produced in early China were destroyed during the 10th–13th centuries AD, a factor which led to the widespread attempts to copy ancient bronzes. A much reprinted treatise, the 'Hsüan Ho Po Ku Tu Lu' by Wang Fu, concerning ancient bronzes and containing illustrations, provided the inspiration for many of these copies. The custom of artificially patinating bronzes in imitation of the finish of the buried artefacts that did survive from the Shang and Chou periods probably first occurred at this time in the 12th-century Chin period, as part of the attempt to recreate the lost heritage of these earlier periods. The practice of copying ancient bronzes continued in subsequent periods, notably in the Ming dynasty when much high quality bronzework with fine patination was produced, and more recently in the Ching dynasty particularly in the mid-eighteenth century. Little information is available regarding the processes that were used, but they probably included controlled burial in the ground and treatment with alkaline chemicals. One recipe that has been recorded consisted of a mixture of cinnabar, verdigris, alum and sal ammoniac, which was pasted onto the surface of the object, which was subsequently heated gradually and uniformly.[14]

A number of the key developments in Chinese metalworking occur during the late Chou period.[15] At that time the bronze-worker became less concerned with scale, and devoted more attention to surface treatment. Consequently during this period, there is increasing sophistication in the preparation of moulds for casting, and in subsequent finishing techniques. Mercury was first distilled in China at this time and the process of fire-gilding, which probably originated in China, passed to Persia and then to Egypt and the Roman world.

Japanese metalworking and colouring

Arguably the finest tradition of metal colouring and patination was that of the Japanese. Their metallurgy probably originated in China and Korea, but the development and use of non-ferrous copper-based alloys, particularly during the 15th and 16th centuries, was unrivalled. From an early date they were able to produce very high grade copper sheet using the simplest types of furnace.[16] The casting technique was unusual in that the ingots were cast under water. Canvas sacks draped over wooden moulds were lowered into hot or boiling water and the molten copper poured into them.[17] The metallurgical reason for this procedure is not clear, but the process has the effect of imparting a red oxide colour to the surface of the metal which is very tough and resistant to tarnish, and it is possible that this may have provided one of the stimuli to the later development of non-ferrous metal colouring.

The early metalwork of Japan was much influenced by the arrival of Buddhism, which was heralded in 552 AD when Buddhist images were sent by the king of Korea to the Japanese emperor. This stimulated the application of metalworking skills to a wide range of religious objects including statues of the Buddha with aureoles that were fashioned from pierced and worked sheet, *sharito* or reliquaries, and temple lanterns. Bronze bells and gongs such as the *kei* or *kyo*, *egoro* or censers, and water vessels termed *kundika*, were also made. The stylistic influence of the Korean images was also augmented by that of India during the Asuka period (673–710 AD), which is thought to have come to Japan via China. Bronze statues of a very large scale were made during the 8th century using a leaded tin bronze, *karakane*, which was often cast in situ and joined by soldering with an alloy of tin and lead. Mercury gilding was used on bronze statuary of this type, a process which the Japanese probably acquired from China where mercury was well known.[18]

The increasing influence of Zen Buddhism towards the end of the 15th century had the effect of changing the status of the minor crafts to a high art form, and metalwork was held in particular esteem. The accompanying

12. The most recent and fullest account to date is provided by Chase and Franklin, Bib. 89. See also Chase, Bib. 88. Earlier discussions of the nature of the black patina include Gettens, Bib. 150; Collins, Bib. 94; Karlbeck, Bib. 204; and Yetts, Bib. 401.

13. Pattern-etched surfaces are discussed in the article by Chase and Franklin, Bib. 89.

14. This recipe is recorded in Fortnum, Bib. 138. The object is said to have been finished by washing and polishing, when cool.

15. The general background to the late Chou period is summarised in a short section in the article by Chase and Franklin, Bib. 89. Other aspects of the period mentioned, include the use of the lost-wax casting technique, the development of glass making, and the production of iron.

16. The techniques of preparing and casting copper and bronze are discussed in Gowland, Bib. 156, 157. Accounts are also given in Brinckley, Bib. 73.

17. The process is briefly described in Tylecote, Bib. 370. The process is described in more detail for the case of silver in Gowland, Bib. 156. (This article is also included in Hickman, Bib. 180). An interesting description is given in early editions of Spon, Bib. 355 (fifth series; the account is omitted from later editions).

18. Japanese methods of gilding are discussed by Moran, Bib. 268.

19. Accounts of the nature and development of swords and sword furniture are given by Robinson, Bib. 304; Compton, Bib. 95; Hawley, Bib. 176; Inami, Bib. 193; and Tanimura, Bib. 362. More specialised works include Wakayama, Bib. 386, on the detailed nomenclature, and Kuni-Ichi, Bib. 213, on the scientific study of Japanese swords.

20. Japanese alloys and their colours are described in Roberts-Austen, Bib. 301, 302; Gowland, Bib. 156; Uno, Bib. 371; Pijanowski, Bib. 281; and Pumpelly, Bib. 294.

changes in the nature of the military class and the increased importance of the Japanese sword encouraged the metal craftsman to apply his skills to the various items of sword furniture, and in particular to the *tsuba* or sword guard.[19] In the fifteenth century these were usually simple forged iron discs, which were sometimes pierced or inlaid with brass or copper to provide some decorative treatment. The years of peace that followed the feudal strife of the sixteenth century, encouraged the development of more decorative *tsuba* which culminated in the use of non-ferrous metals and alloys. A variety of alloys were developed which were intended for patination, and which when used in conjunction with gold, silver (*gin*) and copper (*agakane*), provided a range of subtle and beautiful colour contrasts.[20] The more important alloys that were used included *shakudo*, which was predominantly copper but containing up to 5 percent gold and sometimes a little silver, which produced a lustrous purple-black hue when pickled;

Tsuba, 19th century, showing contrast between gold incrustation and patinated *shakudo* ground.

Tsuba, 19th century, showing variously textured (nanako) ground which has been subsequently patinated.

shibuichi, an alloy of copper and silver which takes on various colours from olive-brown to a light grey, depending on the proportions of the constituent metals; and *kuromi-do*, an alloy of copper with a very small addition of arsenic, which takes on a dark brown to black colour after pickling. The yellow bronze known as *sentoku* was also used and acquired a chrome yellow colour after patination. A restricted range of pickling solutions were used, and the subtle colour combinations achieved by the careful selection of metals and alloys.[21]

The essentially simple pierced silhouettes that were first used on decorated *tsuba* in the fifteenth century were gradually supplemented by the use of brass inlay and a more sculptural approach to the carving of the disc, which was subsequently replaced by the style of the famous and influential Goto family of metalworkers. They favoured the use of *shakudo* which was textured with a fine *nanako* or 'fish-roe' surface, as a ground for relief sculpture in gold. A greater range of alloys, and therefore colours, was used by the Nara school in the late 17th century who were influenced by the subject matter and the style of the Kano painters. New techniques of engraving and the influence of brush stroke styles provided the basis for the Yokoya school. The general tendency was for an increasingly pictorial style coupled with incrustation of ever greater richness, which extended into the 19th century.

Although the skill of the Japanese metalworker tends to be associated with the *tsuba* and other sword furniture, the more domestic articles often made from bronze were also finely wrought and decorated.[22] In addition there are the *okimono*, which were items not intended to have a practical use, but which were rather designed and made for a particular place. Many of these are incense burners or vases, or figures of gods, dragons and other supernatural creatures.

In addition to the techniques and alloys which are specific to the Japanese tradition, the more generally used processes, such as the surface enrichment of alloys by cementation or parting action, were also employed.[23] Coinage made from gold-silver alloys, which contained enough silver to make them white in appearance, were surface-enriched to produce a pure gold finish. A paste composed of iron and copper sulphates, potassium nitrate, calcined sodium chloride, and resin was applied to the surface, and the coins then heated to redness on a grating over a charcoal fire. They were then immersed in a concentrated sodium chloride solution, washed and dried. Nevertheless, it is for the use of colouring in conjunction with the specialised alloys that the Japanese tradition is most noteworthy.

The classical world: Greece and Rome

Extensive use of bronze was made in the pre-classical and classical periods of Greece and Italy. The bronze was either cast or wrought from ingots, large statuary initially being made by hammering sheets of metal over wooden formers and then joining the sections by rivetting. Large objects were also cast in sections and then joined by soldering or welding techniques.[24] The casting of bronze was generally confined to smaller statuary however, because the material was expensive and required for functional use in weapons and for domestic purposes. The first cast bronze statuary is credited to Rhoecus and his sons Telecles and Theodorus of Samos, who are reputed to have learnt the art in Egypt. The general influence of the near and middle east on metalworking technique and style during and after the 8th century BC is clearly recognisable, and the art of lost-wax casting becomes increasingly important at this time and was used very extensively during the 7th century. Initially at least, the use of cores does not appear to have been known, and this fact combined with the general scarcity of the material tended to restrict its widespread use to statuettes of a small scale. Bronze was also used for furniture mounts and for architectural purposes, but few examples have survived. At the beginning of the classical period, in about the 6th century BC, statuettes began to be made in larger quantities and became progressively more important until the Romans finally engaged in the commercial large-scale production of these items. Statues of the gods and portraits of well known philosophers and poets were collected by wealthy connoisseurs, and as a result bronzes were made and finished with increasing care.[25] The preference for small statuary and domestic sculpture is particularly important in the development of metal colouring. Early public and architectural sculpture was generally painted, often with bright colours, which suited their symbolic and public use in almost a theatrical

21. Japanese colouring is discussed in Pijanowski, Bib. 281, 282; Uno, Bib. 371. Collected recipes and procedures for Japanese colouring are given in Niiyama, Bib. 273.

A detailed account of the metalworking techniques and related metallurgy is given in Savage and Smith, Bib. 315.

Accounts of practical methods for making and colouring laminated metals (*Mokume-Gane*) are given in Pijanowski, Bib. 281, 282.

See also Appendix 1, Japanese alloys and colouring.

22. Apart from the accounts of these in general books on Japanese craft, eg M. Feddersen, 'Japanese Decorative Art', London, 1962, which provides a good introduction, see Smith and Hawthorne, Bib. 348, and Watanabe, Bib. 388, on Japanese 'magic mirrors'. Also Nagago, Bib. 270, which gives superb illustrations of Japanese tea kettles.

23. These techniques were in general use in a number of cultures, see note 7, page 10.

24. The background to the technology of mining and metallurgy in the Greek and Roman world is described in Healey, Bib. 177; Forbes, Bib. 136, 137, also deals with the pre-classical and early classical periods. Much useful background material, both of an historical and of a technical nature, is included in Royal Academy, Bib. 310.

25. A general introduction to classical bronzes is given in Savage, Bib. 316.

Bronze charioteer with fine natural patina. C. 470 BC, found at Delphi.

26. The description occurs in book XVIII of the Iliad.

27. Scattered references occur in both the Iliad and the Odyssey.

28. This interpretation is given by Yapp, in his catalogue to the metalwork section of the Exhibition of 1878, entitled 'Metalwork. Art Industry.'

29. These comments are reported in Michel, Bib. 266. Research carried out in Berlin by the Society for the Encouragement of the Arts, in 1864, has confirmed that a better patination is obtained if the surface of the bronze is very lightly oiled and rubbed regularly. The results are summarised in Hiorns, Bib. 182; and more briefly in Michel, Bib. 266.

30. Plinius Secundus, Bib. 288. Pliny's treatment of chemical matters is described and discussed in Bailey, Bib. 54. Pliny is the major source among classical authors, although many of his passages are difficult to interpret.

31. Gilding in the Greek and Roman world is described and discussed in Oddy, Bib. 276. Pliny's description of gilding is discussed, and a new interpretation of his remarks given, in Vittori, Bib. 385.

sense. Later sculpture on a large scale was often set up in the open air and could be allowed to colour naturally by the effects of the atmosphere. The smaller-scale items, which were collected for personal or domestic use, could not be relied upon to achieve a good natural patina, and were of a more intimate nature which tended to discourage the use of painted finishes. Although patination of larger sculpture was probably carried out, the appreciation of patinated finishes and subtlety of bronze colours was almost certainly stimulated more by the increase in quantity and quality of work on a smaller scale.

Direct references to methods of colouring are generally scarce in classical sources. If this were an isolated omission, then one might justifiably surmise that artificial patination was seldom carried out. However, direct accounts of workshop practice were generally little recorded, even in fields where there was an undeniable wealth of activity and sophistication. Some accounts of metalworking practice do occur in the work of early authors, usually in the context of a historical or mythical narrative. Homer, for example, describes the making of armour and a shield for Achilles by Hephaistos,[26] which were forged from bronze and elaborately decorated with a variety of metals, enamel and other coloured effects including black patination. Further references also occur in Homer[27] to the blue colour of a metal that was used for decorative purposes on the interior walls of a palace and also for a shield. This is interpreted by Yapp[28] as a reference to copper which was heated and then immersed in sulphurous water, which does produce bluish steely colours, or black, depending on the precise conditions. Although accounts of this kind are not necessarily factual descriptions of actual artefacts or their production, they do give an indication of the decorative treatments and materials that were familiar to authors of the time and can be presumed to have been in use to some extent.

More direct accounts also occur in the work of later classical authors. Plutarch, for example, discusses natural patination and includes comments on the geographical locations and climate which give rise to particularly fine patination. Both Aristotle and Plutarch comment on the fact that improved natural patination can be obtained on bronze, if the surface of the object is rubbed with oil or other greasy materials.[29] There are also the more extensive accounts of various aspects of metalworking and finishing that are given by Pliny in his Natural History.[30] He describes the use of additions of metals to bronze, including lead which he generally appears to have regarded as a form of adulteration although it does allow the metal to flow more freely and to achieve more faithful impressions, and produces a good brown patina. He also discusses the effects of additions to the furnace charge on the colour of the bronze that is produced. In a further passage he describes the attempt by the sculptor Aristonidas to produce a blush of colour to the face of a bronze statue of Athamas, with the use of iron. It is not clear whether the iron was used as an addition to the melt, or as is more likely, iron salts were used to colour the bronze after casting. A number of other coloured bronzes are mentioned, including *hepatizon* which was liver-coloured, *aes deliacum*, a light coloured bronze which originated from the island of Delos, and a golden bronze which was thought to contain gold but was more likely to have been a form of brass. Philostratus mentions a bronze which was black in colour and called *aes nigrum*.

One of the major developments in metalworking advanced by the Romans was in the refinement of methods of joining metals by soldering, welding and brazing. They developed specialised solders by alloying and used both soldering irons and blow-pipes together with resin fluxes. Although these techniques were originally developed in the context of producing equipment for their vast armies, they were also used to produce better everyday domestic articles in both plain and decorated bronze. The knowledge of alloys acquired in this way also led to refinements in processes such as the tinning of copper vessels, which enabled them to be used for food. Pliny notes that tinned vessels gave a less disagreeable flavour, and that tinning prevented the formation of verdigris. The tinning that was carried out in the Gallic provinces was reputedly the most advanced, and skilfully tinned vessels were passed off as silver. The pewter alloy made from equal parts of lead and tin was also used in imitation of silver, and was appropriately called *argentarium*. Gilding was also practised by the Romans. It is likely that they used both gold-leaf, which is mentioned by Pliny in a notoriously obscure passage, and mercury gilding processes.[31] Mercury was certainly well known to the Romans who used it in the refining of gold.

The middle east and mediaeval Europe

The middle east, which had been a centre for the early development and dissemination of metalworking, again assumed great importance in the period which followed the decline of the Roman empire. In the first place it continued to provide links with both the far east and the west, and craftsmen from Persia, for example, carried their skills both to China and to Byzantium. The embossed and inlaid work of the Sassanian period (227–641 AD), represented by ewers and rhytons decorated with foliage and animals, were equally admired in China and the west, and exerted a considerable stylistic influence on the work of both cultures. The other aspect of middle-eastern influence was less direct but in the long term, far more profound historically. The Arabs were the heirs to the accumulated knowledge of the classical period, represented by the great library of Alexandria, and were instrumental in the development and dissemination of the theoretical and scientific work of the Greeks and the Alexandrians. Traditionally there had been little connection between the speculations of the philosopher-scientists, and the practitioners in most fields. Metalworking was essentially a pragmatic affair with little systematic underpinning, while the philosophers' speculations on the nature of substance were remote from practical application. During the period beginning in about the first century AD, attempts began to be made to systematically describe the nature and properties of materials, and to relate these to ideas about the nature of substance. Although these approaches were compromised to some extent by religious and mystical dogma, the Arabs who inherited the results exploited the methods and approaches in the development of a science which was based on observation and led to a wealth of technical knowledge. The sphere in which investigation was most obscured by mysticism and philosophical dogma, alchemy, is probably the one which is of greatest relevance to the development of aspects of metalworking such as colouring. At its best it is represented by Geber (*ca.* 850), who although working with a set of assumptions on the nature of substance derived from the Greeks, regarded alchemy essentially as a matter of experimental research and systematisation. As a result of his work he was able to describe improved methods for melting, distillation, crystallization etc; and understood the preparation, purification and use of a wide range of substances including acids, metal salts and sulphides. More commonly, however, the work carried out by alchemists was rooted in a mystical tradition extending back to the Egyptians which vitiated the careful observations that they undoubtedly made.[32] In addition their work tended to become increasingly shrouded in secrecy and was presented in an allegorical form which made it almost totally impenetrable.

Islamic science and Arabic translations of Greek and Roman works reached the west via Moorish Spain, where they were often translated into Latin by Jewish scholars, providing an important stimulus to science and philosophy in the late middle ages. This also included much information of a technical nature in a number of fields, including metallurgy and metalworking, and chemistry.

The alchemical tradition, which was taken up in the west, continued through the medieval period until the 17th century, and although it did provide useful methods and observations relating to the manipulation of metals, these tended to be unsystematic and isolated, and buried in an occult literature. The alternative more direct technical accounts are few and far between, represented by early manuscripts such as the Leyden Papyrus (3rd century AD),[33] and later works such as the Lucca manuscript (early 9th century)[34] and the Mappae Clavicula (9th–12th century).[35] These include recipes for making pigments and dyes, the preparation of complex coloured alloys that imitate gold, and the use of arsenic compounds to colour copper. The later Mappae Clavicula includes recipes for colouring silver with sulphides, the methods of gilding, and the application of niello. The problem with these manuscripts is that they tend to be incomprehensible because of the extensive alterations that have occurred when they were transcribed, and because they represent second- or third-hand accounts of earlier records, rather than accounts of contemporary practice. The first workshop manual to appear which gives a generally clear account of the procedures used was the *De Diversis Artibus* of Theophilus, in the 12th century.[36] He has been identified as Roger of Helmershausen, a German Benedictine monk who was a noted metalworker of the time. His writings bear all the hallmarks of a first-hand account, and include procedures for

32. See Lindsay, Bib. 236, which includes a large number of typical descriptive passages from alchemical works, and includes an extensive bibliography.

33. See Caley, Bib. 80.

34. See Hedfors, Bib. 178.

35. See Phillip, Bib. 280; and the annotated translation by Smith and Hawthorne, Bib. 347.

36. See Theophilus, Bib. 364, and translation with technical notes by Smith and Hawthorne, Bib. 349.

working gold, silver and the base metals, preparing pigments, gilding, stone-setting, enamelling, and give an account of methods for colouring copper. There is no other comparable book on workshop practice until the renaissance.

The Renaissance tradition

The development of metal colouring in the Renaissance is inseparable from the development of cast bronze sculpture in general, and from small bronzes and statuettes in particular.[37] It was on objects of this type and scale that great care and attention was given to achieving rich and varied qualities of surface and finish. Techniques were evolved in which layers of varnish were built up on patinated ground colours, and by skilful control of the densities and hues of each, a great variety of individual effects were produced. The development of these techniques, which were not generally suitable for outdoor use, coincided with an increasing interest in the small domestic bronze.

The 15th and 16th centuries saw the age of humanism in Italy, and the enthusiasm for all things classical that accompanied it created a new interest in Greek and Roman art and artefacts. Connoisseur-collectors were eager to acquire classical antiquities, and the market for classically inspired sculpture flourished. Classical themes such as the heroic struggle, the equestrian figure, and the standing nude figure were the typical subjects for pieces of this type. Occasionally these statues doubled as useful domestic objects and were so arranged that they could function as salt cellars, inkwells or candlesticks. More commonly, these useful domestic bronzes appeared in the guise of grotesques or fabulous beasts. Door knockers, andirons, dishes, inkwells, and both hanging lamps and table lamps, were all items that were treated in this fashion. A closely associated treatment was used for items such as boxes and containers, which were made by casting directly from nature. Items such as containers that were cast in the form of crabs, frogs and lizards were very popular at the turn of the 15th century.

Box in the form of a crab, probably cast from life. Patinated and varnished bronze. Padua, late 15th or early 16th century. Attributed to Riccio.

Both statuettes, and objects of the kind just described, were patinated and very often also varnished. The practice of patination, with or without the subsequent use of layers of varnish, seems to have been to some extent a matter of local tradition. Both Donatello and Bertoldo, for instance, preferred unvarnished brown patination, and this was typical of the Tuscan School at the time. Sculptors of the Venetian School, on the other hand, would typically employ varnish.

The choice of colours used was of course a matter for the discretion of the individual sculptor. Vasari, writing in the middle of the 16th century,[38] comments 'This bronze which is red when it is worked assumes through time by a natural change a colour that draws towards black. Some turn it black with oil, others with vinegar make it green, and others with varnish give it the colour of black, so that every one makes it come as he likes best...' Once an effective colour or tone had been developed by an individual or a studio, it would presumably be taken up and used regularly

37. A useful summary of this period is given by Savage, Bib. 316. Accounts of small Renaissance bronzes are also given in M.G.C. Dupre, *Small Renaissance bronzes,* London, 1970; J. Montagu, *Bronzes,* London, 1972; J. Pope-Hennessy, *Essays on Italian sculpture,* London, 1968; Y. Hackenbroch, *Bronzes, other metalwork and sculpture—The Irwin Untermyer Collection,* London, 1962.

38. See Vasari, Bib. 375.

as a matter of course. It is strange that Cellini,[39] in his treatise on goldsmithing and sculpture, makes no mention of any metal-colouring techniques or recipes other than those intended for gilding. Giovanni Bologna (Gianbologna) is known to have developed a distinctive translucent red colour over a yellow bronze ground, which was taken up by his studio and followers. Antonio and Francesco Susini also used a translucent red which was much admired, as did Giovanni Fonduli. The predominant range of colours used for small bronzes varied from rich dark brown to black. Sculptors such as Andrea Briosco (Riccio) and Jacopo Sansovino used this colour range to great effect, as did the other late masters of the renaissance such as Gianbologna and Baccio Bandinelli. Green patination was used generally for the associations with classical antiquities that it evoked.

Occasionally, green patination was used by sculptors who tried to conceal bad workmanship under an ostensible incrustation of age. There also appears to have been a fashion for a green varnish, designed to imitate natural patination, which was recommended and used by Leoni,[40] during the 16th century.

The general stylistic decline that took place in Italy, France and Germany during the 17th and 18th centuries does not appear to have affected standards of craftsmanship. These remained very high, and the use of bronze for statuary, coins and medals, and domestic objects, proliferated.

Medal depicting Henri III of France. Germaine Pillon, 1575. The making of medals and medallions reached a high level of sophistication during the Renaissance. Celebrated sculptors, such as Cellini and Leoni in Italy and Pillon in France, produced fine portraiture works on this scale.

39. See Cellini, Bib. 85, 86. There is very little contemporary technical material relating to this period, although scattered comments occur in a number of sources. Cellini mentions coloured metal foils, and the colouring of metal foils which were used to back gemstones is described in detail in della Porta, Bib. 291. These methods are also described in a text of the 18th century, Smith, Bib. 351.

40. In a letter to Cardinal Granvella, which accompanied candlesticks which he had coloured by this method, Leoni is keen to point out that the metal surface is perfectly finished, under the varnish. Other craftsmen may not have been so scrupulous.

Although the use of patination of certain kinds was genuinely connected with a desire to emulate antique objects, it was certainly also sometimes used to pass off contemporary work as antique. Paolo Giovio, writing in about 1550, describes such cases. (Some of these accounts by Giovio are quoted in F. Arnau, *Three thousand years of deception in art and antiques*, London, 1961.)

Pluto carrying off Proserpine. Patinated bronze. Louis-Simon Boizot, c. 1786. This small statuette is typical of the re-working of classical themes commonly found in the Renaissance and post-Renaissance traditions.

The French tradition in the 19th century

The taste for classical bronzes extended right through the 19th century, and foundries such as those of Barbedienne, Keller, Susse Frères, E. Colin, Louchet and Thiebaut Frères flourished, producing statuary in great quantity. This was made economically possible by the use of the sculpture-reducing machine, an innovation introduced by Achille Collas who was a friend and partner of Barbedienne. This machine was a type of pantograph which was used to accurately scale down and prepare patterns from the work of such renaissance masters as Riccio and Gianbologna. It is not surprising therefore that colouring enjoyed a revival of interest during this period, particularly in France, and individual foundries became known for

41. See Lacombe, Bib. 224.

42. Sanguine is the name used for a type of red clay or earth which was used for making crayons or chalks. It appears to consist largely of ferric oxide, according to the accounts given in Lacombe.

43. The colours or tints described in French manuals in the late 19th century as 'Chinese' and 'Mordoré', are reported to have been generally produced by a smoking process. The colour nomenclature is however rather inconsistent, and terms such as 'florentine' or 'berbedienne' which originally had quite precise meanings, tend later to be used for a wide range of different coloured finishes.

Statuette of Ratapoil. Patinated bronze. Honoré Daumier (1810–79).

the excellence of the patinas that they could achieve. There was keen interest in the development of new colouring techniques which was reflected in the succession of short-lived fashions for particular coloured finishes that occurred at this time. Lacombe, in a brief review of the period written as an introduction to a book of workshop receipts in 1910, summarises the changing tastes in colouring.[41] He points out that modern trends in colouring really begin in about 1828 when a preference for a florentine tint was made popular by the work of the founder Lafleur. Prior to that date the only commonly used patination was the antique green or water green patina, used to imitate natural weathering, and various attempts to copy the tones of Italian bronzes, particularly those from Florence, which were considered to be particularly difficult to reproduce because of the subtle changes in colour they exhibited after long periods of aging. Lafleur's finish involved coating the object with copper and applying a thin paste of sanguine[42] and plumbago in an alcohol or spirit-based solution, which was removed with a brush after drying. The finish was sometimes toned or darkened by brushing with a trace of dry plumbago. An alternative finish of this general type but with softer tones was introduced by Camus in 1833, and seems to have been preferred for a time. It was produced by an entirely different method which was used extensively in the 19th century, involving the exposure of the object to the effects of hot smoke in a special smoking oven.[43] This technique was sometimes used in isolation, but was also used to modify a surface which had previously been patinated by the application of pastes or solutions.

These brown finishes were succeeded by a wide range of green patinas including the colour known as 'artistic green', a pale ashen green which was often enhanced by the use of yellow pigments which were applied in a medium of oil of lavender. The widespread use of pigments which were applied to simply produced ground colours, to give a greater richness and variety of coloured effects, was probably a response to the increasing demand for novelty at a time when bronzes were fashionable items. Masselotte, who was Deniere's head craftsman, produced many finishes which were much in demand. He varied the colouring obtained, by using a variety of yellow and red pigments to highlight both green and brown ground colours. These were generally bound to the surface using oil-based varnishes, a technique which marks an important change in approach to coloured finishes. While the production of patinas by the chemical alteration of the surface of the metal continued to be practiced, the development of colouring at this time was accompanied by the increased use of superficial pigments which were bound to the surface with waxes, varnishes and lacquers. Although the sparing use of pigments to modify the colour or tone of a chemically induced patina is capable of producing subtle and lasting results, the use of pigments was developed to a point where it had little connection with true patination and was more akin to superficial finishing by means of coloured lacquers. Bronze powders were included in the finishes to give multi-coloured reflective effects, of which a typical example of the period was the use of a greenish powdered bronze which was lightly scattered over parts of the surface to create a variegated appearance which was likened to a 'pigeon's neck'.

Other finishes that were commonly used were based on black and brown grounds, some of which were relieved with powdered tin to produce an effect that resembled armour, and which was referred to as 'iron bronze'. Black grounds were also relieved by abrasion to reveal near copper-coloured highlights, a finish which was termed 'fly's wing bronze' by manufacturers. Variations were produced by lightly toning the surfaces with sanguine to produce a rosy hue. Finishes such as these were much used on domestic articles such as embossed table lamps, which were produced commercially in large numbers in the latter part of the century, because their relieved surfaces proved suitable to objects that were likely to be handled.

The increasing use of sandcasting to produce small bronzes, the process of electrotyping, and the development of mechanical forming processes for sheet metalwork, such as spinning and presswork, resulted in the commercial production of large quantities of domestic articles and small decorative items. In most cases the use of traditional colouring techniques, which were labour-intensive and required considerable skill, were not appropriate to these newer methods of production. The possibility of colouring by immersion, where batches of objects could be dipped in baths of solutions, rather than being worked individually by hand, was investigated particularly in Germany in the late 19th and early 20th century.[44] A number of processes were developed and patented, predominantly for the production of black and brown finishes, and some of these are still in commercial use.

The animaliers, Art Nouveau, and Art Deco

Lacombe, in his brief survey of 19th century colouring, ends by mentioning A. L. Barye the animal sculptor who was able to produce 'beautiful variegated natural water greens in his own workshop'. This phrase reminds us that it was normally the practice for artists to entrust the colouring of pieces to the foundry. Barye was an exceptional talent and liked to have control over all aspects of the production of his pieces. He is known to have spent much time experimenting with different colours, and was able to produce many variegated tones ranging from dark velvet greens to reddish brown colours and black. Barye was the leading exponent of the 'animalier' tradition, which flourished in the 19th century, specialising in the realistic portrayal of animals.[45] Other notable animaliers included Emmanuel Fremiet, Pierre Jules Mene and the Bonheur family—Isidore Jules, Marie Rosalie and their brother-in-law Hippolyte Peyrol. A later but more impressionistic animalier style is represented by the work of Rembrandt Bugatti. All used patinated bronze as their medium of expression, and the wide range of colouring techniques available in the French tradition are represented in their work.[46]

Goat with hare. Patinated bronze. A.L.Barye (1796–1875).

The use of bronze was in no way diminished with the introduction of the 'New Art' in the last decade of the 19th century. Indeed the Art Nouveau philosophy, which encouraged the integration of sculpture with the decorative arts, interior design and architecture, resulted in a fruitful cross-fertilisation of styles and techniques. The sculptural style was adapted to all manner of domestic objects including lamps, ashtrays, inkwells, vases, firedogs and candlesticks. Celebrated sculptors were happy to turn their attention to the design of such items. Raoul Larche, Maurice Bouval, Theodore Rivière and Louis Chalon all made use of bronze in this way, often contrasting the patinated surfaces with areas of gilding, enamelling and semi-precious stones. The female form was the dominant theme running through Art Nouveau sculpture and this was to continue in the Art Deco style, at least in the area of small statuettes. The bronze figures made by such sculptors as Ferdinand Preiss, Demetre Chiparus, Gerdago, Otto Poerzl, Pierre le Faguays, Josef Lorenzl, Alexandre Kelety and Bruno Zach, were usually patinated or part-patinated.[47] The fashion for objects made from several materials continued and ivory was much used in conjunction with bronze and semi-precious stones, the ivory mostly intended to imitate flesh. Mixed metal techniques were also used on these pieces, the bronze often being parcel-gilt or inlaid with silver, and subsequently patinated to give colours ranging from golden brown to green and black.

Contemporary with the production of these bronze statuettes is the dinanderie,[48] or beaten holloware, which received a brief but brightly creative revival at the hands of a small group of fine metalworkers living in

44. In the late 19th century the German colourist Puscher experimented with the use of 'lustre washes' including thiosulphates. These were later developed by Beutel, Bib. 65; and Gross, Bib. 160. In Britain further tests were carried out by Holt and Ward, Bib. 187; some of which have been repeated more recently, see Gainsbury, Bib. 145.

Alkali oxidising agents were investigated by Groschuff, and are reported in Krause, Bib. 210, 211; he also investigated the use of potassium permanganate and copper sulphate.

Collected recipes and techniques are found in Buchner, Bib. 76; Beutel, 65; Krause, 210; and the Deutsches Kupfer Institut, Bib. 112. An English translation of Krause is available, see Krause, Bib. 211.

45. For a history of the animalier tradition, see Mackay, Bib. 242.

46. For collected recipes and techniques used in France, see Debonliez and Malpeyre, Bib. 108; and the revised and augmented version by Lacombe, Bib. 224. See also Michel, Bib. 266; and in particular the revised edition of 1931. A number of other French references occur in the bibliography.

47. The styles of art deco sculpture and metalwork are described in Victor Arwas, Art Deco, London, 1980; and in Alain Lesieutre, The spirit and splendour of Art Deco, London, 1978.

48. Dinanderie is the generic term for fine metalwork in brass and bronze, and is named after Dinant, a town in Belgium which from the 10th to the 15th century acquired a high reputation for producing and exporting fine metalwork. The name has been re-applied to certain kinds of modern work in beaten copper and brass.

France at this time. Foremost amongst them was Jean Dunand. Born in Geneva in 1877, Dunand set out to become a sculptor and was awarded a scholarship in 1897 to study in Paris under Jean Dampt. Whilst there he developed an interest in the decorative arts and began learning the techniques of fine metalwork. He also studied with the coppersmith Danhauer in his native Geneva and in 1905 began exhibiting his dinanderie. These consisted predominantly of tall vases made in one piece by hammer-raising sheets of copper and bronze. He used inlays of various metals, including silver and gold, to decorate these pieces and invariably patinated them to bring out the contrast between the metals and generally to enhance the surface. Dunand also became an expert in the use of lacquer and occasionally combined patinated inlay with lacquer in his extraordinary vases. He established a thriving business in metalwork and lacquer, and gave courses in sculpture, silversmithing and chasing. One of his pupils, Claudius Linoissier, later settled in Lyons and produced a wide range of dinanderie vases, bowls and plates which combined geometric inlay with

Vase. Copper, part silvered, part patinated, and part lacquered black. Jean Dunand. Paris, c. 1925.

chemical and heat-induced patination. Other dinandiers who peopled this unique 'school' of fine metalworking included Paul Louis Mergier, Jean Serrière, Armand-Albert Rateau, Luclanel who worked for Christofle, Berthe Cazin, Gaston Bigard and Edouard Schenk.

Britain: 19th century to the present day

In Britain, the use of bronzing and colouring techniques for domestic metalwork appears to have arrived from the continent by way of the Great Exhibitions of 1851 to 1862. Examples of patinated metalwork from both Germany and France were exhibited in the exhibition of 1862, and judging by the reports that appeared in the art journals of the day, the British craftsmen visiting the exhibition were inspired by what they saw. It also became the practice of large firms, such as Elkingtons, to employ French artist-craftsmen solely for the purpose of creating showpieces for international exhibitions such as these.[49] Leonard Morel-Ladeuil and Emile Jeannest were two such French craftsmen, contracted for a number of years to produce extraordinary pieces for Elkingtons. They had both trained as sculptors in Paris and it is more than probable that they brought with them the French enthusiasm for bronzing and colouring, if not the actual techniques and recipes for achieving them. Elkingtons was a Birmingham firm, and it was here that a considerable industry in the bronzing and colouring of copper and silver articles such as boxes, dishes, trays, clocks, furnishing fittings and jewellery was carried on at least until the 1920's if not later. The character of these objects owed much to the aesthetic of the Arts and Crafts movement, originated by William Morris and his circle. The warm, earthy bronzing colours seemed an appropriate method of finishing their simple, almost rustic forms, where the hammer marks were deliberately left on the surface, showing evidence of hand craftwork.

One firm that made such items was A. Edward Jones Ltd, and is described by Glennys Wild in the catalogue notes to the exhibition celebrating the work of this Birmingham firm of silversmiths.[50] In her essay she mentions the colouring process evolved by R. Llewellyn Rathbone, later to become head of Sir John Cass college in London, and which was developed by one of his employees, a craftsman named Frank Salthouse. The process seems to have been practiced in an atmosphere of secrecy, Salthouse working in a room separate from the other craftsmen. From this room, he would emerge at intervals bearing objects coloured yellow, brown, black, green and even blue.

The jealous guarding of techniques and recipes typifies the attitude of the practitioners of colouring and bronzing, and it is very difficult to detect

49. In the catalogue to the exhibition marking the bi-centenary of the Birmingham assay office, the development of silversmithing and the role of companies such as Elkingtons, are described. *Birmingham gold and silver, 1773–1973*, City Museum and Art Gallery, Birmingham.

50. Exhibition catalogue, *A. Edward Jones—metalcraftsman*, Glennys Wild, Birmingham City Museum and Art Gallery.

Animal head. Patinated bronze. Henry Moore, 1951. The tradition of using bronze as a medium for sculpture continues in contemporary work.

51. The major collections of recipes and procedures published in England are Hiorns, Bib. 182; and Field and Bonney, Bib. 124. See Fishlock, Bib. 134, for a more recent account of commercial colouring procedures.

Metal colouring recipes are also included in the collections of workshop receipts, represented in England by Spon, Bib. 355; and in America by Hiscox, Bib. 184.

the passage of knowledge between art and commerce. One practical attempt to disseminate knowledge of these processes was the course set up by Arthur Hiorns, who was head of the metallurgy department of Birmingham Technical Institute at the turn of the century.[51] However, architectural bronze and brass workers, sculptors and decorative metalworkers, who all used metal colouring techniques, seem to have operated in comparative isolation from one another.

With the exception of the exuberant work of the thirties in Europe, metal colouring in the present century has tended to be a rather neglected area. The radical changes that had occurred in sculpture, and the influence of mass production on design in fine metalwork, shifted attention away from techniques such as bronzing and patination. More recently however, there has been a renewed interest in the use of contrasting metals, oxidised finishes and Japanese techniques, that together indicate a growing interest in the potential of the decorative use of colour in metalwork, that is well exemplified in many earlier traditions.

METAL COLOURING TECHNIQUES

1 Surfaces for colouring

Metal colouring, bronzing and patination are not processes that can be considered in isolation from the overall surface quality of an object. It is important that the surface to be coloured is carefully prepared, whether finely polished, abraded or textured. Generally speaking colouring will not conceal defects in the surface. Often the quality of the finished surface is enhanced by preparation involving the use of texture, and many effective coloured finishes use the contrast between coloured areas and areas of exposed metal as the basis for decorative treatments. In addition, features of the surface that are undetectable in a polished finish, such as the grain structure, may be revealed by the process of colouring.

Structure and surface

One inherent difference that is of significance in colouring is the difference between the nature of rolled sheet materials and cast materials. We are used to being able to obtain sheet materials in stock thicknesses, with good surface finishes, carefully controlled compositions and a resulting high level of consistency in working properties. These characteristics of modern sheet metal stock have been carefully developed by manufacturers in order to provide industry and the crafts with materials which can produce controlled and predictable results. On the other hand we are also used to the idea that with cast materials, although the composition of the ingots used can be accurately specified, the quality of the result depends to a large extent on the particular casting process used, the design of moulds, and the skill with which the process is carried out. However when cast and sheet materials are ground and polished they may appear to be closely similar. Although metals presented in this form appear homogeneous to the eye, they do in fact have a distinct grain structure, whose exact form depends on the processes which the material has undergone. Modern production processes result in sheet materials that have a very fine even grain structure, of a scale that makes the grain invisible to the naked eye, even after suitable surface treatment.

Grain-enhancement

In the case of castings, however, the grain structure will vary according to the flow of material in the mould, and the differential rates of cooling of various parts of the casting. Often the individual grains are of a size that would make them visible to the naked eye if suitable surface treatments were applied. This occurred in many of the samples tested, both in cast yellow brass and in cast bronze, grains sometimes measuring several millimetres in diameter. The grain structure is not visible in freshly cast material, or after grinding and polishing operations, but the action of some of the colouring agents used will differentially etch the surface of adjacent grains, making them a clearly visible feature of the metal surface. Usually where this occurs, the colouring effects of the solutions further enhance this grain structure. These effects are referred to in this book as 'grain-enhancement', and the solutions and procedures that have given rise to them are recorded in the descriptions accompanying individual recipes. The degree to which they will occur when objects are coloured depends largely on the characteristics of the particular casting involved. Generally speaking, grain size and distribution in a casting will not be even but will vary, sometimes abruptly, according to the design of the object.[1] In particular, differences in grain size and texture density will tend to occur where there

1. An example of a grain-enhanced cast brass surface with an abrupt change in grain size. The change occurs where a large runner fed the back of the large cast plate from which the sample was cut.

are significant changes in section, or again, where a substantial runner feeding the casting has been removed. Although no trace of surface differences may be visible when the surface is ground smooth or polished, subsequent colouring with the appropriate solution will reveal them. One reason for this is the difference in rates of cooling of the molten metal in the mould, the more massive parts of the casting taking significantly longer to cool than thin sections. The nature of the grain-enhancement that is obtained also depends on the flow pattern of the molten metal, more evenly distributed grain structure occurring in moulds of a design which allows the material to flow freely and evenly. Working the surface of a finished casting by hammering and reheating will also affect the grain size, through a phenomenon known as recrystallization, but this is very difficult to use as a means of controlling grain size, as the effects are only visible after the object has been coloured or etched with a suitable agent.

As-cast surfaces

More obviously apparent than the grain structure that is revealed on smooth surfaces during colouring, there are of course the immediately visible surface qualities and textures that are produced by the contact of the metal with the surface of the mould. The precise nature of these will depend on the method of casting, the texture of the investment material in the case of lost-wax casting and the texture imposed by the sand in sand casting, for example. These relatively rough and porous surfaces generally respond well to colouring treatments, and often differ slightly in colour or tone from the results produced on ground and polished surfaces. The contrast between areas of a surface that are left as-cast and areas that have been machined, or ground and polished, clearly provides the basis for a wide range of decorative treatments.

Porosities and inclusions

In addition to the textures that are inherent in particular casting processes rather than in the structure of the metal itself, the occurrence of defects such as porosities and inclusions also needs to be considered. Although the careful use of appropriate casting methods will help to minimise the occurrence of holes caused by trapped gas, and clean working methods will reduce the likelihood of particles of foreign matter being trapped in the melt, it is virtually impossible to produce castings that are entirely free of these defects. Although in many cases surfaces in the as-cast state will appear to be unblemished, grinding and polishing will reveal porosities and inclusions clearly. Traditionally these small blemishes have been corrected by using a variety of chasing tools to close the surface by spreading the surrounding metal. Visible blemishes in as-cast surfaces that are to be left unground are similarly blended into the surrounding texture by the use of matting tools and the like. When cast surfaces are worked in this way, the localised working can effectively alter the structure of the metal at that point, which may be revealed as a difference in tone or surface quality when the object is subsequently coloured. The extent of chasing or matting should therefore be considered in relation to subsequent colouring, and may need to be more extensive than a particular porosity demands, in order to achieve a unity of surface finish.

Abrasive finishing

A combination of the varieties of surface finish that arise out of the nature of the material and of the techniques employed has often been used to great effect both in sculpture and in the decorative arts. A particularly refined use of these effects in conjunction with colouring is exemplified by the decorative sword furniture produced by Japanese craftsmen. The full range of matted finishes contrasting with chased or smoothly ground areas are to be found in many of the *tsuba* or sword guards that they produced. The particularly fine grain-enhanced surfaces that are found on some *tsuba* are the result of an approach to surface finishing that is characteristic of Japanese metalwork of this kind, and that differs from the traditional approach to finishing in western metalworking. A series of graded abrasives were used to refine the surface gradually to an extremely fine matt finish, without the use of mechanical polishing techniques or buffing.[2] The result of this finely controlled abrasion was to produce a smooth finish whose structure remained undistorted even on a microscopic scale, and which when mildly etched and coloured gave rise to very fine surfaces. The more

2. The finishing techniques used are similar to the surface preparation that is carried out by metallurgists when preparing metal samples for metallographic examination of their fine structure.

forceful mechanical polishing techniques, that were often used to finish plain surfaces in the western tradition, tend to cause a superficial flow of material that can entirely mask the grain structure in some cases. The use of a sequence of abrasives, without a final mechanical polish, also appears to have the effect of providing a better surface for the action of colouring agents. In general it is also advantageous to use a fine abrasive immediately prior to colouring, where traditional polishing sequences are employed. It also appears to be the case, judging from the tests carried out, that even and unblemished colouring is most difficult to obtain on buffed and highly polished surfaces.

Imposed texture

Clearly it is a small step from the controlled use of the various inherent surface qualities and textures that arise out of the nature of a material and its method of working, to the use of imposed textures. In casting, texture of a scale that is coarser than that of the mould material can be introduced into the object by working on the wax or pattern, but perhaps of more importance are textures that are imposed on the metal surface directly. This is particularly true of sheet materials, which do not offer the range of surface qualities that are available with castings. The grain structure of rolled sheet is generally so fine that the mild etching that can so enrich a cast surface often has little visible effect. This means that the plain coloured surfaces produced on sheet materials can seem comparatively lifeless. The use of textured surfaces[3] of various kinds in conjunction with colouring techniques significantly enriches the range of finishes that are available for these materials.

3. Examples of some of the textured finishes that can be effectively used in conjunction with colouring techniques. The textures shown are produced by hammering, A and B; punching, C and D; and etching, E and F.

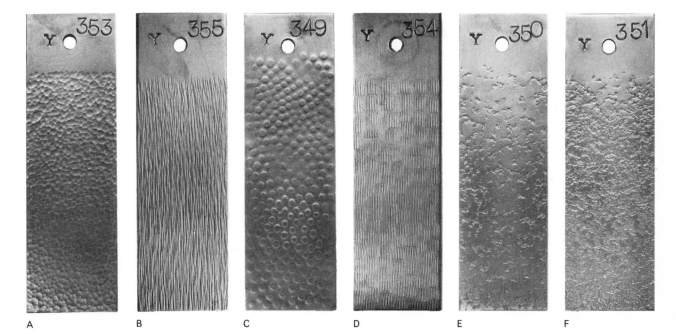

A   B   C   D   E   F

In addition, the use of texture can considerably improve the wearing properties of finishes,[4] by providing areas which are less exposed to continuous direct contact than plain surfaces. A wide variety of techniques can be used, including machining and engraving, hammered and punched finishes, sandblasting and etching. The effects produced can be used in conjunction with the technique known as relieving, where the raised portions of a texture are abraded, while the recesses are left fully coloured.

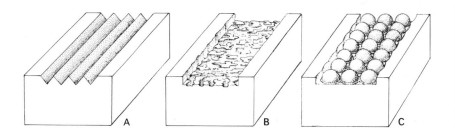

4. Diagrammatic views of the use of various textures to provide coloured surfaces in which the finish is protected from excessive wear. In A, which could be produced by machining or engraving, the peaks provide a raised contact surface, while the colour is retained on the walls of the grooves. In B, produced by etching, an area of the surface is lowered by chemical action and textured to provide a 'key' for the colouring. In C, which could be produced using a small hollow punch, the colour is retained in the crevices and on the walls of the domed protrusions, while the tops act as a contact surface which can either be relieved or allowed to wear naturally.

Fabrication

Metal objects are usually designed to be constructed from a number of parts. The complexity or sheer scale of a bronze sculpture, for example, may require that it is cast in sections which must subsequently be joined together by welding. If a filler rod of the same material as the sculpture is used, then the join can be cleaned up, and coloured over during patination. In the case of much silversmithing and fine metalwork, objects are constructed from sheet materials, often as a series of components that are then joined together using a range of silver solders. Copper and brass objects are similarly constructed using brazing alloys and solders. The brazing alloys and solders used differ in composition from the metals that they are used to join, since they are formulated to obtain specifically useful properties such as low flow temperatures. Consequently the colours produced on these joining materials will differ from those produced on the metals themselves, the differences tending to be most marked on plain unvariegated surface colourings. It is important to consider colouring as an integral part of the design of the whole object so that these factors are taken into account. In many cases joints can be positioned or designed in such a way that the solder line is concealed. Where exposed seams are unavoidable, it is often possible to use the technique of burnishing or chasing over the joint. Where this procedure is not possible, the entire piece may be plated and a colouring technique suitable for plated surfaces selected accordingly.

2 Degreasing and cleaning

Adherent, even and spot-free colouring can only be obtained on clean metal surfaces that are entirely free of grease. Any residual grease, including fingermarks, will inhibit colouring locally and give rise to unevenness, marks and spots that mar the finish. The polishing stages in the surface preparation of metals normally includes the use of grease-bound polishing compounds, such as tripoli or rouge, which must be scrupulously removed. General degreasing and the removal of these compounds can be carried out by a number of methods including treatment with alkaline degreasing agents or with organic grease-dissolving agents. In commercial applications the metal is either immersed in hot alkaline degreasing agents, or sprayed with them. Organic grease-dissolving agents, on the other hand, are often used in the form of a vapour which is allowed to condense on the object.

Degreasing

In the workshop simpler methods can be used effectively. Initial degreasing can be carried out by bristle-brushing the surface with organic solvents or with the equivalent commercially available polishing compound removers.[5] Generally speaking, at least two cleaning operations should be carried out with one of these agents; the first to remove bulk residues, and the second to provide a thorough degreasing using fresh solution. The surface should be thoroughly scrubbed with the solution using a bristle-brush, ensuring that any holes, hollows, crevices or textured areas are scrupulously cleaned out. In the case of castings, particular attention should be paid to porosities and inclusions which tend to trap small quantities of grease that can ooze onto the surface when colouring is carried out.[6]

Cleaning

After the initial thorough degreasing a further cleaning stage is required. A non-ammoniated emulsion cleaner which is water-soluble is suitable for this purpose. It provides the transition between the organic solvents used for degreasing and the water used for final cleaning, which are generally incompatible. Cleaning is again carried out by scrubbing the object with the solution, using a clean bristle-brush. When the surface has been thoroughly treated, it is washed under cold running water.

Pumicing

Finally the object can be lightly bristle-brushed with pumice and a little water, and again washed under cold running water, immediately prior to colouring. This provides a final mechanical cleaning stage, helps to reduce surface tension and assists the action of colouring agents. It should not be carried out where a very reflective gloss finish or a lustre colour is required, but is desirable in all other cases, including gloss finishes in general.

5. See *Safety: chemicals and their hazards,* final paragraph. Aqueous solutions are available, but immersion in a hot solution is generally required.

6. In this test sample, the small porosities have retained some grease which was not cleared during degreasing and cleaning operations. When immersed in the hot colouring solution, the grease has oozed out and prevented colouring in the regions surrounding the porosities.

Timing

Final polishing operations, degreasing and cleaning should be carried out just prior to colouring to minimise the risk of tarnishing. If the object needs to be left for a short time, for instance while one attends to the colouring bath, then it should be left under cold running water in such a way that the whole surface remains immersed.[7] It is not advisable to leave objects in degreasing or cleaning solutions, or to leave their surfaces wet with these agents, for any length of time. Differential drying tends to occur as they evaporate, sometimes leading to soiling or tarnishing and patchiness of the finished surface. If tarnishing does occur at any stage during cleaning, then it is necessary to dry the object and repeat the finishing and cleaning cycle, beginning at the final polishing stages.

Ultrasonic cleaning

Surfaces can also be cleaned ultrasonically. This technique is particularly useful for cleaning small complex articles that are difficult to clean manually. The process is carried out in three stages as for manual cleaning, beginning with a degreasing agent followed by a non-ammoniated cleaner, and finishing with water. These solutions can be placed directly in the ultrasonic tank, but it is usually more convenient to contain them in glass vessels which are successively placed in the tank, so that continual emptying and cleaning of the tank is avoided. The tank is filled with water and the glass vessel suspended in it, so that it clears the side walls and bottom. The articles to be cleaned are suspended in the cleaning solutions in the glass vessel, and the tank is then switched on for the appropriate length of time.[8] The particular solutions to use, and timings, are best determined in consultation with the manufacturer of the ultrasonic equipment used. The articles are cleaned by the effect of the high frequency sound produced by the equipment, which activates the solution at the surface of the object through a phenomenon known as cavitation.

Cleaning of equipment

Whether manual or other techniques are used, it should be borne in mind that any equipment such as suspension rods or supporting cradles that is used in conjunction with objects being coloured, or is immersed in colouring baths with them, should also be clean and thoroughly degreased to prevent contamination. Immersion vessels should obviously also be grease-free and clean prior to use.

3 Immersion colouring

A wide variety of colours and surface finishes are obtained by totally immersing the object in a solution of chemicals, at temperatures ranging from cold to boiling, for varying lengths of time.

Surface preparation and cleaning

The object should be thoroughly clean and degreased prior to colouring. It is essential to ensure that any dirt or grease that may have become trapped in holes, hollows or angles during polishing operations is removed. Any residual grease will tend to be released when the object is immersed and may contaminate surfaces and prevent colouring.[9] Particular attention should be paid to porosities and inclusions in cast surfaces.

Generally speaking the surface should be bristle-brushed with pumice prior to immersion, to provide a slight 'tooth' which assists the action of the chemical agents and helps to produce a more even and tenacious finish. In some cases where the surface effect depends on a high gloss finish, lustre colours for example, this should not be carried out. In the majority of cases, the solutions used will tend to etch the surface to some degree, and there is no advantage in starting with a high polish. The extent to which particular solutions may be expected to alter the surface quality is indicated in individual recipes.

Small objects

Small objects, such as articles of jewellery, can be immersion coloured in glass vessels. Pyrex glassware, such as that used in chemical laboratories, is well suited for this, and colouring involving quantities of solution of up to

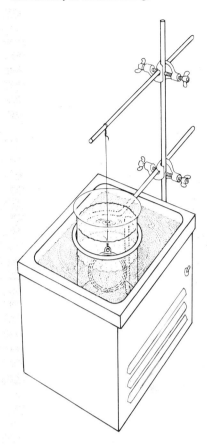

7. Immersion in a cold, weak solution of potassium bi-tartrate can be used as an alternative. This is often the method used commercially in metal-finishing.

8. A suitable arrangement for ultrasonically cleaning small articles. The article being cleaned is suspended in a cleaning solution, contained in a glass beaker. This is in turn suspended in the ultrasonic tank, which contains water. It is preferable to avoid contact between the beaker and the bottom or walls of the tank.

9. An example of a streaky surface caused by a small amount of grease, trapped in a suspension hole, which has flowed on to the surface on immersion.

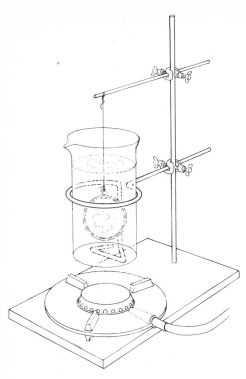

10. Cradle made from mild steel rod, held in a retort stand, used to support and retain a colouring vessel for small articles.

11. Colouring procedures which were found to give rise to severe bumping in tests are indicated in the notes accompanying individual recipes.

about one or two litres can be safely carried out in beakers. These may be heated over a gas ring. The beaker should be positively supported and retained, and not simply placed on a gauze or directly on the ring. A suitable supporting cradle can easily be made from mild steel rod, which can be clamped into a metal retort stand.[10]

Bumping

In some cases, when solutions are heated, the liquid in the glass vessel may be subject to a violent shuddering, known as 'bumping', as trapped air is expelled. This can often be alleviated by the addition of a few fragments of porous pot in the bottom of the vessel. Sometimes dense precipitates are formed as solutions are heated, which can give rise to a heavy sediment in the bottom of the vessel also causing severe bumping. In extreme cases this will not be alleviated by the addition of porous pot, and colouring should then be carried out in a copper vessel rather than glass.[11]

Larger objects

Larger objects cannot safely be coloured in glass vessels. With the increasing quantities of solution required the potential dangers arising from a breakage become too great. Many of the alternatives suggested in the literature are only suitable for a proportion of the colouring solutions that are used. The enamelled vessels that are often suggested cannot be used with some oxidising agents, for example potassium permanganate, which attacks the enamel. Lead-lined steel tanks, which are often used for the acid pickling of metals, are only suitable for a proportion of the solutions involved, and in particular cannot be recommended for use at higher temperatures. Rubber-lined steel tanks are also unsuitable for use at higher temperatures and are seriously affected by strong oxidising agents which are regularly used in immersion colouring. Steel tanks internally coated with PVC can be used for a wider range of solutions but cannot be recommended for the full range of chemicals employed, particularly at higher temperatures. Steel tanks and galvanised tanks cannot be used because of the undesirable electrochemical effects that are produced. Although copper-plated steel tanks can be used they are generally not sufficiently resistant for long term use.

There are two chief types of tank material that are suitable for the full range of colouring solutions generally employed: copper and polypropylene.

Copper tanks

Copper tanks can be purchased, or alternatively fabricated to suit the particular needs of a workshop. The only disadvantage with this material is that since the colouring agents employed are intended for cuprous metals, the tank will also be coloured. Although this does not adversely affect the colouring of objects directly, it does mean that the solutions used will be depleted in colouring the tank surface. The tank will therefore need to be 'worked in' with some of the colouring solution, and washed, prior to use. Additionally, in some cases the colouring of the tank surface produced by one solution will be stripped when a different solution is subsequently used, and may cause contamination of the new solution. The precise effects are difficult to predict and it is therefore usually necessary to clean the tank surface back to the metal when a solution of a different type is to be used. Nevertheless copper tanks are probably the most suitable for immersion colouring in the average workshop, since they can be fabricated fairly easily to meet particular requirements and can be externally heated.

Polypropylene tanks and immersion heating

The alternative is to use a polypropylene tank or a polypropylene-lined steel tank. These are readily available and can be obtained to order over a wide size range. Polypropylene is inert with respect to the chemicals used in immersion colouring and is not subject to the 'working in' problems associated with copper tanks. The major disadvantage with this material is that, although it will withstand the solution temperatures involved, tanks made from it cannot be externally heated. Some form of immersion heater is required. The majority of immersion heaters are jacketed externally with steel and are therefore unsuitable for use in colouring cuprous metals because of the electrochemical effects produced. Lead-coated and titanium versions are available but these are also subject to some limitations in respect of the range of solutions for which they may be used. Silica-glass cased heaters are suitable for a wider range of solutions but cannot be used for those containing caustic alkalis or concentrated phosphoric acid. They are also rather fragile in use and are generally only available in a straight form which may imply the use of deep tanks in order to achieve the temperatures required, resulting in the need to use uneconomically large quantities of solution for colouring.

The most suitable type of immersion heater available is cased in Teflon, which is resistant to the full range of chemicals normally encountered in immersion colouring and produces no electrochemical effects. Heaters of this type can be obtained either in the form of a straight coil, or as a spirally coiled flat plate.[12] Both forms are supplied with a non-heating support section, which can be bent to fit the tank available. The flat-plate type appears to be particularly suitable for many immersion colouring applications; sited near the bottom of a suitably sized tank, it allows for a more economic use of solutions, and produces convection currents rising evenly through the tank. Sufficient clearance should be allowed between the heater and the bottom of the tank to allow for the sedimentation of dense precipitates.

Tanks of either type should be provided with a lid, as this significantly improves heating-up times and prevents excessive evaporation of solutions.

Tank size and heater rating

The size of tank for a particular situation should be carefully considered. Although the choice of a large tank may appear to offer greater flexibility in the range of objects that may be coloured, this may well also imply the use of large quantities of solution for colouring relatively small items, which can prove to be uneconomical. A sufficient depth of solution must be allowed to enable the object being coloured to be kept well clear of the heater or the bottom of the tank, and to ensure that it remains well covered. Some solution will be lost by evaporation, particularly in the case of boiling solutions, and this should be taken into account. In many situations it may be better to invest in more than one tank, so that solutions can be used economically.

Where solutions are heated with an immersion heater, the tank form and size should be considered in relation to suitably rated heaters. Immersion heaters are available in a wide range of sizes and power ratings, and consultation with the manufacturer is recommended in determining the correctly rated heater for a particular tank size.[13] The heater should be capable of bringing the volume of solution to boiling point and maintaining a steady rolling boil while the object is immersed. The heating-up time should not be too prolonged, as in some cases the slow rise in temperature over a long period of time can reduce the effectiveness of the solutions used. Generally speaking, the heater should boil the solution within about one hour or preferably less. As an approximate guide, it was found in practice that a three kilowatt heater was suitable for use with from twenty to fifty litres of solution.

Suspension of the object

The object to be coloured is suspended or supported in the solution during colouring, and should be kept well clear of the walls of the containing vessel. Various means of suspension can be used including wires, rods, nylon filaments and plastic or metal cradles. The materials employed should be capable of withstanding the corrosive nature of the chemicals that are used and the temperature of the solution. If metal suspension wires or supports are used then these should be of copper or brass since other metals, particularly iron and steel, will give rise to electrochemical effects that may affect the colouring. As an alternative, supporting structures which are to stand in the tank during colouring can be constructed from a plastic such as polypropylene, which is resistant to the various corrosive chemicals used and will withstand the temperature of boiling solutions without deforming or sagging if sufficiently substantial sections are used. Nylon filaments can also be used for the suspension of smaller articles.

One method of suspension which has been found to be particularly effective in practice for use with objects constructed from sheet materials consists of attaching a copper or brass wire or rod to the object with lead solder.[14] This gives positive control of the object while in solution, and provides a means of manipulating the object during subsequent washing, drying and finishing operations which does not involve direct handling of

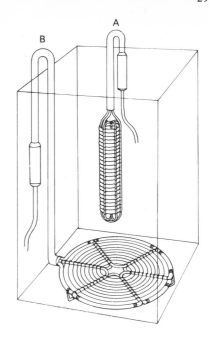

12. Immersion heaters of the straight coil type, A, and in the form of a coiled flat plate, B. The tanks and immersion heaters discussed in this section can be obtained from the suppliers of equipment to the electroplating and metal-finishing trades.

13. It is essential to ensure that immersion heaters are properly earthed. They should also be fitted with an automatic earth leakage circuit breaker. These are not normally supplied with heaters but are readily available and are essential for the safe use of heaters in this application. Immersion heaters should never be switched on unless they are immersed, or damage will rapidly be caused. A thermostat can also be fitted to control the temperature of the solution.

14. An example of a supporting rod used to position the object while in the immersion tank. The rod has been designed so that it ensures adequate clearance at the bottom. The object is prevented from rotating in the tank by the lateral rod which engages the lip of the tank.

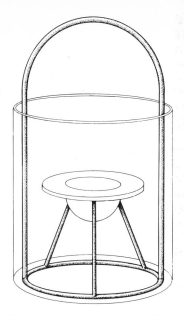

15. An example of a supporting cradle fabricated from brass rod, similar to that used for the cast bowls illustrated in the colour plates. The bowl is held under the rim by a three point contact. The cradle ensures all round clearance and provides a means of manipulating the object during colouring and washing.

16. An example of a surface defect caused by the suspension of the sample too close to the vertical support of a polypropylene suspension tree.

the newly coloured surface. The use of lead solder for this purpose has not given rise to any adverse effects on colouring in the tests which have been carried out. These wires or rods should not be re-heated to melt the solder, when removing the object, as this will cause discolouration.

In the case of castings, which are relatively heavier, or where a small soldered joint is insufficient, a blind hole can be drilled and tapped to accept the threaded end of a brass rod. Alternatively, supporting cradles can be devised and constructed for particular objects.[15]

Clearance

It is essential that suspension wires and supporting cradles are kept clear of the surface of the object, except at the selected point of attachment, and that the object is kept well clear of the walls of the vessel. Anything lying in close proximity to the surfaces being coloured will give rise to local colour variations, taking the form of surface marks or streaks that are generally lighter in colour than the remainder of the surface.[16] Sufficient clearance should also be provided between the object and the bottom of the tank, so that it remains clear of any sediments or settled precipitates that are produced during colouring. If the object is partly covered by these, then local colour variations tend to be produced.

Monitoring colour development

Whichever means of suspension or support is adopted, it should be possible to remove and replace the object quickly and easily during colouring, so that it may be examined. When monitoring is carried out, the object should be rapidly transferred to a container of water and rinsed by agitation. It may then be examined and replaced if necessary. It is not advisable to examine the object directly after removal from the colouring vessel, since the surface of the hot object dries very rapidly as the colouring solution drains from the surface, which can give rise to streaks and discolourations that are difficult to remedy. In the case of hot or boiling immersion, the water used for intermediate rinsing should be 20–30°C lower in temperature than the colouring solution, while in the case of cold immersion the rinsing water should be cold.

If the surface develops unevenly during colouring, then the surface of the object can be lightly bristle-brushed with water or with pumice and water, as necessary, prior to re-immersion. This should not be carried out too hastily, as the initial stages of many immersion colouring procedures tend to produce slightly uneven results.

Timing

There are a number of factors which will affect the length of time taken for immersion colouring. Generally speaking larger objects take longer to colour than small articles. This is partly due to the fact that the larger quantities of solution required for objects of any size will take significantly longer to change temperature, than small quantities that can be rapidly brought to the boil. When cast objects are placed in the colouring solution they may take some time to warm up, and often significantly reduce the temperature of the solution, which may then take some time to come back to the boil. This will tend to retard the colouring, but may be alleviated to some extent if the casting is placed in hot water for a few minutes immediately prior to immersion. The hot washing tank can be used for this purpose. It is inadvisable however to leave the object in hot water for extended periods, as the surface will tend to tarnish fairly rapidly.

Perhaps the most significant factor affecting timing is the depletion of the solution as the immersion tank is coloured. This problem does not arise in the case of polypropylene tanks. In the case of copper tanks, which will be coloured by the solutions placed in them, the surface area of the tank may be significantly larger than the object that is placed in it. The tank will tend to be preferentially coloured and the solution contained within it may be depleted rapidly, to the extent that the object is coloured only very slowly. This problem can be overcome to some extent by working-in the tank prior to colouring. The tank is first coloured with some of the solution, which is then removed. Fresh solution is placed in the tank and brought up to temperature, prior to the immersion of the object. Although this generally improves the situation, the solution may still be steadily depleted as the tank coating gradually builds. In practice it was found that although timings were significantly affected, sometimes taking four or five times as long as for the colouring of small articles in glass vessels, the results produced were not adversely affected. Colour development should be monitored visually and the timings quoted in recipes, which relate to the colouring of small articles (normally in glass vessels), used as a guide to the general pattern of colour development.

Trapped air and precipitates

Objects which have hollows or concavities should be carefully oriented and monitored during colouring. These may tend to trap pockets of air as the object is immersed, resulting in patchy or uncoloured areas. The simple remedy is to ensure that the object can be oriented in the solution so as to release the trapped air. Hollows and concavities will also tend to collect precipitates, which can cause local differences in colour if they are allowed to accumulate. Correct orientation of the object in the tank may solve the problem, but a position which will prevent the accumulation of precipitates is likely to encourage trapped air. If this is the case then either provision should be made to enable the object to be repositioned while in solution, or the object should be positioned initially to prevent the trapping of air, and removed from time to time for rinsing to clear accumulated precipitates.

Washing, drying and finishing

When colouring is complete, the object should be removed to a tank of water for thorough rinsing. Preferably a sequence of tanks should be used so that very thorough rinsing is ensured. In the case of colouring by cold immersion, cold water should be used for rinsing. After colouring by hot or boiling immersion, the initial rinsing should be carried out in water that is 20–30°C cooler than the temperature of the colouring bath. Subsequent rinsings are carried out in progressively cooler water, and the object is finally rinsed under running cold water. The object is then either dried in sawdust or allowed to dry in air before wax finishing or lacquering.

Emptying tanks and disposal

If the tank used is of such a size that it cannot be emptied directly, then it may be emptied using either a siphon or a pump. Moderately sized tanks can be emptied using a siphon fitted with a priming bulb and a stop-cock, allowing solutions to be decanted safely for storage or disposal. Larger tanks will probably require a pump. Magnetically coupled pumps with polypropylene fittings are available in a wide range of sizes and flow capacities which are suitable for handling the range of chemical solutions used in colouring and which will withstand hot solutions.[17]

17. Siphons and pumps of the type described in this section are available from the suppliers of equipment to the electroplating and metal finishing trades.

If dense precipitates are formed during colouring, then these should be allowed to settle before the solution is siphoned or pumped to waste. The residues can be transferred to a plastic container and allowed to dry out before disposing of them as normal refuse. Exceptions to this general rule and further comments on the disposal of solutions are given in the chapter devoted to safety. Heavy precipitates should not be run to waste through normal sinks, unless a special trap is provided for the purpose which may be emptied from time to time, as they will tend to accumulate.

4 Application techniques

Many traditional finishes, particularly those associated with the production of a green patina on bronze sculpture, are produced by the direct application of solutions to the surface of an object. Colouring is seldom accomplished in one application, but characteristically involves a cycle of sparing applications and periods of drying, for a matter of days or even weeks, until the desired colour has developed. The technique is of particular relevance in the colouring of large sculptures, which it would be impractical to colour by immersion, but is equally useful for colouring smaller articles. It is also very economical in the use of chemicals.[18]

18. It should be noted that although the recipes and techniques quoted in this book are specified in terms of relatively pure chemicals, metalworkers and sculptors have traditionally employed crude or naturally occurring substances for colouring. Acetic acid, for example, was used in the form of crude malt or wine vinegar, and urine, which yields ammonia when it becomes stale, was also used. In addition naturally occurring minerals such as common salt, alum, or copiapite (basic ferric sulphate) and plant products, such as the radish pastes used as a pre-treatment in Japanese colouring, were also used. This 'vernacular' aspect of colouring and patination continues at the present time, vinegar and urine being commonly used by sculptors in the patination of cast bronze.

Basic procedure

The surface of the object to be coloured must be thoroughly grease-free prior to treatment. Even small traces of grease will stubbornly resist colouring, and are difficult to eliminate without marring the surface once colouring is under way. Colouring is generally easier to achieve on as-cast rather than polished surfaces. Polished surfaces may be bristle-brushed with pumice and then thoroughly washed in cold water. This provides a slight 'tooth' to the surface, which assists in breaking down surface tension and obtaining a more positive effect when the colouring solution is initially applied. It also provides a final mechanical cleaning stage which helps to ensure that any surface contamination is removed, immediately prior to colouring.

Although surfaces must be thoroughly degreased and should not be handled after cleaning, firmly adherent brown or black oxide layers that occur on castings need not be removed as these provide a ground on which a green or blue-green patina will develop. However it should be borne in mind that if the oxide layers only partially cover a surface then a different colour will probably occur on the oxide layer than the one that occurs on areas of originally clear metal. This factor has sometimes been used to obtain differences in tone on cast bronze sculpture. An oxide ground is sometimes abraded back on raised areas and left undisturbed in hollows, providing a colour difference after patination which can accentuate or complement the form.

The solution is applied to the surface by one of the methods outlined below, and the surface left barely moist. Any excess solution should be removed. The surface is then left to dry thoroughly. Initial applications should be left to dry for at least a day, and may require longer. When thoroughly dry, the surface is rubbed with a dry soft cloth prior to the next application of solution. This removes any loose residue that is left as the solution dries out and helps to consolidate the surface. As treatment continues, applications can be carried out more frequently, but it is always essential to ensure that the surface is thoroughly dry and well rubbed down with a soft dry cloth before a further application.

Application by dabbing and wiping

A good method of applying the solution is to dab or wipe it onto the surface with a soft cloth. In many cases it may be necessary to dab the surface vigorously with the moist cloth for the initial applications, in order to help break down the surface tension and to ensure that adequate wetting takes place. The surface should be left only very sparingly moist. One way of ensuring this is to apply the solution with one cloth and then to wipe over the wet surface with a second, nearly dry cloth, until only the very slightest glaze of moisture is apparent. The object is then left to dry thoroughly. When completely dry it is rubbed with a soft dry cloth, as described above, prior to the next application. This sequence of sparing applications and periods of thorough drying is repeated until the patina has fully developed. A very gradual development is essential to a well patinated and coherent surface. The temptation to hasten the process should be resisted. More liberal applications may appear to increase the development of an opaque layer more quickly, but the results are often superficial and the surface tends to break down when further solution is applied. Forced drying by warming the object on a hotplate or with hot air has a similar effect, producing a superficial layer of dried chemicals which tends to have little adherence.

In many cases, the finished surface consists of a green or blue-green patina on a ground colour which may be a shade of ochre, brown or black. Sometimes, when the solution is applied very sparingly by wiping, the ground colour alone tends to develop and the green or blue-green colour is slow to form. When this occurs, the solution may be applied less sparingly when the ground colour is well developed.

When the patina has developed fully, and treatment is complete, the object should be left to dry for several days at least and rubbed down carefully with a soft dry cloth prior to finishing with wax.

Application by brushing

An alternative method of application, favoured by French colourists of the nineteenth century, is carried out by applying and thinning the solution with soft bristle-brushes. The solution is first brushed on so that the whole surface of the object is wetted. A dry brush is then used to thin out the solution on the surface with light rapid strokes in all directions. When the brush becomes too moist to be effective a further dry brush is used, and is replaced as necessary until the whole surface is nearly dry. The object is then left to dry completely before the next application. The brushing application was often carried out while the object was gently warmed on a hotplate. This technique tends to produce a surface with a powdery appearance and although it can produce results more quickly than by wiping, requires considerable practice to perfect.

Application by spraying

A more recent method of application consists of spraying the surface of the object with a fine mist of solution, and allowing it to dry before a further application. This process is repeated until the patina has developed. No specialised spraying equipment is required, as a fine mist can be obtained with readily available hand-held pistol grip sprays as used by gardeners for insecticides. These are made of plastic that is resistant to the range of chemicals that are suitable for use in spray treatments. The fine-bore tubing and filter used in these hand sprays should be thoroughly cleaned after use.

The success of the technique depends to a large extent on the care with which it is carried out. If the spray is applied unevenly or too liberally, then the solution tends either to pool or stream, depending on the orientation of the surface. This results in uneven streaks and patches in the final patina. Ideally the surface should become coated with a mist of fine droplets. Some specialised chemical compounds are available which can help to achieve this. Very small additions of an organic silicon compound (a silicone) or of trimethylhexanol have the effect of retracting the solution into droplets of an even size, and tend to prevent pooling and streaming. Additions of these chemicals have been used commercially in the artificial patination of copper roofs, and results are reported to be good.

The chief disadvantage of spraying as a technique, apart from the difficulty of obtaining an even fine coating, is that it presents a significantly greater health hazard. Fine airborne mists of chemicals are easily inhaled and absorbed, and for this reason the more toxic or corrosive chemicals should never be applied by spraying. It is also the case that chemicals which can be applied relatively safely by other methods may be harmful if inhaled in the form of a spray. It is essential therefore when using this technique to ensure that there is adequate localised extraction to draw away the airborne mists, and to wear a nose and mouth face mask fitted with the correct filter.

Application by dipping

Smaller articles may conveniently be coated with solution by a short immersion. The article is then removed and the surface either wiped or brushed, as described above, until very sparingly moist, and allowed to dry.

Timings

Precise timings for patina development and drying cannot be given. The actual time taken for a patina to develop will depend to a large extent on the scale of the object being coloured and the general atmospheric conditions prevailing, particularly the level of humidity in the air. A humid or damp atmosphere can assist in the development of a green or blue-green patina, but will also tend to increase the required drying times. The timings recorded in individual recipes are for small scale test surfaces, and give an indication of the length of time taken for patina development in the tests that were carried out. With larger objects and variant levels of humidity, colour development can take several times as long.

Wetting problems

In some cases a solution may fail to wet the surface of the metal evenly, but will rather retract into pools or runs, producing streaks or spots of colour on a predominantly bare metal surface. This problem, which occurs most frequently on highly polished surfaces, can often be overcome by a combination of simple means. Small quantities of commercial wetting agents or of methylated spirit can be added to the solution, which will help to improve the contact between solution and surface. In addition the metal surface can be bristle-brushed with pumice to provide a slightly 'toothed' finish which will assist the solution to spread. The initial applications of solution can also take the form of a vigorous dabbing or pounding of the surface with a soft cloth pad moistened with the solution, which helps to break up the liquid into droplets and enforces a more distributed contact with the surface. A combination of these techniques will normally alleviate wetting problems.[19]

If the surface stubbornly resists wetting, then a thin ground of colour can be applied using a different chemical, and the patinising solution then applied in the normal way. The ground may be conveniently produced in

A   B

19. When a solution fails to wet the surface it retracts into pools and produces patches of colour, A. If the surface is then pounded with a soft cloth the action produces better initial contact and spread, B, and assists subsequent applications.

two ways. A single application of a different patinising solution containing a high concentration of ammonium salts or ammonia can be used to provide an initial 'bite' to the surface. Alternatively the surface can be firmly rubbed with a cloth moistened with paste comprised of zinc chloride, ground with half its weight of copper sulphate and a little water. This provides a slight but even colouring, to which the patinising solution can be applied in the normal way.

Pooling

Some pooling tends to occur even when a surface is only barely moist. In particular the solution will accumulate just behind sharp edges or changes in direction on both horizontal and vertical surfaces. It will of course also build up in any hollows that are present. There is no special remedy for this; the excess moisture should simply be wiped away.

Pooling will also tend to occur around the base of an object where it is in contact with a surface. If this proves to be a problem with a particular object, then the object can be placed on a gauze or mesh over a tray, so that there is less opportunity for the solution to accumulate.

It is important to remove excess moisture from areas where it tends to accumulate, as any accumulation will tend to produce patches of thicker and less adherent patina, which tend to break down unevenly when further solution is applied.

Finishing

It is essential to allow ample time for the surface to dry out thoroughly after treatment, prior to wax finishing. Any moisture trapped beneath the wax will give rise to imperfections that are difficult to remedy. Some green or blue-green patinas, particularly those produced with solutions containing higher concentrations of chlorides, have a tendency to 'sweat' moisture intermittently for some time. Although this will be minimised by very sparing applications of solution, drying time should always be generous. When patina development is complete, and the surface is thoroughly dry, the object is finally rubbed down with a dry cloth and then wax finished in the normal way.

5 Scratch-brushing, bristle-brushing

Procedure

In some cases solutions are applied to the surface of the metal using a bristle-brush or a brass scratch-brush. This enables the surface to be worked to some degree in the presence of the solution. A bristle-brush is generally used to keep the solution in motion on the surface and to even its action, without directly affecting the surface itself. A brass scratch-brush is used where a more positive action is required which will activate the surface to some extent so that it is receptive to the effects of the solution. Both of these techniques are best carried out using small circular motions of the brush, except where a texture or grain dictates otherwise.

Sulphide solutions

These techniques are of particular importance in the development of colours using sulphide solutions. In these cases the bristle-brush is used to lay the solution on to the surface and to clear any loose superficial deposits that are produced. The brass scratch-brush is more commonly used and is an essential component in the production of many finishes, where it is employed to work the solution into the surface of the metal and also in many cases to burnish the result, often with the use of hot or cold water once the initial colour has been established. The details of these procedures are given in the notes that accompany individual recipes.[20,21]

Commercial sulphide solutions

Sulphide solutions are often used commercially to produce black, steel-grey and bronzing colours on the cuprous metals and silver. In many cases proprietary solutions are available which have been developed to produce this particular range of colours. They generally produce good results and are probably the simplest and most economical method of obtaining these finishes. Many of them also involve the use of scratch-brushed and bristle-brushed application techniques.

20. An additional technique has been reported for producing variegated finishes on copper. This involves rubbing the surface of the metal with a lump of potassium polysulphide, under water. The partially adherent black colouring produced is then scratch-brushed with a very dilute solution of the polysulphide. The process may be repeated, as necessary, to develop the variegated finish.

21. It is also reported that even and highly burnished sulphide finishes can be produced by barrelling objects in a barrel polisher containing a solution of ammonium sulphide. The barrelling beads should be thoroughly washed and treated with a rust inhibitor after use.

6 Applied pastes

A restricted range of chemicals have traditionally been used in the form of pastes, to colour objects on a variety of scales. The most commonly used are the statuary pastes based on sulphides, which are used to produce bronzing finishes. These are discussed separately below. The remainder tend to produce variegated surfaces of various colours, which often include the development of some green or blue-green patina. Pastes of this type were investigated in the research as a possible means of developing patination in cases where the direct application of solutions failed to wet the surface of the metal. Two basic approaches were tried. Patinating solutions were thickened into pastes, using inert media such as clays, gelatin or flour to provide the body. These generally failed to produce patination, probably because the resulting pastes tended to exclude air from the surface. The second approach was to take the solid ingredients of patinating solutions and to add only very small quantities of water, so that concentrated thick pastes were produced. These did produce positive results, but they were markedly different from those produced by the application of the equivalent patinating solutions, tending to result in very variegated surfaces.

Preparation of pastes

The solid ingredients should be ground to a fine consistency using a pestle and mortar. The liquid components are then added very gradually, with continued grinding. The additions must be made very gradually as the paste can easily become too thin. A thick creamy consistency is the most suitable, as thinner pastes tend to drain from vertical surfaces and to pool in concave surfaces and hollows.

Application

There are two basic approaches to the application of pastes, which tend to give rise to slightly different results. They can be applied sparingly by wiping with a soft cloth and allowed to dry; the applications being repeated a number of times as necessary until the surface has developed. The depth of colour and precise surface quality obtained will depend on the period of treatment. Alternatively, the paste can be applied thickly by brushing, and left for a longer period to dry out completely. The residual dry paste is then cleared by rinsing in cold water, revealing the underlying colour. This method tends to encourage the formation of variegated green or blue-green patinas on coloured grounds of various kinds. Variations in the surface quality obtained can be made by altering the method of application slightly. The paste can be stippled into place, applied locally in patches so that the surface is gradually built up, or the applied paste can be 'raked' so that lines are produced that nearly break through to the metal surface.

In some cases a paste may fail to dry out, and gradually becomes more liquid. If this occurs the residue should be rinsed away in cold water, and the application repeated with a paste of a thicker consistency. Whichever method of application is adopted, the object must be allowed to dry out thoroughly for some time before finishing is carried out.

Sulphide pastes

Sulphide-based pastes are commonly used for the colouring of statuary and architectural bronzework. These can be applied in the same way as the pastes described above, but a particular method suited to sulphides can be used to produce a more even colouring. The paste is mixed to a thick creamy consistency and brushed onto the surface of the object, so that it is quickly covered with an even thin layer. After a few minutes the residue can be gradually brushed off with a stiff bristle- or nylon brush, which is frequently dipped into a mixture of the dry ingredients of the paste, so that the surface is dried as it is burnished. When the paste is initially brushed off it will appear to have hardly coloured the underlying metal, but vigorous burnishing with the brush loaded with dry powder quickly brings up the colour. The surface should finally be burnished with a dry brass scratch-brush charged with a little dry powder.[22]

Commercial bronzing pastes

As the technique of bronzing with sulphide pastes is commonly employed commercially in the colouring of statuary and architectural metalwork, a

22. In addition to the pastes noted in the recipe section, it is reported that chocolate-brown to red-brown colours can be obtained with a paste made from three parts of antimony pentasulphide to one part of ferric oxide, mixed with a little ammonia or ammonium sulphide solution and applied in the manner described in this section. Antimony pentasulphide is sometimes referred to as 'gold sulphur'; this should not be confused with references to 'gold sulphide'.

number of proprietary pastes are available. These are formulated to provide a range of bronzing colours. The results obtained are generally very good, and they are probably the most convenient and economical means of obtaining this type of finish.

7 Torch technique

This method of colouring, which involves stippling solutions onto metal surfaces that are heated with a blow-torch, is well known to sculptors. A wide range of green and brown colours can be produced by this method, but the quality of the finished result depends largely on the skill and artistry of the patinator. The technique is particularly suitable for castings, and is applicable to both small and large objects.

Method

Colour is gradually built up by alternately heating a small area of the metal with an oxidising flame, and then applying the solution to the heated area. The solution is generally applied either by stippling with a paint brush or with a cloth.[23] In some cases it is necessary to work the solution into the surface with a brass scratch-brush.[24] It is essential to apply the solution sparingly, whichever application technique is adopted. The solution should not be flooded onto the surface.

Temperature

The temperature of the metal surface to which the solution is applied is critical. It cannot be measured directly, but can be accurately gauged by the way in which the solution reacts as it is applied. The solution should sizzle as it is stippled onto the surface. If the temperature is slightly too hot then it will tend to spit, and at higher temperatures will spit violently and fail to wet the surface at all. If the temperature is too low, then the solution does not steadily sizzle, but simply cools the metal surface. The correct working temperature is not difficult to assess after a little practice, but considerable skill is required in sustaining the right surface conditions as the colour is gradually developed on an object.

The relative frequency of periods of heating and applications of solution will depend to a large extent on the nature of the object being coloured. A large casting will tend to retain heat for some time and will require less frequent but more sustained periods of heating. A thin sheet material, on the other hand, gains heat quickly but is rapidly cooled as the solution is applied, and requires a more frequent alternation of flame and solution. A satisfactory patina is only produced when the solution reacts with the hot metal surface at a positive but steady sizzling temperature.

Fumes and vapours will inevitably be produced as solutions are applied to the hot metal. These must not be inhaled. Adequate ventilation must be provided, preferably in the form of localised extraction, and a nose and mouth face mask fitted with a filter suitable for acid vapours should be worn.

Finishing

When patination is complete, the object should be left to cool and dry out thoroughly before wax finishing. It is sometimes suggested that wax should be applied while the object is still hot, as this facilitates its application, but this practice is not recommended. Minute quantities of reactants can become trapped beneath the wax, causing imperfections which are difficult to remedy. It is generally better to allow the object to cool completely and dry and then apply the wax in the normal way, after moistening with pure turpentine.

8 Heat colouring

Heat has been traditionally used to produce a restricted range of colours on metal surfaces, without the use of chemicals. The application of heat induces the formation of the natural oxides of the metal, producing thin layers at lower temperatures which yield interference colours or bronzing finishes, and opaque colours ranging from orange through red to brown or black with the thicker films produced at higher temperatures. Many of these

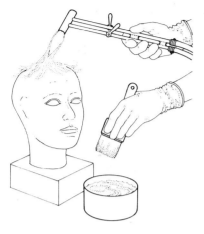

23. Alternate heating and application, using a blow-torch and brush.

24. Where this technique is necessary, it is indicated in the notes to individual recipes.

colours will be familiar as the transitory and variegated colourings that occur when annealing operations are carried out. Although there are a few standard finishes that are produced by heat alone, such as the Bronze Metal Antique finish which is used with bronzes having a high copper and tin content, the technique has generally been little used recently. This is probably because it is difficult to produce even and standardised colours, as the process requires considerable care and skill if particular colours are to be obtained.

Preparation and cleaning

It is essential, when preparing an object for heat colouring, to ensure that the surface is thoroughly degreased and clean, and is not handled after cleaning. Any slight surface contamination will tend to cause imperfections in the final finish which may be indelible, particularly in the case of the bronzing and interference colours produced at lower temperatures.

Colouring

Although colouring can sometimes be carried out using a torch, it is generally preferable to use a kiln or muffle furnace, so that the whole object is more evenly heated. On copper and bronze a range of interference colours is produced at lower temperatures, similar to the temper colours produced on steel, ranging from pale golden yellow, through orange and greenish tints to red. The most distinctive colours are produced on highly polished copper surfaces. Colour development is best monitored visually in practice, as timings will vary with the particular kiln used.

At higher temperatures opaque colours are produced, the most even finishes being black or dark brown. On cast bronze these opaque colours are generally even and adherent. On sheet copper, on the other hand, the black layers tend to flake as the metal cools revealing underlying variegated colourings whose precise form is difficult to predict or control. Although these variegated surfaces are often interesting, it should be borne in mind that they occur at temperatures above the annealing point of the metal, and will therefore have only a restricted application in practice.

Brass is generally less easily coloured by heat alone. Although yellow brass castings can be coloured in a kiln at higher temperatures to produce opaque brown, grey or black finishes, the surfaces produced tend to be variegated and are somewhat unpredictable. Rolled sheet brass is very difficult to colour by heating, producing only very pale lustre colours even at high temperatures, and the technique cannot be recommended for use with this material.

Additional techniques

One particular technique has been used to obtain red and purplish red colours with copper and bronze. The object to be coloured is heated to red heat in a kiln or muffle furnace and then rapidly plunged into boiling water. The technique is clearly limited to objects of a certain scale, particularly since it must be carried out rapidly as it is important to lose as little heat as possible during the transfer from the heat source to the boiling water. A torch can be used to heat the object, but it is generally more difficult to obtain an even colouring by this method, except with small articles. If a torch is used with sheet materials, then it should be noted that more even colourings are produced on the reverse of the face that is heated.

A method of colouring is sometimes used that is particular to the technique of raising copper and gilding metal objects by hammering. After the metal has been worked for some time, it becomes work-hardened and must be annealed by heating with a torch, prior to further hammering. Annealing produces oxide layers on the surface which are normally removed by pickling the object in dilute acid. If the pickling is omitted, then the oxide layers will gradually build up and be modified as the object is successively hammered and annealed. The disposition of colours cannot be controlled, but the technique produces very variegated surfaces which will include many of the colours typically seen in annealing.

9 Vapour technique

Metal surfaces can also be coloured by exposing them to the action of vapours or gases for varying lengths of time. The idea of producing patinas by sealing objects in containers in the presence of particular vapours or gases arose out of the desire to simulate the development of natural patinas by attempting to copy the atmospheric conditions that were supposed to have produced them. The technique is seldom recorded as a practical procedure in the literature, although it has been used by sculptors, and one or two particular recipes are noted in manuals relating to sculpture techniques.

Procedure

Essentially the procedure consists of placing the object to be coloured in a sealed container, in which a solution is present which will liberate a vapour or a gas. The solution is normally placed in the bottom of the container and the object suspended or supported above it in such a position that will ensure the free circulation of vapour around the surfaces that are to be coloured. The container is then sealed and left for some time. Subsequently the container is opened, and the object removed and examined. If further colour development is required then the object is replaced for a further period; if colour development is satisfactory then the object is washed in cold water and dried and finished in the normal way. The technique is not complicated to operate where the vapours used are relatively harmless, but in cases where potentially harmful vapours are used the procedure must be carefully planned. The container must be thoroughly sealed, so that there is no leak of vapour into the atmosphere. In addition, it should be possible to arrange the equipment in such manner that the solution can be introduced into the container, and the vapour and the solution removed from the container, without opening the vessel. An illustration of a suitable arrangement is given in the accompanying diagram.[25] The object is

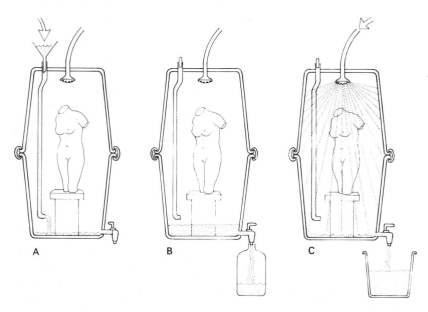

25. Sequence of operations for the vapour technique, in the case of strong vapours. The solution is introduced into the sealed container via a funnel, A. The funnel is removed and the object left. Before opening the container the solution is drained, B. The vapour is cleared by operating the water spray, the water being drained at the bottom, C. When clear of vapour, the container is opened and the object removed.

A   B   C

suspended or placed on a stand in the container, which is then sealed. The solution is introduced into the bottom of the container via a rubber or plastic tube which is then closed with a bung. When the object has been left for an appropriate length of time and needs to be removed, the solution is drained from the bottom of the container by means of the tap provided. The container will still be full of the concentrated vapour, and this is removed by spraying cold water into it from the spray that is set into the lid. The water will gradually remove the vapour from the air in the container, and may be drained via the tap at the bottom. When the spray has been operated for some time, the container may be opened to the atmosphere, and the object removed.

The technique is generally little recorded in the literature on metal colouring, with the exception of the use of ammonia, either in isolation or in conjunction with concentrated acetic acid, and the slow development of a blue-green patina in a damp atmosphere in which carbon dioxide is slowly liberated by the action of hydrochloric acid on marble chips. The tests carried out in research into this technique have largely been concerned with refining the practical procedure, rather than seeking to extend the list of possible solutions, since this appears to be the key to any development of it as a practical possibility. In so far as the procedure outlined above provides a workable method of dealing with strong vapours, a number of alternative colouring agents can be considered, for example nitric acid, ammonium sulphide, ammonium hydrosulphide, and aqueous solutions of potassium polysulphide.

Stress corrosion cracking

In some tests carried out using the concentrated vapours of ammonia solution and acetic acid, a destructive corrosion phenomenon occurred with the rolled sheet yellow brass samples that were tested. Suspension in these vapours caused fine hairline cracks to appear at right angles to the edges of the test samples, and also radiating from the drilled suspension hole. The cracks correspond to areas of high stress in the material, where it had been worked by shearing with a treadle guillotine and bored using a pillar drill. The cracks worsened with continued vapour treatment, penetrating the thickness of the material, and causing total fracture and crumbling of the metal in the region of the cracks.[26] This phenomenon, which is known as stress corrosion cracking, can occur with other materials but is most common with brass. The vapour technique should clearly be used with caution in the case of this material.

10 Particle technique (sawdust technique)

Richly stippled and textured surfaces with variegated colours can be produced by using a technique in which a moisture-retaining medium is wetted with a colouring solution, and left in close contact with the surface for some time. Although a wide range of moisture-retaining media can be used, sawdusts and wood shavings were found to be particularly effective.

Preparing the medium

The medium used must be evenly moistened with the solution. This can be most easily carried out by containing the sawdust or wood shavings in a plastic tray or bucket, and gradually adding small quantities of solution which are thoroughly worked in by hand. It is obviously essential to wear gloves when carrying out these operations. The degree to which the sawdust or shavings are moistened is very important. If the medium is too dry then the effect will be limited to localised points of contact and the solution will tend to dry out and cease to have any effect before the surface is sufficiently developed. It is more likely, however, that the medium will be made too moist, allowing local pools of solution to accumulate on the metal surface, causing unevenness, patchiness and considerable variation in surface development over the object being coloured. As a guide to the correct degree of moistness, two general indications can be used. If a handful of the prepared sawdust is squeezed as hard as possible, then it should not yield any drops of solution. Secondly, if the prepared medium is tamped on to a smooth non-absorbent surface, such as a sheet of plastic, then it should leave only a slight trace of dampness which reflects the texture of the medium. It should not significantly wet the surface, or produce distinct drops or smears of solution.

Solutions

A wide range of chemical solutions can be used to produce a variety of effects. The examples given in the recipe section indicate the effects produced with a number of different types of colouring agents. Other solutions can be used, but it should be borne in mind that not all solutions can be safely mixed with organic materials such as sawdust and wood shavings. Certain classes of chemicals that are generally used as colouring agents and appear in this book under other colouring techniques, but which are not recommended for use with organic materials, are indicated in the chapter on safety. The chemical safety literature should be consulted and expert advice sought if the use of other chemicals is contemplated.

Laying-up the object

Objects are most easily coloured by packing them in a container with the prepared medium. Plastic boxes, buckets or dustbins can be used. Alternatively, cardboard or wooden boxes can be used to provide support for plastic bags which are used to contain the object with the medium. The

26. Stress corrosion cracking in a rolled sheet yellow brass sample caused by the action of ammonia vapour and the vapour from glacial acetic acid, used in succession.

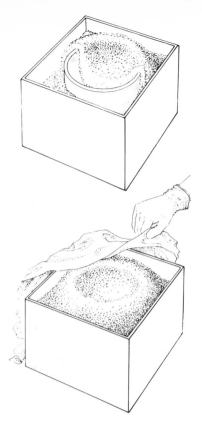

27. The object is embedded in sawdust by placing it in a container and adding sawdust by sprinkling and tamping down. When the object is evenly covered, the container is covered with a plastic bag to prevent excessive drying of the medium at the top of the container.

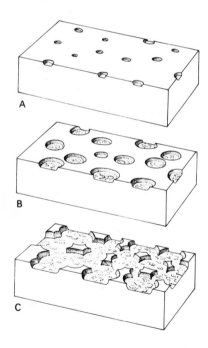

28. Typical surface development obtained with a particle technique, using a fine granular material. The surface is initially etched at the points of contact with the medium, A. As treatment progresses the etched points gradually expand, B, until a stage is reached where the bulk of the surface is etched and points of unetched surface remain, C. Colouring generally occurs both on the etched and unetched parts. Continuation of the treatment will result in a totally etched surface.

particular method used will obviously depend on the scale and type of object being coloured, but it is generally advantageous to use a container which approximates to the form of the object, so that a reasonably even depth of medium around the object is achieved. The prepared sawdust is sprinkled into the container and lightly tamped down to form a bed. The object is placed on this and further prepared sawdust added, by sprinkling and tamping, until the object is completely and evenly covered.[27]

Colour development and monitoring

Although some solutions produce mottled effects or stippled colours, without significantly altering the surface quality of the metal, the majority give rise to selectively etched surfaces. Typically, a granular sawdust will produce localised points of action where the individual grains are in contact with the metal surface. The areas surrounding these points where etching takes place gradually become coloured by the effect of the moist atmosphere of solution present, but are generally not etched initially. If the object is left in the medium, then the etched points tend to extend outwards from their centres, and the remaining surface becomes increasingly affected. A stage is usually reached where the etched areas predominate, leaving a stipple of coloured points which are not markedly etched. Finally the whole surface may be etched back, as these remaining points are progressively lost.[28]

Interesting surface effects are produced at all stages, and it is clearly a matter of personal judgement to determine the stage at which a particular process should be halted. The difficulty is that the object is totally covered and the changes cannot be directly observed. There are two approaches to monitoring surface development in this situation. Probably the most generally useful method is to use one or more samples, which are packed with the object being coloured and which can be removed from time to time to check progress. Alternatively, the object itself can be removed and repacked from time to time, until the surface has developed satisfactorily. The disadvantage with this method, apart from the inconvenience of repacking the object several times, is that the sharp quality of a textured surface may tend to be lost as the object is repeatedly repacked and new areas of the surface are brought into close contact with the granular medium. If the object is removed, then the medium should be thoroughly mixed again before it is repacked, to ensure an even redistribution of moisture throughout.

Removing the object

If the medium is allowed to dry out completely before the object is removed, then particles of medium tend to become bound to the surface as the residual moisture in contact with the surface dries out. These particles are often difficult to remove. It is preferable to remove the object before the medium has completely dried out. If colour development is insufficient when this stage has been reached, the object will need to be repacked with fresh medium and left for a further period. When the surface has developed satisfactorily, the object is removed and any loose material cleared by washing under cold water. A bristle-brush can be used if necessary. It is then allowed to dry out thoroughly, and brushed with a soft dry bristle-brush before wax finishing or lacquering.

Variables

The exact finish obtained will depend on a number of factors, the most significant of which is the degree of coarseness or fineness of the medium used. Fine dusty materials should be avoided as these tend to aggregate into lumps when they are moistened, resulting in unevenness in the surface. A distinctly granular sawdust generally produces good results, giving stippled or closely textured surfaces, and giving rise to the most marked etching effects. Alternatively, very fine wood shavings can be used, such as those produced by a planing machine set to a very fine cut. These also produce stippled or closely textured surfaces, but the etching effects tend to be less marked. Coarse shavings and shredded wood fibres such as those that are used as packing materials can be used to obtain larger-scale, more open textures.

Generally speaking, the coarser materials require significantly longer times to produce their effects. With these materials there is less extensive contact with the metal surface, the colouring depending to a greater extent on the humid atmosphere of the solution in close proximity to the surface.

The finer materials are capable of retaining large quantities of solution because of the much increased surface area that they offer, and maintain a closer direct contact with the metal surface. The action produced by wood fibres and particles appears to have a more positive effect than could be expected if they were simply passive carriers for the solution. They appear rather to act as centres for corrosive action. This action tends to be more pronounced where these points of contact are closely spaced, than where they are sparsely located.

The degree to which the medium is moistened can also significantly affect the timing. Longer times may be required both when the medium is too dry and when it is too wet. It is important to maintain the optimum moisture level, in order to achieve the required surface development without the need to repack the object a number of times.

Types of sawdust

Different types of wood contain quantities of various chemicals that may be liberated when their sawdust or shavings are moistened. Traditionally, the sawdust of woods such as box, or to a lesser extent beech, have been used as drying agents by the metal finishing trades, as they do not produce any tarnishing effect on copper or silver in drying operations. On the other hand softwoods have generally been avoided as they are known to yield resinous substances. In colouring tests carried out using the sawdust technique, a number of different types of sawdust have been used including beech, mahogany and softwoods. In this application no appreciable difference could be observed in the effects obtained, the particle size appearing to be the key factor rather than the type of wood used. It is nevertheless possible that in certain cases variant effects may be produced if different types of sawdust are used.

29. A technique which is reported to have been used commercially at one time, for producing even colouring, consisted of tumbling the object in sawdust which had been moistened with a colouring solution.

Extensions of the technique

The technique can be elaborated in various ways to produce additional effects.[29] The sawdust or shavings can be applied to a surface which has been provided with a ground colour produced by another technique. Alternatively, different solutions can be used in succession, possibly with media having markedly different degrees of coarseness. For example, a granular sawdust might be used to provide a finely textured ground to which coarse shavings impregnated with a second solution are applied, to superimpose a bolder texture.

Localised effects can be produced by masking off areas with resists such as stopping-out varnishes, self-adhesive tapes or films, or wax. Alternatively, on larger scale objects where part of the surface is to be treated, the area to be coloured can be walled off with modelling clays or waxes, providing a localised container for the prepared media.[30]

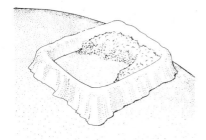

30. Modelling clay used to isolate an area and contain the sawdust, when colouring part of a surface. When the sawdust is in place it is covered with a plastic sheet to prevent excessive evaporation.

11 Appressed materials (cotton-wool technique, cloth technique)

An alternative range of media for retaining colouring solutions in contact with metal surfaces is represented by materials such as cotton-wool and cloth. These tend to act in a different manner from particle media, in that they hold a thin layer of solution in contact with the surface. The precise effects obtained vary, depending on the type of material used and the openness of the weave, but generally the results are variegated.

Procedure

Strips or pieces of cotton-wool or cloth are wetted with the solution and applied to the surface of the object. A stiff bristle paint brush is used to stipple the material into place, so that pockets of trapped air are forced out and close contact with the surface is ensured. The material is applied in small pieces until the whole surface of the object is covered.[31]

Removing the material

If the cotton-wool or cloth is allowed to dry out completely, then the fibres tend to become bound to the surface as the solution dries. These are difficult to remove. It is preferable to remove the applied materials before they are completely dry, so that this is avoided. When the applied materials have

31. Strips of cloth are moistened and stippled into place using a stiff brush, to exclude air and ensure close contact.

32. An example of a coloured and textured finish produced with a moistened cotton cloth with an open weave.

been removed, the object is rinsed in cold water and allowed to dry thoroughly. It may then be wax finished or lacquered as necessary.

Materials

The results obtained depend to a large extent on the particular material used. The majority of examples given in the recipe section involve the use of cotton-wool, which produced the most distinctive results in tests. As a medium, it retains a greater quantity of solution at the surface than woven cloths, tending to produce overall etching effects and variegated surfaces that are smooth to the touch. It is important to ensure that pockets of air are eliminated and that the cotton-wool is in close contact with the surface when the object is prepared. On vertical surfaces and concavities, the solution will drain and pool if the cotton-wool medium is liberally wetted. In these cases it is better to use the solution sparingly so that pooling is avoided, and to repeat the application as necessary.

Woven materials such as cotton cloths tended to produce different results. Relatively smaller quantities of solution are held by these materials, and the results are more akin to those produced by the direct application of solutions. Ground colours are produced first, with blue-green patinas subsequently appearing as the medium dries out. The colours produced generally take on the texture of the applied cloth.[32] In practice it was found that light pure cotton cloth was readily appressed to metal surfaces, where as heavier cloths and those containing a proportion of man-made fibres were more springy and difficult to maintain in close contact with the metal.

Solutions

A wide range of chemical solutions can be used with cotton-wool to produce a variety of effects. The cloth technique can also be used with a wide range of solutions, but is probably most applicable in the case of solutions which are used to produce blue-green patinas. The examples given in the recipe section indicate the effects produced with a number of different types of colouring agent. Other solutions can be used, but it should be borne in mind that some chemicals cannot be used safely with cotton-wool or cloth. Certain classes of chemicals that are generally used as colouring agents and appear in this book under other colouring techniques, but which are not recommended for use with these materials, are indicated in the chapter devoted to safety. The chemical safety literature should be consulted and expert advice sought if the use of other chemicals is contemplated.

Extensions of the technique

The technique can be elaborated in various ways to produce additional effects. Cloths or cotton-wool can be used to modify a surface or part of a surface that has previously been provided with a ground colour using another technique. Alternatively, different solutions can be applied in succession possibly using a combination of finely woven and open textured materials. In addition resists of various kinds such as stopping-out varnishes, self-adhesive films or tapes, or wax, can be used to produce localised effects. A wide range of surface finishes can be achieved with the skilful use of a combination of techniques.

In addition to the materials noted above, absorbent paper can also be used as a solution retaining medium and can be applied in the same way as cloth. Alternatively, absorbent papers can be crumpled and packed around the object, in a manner similar to that used in particle techniques, producing distinctive textures on ground colours that are prepared using other techniques.

Further effects can be obtained if loose fibrous materials, thread, string or similar alternatives are used in place of woven materials or cotton-wool.

12 Partial plate colouring (bi-metallic colouring)

Richly variegated surfaces and effects combining two or more colours can be produced using a technique developed in the course of the research, which exploits the potential of colouring procedures that produce different colours on different metals. Essentially, the technique consists of colouring a partially plated surface, and is applicable to copper plate, silver plate and also to combinations of plated finishes. Most of the available methods of colouring can be used, so long as a different colour is produced on the plated surface than on the underlying metal.[33] Bi-metallic colouring effects that were particularly noteworthy in tests have been included in the recipe sections headed 'Partial plate colouring'. These sections are not exhaustive, but provide a working basis for a technique that clearly has a far wider potential.

Basic procedures

The colouring methods adopted are those used for colouring in general, the particular effects obtained in this case depending on the precise way in which the plated surface is prepared prior to colouring. A number of basic approaches are available.

1 The article can be part-plated initially, by masking with a resist, so that selected areas of the underlying metal remain unplated.
2 The article can be wholly plated and portions of the underlying metal then exposed using abrasive techniques such as emery papers and abrasive mops or wheels, or by cutting through using a variety of methods such as engraving, machining, filing or sgraffito.
3 The article can be wholly plated, selectively masked to protect areas of the plate that are to be preserved, and then sandblasted to expose the underlying metal.
4 The article can be wholly plated and then etched to expose the underlying metal. The areas of plating that are to be preserved are protected with traditional 'stopping-out' resists that are used in etching.

Once the surface has been prepared using one or more of these techniques, the article should be thoroughly degreased and cleaned prior to colouring.

Selecting a preparation method

The simplest effects that can be obtained are produced with a single plated layer on an underlying metal. The surface quality and the edge quality of the coloured areas will depend on the method used when the plated article is prepared. If the article is part-plated initially, then the polished finish on both the plate and the underlying metal will be preserved, and the edges of the coloured areas will be sharply defined.[34] Abrasive techniques which gradually work through the plating will tend to produce more variegated colourings with less precise edges.[35] Etching or sandblasting can be used to produce effects of both kinds, depending on the particular way in which these processes are carried out. They will also produce a textured finish on the exposed underlying metal. The selection of preparation procedures will clearly depend on the type and quality of surface effect desired for a particular object, and on the accessibility of the surface to the various available preparation methods.

Colour

The greatest colour contrast tends to be produced with silver-plate, where the plate remains unaffected and the underlying copper, gilding metal or brass is coloured. Subtler colour combinations are produced where both the plated areas and the underlying metal are coloured. In selecting a colouring method, the relative times taken for colours to develop on the different metals should be taken into account. Widely divergent timings and closely similar timings can both be used to advantage. If, for example, an underlying metal quickly develops a colour while the plated metal takes some time to be affected, then the contrast between the uncoloured plate and the coloured ground can be preserved. If, on the other hand, distinct colours are required on both surfaces then closely similar timings are desirable.

Multi-colour techniques

The technique is not limited to simple two-colour effects, but rather produces its richest possibilities where more than one plated layer is used. Gilding metal or brass, for example, may be copper-plated and silver-plated prior to surface preparation and colouring. The same range of methods can be adopted—partial plating in the first instance, abrading through the plate, or combinations of these techniques. Some of the richest effects obtained in the course of research were produced by copper-plating and then silver-plating on a gilding metal or brass surface, and gradually exposing the layers

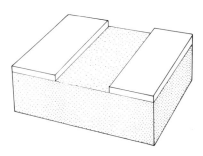

33. With the exception of torch techniques and colouring by heating in a kiln, which should not be used with plated surfaces.

34. An example and diagram of a hard-edge partial plate effect, produced by cutting through the plate, in this case silver-plated gilding metal. Similar hard-edge effects can be produced by stopping out prior to plating. Commercial stopping-out varnishes are available for this purpose.

35. This is of particular value where more than one plated layer is used. See 'Multi-colour techniques', below.

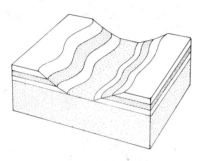

36. An example and diagram of a multi-coloured surface produced by the exposure of successive layers by abrasion prior to colouring.

by abrasion, subsequent treatment by immersion producing differential colouring of the exposed layers.[36]

The technique can be further extended by alternately plating with silver and copper, to build up a sequence of layers that are then exposed by gradual abrasion. Parcel-gilt, which is unaffected by the majority of colouring agents, can also be used to provide an additional contrasting metal finish. A combination of partial plating and subsequent etching, sandblasting and abrasive treatments can be used to achieve surfaces with both hard edged and variegated coloured areas.

Monitoring colour development

In some cases, colouring agents may tend to erode the plated areas at their exposed edges, particularly if these have been feathered by abrasion. Although this does sometimes occur, it is a problem that can generally be avoided by careful monitoring of the article during colouring. It is advantageous to prepare one or two samples that are plated with the article being prepared for colouring, so that tests can be carried out prior to colouring the article itself.

Commercial plating

It is essential to provide clear and precise instructions for the plater when articles are sent to be plated commercially. Commercial electroplating firms will generally adopt standard plating sequences, including 'flash' coatings, unless clearly advised to the contrary. The plated layers should be substantial and it should be made clear that 'flash' plating is not sufficient.

Cleaning and polishing

If a surface is produced in which uncoloured areas of metal occur, then care must be taken in any subsequent cleaning. The usual proprietary cleaning agents and abrasives that are used to remove tarnish and to clean both silver and copper will generally affect any coloured areas. Great care should be taken when cleaning is carried out. As an alternative to wax finishing, a low temperature stoving lacquer can be used. The problem does not arise in the case of surfaces where both the plated areas and the underlying metal are coloured, as the whole surface can then be wax finished in the normal way and re-waxed as necessary.

13 Finishing

The majority of the colouring procedures used are concluded by thoroughly washing the article, when colouring is complete, to remove all traces of chemicals from the surface. The washing sequence normally ends with a thorough rinsing in cold water. The object must then be dried. It is generally inadvisable to simply leave the wet object to dry, as this can sometimes lead to spots or fringes as isolated patches of moisture evaporate, particularly on plain surfaces coloured by immersion. Surfaces can be dried by dabbing with a soft absorbent material such as a paper towel or a soft dry cloth. Alternatively they may be dried in sawdust. The sawdust used should be fine and dry, and sieved to remove any chippings or splinters of wood that may be present. Softwood sawdusts from pine, deal and other resinous woods should be avoided as they may exude acid resins which can mark coloured surfaces. Traditionally boxwood and beech sawdusts have been used as drying agents in metal finishing, and these and other non-resinous hardwoods are generally satisfactory. In commercial applications drying is normally carried out in a stationary or rotating barrel in which the sawdust is heated. In the average workshop slightly simpler methods may be adopted. Small articles, such as items of jewellery, can tumbled by hand in a closed container of sawdust. In the case of larger objects the sawdust can be sprinkled directly onto the surface until it dries. Initially the sawdust will tend to stick to the moist surface of the piece, but as further sawdust is continuously added this will gradually fall away as the surface is dried out. When the dust no longer adheres to any part of the surface, the object can be lightly brushed with a soft brush or dusted with a soft cloth prior to wax finishing or lacquering.

Relieving

If a relieved surface is required, then relieving should be carried out at this stage, prior to wax finishing or lacquering. Relieving is the term used to describe the gentle abrasion of a coloured surface which is used to lighten the tone or provide a gradation in tone, or to selectively expose the underlying metal.[37] It is normally carried out using a bristle-brush with a fine abrasive powder such as whiting or fine pumice, or with combinations of these. Alternatively it can be carried out with a brass scratch-brush. It is a technique that calls for considerable care and skill. Relieving was at one time frequently used as an after-treatment in the colouring of bronze sculpture, particularly on pieces where areas of deep texture or relief modelling were prominent. It was normally carried out on the raised areas of a piece, which were highlighted by relieving, while the hollows and recesses were left fully coloured.

Pigments

Although the use of pigments falls outside the general scope of this book, it should nevertheless be mentioned that coloured or patinated surfaces were often modified by the use of pure pigments in the form of finely ground powders, particularly in the French tradition of colouring bronze sculpture. These were applied to the coloured surface by taking a little of the powdered pigment on a soft bristle-brush and lightly brushing the object. Plain ground colours could be modified in this way to achieve subtle changes of colour and tone. Plumbago was often used to produce a steely or leaden effect on certain types of surface, while pigments such as chrome yellow were used to alter the colour of a green or blue-green patina, or to add warmth of tone to a plain brown ground. After the surface had been lightly and rapidly buffed with soft cloths or brushes, it was usually wax finished. Although a limited and selective use of this technique can enhance the quality of a surface and yield lasting results, the use of wax to bind various loose materials to the surface clearly cannot be generally recommended.

Wax finishing

When colouring is complete and the object is thoroughly dry, the coloured surface will need to be sealed. Thorough drying before sealing with waxes, or lacquers and varnishes, is essential. If even minute traces of moisture become trapped they will give rise to spots or opaque cloudy patches in the finished surface, which are difficult to remedy.[38]

A traditional method of finishing bronze sculpture, which is still commonly used, is to apply a natural wax such as beeswax, which not only seals the surface and makes it more resistant to wear, but also brings out the full colour of the patina. Pure beeswax is rubbed on with a soft cloth that has been moistened with a little pure distilled turpentine to soften it, and then rubbed and burnished with a soft cloth, or with a soft bristle-brush in the case of as-cast or textured surfaces. This finish is normally found to be satisfactory for sculpture and decorative metalwork. Harder waxes are available, including commercial preparations of carnauba wax, which can be applied with a cloth and which harden considerably after drying for some time. The more recently developed silicone polymer waxes and sealants are probably more suitable for objects that are likely to receive constant handling. They have been developed for the protection of painted surfaces and provide very durable and hard finishes.[39] They are applied with a cloth, and burnish up well after drying, bringing out the full colour of the patination as do natural waxes. Some silicone polymer sealants also contain ultra-violet inhibitors.

Wax finishes are normally maintained simply by burnishing the surface with a soft cloth from time to time, and adding a little additional wax very occasionally, as necessary.[40]

Lacquers and varnishes

The traditional alternatives to wax, commonly used on sculpture and a wide range of domestic articles, are the lacquers and varnishes. These were originally spirit lacquers based on natural resins, that were applied by dipping or brushing and usually air-dried. The natural resins used tend to have an inherent golden or reddish colour which was sometimes augmented, or even totally altered, by the addition of dyes. Air-drying lacquers and

37. An example of a plain surface that has been relieved by bristle-brushing with pumice, to produce a change in tone.

38. This sample was wax finished prematurely. Some slight traces of moisture in the surface, which had not dried out, have caused patches of opacity in the finish. In extreme cases the surface can blister and lift in small patches.

39. It is essential to ensure that the wax used contains no abrasives or cutting agents. Waxes should be applied lightly and allowed to dry thoroughly before bringing up the finish with a soft cloth.

40. Lemon oil and oil of lavender were used to finish patinated bronze in the French colouring tradition. Waxes and oils still have some application in the protection of bronzed finishes. Ornamental bronze work in some buildings is still maintained by regular cleaning, brushing and re-oiling with lemon oil, or linseed oil in some cases.

varnishes based on natural resins are still used, particularly in the brushed finishing of larger objects.

More recent air-drying lacquers have been based on nitro-cellulose or synthetic resins, with a wide range of additions to meet specific requirements. Harder and more durable finishes can be obtained using stoving lacquers based on a range of synthetic resins such as urea-formaldehyde, melamine-formaldehyde, acrylic and epoxy resins. These may be stoved at various temperatures, and it should be noted that only the low temperature formulations and stoving schedules are generally suitable for metal colouring applications. In many cases the equipment required, and the safety regulations that must be adhered to, imply that the process will need to be carried out commercially or in specialised workshops. The process of lacquering is sensitive to factors such as dust, ambient temperature, and air humidity, and these need to be carefully controlled to obtain good finishes.

Additional alternatives include the use of brushed-on polyurethane finishes and two-part epoxy coatings of various kinds. The range of air-drying lacquers, stoving lacquers and varnishes available is very wide, and it is advisable to consult the manufacturers when selecting a finish for a particular application. The choice of finish will depend chiefly on the scale of the object, the surface quality and degree of resistance required, and the equipment and facilities available for carrying out the application.

Although tough finishes are obtained with this range of materials, their visual quality depends to a large extent on the application technique used and the care with which they are applied. While it is a relatively simple matter to obtain a moderately thick coating evenly using brushed application techniques, the thickness of the resulting film may tend to obscure the colouring or patination rather than enhance it. In addition the quality of the film is generally more difficult to control on edges and sharp changes in surface direction, tending to make it particularly obtrusive at these points. The very fine coatings that are desirable in the case of colouring and patination require considerable skill to achieve, and may involve the use of application techniques such as electrostatic spraying which can only be carried out commercially in specialised workshops.

Lacquers and varnishes tend to yield high gloss finishes which do not necessarily complement the patination. If less reflective satin finishes are required then the gloss finish produced by natural resin lacquers can be taken down by careful light rubbing with a fine wire wool charged with a little beeswax. Similar techniques can be used with many of the more recently developed lacquers and varnishes, but methods can vary according to the particular material involved, and the manufacturer should be consulted for the correct procedure to apply in a particular case.

Benzotriazole

Benzotriazole is a compound which has been extensively used to prevent the tarnishing and corrosion of copper and its alloys. In recent years it has also been used by conservators to treat copper and bronze antiquities, including artefacts covered by a variety of corrosion products or patinas. Objects are treated by immersing them, for varying periods of time, in cold solutions of benzotriazole in water or alcohol. The treatment is reported to provide effective inhibition, preventing surface deterioration due to atmospheric and other factors. Although the treatment has not been used specifically in metal colouring applications to date, a combination of treatment with benzotriazole and subsequent wax or lacquer finishing may be of interest in particular cases and worthy of further investigation.

THE PLATES

LIBRARY, JACOB KRAMER
COLLEGE, LEEDS

PLATE I

49

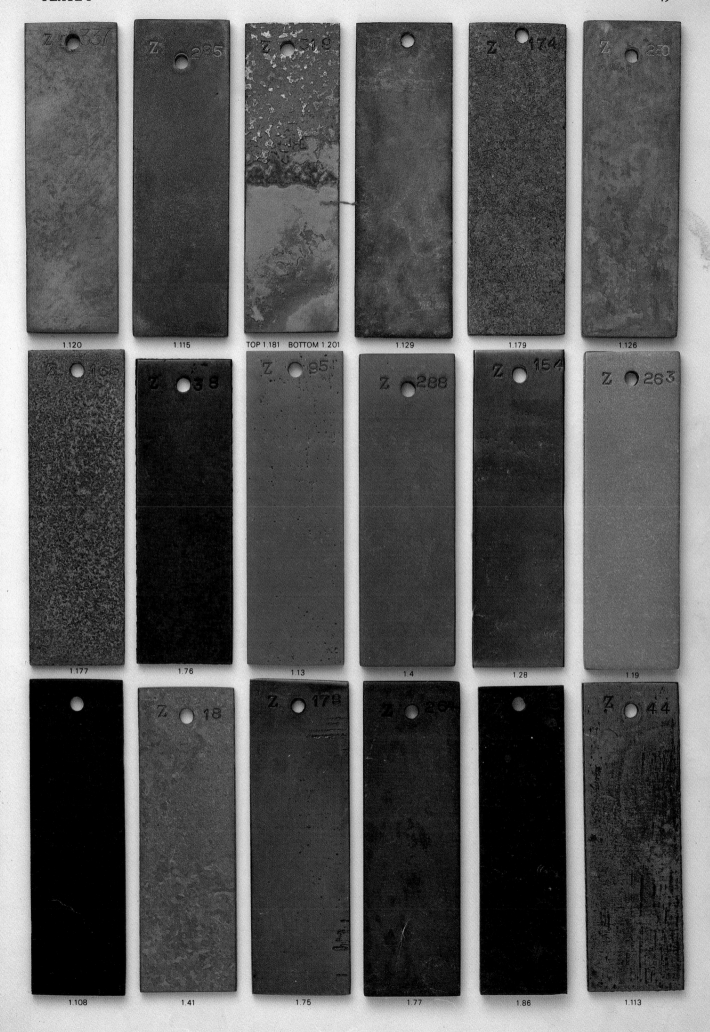

1.120   1.115   TOP 1.181 BOTTOM 1.201   1.129   1.179   1.126

1.177   1.76   1.13   1.4   1.28   1.19

1.108   1.41   1.75   1.77   1.86   1.113

PLATE II

50

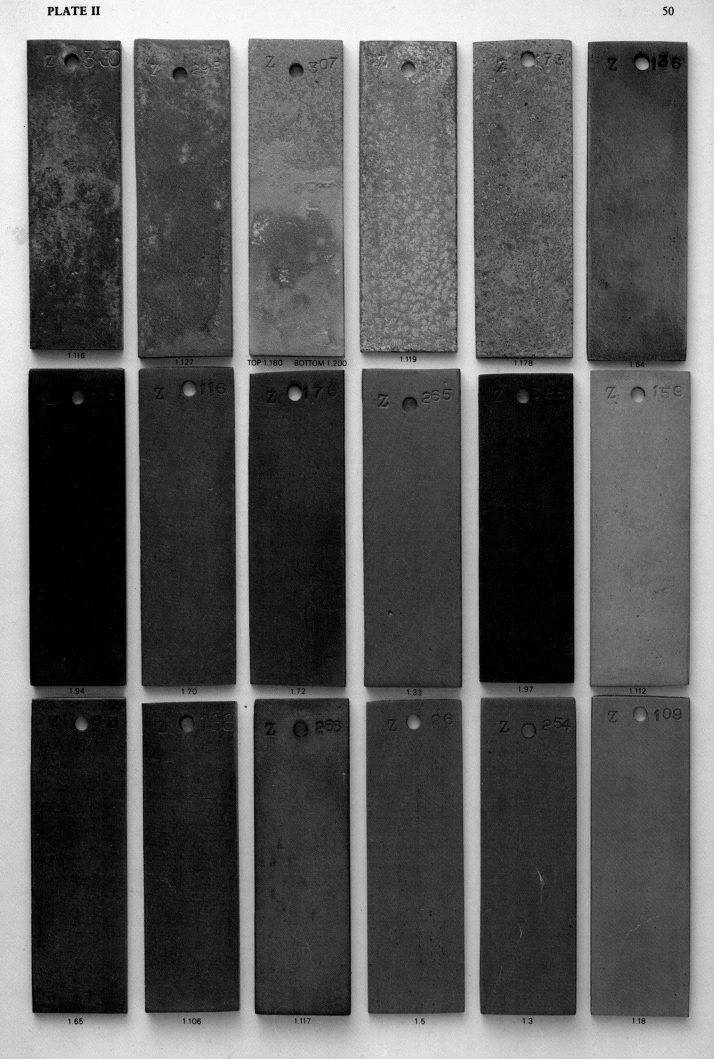

1.116

1.127

TOP 1.180   BOTTOM 1.200

1.119

1.178

1.64

1.94

1.70

1.72

1.33

1.97

1.112

1.65

1.106

1.117

1.5

1.3

1.18

PLATE III   51

2.10   2.53   2.32   2.149   2.69   2.137

2.146   TOP 2.175   BOTTOM 2.196   2.112   2.90   2.128   2.52

2.40   2.46   2.129   2.117   2.147   2.3

PLATE IV

52

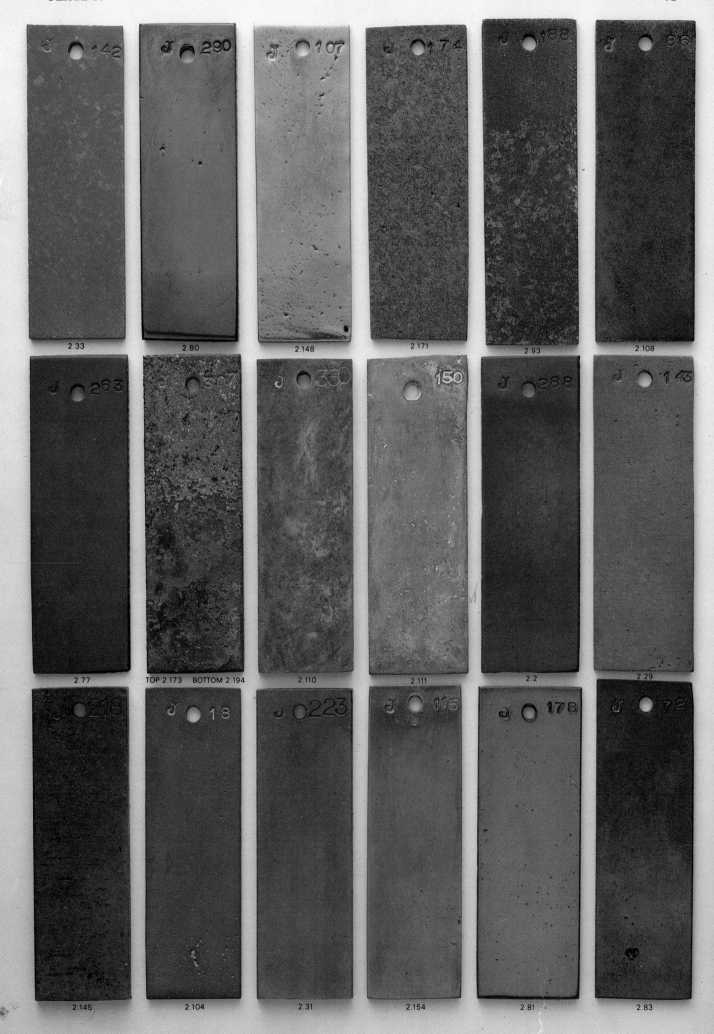

2.33 2.80 2.148 2.171 2.93 2.108

2.77 TOP 2.173 BOTTOM 2.194 2.110 2.111 2.2 2.29

2.145 2.104 2.31 2.154 2.81 2.83

PLATE V 53

3.93   3.89   3.160   3.186   3.177   3.38

3.105   3.140   3.148   3.142   3.31   3.20

3.35   3.128   3.129   3.5   3.137   3.127

PLATE VI 54

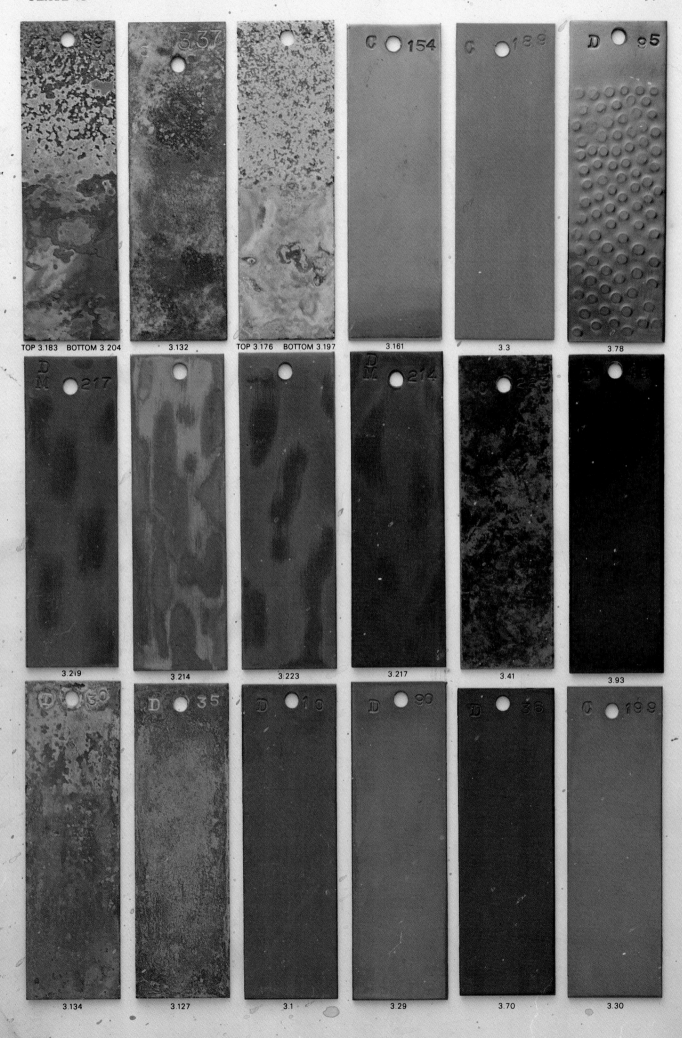

TOP 3.183 BOTTOM 3.204   3.132   TOP 3.176 BOTTOM 3.197   3.161   3.3   3.78

3.219   3.214   3.223   3.217   3.41   3.93

3.134   3.127   3.1   3.29   3.70   3.30

PLATE VII 55

4.123 4.139 4.98 4.124 4.71 4.11

4.52 4.1 4.9 TOP 4.183 BOTTOM 4.204 4.173 4.33

4.172 4.97 4.76 4.152 4.175 4.136

PLATE VIII

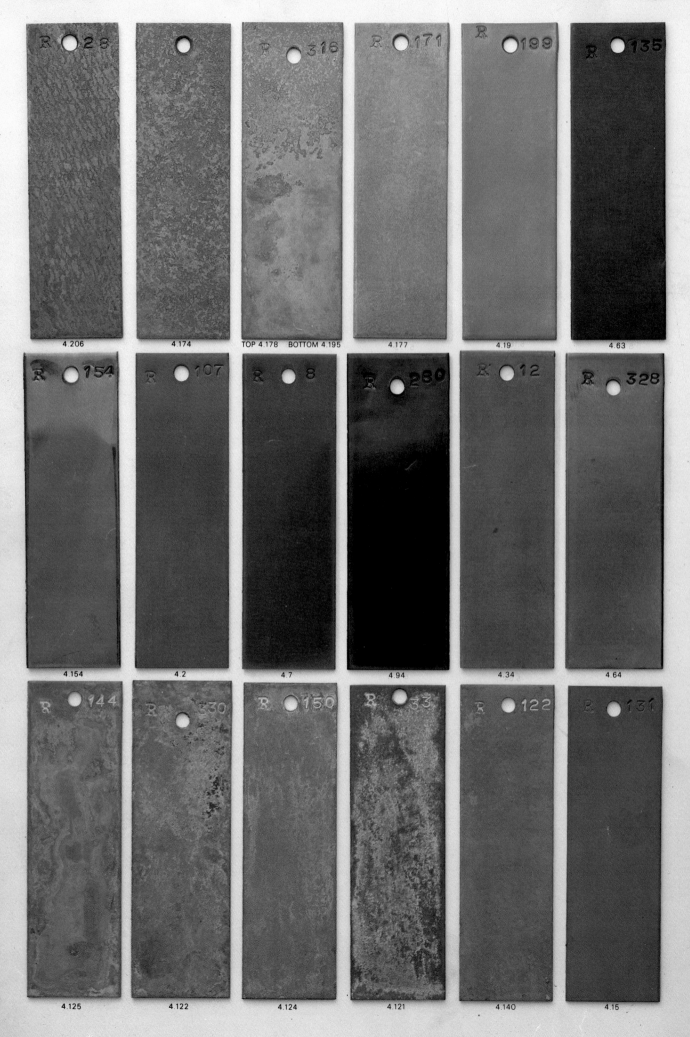

4.206   4.174   TOP 4.178   BOTTOM 4.195   4.177   4.19   4.63

4.154   4.2   4.7   4.94   4.34   4.64

4.125   4.122   4.124   4.121   4.140   4.15

PLATE IX

57

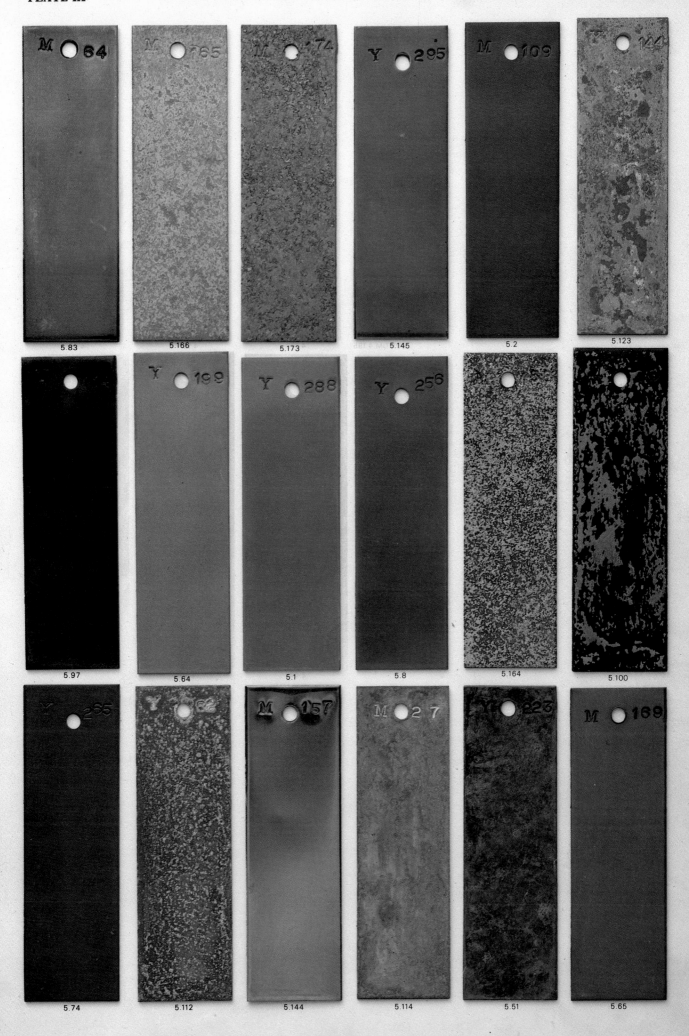

5.83

5.166

5.173

5.145

5.2

5.123

5.97

5.64

5.1

5.8

5.164

5.100

5.74

5.112

5.144

5.114

5.51

5.65

PLATE X

58

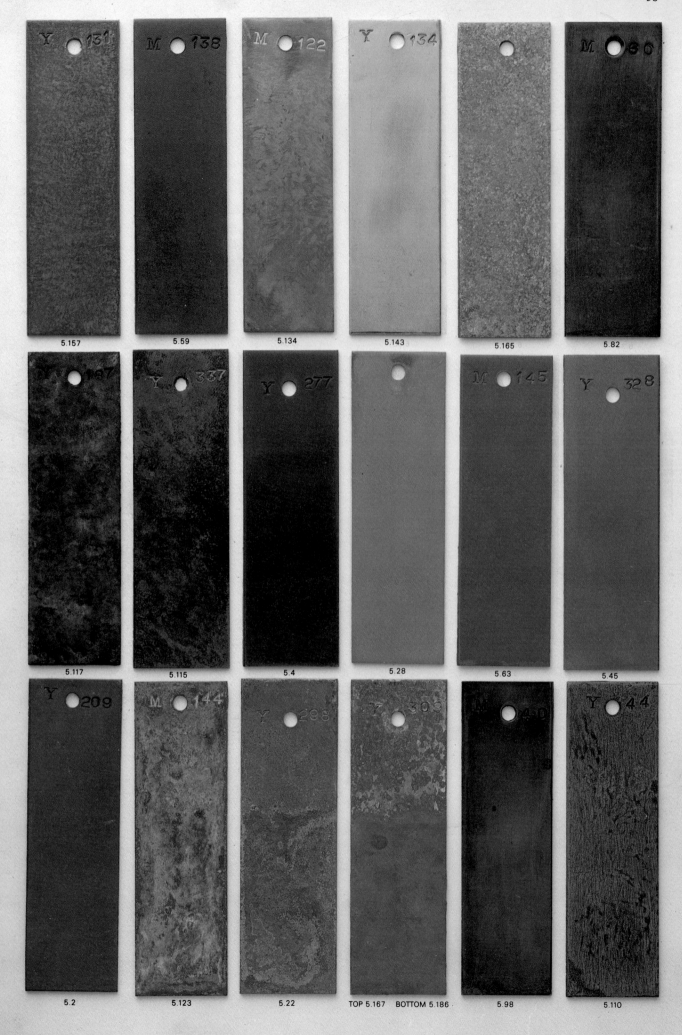

5.157

5.59

5.134

5.143

5.165

5.82

5.117

5.115

5.4

5.28

5.63

5.45

5.2

5.123

5.22

TOP 5.167   BOTTOM 5.186

5.98

5.110

PLATE XI

59

6.33   6.35   6.30   6.34   TOP 6.39   BOTTOM 6.48   6.36

6.2   6.22   6.1   6.25   6.27   2.3

6.63   6.68   6.59i   6.59ii   6.60   6.67

PLATE XII 60

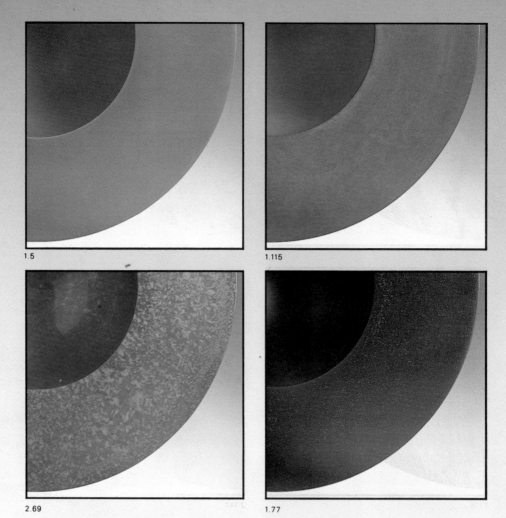

1.5

1.115

2.69

1.77

2.2

PLATE XIII 61

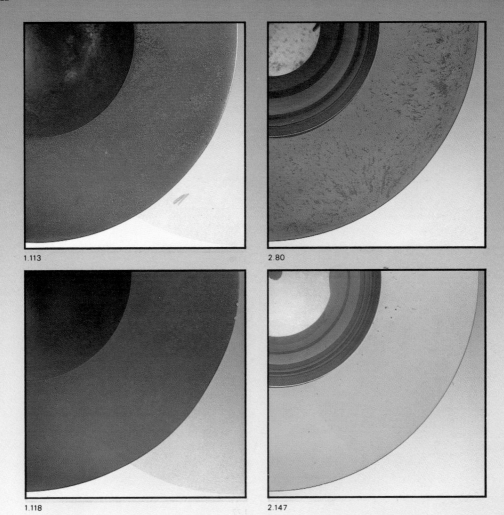

1.113

2.80

1.118

2.147

1.27

PLATE XIV

62

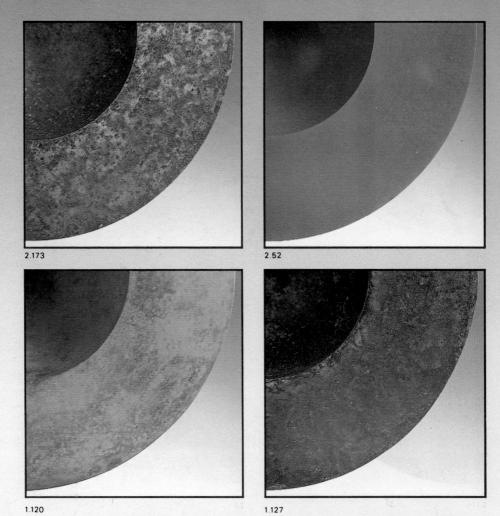

2.173

2.52

1.120

1.127

1.176

PLATE XV   63

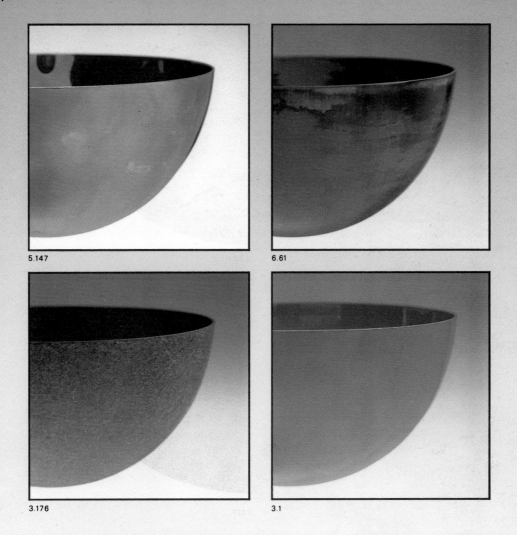

5.147

6.61

3.176

3.1

5.2

PLATE XVI

64

6.62

3.211

3.45

4.96

3.177

THE RECIPES

The recipes are grouped into six main sections, dealing with cast bronze, cast yellow brass, copper and copper-plate, gilding metal, yellow brass, silver and silver-plate. Within each section, the recipes employing the more usual techniques of metal colouring are grouped according to the colour and surface finish that is obtained. These are followed by recipes involving techniques developed during the research, that produce very variegated colours and textures, and are grouped according to the technique used. A list of recipes that were not successful in tests is included in an appendix for information.

Each recipe gives a list of ingredients, outlines the procedure to be used, and provides information about stages in the process of colouring. Detailed information about the techniques employed and the equipment required are given in the chapter on metal colouring techniques, to which the reader should refer when selecting a recipe and planning the colouring of a particular piece. The reader is also strongly advised to consult the chapter on safety at this time. Although specific hazards are indicated in the notes accompanying each recipe, the more general safety considerations are as important and must not be ignored.

The difficulties that may be encountered during colouring are given in the notes accompanying individual recipes. Recipes are described as 'not recommended' if they proved to be particularly intractable in tests, and are likely to be problematic in colouring objects. Those recipes, on the other hand, which readily produced consistently good results are marked with an asterisk.

There are various factors in the case of each colouring technique which will affect the timings quoted in recipes. One of the more obvious is the question of scale. Where a surface is worked with a scratch-brush or a torch, the time taken will clearly depend on the size of the object. Timings, where quoted, relate to the colouring of small surfaces. In the case of immersion colouring, the timings given are for small objects where one or two litres of solution are involved. It should be noted that as the size of object is increased, and larger quantities of solutions are used, timings tend to become progressively extended; an object requiring ten or fifteen litres of solution taking up to two or three times as long to colour. In the case of particle techniques, and in the use of appressed materials, timing will be significantly affected by the coarseness of the medium and the degree to which it is moistened.

Each procedure has variable factors of this kind, which are discussed in the particular section relating to it in the chapter on Metal Colouring Techniques. Generally speaking colour development should be monitored visually and the timings quoted used as a guide to the general pattern of colour development.

Marginal notes dealing with hazards and safety precautions are marked with a triangle ▲.

Recipe selection

Selecting a recipe for a particular application obviously involves rather more than simply choosing a colour. In addition to the particular design problems that may be associated with specific objects, two important general considerations should be borne in mind when planning pieces that involve bronzing, colouring or patination.

Firstly it should be noted that chemically induced finishes are generally unsuitable for use on articles for foodstuffs, and *we cannot recommend that the finishes given in this book be used on surfaces that will come into direct contact with food or drink.*

Secondly, handling and wear are factors that should be taken into account. Finishes can vary from deeply etched surfaces to light blooms of colour, and the choice of a recipe will clearly depend on the particular application. Some finishes may not be suitable for functional use in a particular case, or may need to be used in conjunction with textured rather than plain surfaces, for example where constant handling is expected. It is advisable to carry out a small scale test before embarking on the colouring of a piece, where investment in a considerable quantity of chemicals may be involved. This will help to ensure that the surface quality and toughness produced are appropriate to the proposed application.

The principle applications for metal colouring, bronzing and patination are, as they have traditionally been, in the fields of decorative fine metalwork and sculpture.

1 Cast bronze

The particular bronzes to which the recipes and results refer are LG3 (copper 86%, tin 7%, zinc 5%, lead 2%) and LG4 (copper 87%, tin 7%, zinc 3%, lead 3%). These bronzes are commonly used by art bronze foundries in Great Britain. Bronzes with a significantly different composition will in many cases yield variant colourings.

The main colour heading for each recipe refers to the results obtained on polished surfaces, unless otherwise stated. Reference is made to the colour and surface finish obtained, and to grain enhancement where this has occurred. The results obtained on rough or as-cast surfaces are indicated in the accompanying marginal notes.

	Recipe numbers
Red Red-purple/red-orange/red-brown	1.1 – 1.18
Brown Light bronzing/brown/dark brown	1.19 – 1.96
Black Black/dark slate/grey	1.97 – 1.114
Green patina	1.115 – 1.153
Lustre colours	1.154 – 1.175
Special techniques	1.176 – 1.213

1.1*   Red Satin/semi-matt

Copper nitrate	80 gm
Water	1 litre
Ammonia (.880 solution)	3–5 cm³

Boiling immersion (Thirty minutes)

The article is immersed in the boiling solution. After five minutes a pink tinge is apparent, followed by a gradual and even build-up of colour. After thirty minutes the article is removed and washed thoroughly in hot water before drying in sawdust. When dry the surface has a characteristic greyish bloom which becomes nearly invisible after wax finishing, and is only apparent when viewed from oblique angles.

Increasing the quantity of ammonia added, up to about 10 cm³/litre, produces a brighter colour, but this is accompanied by a decrease in adherence and a tendency for darker marks and irregularities to appear.

Reducing the ammonia content to below 3 cm³/litre tends to produce results similar to those obtained with concentrated copper nitrate alone. Decreasing the overall concentration of the ingredients also has this effect.

1.2   Light reddish-purple Semi-matt/semi-gloss

Copper nitrate	200 gm
Water	1 litre

Boiling immersion (Twenty minutes)

The article is immersed in the boiling solution. After five minutes the surface darkens and acquires a pink tinge, gradually becoming brown as immersion continues. A drab grey layer forms during the later stages of immersion. After twenty minutes the article is removed and washed in hot water. The grey layer washes away revealing the light reddish-purple beneath. The article is washed thoroughly and dried in sawdust. When dry, the surface exhibits a characteristic grey bloom, which becomes almost invisible when the article is wax finished.

Tests carried out using lower concentrations of copper nitrate produced similar results. These were lighter in colour, pale orange with a purple tinge. At lower concentrations there is more variation in colour over the surface.

Tests carried out using concentrated hot solutions (not boiling) produced similar results, but gave rise to streaks and bare patches.

Tests carried out with prolonged immersion in cold concentrated solutions (36–48 hours; 200 gm/litre at 15–20°C) produced similar colours, but the surfaces were marred by silvery streaks.

1.3*   Reddish-purple Satin   *Pl. II*

Copper sulphate	25 gm
Water	1 litre

Boiling immersion (Ten to thirty minutes)

The object is immersed in the boiling solution. After two or three minutes the surface takes on a reddish-orange colour which gradually darkens with continued immersion. When the surface has attained a reddish-purple colour, which may take from ten to thirty minutes, it is removed and washed thoroughly in hot water. After drying in sawdust it is wax finished.

A reddish-purple colour is produced on both rough and as-cast or polished surfaces.

Tests have shown that the time taken for the full colour to develop is very variable, and in some cases may be longer than thirty minutes.

Tests with more concentrated solutions of copper sulphate tended to produce lighter and more orange results on polished surfaces, although the colour obtained on rough and as-cast surfaces was consistently reddish-purple at a wide variety of concentrations.

1.4*   Rich orange-red Satin   *Pl. I*

Copper sulphate	120 gm
Ammonia (.880 solution)	30 cm³
Water	1 litre

Boiling immersion (Twenty minutes)

Immersion in the boiling solution produces lustre colours after a minute, followed by the development of an orange-pink colour which is well developed after seven or eight minutes. Continued immersion produces a gradual and even deepening of the colour. After twenty minutes the object is removed, washed thoroughly in hot water, and dried in sawdust. When completely dry it is wax finished.

A more purple colour is produced on rough or as-cast than on polished surfaces.

▲ A dense precipitate is formed in the solution which tends to sediment in the immersion vessel, causing 'bumping'. Glass vessels should not be used as the bumping may be severe.

▲ Preparation of the solution and colouring should be carried out in a well ventilated area to prevent irritation of the eyes and the respiratory system by the ammonia vapour. Adequate protection should be worn to safeguard the skin and particularly the eyes from accidental splashes of this corrosive alkali.

1.5*   Red-orange terracotta Gloss   *Pl. II, Pl. XII*

Copper acetate	70 gm
Copper sulphate	40 gm
Potassium aluminium sulphate	6 gm
Acetic acid (10% solution)	10 cm³
Water	1 litre

Boiling immersion (Fifteen minutes)

Immerse the object in the boiling solution. The colour development is gradual and even throughout the period of immersion, producing purple colours after about five minutes. The object is removed after fifteen minutes, washed thoroughly in hot water and dried in sawdust. When completely dry it is wax finished.

The colour obtained on rough and as-cast surfaces is reddish-purple.

1.6*   Purplish red Satin/semi-matt

Copper sulphate	25 gm
Ammonia (.880 solution)	3 cm³
Water	1 litre

Boiling immersion (Thirty minutes)

Immersion in the boiling solution rapidly produces lustre colours within a minute, which quickly change to a more opaque orange colour after three minutes. Continued immersion produces a gradual darkening of the surface to reddish-brown. After about thirty minutes, when the colour is fully developed, the object is removed, washed in hot water and dried in sawdust. When dry it is wax finished.

A more purple colour is obtained on rough or as-cast than on polished surfaces.

Increasing the ammonia content tends to inhibit colour development. At a rate of 10 cm³/litre of ammonia, with the stated concentration of copper sulphate, no colour is produced.

1.7   Red on golden bronze (variegated) Satin/semi-matt

Copper sulphate	125 gm
Ferrous sulphate	100 gm
Acetic acid (glacial)	6.5 cm³
Water	1 litre

Boiling immersion (Thirty to forty minutes)

The object is immersed in the boiling solution. The surface darkens gradually during the course of immersion, initially tending to be drab and rather uneven, but later becoming more even and generally darker and richer in colour. When the colour has developed, after thirty to forty minutes, the object is removed and washed in hot water. It is dried in sawdust, and wax finished when dry.

A variegated purple and ochre is produced on rough or as-cast surfaces. Tests on polished surfaces gave variable results. The golden bronze ground was consistently produced, but there was variation in the extent to which the red colour developed.

▲ Acetic acid in the highly concentrated 'glacial' form is corrosive, and must not be allowed to come into contact with the skin, and more particularly the eyes. It liberates a strong vapour which is irritating to the respiratory system and eyes.

1.8   Rich red-purple Semi-matt

| Copper acetate | 80 gm |
| Water | 1 litre |

Boiling immersion (Forty-five minutes)

Immerse the object in the boiling solution. After five minutes a pink/orange colour develops on the surface, which gradually darkens as immersion continues to produce a deep lilac tinged with orange after twenty minutes. After forty-five minutes the object is removed, washed thoroughly in hot water and dried in sawdust. When completely dry it is wax finished.

On rough and as-cast surfaces a darker and more even purple is produced. On polished surfaces there is some variation in colour over the surface.

After about fifteen minutes boiling, a green precipitate is formed which becomes more abundant and grey in colour as boiling continues. This should be washed from the surface of the object occasionally, either by agitating the object in the colouring vessel, or by transferring briefly to a rinsing tank of hot water.

1.9   Reddish-purple (variegated) Semi-matt

Object heated and plunged into boiling water

The object is heated evenly to a full red/orange colour using a blow torch or preferably a kiln or muffle furnace. It is then immediately plunged into a bath of turbulently boiling water. When it has cooled to the temperature of the water, it is removed from the bath and washed in cold water. It may then be dried in sawdust and wax finished when dry.

The temperature of the object as it enters the water appears to be critical. The equipment should be arranged so that as little heat as possible is lost during the transfer from heat source to immersion in the boiling water.

Although good results are readily achieved with small-scale items and thin materials, as the scale and wall thickness increases so it becomes more difficult to obtain an even colour.

1.10   Terracotta Satin

Copper acetate	50 gm
Copper nitrate	50 gm
Water	1 litre

Boiling immersion (Thirty minutes)

Immersing the object in the boiling solution produces lustrous golden colours after one minute, which gradually change to more opaque matt pink/orange tints after five minutes. Continued immersion produces a gradual darkening to a terracotta. After thirty minutes the object is removed, washed in hot water and dried in sawdust. When completely dry it is wax finished.

On rough and as-cast surfaces the colour produced is reddish-purple.

1.11   Orange/orange-brown Semi-gloss

Copper sulphate	125 gm
Sodium acetate	12.5 gm
Water	1 litre

Boiling immersion   (Ten to fifteen minutes)

An even reddish-purple colour is produced on rough or as-cast surfaces.

After one minute of immersion in the boiling solution a slightly lustrous orange/pink colour develops on the surface of the object. This gradually darkens tending to become brown as immersion continues. When the colour has developed, after ten to fifteen minutes, the object is removed and washed in hot water. It is wax finished after thorough drying in sawdust.

1.12*   Orange-brown (slight petrol lustre) Semi-gloss

Copper sulphate	60 gm
Copper acetate	10 gm
Potassium aluminium sulphate	25 gm
Water	1 litre

Boiling immersion   (Thirty minutes)

A reddish-purple colour is obtained on rough or as-cast surfaces.

The object is immersed in the boiling solution, producing a gradual darkening of the surface to a drab ochre/brown colour after ten minutes. A slight lustre may also be apparent. Continued immersion produces further darkening, although colour development is slow. After thirty minutes the object is removed, washed in hot water and dried in sawdust. When dry, it is wax finished.

1.13*   Orange-brown Gloss   *Pl. I*

Copper acetate	25 gm
Copper sulphate	19 gm
Water	1 litre

Hot immersion   (Fifteen minutes)

On rough and as-cast surfaces the colour obtained is brick red.

The article is immersed in the hot solution (80°C). The colour develops gradually to produce a light terracotta after about five minutes, which darkens evenly as immersion continues. After ten minutes the surface is orange-brown in colour. The article is removed after fifteen minutes immersion, washed thoroughly and dried in sawdust. When dry, the surface has a characteristic grey bloom which is invisible after wax finishing.

1.14   Reddish-brown Semi-matt

Copper sulphate	25 gm
Lead acetate	25 gm
Water	1 litre

Hot immersion   (Thirty minutes)

A mid-brown colour is produced on rough or as-cast surfaces.

Tests showed that a significant brightening of the colour to a more reddish-brown could be obtained by the addition of a small quantity of ammonium chloride. This should be added at a rate of 2.5 gm/litre at the end of immersion, which is then continued for an extra three or four minutes.

▲ Lead acetate is highly toxic, and every precaution should be taken to prevent inhalation of the dust during weighing and preparation of the solution. Colouring should be carried out in a well ventilated area. The fumes must not be inhaled.

The object is immersed in the hot solution (80°C) producing lustre colours after about one minute. Continued immersion produces a gradual change to opaque colours that slowly become more brown. After thirty minutes the object is removed and washed in hot water. After thorough drying in sawdust, it is wax finished.

1.15   Thin orange on grain enhanced ground Semi-matt

Copper sulphate	6.25 gm
Copper acetate	1.25 gm
Sodium chloride	2 gm
Potassium nitrate	1.25 gm
Water	1 litre

Boiling immersion   (About one hour)

The colour on polished surfaces tends to have a thin and translucent appearance, and shows signs of being fragile.

Variegated brown colours are produced on as-cast surfaces.

▲ Potassium nitrate is a powerful oxidising agent and must not be allowed to come into contact with combustible materials.

The article is immersed in the boiling solution and after a few minutes the surface darkens and becomes more opaque as the colour develops. The surface develops gradually as immersion continues, and after about one hour the article is removed and washed in hot water. It is dried in sawdust and may be wax finished when completely dry.

1.16   Greyish bloom on light orange Semi-matt

    A Potassium nitrate   100 gm
        Water   1 litre

    B Copper sulphate   25 gm
        Ferrous sulphate   20 gm
        Acetic acid (glacial)   1 cm^3

    Boiling immersion (Thirty minutes)

The object is immersed in the potassium nitrate solution for a period of about fifteen minutes, which produces a slight brightening of the surface. The object is removed to a bath of hot water, while the copper sulphate, ferrous sulphate and acetic acid are separately added to the colouring solution. The object is re-immersed in the boiling solution, which gradually produces a greyish-brown colour, and removed after about fifteen minutes. It is washed in hot water and allowed to dry in air. When dry it is wax finished.

An even purple colour is produced on as-cast surfaces.

▲ Potassium nitrate is a powerful oxidant and must be prevented from coming into contact with combustible materials.

▲ Glacial acetic acid is very corrosive and must be prevented from coming into contact with the eyes and skin.

1.17★   Orange-brown (slight petrol lustre) Semi-gloss

    Copper sulphate   100 gm
    Copper acetate   40 gm
    Potassium nitrate   40 gm
    Water   1 litre

    Boiling immersion (Ten minutes)

The article is immersed in the boiling solution, which produces drab orange colours after one minute. These gradually darken and become more opaque with continued immersion. The article is removed when the colour has developed, after about ten minutes, and washed in hot water. After thorough drying in sawdust, it is wax finished.

An even reddish-brown colour is produced on rough or as-cast surfaces.

▲ Potassium nitrate is a powerful oxidising agent and must not be allowed to come into contact with combustible material.

1.18★   Yellowish-orange (slight petrol lustre) Satin/semi-gloss   *Pl. II*

    Copper sulphate   25 gm
    Ammonia (.880 solution)   3–5 cm^3
    Water   1 litre

    Boiling immersion (Two to three minutes)

Immersion in the boiling solution rapidly produces lustre colours within a minute. These quickly change to a more even and opaque orange colour. The object is removed when this colour has developed evenly, after two or three minutes. It is thoroughly washed in hot water, dried in sawdust, and wax finished when dry.

A warm brown colour is produced on rough or as-cast surfaces.

Increasing the ammonia content beyond the stated amount tends to inhibit colour development. At a rate of 10 cm^3/litre of ammonia, with this concentration of copper sulphate, no colour is produced.

1.19★   Golden-yellow (grain enhanced) Satin   *Pl. I*

    Copper sulphate   50 gm
    Ferrous sulphate   5 gm
    Zinc sulphate   5 gm
    Potassium permanganate   2.5 gm
    Water   1 litre

    Boiling immersion (Fifteen minutes)

The object is immersed in the boiling solution. The dark brown layer that forms on the surface should be removed after one or two minutes, by removing the object and bristle-brushing under hot water. It is then re-immersed and the procedure repeated after two minutes if necessary. Immersion is then continued to about fifteen minutes, after which the object is removed and washed in hot water, using a bristle-brush if necessary. After thorough washing, it is dried in sawdust, and wax finished when completely dry.

A reddish ochre colour is produced on rough or as-cast surfaces.

1.20   Dull gold (or variegated brown on dull gold) Semi-matt

Potassium permanganate	2.5 gm
Copper sulphate	50 gm
Ferric sulphate	5 gm
Water	1 litre

Boiling immersion (Fifteen minutes)

The article is immersed in the boiling solution, and the surface becomes coated with a superficial dark brown film, which should be removed by bristle-brushing in a bath of hot water. The article is re-immersed in the boiling solution and again becomes coated with dark brown.

If a plain brown surface is required, the brown film should be removed every few minutes as immersion continues. After about fifteen minutes the article is removed and washed thoroughly in hot water. After drying in air it may be wax finished.

If a variegated surface is required, then the brown film should be left on the surface as immersion continues. After fifteen minutes, the article is removed, and allowed to dry in air for a few minutes. The surface is then gently brushed with a dry bristle-brush, which will partially remove the brown layer, leaving a variegated surface. The article should be wax finished.

A slight grain-enhanced effect is produced on polished surfaces. The variegated effect is difficult to obtain.

Similar colours are produced on rough or as-cast surfaces.

The solution should not be boiled for any length of time, prior to the immersion of the article, or the dark layer tends not to be deposited and the variegated effect not produced.

1.21   Light bronzing Gloss

'Rokusho'	5 gm
Copper sulphate	5 gm
Water	1 litre

Boiling immersion (About one hour)

The article is immersed in the boiling solution and after a few minutes the surface darkens and becomes more opaque as the colour develops. The surface develops gradually as immersion continues, and after about one hour the article is removed and washed in hot water. It is dried in sawdust and may be wax finished when completely dry.

A mid-brown colour is produced on as-cast surfaces.

Tests involving the addition of potassium aluminium sulphate (2–3 gms/litre) produced virtually no colouring on polished surfaces.

For description of 'Rokusho' and its method of preparation, see appendix 1.

1.22   Lightly bronzed grain-enhanced surface Satin

A	Copper nitrate	20 gm
	Ammonium chloride	20 gm
	Calcium hypochlorite	20 gm
	Water	1 litre
B	Water	

Cold immersion (A)   (Ten minutes)
Cold immersion (B)   (Twenty hours)

The article is immersed in solution A for about ten minutes, which produces a brightening of the surface to a slightly golden colour. The article is rinsed and transferred to a container of cold water, in which it is left for about twenty hours. The article is then removed and dried in sawdust. When dry, it may be wax finished.

The initial immersion produces a red-gold colouration.

A yellowish or greenish-brown colour is produced on as-cast surfaces.

▲ Care should be taken to avoid inhaling the fine dust that may be raised from the calcium hypochlorite (bleaching powder). It should also be prevented from coming into contact with the eyes or skin.

1.23   Mid-brown Gloss

A	Ammonium carbonate	180 gm
	Copper sulphate	60 gm
	Acetic acid (6% solution)	1 litre
B	Oxalic acid	1 gm
	Ammonium chloride	0.25 gm
	Acetic acid (6% solution)	500 cm³

Hot immersion (Fifteen minutes)

Solution A is boiled and allowed to gradually evaporate until its volume is reduced by half. Solution B is then added and the mixture heated to bring the temperature to 80°C. The article is immersed in the hot solution, the colour gradually darkening until it becomes brown after ten minutes. After fifteen minutes, when the colour has fully developed, the article is removed and rinsed carefully in warm water. The surface should not be disturbed at this stage. It is allowed to dry in air, and wax finished when completely dry.

A slight bloom may occur on polished surfaces during the latter stages of immersion. This becomes invisible after wax finishing.

▲ The ammonium carbonate should be added to the acetic acid in small quantities. The reaction is effervescent, and if large amounts are added the effervescence is very vigorous.

▲ Preparation of the solution, and colouring, should be carried out in a well ventilated area. Acetic acid vapour will be liberated when the solutions are boiled.

▲ Oxalic acid is harmful if taken internally, and irritating to the eyes and skin.

1.24   Mid-brown bronzing Semi-matt

Potassium sulphide	2 gm
Ammonium sulphate	0.5 gm
Water	1 litre

Cold scratch-brushing (Several minutes)

A mid-brown bronze colour with a semi-matt finish is produced on polished surfaces. A darker and more opaque colour is produced on as-cast surfaces.

The article is directly scratch-brushed with the cold solution, and worked for several minutes, alternating with hot and cold water. The technique will gradually dull the surface as a pale bronzing colour develops. When the colour and surface finish have developed, the article is washed in cold water and allowed to dry in air. When dry, it may be wax finished.

1.25   Mid-brown bronze Semi-gloss

Barium chlorate	52 gm
Copper sulphate	40 gm
Water	1 litre

Hot immersion (Fifteen minutes)

An even mid-brown is produced on rough or as-cast surfaces.

The lustre colours produced in the early stages of immersion are generally even, but are less bright than the equivalent colours produced by other methods.

▲ Barium compounds are poisonous and harmful if taken internally. Barium chlorate is a strong oxidising agent and should not be allowed to come into contact with combustible material. It should be prevented from coming into contact with the eyes, skin or clothing.

Immersion in the hot solution (70˚C) produces an immediate golden lustre colour, which changes gradually to pink/green and then to a brownish lustre after about three or four minutes. Continued immersion produces a greater opacity of colour and a gradual development to a yellow-brown or orange-brown colour. When the colour has fully developed, after fifteen minutes, the object is removed and washed in hot water. It is wax finished, after thorough drying in sawdust.

1.26*   Mid-brown (slight petrol lustre) Semi-gloss

Copper sulphate	25 gm
Nickel ammonium sulphate	25 gm
Potassium chlorate	25 gm
Water	1 litre

Boiling immersion (Twenty minutes)

A yellow-green patina on a brown ground is produced on rough or as-cast surfaces.

▲ Nickel salts are a common cause of sensitive skin reactions.

▲ Potassium chlorate is a powerful oxidising agent and should not be allowed to come into contact with combustible material. The solid, and solutions, should not be allowed to come into contact with the eyes, skin or clothing.

Immersing the object in the boiling solution produces an immediate lustrous mottling of the surface. Continued immersion induces a change to a more even orange colour which darkens gradually. After twenty minutes the object is removed and washed in hot water. After thorough drying in sawdust it is wax finished.

1.27   Light orange-brown (slight grain-enhancement) Semi-gloss   *Pl. XIII*

Ammonium carbonate	50 gm
Ammonium chloride	5 gm
Copper acetate	25 gm
Oxalic acid (crystals)	1 gm
Acetic acid (6% solution)	1 litre

Boiling immersion (One hour)

A light greenish-brown colour is produced on rough or as-cast surfaces. The colour finish produced on polished faces tends to be slightly variegated or streaky.

▲ The ammonium carbonate should be added to the acetic acid in small quantities. The reaction is effervescent and if large amounts are added, the effervescence becomes very vigorous.

▲ Oxalic acid is harmful if taken internally. Contact with the eyes and skin should be avoided.

The object is immersed in the boiling solution and an uneven orange-yellow colouration with traces of green is produced after a few minutes. The surface gradually darkens with continued immersion producing a reddish-brown colour. After about one hour, the object is removed and washed in hot water. After thorough drying in sawdust, it may be wax finished.

1.28   Orange-brown Semi-gloss   *Pl. I*

Copper sulphate	6.25 gm
Copper acetate	1.25 gm
Potassium nitrate	1.25 gm
Water	1 litre

Boiling immersion (About one hour)

Variegated brown colours are produced on as-cast surfaces.

▲ Potassium nitrate is a powerful oxidising agent and must not be allowed to come into contact with combustible material.

The article is immersed in the boiling solution and after a few minutes the surface darkens and becomes more opaque as the colour develops. The surface develops gradually as immersion continues, and after about one hour the article is removed and washed in hot water. It is dried in sawdust and may be wax finished when completely dry.

1.29*   Mid-brown Gloss

Ammonium acetate	50 gm
Copper acetate	30 gm
Copper sulphate	4 gm
Ammonium chloride	0.5 gm
Water	1 litre

Boiling immersion   (Seven to ten minutes)

A dark grey-brown is produced on rough or as-cast surfaces. The mid-brown that is produced on polished surfaces tends to be slightly lustrous.

After one or two minutes of immersion in the boiling solution, the object takes on an olive colouration, which gradually darkens to a more ochre colour as immersion continues. The full colour tends to develop in about seven minutes, although it may take a little longer. Prolonging the immersion beyond nine or ten minutes tends to cause the colour to become more pallid, and does not improve surface consistency. The object is removed and washed in hot water. After thorough drying in sawdust, it is wax finished.

1.30   Orange-brown (slight grain enhancement) Semi-matt

Copper nitrate	200 gm
Water	1 litre

Applied liquid   (Twice a day for five days)

A thin layer of blue-green patina on a dark brown ground is produced on rough or as-cast surfaces. On polished surfaces no patina is produced if the solution is applied very sparingly.

The solution is initially dabbed on vigorously with a soft cloth, to ensure that the surface is evenly 'wetted'. Subsequent applications should be wiped on, and used very sparingly, to inhibit the formation of green patina. After each application the article is left to dry in air. This procedure is repeated twice a day for about five days. The article is then left to dry thoroughly in air, and wax finished when completely dry.

1.31   Greenish-brown Matt

Copper sulphate	120 gm
Potassium chlorate	60 gm
Water	1 litre

Hot immersion   (Fifteen to twenty minutes)

A similar but darker and more grey colour is produced on rough or as-cast rather than polished surfaces.

▲ The dense precipitate formed in the later stages of immersion tends to sediment and cause 'bumping'. Glass vessels should not be used. The temperature of the solution should not be raised above 80°C, as the formation of precipitate is increased.

▲ Potassium chlorate is a powerful oxidising agent and must not be allowed to come into contact with combustible material. Contact with the eyes, skin and clothing must be avoided.

The object is immersed in the hot solution (70–80°C). A rapid series of lustre colours is produced, which become more opaque and darken to an orange-brown colour. After about ten minutes a greenish layer begins to be deposited on the surface, as a dense precipitate is formed in the solution. After about fifteen to twenty minutes, when the ground colour is a reddish-brown and the greenish layer is even, the object is removed. It is washed and bristle-brushed under warm water, which removes any loose material, leaving an even greenish colour which appears integral with the brown ground. After drying in sawdust, it is wax finished.

1.32   Reddish-brown (variegated) Semi-gloss

Ferric nitrate	10 gm
Water	1 litre

Torch technique

A darker variegation can be obtained by finally dabbing with a stronger solution (ferric nitrate, 40 gms/litre).

▲ The fumes evolved as the solution is applied to the heated metal should not be inhaled. Adequate ventilation must be provided and a nose and mouth face mask worn, fitted with the correct filter.

▲ Ferric nitrate is corrosive and should not be allowed to come into contact with the eyes or skin. A face mask or goggles should be worn to protect the eyes from hot splashes as the solution is applied.

The object is heated with a blow torch and the solution applied sparingly with a scratch-brush, using small circular motions and working across the surface. The metal gradually darkens to a reddish-brown and the surface can be finished by lightly dabbing with a cloth dampened with the solution. When treatment is complete and the object has been allowed to cool and dry thoroughly, it may be wax finished.

1.33   Mid-brown (fine green patina) Matt/semi-matt   *Pl. II*

Potassium chlorate	50 gm
Copper sulphate	25 gm
Nickel sulphate	25 gm
Water	1 litre

Boiling immersion (Thirty to forty minutes)

After two or three minutes immersion in the boiling solution an orange colour develops on the surface of the object, which gradually darkens with continued immersion. The full colour develops in about thirty minutes, together with a green tinge which begins to appear after about twenty minutes. If this green colour is to be kept to a minimum, the object should be removed after thirty minutes. Colour development and surface quality are generally better with a longer immersion, the green tending to enhance the surface. The object is removed after about forty minutes, washed in warm water, and either dried in sawdust or allowed to dry in air. When dry, it is wax finished.

On polished surfaces the patina is very fine and integrates with the brown ground to produce a unified olive-green appearance. On rough or as-cast surfaces both the ground colour and the patina are darker.

▲ Potassium chlorate is a powerful oxidant and must not be allowed to come into contact with combustible material. The solid or solutions must be prevented from coming into contact with the eyes, skin or clothing.

▲ Nickel compounds are a common cause of sensitive skin reactions. Precautions should be taken to prevent inhalation of the dust, during preparation of the solution.

1.34*   Ochre Matt/satin

Copper sulphate	125 gm
Sodium chlorate	50 gm
Ferrous sulphate	25 gm
Water	1 litre

Boiling immersion (Five minutes)

Immersion in the boiling solution produces a slight etching effect and gradual darkening of the surface. After five minutes when the light ochre colour has developed, the object is removed and washed in hot water. After drying in sawdust, it is wax finished.

On as-cast and rough rather than polished surfaces a darker ochre ground is produced, overlaid with some greyish patina.

For results obtained with prolonged immersion in this solution see recipe 1.130.

▲ Sodium chlorate is a powerful oxidising agent and must not be allowed to come into contact with combustible material or acids. Inhalation of the dust must be avoided. The eyes and skin should be protected from contact with either the solid or solutions.

1.35   Light brown Satin

Potassium chlorate	50 gm
Copper sulphate	25 gm
Water	1 litre

Hot immersion (Fifteen or twenty minutes)

The object is immersed in the hot solution (80°C) and is initially coloured with an uneven golden lustre finish which changes to a more opaque yellow-green. Continued immersion darkens the colour to orange and then finally to a more brown colour. After fifteen or twenty minutes, when the full colour has developed, the object is removed and washed in hot water. When it has been thoroughly dried in sawdust, it is wax finished. A green powdery film may develop when the object is dried, which can be brushed away prior to waxing. The wax finish tends to make any residual green bloom invisible.

A darker brown is produced on rough or as-cast surfaces.

Some areas with a matt finish occurred in tests carried out on polished surfaces.

▲ Potassium chlorate is a powerful oxidising agent and should not be allowed to come into contact with combustible material. Contact with the eyes, skin and clothing must be avoided.

▲ A nose and mouth face mask fitted with a fine dust filter should be worn when brushing away any dry powder from the surface.

1.36*   Grey patina on brown Semi-matt

Copper acetate	40 gm
Copper sulphate	200 gm
Potassium nitrate	40 gm
Sodium chloride	65 gm
Flowers of sulphur	110 gm
Acetic acid (6% solution)	1 litre

Hot immersion. Scratch-brushing (Five minutes)

The article is immediately etched on immersion in the hot solution (70–80°C). A greyish colour, typical with chloride containing solutions, develops within two or three minutes. After five minutes the article is removed, and the uneven surface is scratch-brushed with the hot solution to improve the surface consistency. The article is then washed in hot water and allowed to dry in air. When dry the surface is exposed evenly to daylight, causing a darkening of the colour. When the colour has stabilised and no further signs of change can be detected, it is wax finished.

On rough or as-cast surfaces the grey patina develops to an even finish. On polished surfaces the patina is less well developed and forms a translucent grey on the grain-enhanced brown ground.

▲ Potassium nitrate is a powerful oxidising agent and should not be allowed to come into contact with combustible material. It must not be brought into contact with the sulphur in a dry state, but should be completely dissolved before the flowers of sulphur are added.

1.37   Grey-green patina on mid-brown ground Matt

Potassium chlorate	50gm
Copper sulphate	25gm
Ferrous sulphate	25gm
Water	1 litre

Boiling immersion (Twenty minutes)

A greenish-grey patina on a mid-brown ground is produced on rough or as-cast surfaces.

▲ Potassium chlorate is a powerful oxidant and must not be allowed to come into contact with combustible materials. It should also be prevented from coming into contact with the eyes, skin or clothing.

The object is immersed in the boiling solution. The colour develops gradually during immersion, becoming darker and tending to acquire a bloom of greenish-grey patina during the latter stages. If the bloom becomes excessive it should be removed by gently bristle-brushing the surface. When the required colour has developed, after about twenty minutes, the object is removed and washed in hot water. It is dried in sawdust, and may be wax finished when dry.

1.38   Light bronzing with greenish tint Matt

Copper sulphate	10gm
Lead acetate	10gm
Ammonium chloride	5gm
Water	to form a paste

Applied paste (Several days)

On rough or as-cast surfaces, a brown-olive ground is produced which is thinly covered with fine blue-green patina. On polished surfaces, no patina development occurred in the tests carried out.

▲ Lead acetate is highly toxic, and every precaution should be taken to prevent the inhalation of the dust during weighing, and preparation of the paste. A mask fitted with a fine dust filter must be worn when brushing away the dry residue. A clean working method is essential throughout.

The ingredients are ground to a creamy paste with a little water, using a pestle and mortar. The paste is applied to the object with a soft brush, and is left to dry out completely. When completely dry, after several hours, the dry residue is brushed away with a bristle-brush. This process is repeated three or four times. When treatment has finished, the object should be allowed to stand in a dry place for several days to ensure thorough drying. After a final brushing with the bristle-brush, it is wax finished.

1.39   Buff/brown (slight grain-enhancement) Semi-matt

Ferric oxide	25gm
Nitric acid (70% solution)	100cm³
Iron filings	5gm
Water	1 litre

Cold immersion (Fifteen minutes)

A similar buff/brown colour is produced on an as-cast surface.

▲ Preparation and colouring should be carried out in a fume cupboard. Toxic and corrosive brown nitrous fumes are evolved which must not be inhaled.

▲ Concentrated nitric acid is extremely corrosive, causing burns, and must be prevented from coming into contact with the eyes and skin.

The ferric oxide is dissolved in the nitric acid (producing toxic fumes), and the mixture is added to the water. The iron filings are added last. The article is immersed in the solution, which produces an etched surface, and removed after about fifteen minutes. It is washed thoroughly in cold water and allowed to dry in air, and may be wax finished when dry.

1.40   Buff tinged grain-enhancement Semi-matt

Copper nitrate	80gm
Nitric acid (10% solution)	100cm³
Water	1 litre

Hot immersion (Five minutes)

A light buff/grey colour is produced on as-cast surfaces.

▲ Nitric acid is corrosive and may cause severe irritation or burns. It should be prevented from coming into contact with the eyes or skin.

The article is immersed in the hot solution (60–70°C), which causes etching of the surface and a gradual darkening. After five minutes it is removed and washed in warm water, and allowed to dry in air. When dry, it may be wax finished. (Some sources suggest using this as a ground for a greenish patina, produced by applying a solution of ammonium carbonate (100 gms/litre) by wiping with a soft cloth and allowing to dry. Tests for extended periods produced only the slightest traces of a green tinge.)

1.41   Buff grain-enhanced surface Matt   *Pl. I*

Bismuth nitrate	50gm
Nitric acid (25% solution)	300cm³
Water	700 cm³

Cold immersion (Twenty minutes)

As-cast surfaces are also etched by this treatment, relatively flat areas of the surface also being subject to grain-enhancement in tests.

▲ Nitric acid and sulphuric acid are corrosive and can cause severe irritation or burns. They must be prevented from coming into contact with the eyes and skin.

The article is briefly pickled in sulphuric acid (50% solution) and rinsed in cold water. It is then immersed in the strongly acidified bismuth nitrate solution. The surface darkens initially and then gradually acquires a mid-grey layer. After about twenty minutes, the article is removed and washed in cold water, which removes the grey layer and reveals the underlying buff etched surface. When completely dry, it may be wax finished.

1.42★   Orange-brown (grain-enhanced) Semi-matt

Copper nitrate	85 gm
Zinc nitrate	85 gm
Ferric chloride	3 gm
Hydrogen peroxide (3% solution)	1 litre

Applied liquid   (Twice a day for five days)

The cold solution is dabbed on to the surface of the object with a soft cloth, and the object left to dry thoroughly in air. This procedure is repeated twice a day for a period of about five days. An initial lustrous colouration is gradually replaced by a more opaque colour. Traces of blue-green patina may also appear during the later stages of treatment, if the application is not sufficiently sparing, and these should be brushed away when dry if not required. The object is wax finished when completely dry.

A dark brown is produced on rough or as-cast surfaces.

Patina development can be encouraged, particularly on rough or as-cast surfaces, by a less sparing use of the solution in the later stages of treatment.

▲ Hydrogen peroxide must be kept from contact with combustible material. In the concentrated form it can cause burns to the eyes and skin, and when diluted will at least cause severe irritation.

1.43   Smokey brown on enhanced grain Matt

Ammonium chloride	15 gm
Copper nitrate	30 gm
Sodium chloride	30 gm
Potassium aluminium sulphate	7 gm
Acetic acid (6% solution)	1 litre

Boiling immersion   (Ten minutes)

The article is immersed in the boiling solution. After two or three minutes the etching effect of the solution brings out the grain structure of the casting. Continued immersion gradually darkens the enhanced granular surface of the object which is removed after ten minutes, washed, and allowed to dry in air. When dry it is wax finished.

On rough and as-cast surfaces a dull grey-brown, without grain-enhancement, is produced.

1.44   Pale brown grain-enhancement Semi-matt/matt

Copper acetate	120 gm
Ammonium chloride	60 gm
Water	1 litre

Boiling immersion. Scratch-brushing   (Ten minutes)

The object is immersed in the boiling solution and shows signs of surface etching after a minute. The grain structure of the cast surface is revealed, and darkened, with continued immersion. An uneven brown colour is produced. After ten minutes the object is removed and scratch-brushed with the hot solution, which evens the surface, but also lightens the colour. The object is thoroughly washed, allowed to dry in air, and wax finished when completely dry.

An even, dull light brown is produced on rough and as cast surfaces. On polished surfaces, the results are pale and nearly matt.

1.45★   Orange grain-enhancement (slight lustre) Semi-gloss

Copper acetate	7 gm
Ammonium chloride	3.5 gm
Acetic acid (6% solution)	100 cm³
Water	1 litre

Boiling immersion   (Twenty minutes)

The object is immersed in the boiling solution. The surface gradually darkens and a slight etching effect slowly reveals the grain structure of the casting on polished surfaces. When the surface colour is orange, after twenty minutes, the object is removed, washed in hot water, and dried in sawdust. When dry, it is wax finished.

On rough or as-cast surfaces, an even warm brown colour is produced. The orange colour on polished faces is translucent in quality and does not mask the grain-enhancement.

1.46   Brown mottled with pale grey Semi-matt

Copper sulphate	25 gm
Ammonium chloride	25 gm
Lead acetate	25 gm
Water	1 litre

Hot immersion   (Fifteen minutes)

The object is immersed in the hot solution (80°C) producing slight etching effects and pale orange/pink colours. After fifteen minutes, the object is removed washed in hot water and allowed to dry in air. Exposure to daylight causes a darkening of the colour. When the colour has stabilised and the object is completely dry, it is wax finished.

A slight grey bloom on a dull brown and greenish-brown ground is produced on rough or as-cast surfaces.

▲ Lead acetate is very toxic and every precaution should be taken to prevent inhalation of the dust during weighing and preparation of the solution. The vapour evolved from the heated solution must not be inhaled.

1.47   Grey-green tinge on light brown (slightly grain enhanced) Semi-matt

Sodium dichromate	150 gm
Nitric acid (10% solution)	20 cm³
Hydrochloric acid (15% solution)	6 cm³
Ethanol	1 cm³
Water	1 litre

Cold immersion   (Fifteen minutes)

The sodium dichromate is dissolved in the water, and the other ingredients are then added separately while stirring the solution. The article is immersed in the cold solution for about three or four minutes. A superficial layer tends to form which should be removed by bristle-brushing under water. The article is re-immersed for a further ten minutes, and then removed and washed in cold water. (Bristle-brushing at this stage will tend to remove the grey-green that has formed, leaving a light brown colour. Some residual grey-green tends to remain.) It is allowed to dry in air, and may be wax finished when dry.

A grey-green on a light brown ground is produced on rough or as-cast surfaces. The extent to which the grey-green colour remains, on polished surfaces, is unpredictable.

▲ Sodium dichromate is a powerful oxidant, and must not be allowed to come into contact with combustible material. The dust must not be inhaled. The solid and solutions should be prevented from coming into contact with the eyes and skin, or clothing. Frequent exposure can cause skin ulceration, and more serious effects through absorption.

▲ Both hydrochloric and nitric acids are corrosive and should be prevented from coming into contact with the eyes, skin or clothing.

1.48   Brownish grain-enhanced surface Matt

Copper nitrate	300 gm
Water	100 cm³
Silver nitrate	2 gm
Water	10 cm³

Applied liquid   (Several minutes)

The copper nitrate and the silver nitrate are separately dissolved in the two portions of water, and then mixed together. The solution is warmed to about 40°C. The article to be coloured is immersed for a few minutes in a cold solution of sulphuric acid (50% solution), and then washed. The warm colouring solution is applied to the article by wiping with a soft cloth, producing a superficial grey layer which is washed away with warm water, revealing the underlying surface effect. When the article is completely dry, it may be wax finished.

The underlying grey colour that is produced tends to be non-adherent on polished surfaces, leaving an uneven brown staining. A more even mid-brown colour is produced on rough or as-cast surfaces. The procedure cannot be generally recommended.

▲ Silver nitrate should be prevented from coming into contact with the skin, and particularly the eyes which may be severely irritated.

▲ Sulphuric acid is very corrosive, causing severe burns, and must be prevented from coming into contact with the eyes and skin.

1.49   Mid-brown bronzing (spotted with dark grey) Semi-matt

Copper acetate	20 gm
Sodium tetraborate (borax)	5 gm
Potassium nitrate	5 gm
Mercuric chloride	2.5 gm
Olive oil	to form a paste

Applied paste   (Two or three days)

The ingredients are ground to a thin paste with a little olive oil, using a pestle and mortar. The paste is applied to the object, either with a brush or a soft cloth, and is allowed to dry for several hours. When dry, the object is brushed with a bristle-brush to remove the dry residue. This procedure is repeated as necessary. When treatment is complete, the surface is again brushed, and then cleaned with pure distilled turpentine. When this has dried, it is wax finished.

A dull blackish-brown colour is produced on rough or as-cast surfaces. The bronzing effect on polished surfaces is slight, and spotted with fine dark grey specks.

▲ Mercuric chloride is a highly toxic substance. It is essential to prevent any contact with the skin, and to prevent any chance of inhaling the dust. Gloves and masks must be worn. A nose and mouth face mask fitted with a fine dust filter must be worn when brushing away the dry residue.

▲ Potassium nitrate is a powerful oxidising agent and should not be allowed to come into contact with combustible material.

▲ In view of the limited colouring potential, and the high risks involved, this method cannot be recommended.

1.50   Greenish-brown (variegated) Semi-matt

Copper nitrate	200 gm
Zinc chloride	200 gm
Water	1 litre

Applied liquid   (Twice a day for five days)

The solution is wiped on to the article with a soft cloth and allowed to dry in air. When it is completely dry, the surface is gently rubbed smooth with a soft dry cloth to remove any loose dust, and a further application of solution wiped on. This procedure is repeated twice a day for about five days. The solution takes effect immediately, giving rise to an etching effect, and pale cream and pink colours typical with chloride solutions. These colours darken on exposure to daylight. When treatment is complete, the article is left to dry thoroughly in air, and when completely dry, it is wax finished.

A dark brown colour tinged with blue-green was produced in tests on rough and as-cast surfaces. On polished faces, the greenish-brown was variegated with some pink and pale blue traces.

1.51   Dark brown (variegated) Semi-matt/matt

Copper sulphate	25 gm
Sodium chloride	100 gm
Potassium polysulphide	10 gm
Water	1 litre

Hot immersion   (Fifteen minutes)

The object is immersed in the hot solution (60°C), producing a slight matting of the surface due to etching, and the gradual development of dull brown colours. After about fifteen minutes, when the brown has developed, the object is removed. Relieving may be carried out at this stage, using the hot solution with the scratch-brush. The object is then removed, washed in warm water and allowed to dry in air. When dry, it is wax finished.

Gentle relieving with the scratch-brush reveals a lighter grain-enhanced finish on polished surfaces.

▲ The colouring should be carried out in a well ventilated area, as some hydrogen sulphide gas will be liberated.

1.52   Light bronzing (grain-enhanced) Semi-matt

Copper acetate	40 gm
Copper sulphate	200 gm
Potassium nitrate	40 gm
Sodium chloride	65 gm
Acetic acid (6% solution)	1 litre

Cold immersion. Scratch-brushing (Ten minutes)

The object is suspended in the cold solution, producing an immediate etching effect. After five minutes a pale cream or pink colour develops unevenly. After ten minutes the object is removed and scratch-brushed with the cold solution to even the surface. It should then be allowed to dry in air without washing. Any wet areas on the surface should be very gently brushed out with a soft brush, to prevent streaking or pooling. When dry, the object is exposed evenly to daylight, causing a darkening of the colours. When the colour has stabilised the object is wax finished.

A thin grey-brown colour is produced on rough or as-cast surfaces. The colouring effect on polished surfaces is slight, and tends to be uneven. The solution cannot generally be recommended.

▲ Potassium nitrate is a powerful oxidising agent and should not be allowed to come into contact with combustible material.

1.53   Pale yellow-brown (slightly grain-enhanced) Matt

A	Copper sulphate	25 gm
	Ammonia (.880 solution)	30 cm^3
	Water	1 litre
B	Potassium hydroxide	16 gm
	Water	1 litre

Hot immersion (A)	(Ten minutes)
Hot immersion (B)	(One minute)

The article is immersed in the hot copper sulphate solution (60°C) for about ten minutes, producing a grain-enhancement effect and darkening of the surface to a greenish-brown drab. The article is then transferred to the hot potassium hydroxide solution (60°C) for about one minute. It is removed, washed in warm water and allowed to dry in air. When dry it is wax finished.

An even greenish-brown colour is produced on rough or as-cast surfaces. The colouring effect on polished surfaces is slight.

▲ Preparation of solution A, and colouring should be carried out in a well ventilated area, to prevent irritation of the eyes and respiratory system by the ammonia vapour liberated.

▲ Potassium hydroxide, in solid form and in solution, is corrosive to the skin and particularly the eyes. Adequate precautions should be taken.

1.54   Variegated orange-brown (as-cast surfaces only) Matt

Ammonium molybdate	12.5 gm
Ammonia (.880 solution)	2 cm^3
Water	1 litre

Hot immersion   (One hour)

The object is immersed in the hot solution (60–70°C), and the colour develops very gradually on the as-cast surface. A slight bloom occurs on a polished surface. Immersion is continued until the colour has fully developed, and the object is then removed, washed in hot water, and allowed to dry in air. It may be wax finished when dry.

A streaky tarnish occurs on polished surfaces as immersion proceeds. No improvement could be obtained on polished faces by variations in the concentration of the ingredients, temperature, or period of immersion.

1.55   Light brown Semi-matt

Copper sulphate	100 gm
Ammonium chloride	0.5 gm
Water	1 litre

Applied liquid to heated metal (Several minutes)

The object is gently heated on a hotplate and the solution applied either by brushing or with a soft cloth. The object should be heated to a temperature such that when the solution is applied, it sizzles gently. Application causes a slight enhancement of the grain structure of the surface, and some slight colouring. When the colour has developed, the object is wax finished.

A mid-brown colour is produced on rough or as-cast rather than polished surfaces. The colouring effect on polished faces is rather slight. Some traces or a bloom of patina may occur as the surface dries out during heating.

1.56   Reddish-brown (as-cast surfaces only)

Ferric oxide	30 gm
Lead dioxide	15 gm
Water	to form a paste

Applied paste and heat (Repeat several times)

The ingredients are mixed to a paste with a little water. The paste is applied to the surface with a soft brush, while the object is gently heated on a hotplate or with a torch. When the paste is completely dry, it is brushed off with a stiff bristle-brush. The procedure is repeated a number of times, until the colour has developed. After the final brushing, the object may be wax finished.

Little colouring is produced on polished surfaces other than in features such as punched marks. The procedure is suitable for producing a contrast between as-cast and polished areas.

▲ Lead dioxide is toxic and very harmful if taken internally. The dust must not be inhaled. A nose and mouth face mask with a fine dust filter must be worn when brushing away the dry residue.

1.57   Mid-brown (as-cast surfaces only)

Copper nitrate	80 gm
Water	1 litre
Ammonia (.880 solution)	5 cm³

Applied liquid (Twice a day for ten days)

The solution is applied to the surface of the article to be coloured, by dabbing with a soft cloth, leaving the surface sparingly moist. The article is then left to dry. This procedure is repeated twice a day for about ten days. When treatment is complete, the article is allowed to dry for several days and may be wax finished when completely dry.

A mid-brown colour with an olive tinge is produced on as-cast surfaces. Some blue-green patina may also develop in the latter stages. This can be encouraged by a more liberal use of the solution.

A slight yellowish/ochre colour is produced on polished surfaces. A blue-green patina will tend to develop in any surface features such as punched marks or textured areas.

1.58   Brown (as-cast surfaces only) Semi-matt

Copper sulphate	500 gm
Copper acetate	100 gm
Acetic acid (30% solution)	25 cm³
Water	1 litre

Hot scratch-brushing (Several minutes)

The article is either immersed in the hot solution (80°C) for about two minutes, or the solution may be applied liberally with a soft brush, to allow the colour to develop. The surface is then gently scratch-brushed with the hot solution until an even finish is produced. Relieved effects may be obtained by continued working, without the hot solution. The article is thoroughly washed in hot water, and either dried in sawdust, or allowed to dry in air. When dry it may be wax finished.

A thin 'veil' of brown on the dulled metal tends to be produced on polished surfaces, which is completely cut back if overworked with the scratch-brush.

▲ Strong acetic acid is corrosive and should not be allowed to come into contact with the eyes and skin.

1.59★   Pale grey-brown bronzing Semi-matt

Potassium sulphide	2 gm
Ammonium sulphate	0.5 gm
Water	1 litre

Cold scratch-brushing (Several minutes)

The object is scratch-brushed with the cold solution, alternating with hot and then cold water. This process is continued until the colour has developed. Any relieving that is required should then be carried out. The object is washed thoroughly in cold water and allowed to dry in air. When completely dry it is wax finished.

On rough and as-cast surfaces a dull grey-brown colour is produced.

1.60   Mid-brown (as-cast surfaces only)

Copper nitrate	20 gm
Zinc sulphate	30 gm
Mercurous chloride	30 gm
Water	1 litre

Applied liquid   (Twice a day for ten days)

The solution is initially applied by vigorous dabbing with a soft cloth, to overcome resistance to wetting. Once some surface effect has been produced, the solution may be applied by wiping. The surface is allowed to dry completely after each application, which should be carried out twice a day for at least ten days. If the solution is used very sparingly then a patina-free ground colour tends to be produced. A more liberal use of the solution during the later stages of treatment will tend to encourage the formation of a blue-green patina. When treatment is complete, the article should be allowed to dry in air for several days before wax finishing.

Little effect other than a slight uneven brassy tint could be obtained on a polished surface.

In tests, the blue-green patina was found to be difficult to produce.

▲ Mercurous chloride may be harmful if taken internally. There are dangers of cumulative effects.

1.61   Light and dark brown (variegated)   Semi-matt

Copper acetate	30 gm
Ammonium chloride	15 gm
Ferric oxide	30 gm
Water	1 litre

Hot immersion. Hot scratch-brushing   (Ten to fifteen minutes)

The solution is brought to the boil and then removed from the source of heat. The object is then immersed, and becomes coated with a thick brown layer. After ten minutes immersion the object is removed and gently heated on a hotplate, the surface being worked with the scratch-brush. More solution is applied and worked in with the scratch-brush, occasionally allowing the surface to dry off without scratch-brushing. When the colour has developed it is allowed to dry on the hotplate for a few minutes, removed and the dry residue removed with a stiff bristle-brush. When the surface is free of residue it is wax finished.

A reddish-brown variegated with dark brown is produced on rough or as-cast surfaces.

An alternative technique which is more suited to larger objects is to heat locally with a torch and apply the pasty liquid with the scratch-brush, without prior immersion.

▲ A nose and mouth face mask fitted with a fine dust filter should be worn when brushing off the dry residue.

1.62   Pale brown lustre   Gloss

Potassium sulphide	1 gm
Ammonium chloride	4 gm
Water	1 litre

Cold scratch-brushing   (Five minutes)

The object is scratch-brushed with the cold solution for a short time and then washed in hot water. This alternation is repeated several times until the colour develops. The surface is then gently burnished with the scratch-brush in hot water, and finally with cold water. After thorough rinsing and drying in air the object may be wax finished.

A dark grey-brown matt finish is produced on rough and as-cast surfaces.

▲ The procedure should be carried out in a well ventilated area, as some hydrogen sulphide gas is liberated from the solution.

For results obtained with a more concentrated solution, see recipe 1.85.

1.63   Mid-brown   Semi-matt

Potassium sulphide	125 gm
Water	1 litre
Ammonia (.880 solution)	100 cm³

Applied liquid   (Several minutes)

The potassium sulphide is ground to a paste with some of the water and a little ammonia, using a pestle and mortar. This paste is added to the rest of the water, and the remaining ammonia, to form a strong solution. This solution is wiped on to the surface of the object with a soft cloth, producing a non-adherent black layer that is easily washed off. The application is repeated a number of times until the surface exposed as the black film is removed, becomes brown. When the colour is satisfactory, the object is washed in cold water, allowed to dry in air, and wax finished when dry.

A dark brown or black colour is produced on rough or as-cast surfaces.

▲ Preparation and colouring should be carried out in a well ventilated area. Some hydrogen sulphide gas will be liberated by the potassium sulphide solution. Ammonia vapour will also be liberated from the solution. These should not be inhaled. The solution should be prevented from coming into contact with the eyes and skin as it may cause burns or severe irritation.

1.64*   Dark brown (black on brown ground) Matt   *Pl. II*

Nickel sulphate	35 gm
Copper sulphate	25 gm
Potassium permanganate	5 gm
Water	1 litre

Hot immersion (Five to thirty minutes)

The object is immersed in the hot solution (80–90°C) and the surface immediately becomes covered with a dark brown layer. After four or five minutes it is removed and gently bristle-brushed with hot water. The dark layer is removed revealing a brown 'bronzing' beneath. If darker colours are required then the process should be repeated. If, on the other hand a full very dark brown/black colour with a more matt finish is required, the object should be re-immersed, and the dark brown layer will then re-form and should be left on the surface as immersion is continued. After thirty minutes the object is removed, and gently bristle-brushed under hot water. After thorough rinsing, and drying in sawdust, it is wax finished.

A long immersion without removing the dark layer after the initial brushing, produces an even black dusting on a brown ground. Immersion combined with repeated brushing produces a slightly figured surface of dark brown on a golden bronze colour. On rough or as-cast rather than polished surfaces, an even black/dark brown is produced by both techniques.

▲ Nickel compounds are a common cause of sensitive skin reactions. Precautions should also be taken to avoid inhalation of the dust when preparing the solution.

1.65*   Dark brown with grey sheen Satin/semi-gloss   *Pl. II*

Sodium sulphide	30 gm
Sulphur (flowers of sulphur)	4 gm
Water	1 litre

Hot immersion and scratch-brushing (Fifteen minutes)

The solids are added to about 200 cm³ of the water, and boiled until they dissolve and yield a clear deep orange solution. This is added to the remaining water and the temperature adjusted to 60–70°C. The object is immersed in the hot solution and the colour allowed to develop for about ten minutes. It is then removed and scratch-brushed with the hot solution, until the surface is even and the colour developed. The object should be washed thoroughly in hot water and dried in sawdust. When dry, it may be wax finished.

A very dark brown/black colour, which burnishes to a glossy sheen, is produced on rough or as-cast surfaces.

▲ Sodium sulphide in solid form and in solution must be prevented from coming into contact with the eyes and skin, as it can cause severe burns. Some hydrogen sulphide gas may be evolved when the solution is heated. Colouring should be carried out in a well ventilated area.

1.66   Mid-brown bronzing Gloss

Antimony trisulphide	12.5 gm
Sodium hydroxide	40 gm
Water	1 litre

Hot scratch-brushing (Several minutes)

The hot solution (60°C) is repeatedly applied liberally to the surface of the object with a soft brush, allowing time for the colour to develop. The surface is then worked lightly with the scratch-brush, while applying the hot solution and hot water alternately. The treatment is concluded by lightly burnishing the surface with the scratch-brush and hot water. After rinsing it is dried in sawdust, and wax finished when dry.

A light purplish brown is produced on rough and as-cast surfaces.

A darker colour can be obtained on both polished and rough or as-cast surfaces by immersing the object in the hot solution for about fifteen minutes, prior to the brushed application.

▲ Sodium hydroxide is a powerful caustic alkali which can cause severe burns to the eyes and skin, in the solid form or in solution. When preparing the solution, small quantities of the sodium hydroxide should be added to a large quantity of water and *not* vice-versa. When the sodium hydroxide has completely dissolved, the antimony trisulphide is added.

1.67   Light brown bronzing (thin brown on bare metal) Semi-matt

Copper sulphate	30 gm
Potassium permanganate	6 gm
Water	1 litre

Boiling immersion and scratch-brushing (Several minutes)

On immersion in the boiling solution, the object becomes coated with a brownish film, which develops to a dark purplish-brown after one minute. This brown colour film is loosely adherent and sloughs off gradually to reveal uneven underlying grey-brown colours. After three or four minutes the object is removed and gently scratch-brushed with the hot solution. Overworking will cut through to the underlying metal. If the surface is unsatisfactory then the process is repeated. When the desired colour has been obtained, the object is washed in hot water, dried either in sawdust or in air, and wax finished when dry.

An even dark brown is produced on rough or as-cast surfaces. The colour on polished surfaces is rather variable and may take the form of a dull 'figure'. Test results tended to be inconsistent.

The development of the loose brown layer appears to be essential to successful colouring. If the solution is boiled for too long before the object is immersed, then a more dense precipitate is formed in the solution, and coating does not take place.

1.68   Light bronzing (slight lustre) Semi-gloss

Ammonium sulphide (16% solution)   100 cm³
Water   1 litre

Cold immersion and scratch-brushing (Ten to thirty minutes)

The article is immersed in the cold solution for a few minutes to allow the colour to develop, and then removed and lightly scratch-brushed with the cold solution, alternating with hot and cold water. It is then re-immersed for about ten minutes and then removed and scratch-brushed as before. This procedure is repeated until the desired colour is achieved. The article is finally lightly scratch-brushed alternately with hot and cold water. It is allowed to dry in air and wax finished.

A brown or grey-brown colour is produced on rough or as-cast surfaces.

Prolonged working with the scratch-brush on polished surfaces tends to produce a very pale dull finish.

As an alternative to immersion in the cold solution, the solution can be applied continuously to the surface with a soft brush. This tended to produce lighter bronzed finishes in tests.

▲ Ammonium sulphide solution liberates a vapour containing ammonia and hydrogen sulphide gas, which should not be inhaled. Preparation and colouring must be carried out in a well ventilated area. The solution must be prevented from coming into contact with the eyes and skin, as it causes burns or severe irritation.

1.69   Reddish-brown Semi-matt

Sodium chlorate   100 gm
Copper nitrate   25 gm
Water   1 litre

Boiling immersion (Forty minutes to one hour)

The object is immersed in the boiling solution, and the surface is quickly coloured with a dark lustre, which develops to a thin yellowish-orange colour. Continued immersion produces a gradual change to a more opaque orange-red or orange-brown colour. When the colour has developed evenly and to the required depth, after about forty minutes to one hour, the object is removed and washed in hot water. It is dried in sawdust, and may be wax finished when dry.

A slight grey-green patina on a dark brown ground is produced on as-cast surfaces. The grey-green patina tends to occur in features such as punched marks or surface texture, on both polished and as-cast surfaces.

For results obtained with prolonged immersion in this solution, see recipe 1.117.

▲ Sodium chlorate is a powerful oxidant and must not be allowed to come into contact with combustible material or acids. It should also be prevented from coming into contact with the eyes, skin or clothing.

1.70★   Reddish-brown (grain-enhanced) Semi-gloss   *Pl. II*

Copper sulphate   100 gm
Ammonium chloride   0.5 gm
Water   1 litre

Hot immersion (Fifteen minutes)

The object is immersed in the hot solution (70–80°C) producing a pale orange colouration, which darkens gradually with continued immersion. This is accompanied by a slight grain-enhancement due to etching. After fifteen minutes the object is removed, washed thoroughly in warm water and dried in sawdust. Exposure to daylight will cause a darkening of the colour initially. When no further darkening is taking place the object is wax finished.

On rough or as-cast surfaces a dark grey-brown colour is produced.

1.71   Purplish-brown Gloss

Sodium hydroxide   100 gm
Sodium tartrate   60 gm
Copper sulphate   60 gm
Water   1 litre

Hot immersion (Forty minutes)

The object is immersed in the hot solution (60°C) producing dull lustre colours after about one minute. The surface gradually becomes more opaque and the colour slowly turns brown as immersion continues. After about forty minutes, when the colour has developed, the object is removed and washed in hot water and either dried in sawdust or allowed to dry in air. It may be wax finished when dry.

A dull orange lustre with a greenish tinge is produced after about ten minutes.

Some spotting tended to occur on the fully developed purplish-brown colour on a polished surface.

An even brown colour is produced on as-cast surfaces.

▲ Sodium hydroxide is a powerful caustic alkali. The solid or solutions can cause severe burns. It must be prevented from coming into contact with the eyes or skin. When preparing the solution, small quantities of the sodium hydroxide should be added to large quantities of water and *not* vice-versa. The solution will become warm as the sodium hydroxide is dissolved. When it is completely dissolved, the other ingredients are added.

1.72★   Reddish-brown/purplish-brown Semi-gloss/gloss   *Pl. II*

Copper carbonate   160 gm
Sodium hydroxide   80 gm
Water   1 litre

Boiling immersion (Thirty minutes)

The sodium hydroxide is added to the water in small quantities, while stirring. Heat is evolved as it dissolves, warming the solution. When it has completely dissolved, the copper carbonate is added and the solution heated to boiling. The object is immersed in the boiling solution, and gradually acquires a pale grey colour which darkens to a greyish-mauve as immersion proceeds. After about thirty minutes the object is removed and dried in sawdust. When completely dry, it may be wax finished.

An even purple colour is produced on rough or as-cast surfaces.

Similar results can be obtained using more dilute solutions for correspondingly longer times, but these tend to be slightly less even.

▲ Sodium hydroxide is a powerful caustic alkali, and the solid or solutions must be prevented from coming into contact with the eyes and skin. The sodium hydroxide, in small quantities, should be added to the water and *not* vice-versa. Colouring should be carried out in a well ventilated area, as inhalation of the caustic vapours should be avoided.

1.73   Purplish-brown   Gloss

Copper acetate	6 gm
Copper sulphate	1.5 gm
Sodium chloride	1.5 gm
Water	1 litre

Boiling immersion (About one hour)

A variegated purple and brown is produced on as-cast surfaces.

The article is immersed in the boiling solution and after a few minutes the surface darkens and becomes more opaque as the colour develops. The surface develops gradually as immersion continues, and after about one hour the article is removed and washed in hot water. It is dried in sawdust and may be wax finished when completely dry.

1.74   Reddish-brown   Semi-matt

A	Copper sulphate	25 gm
	Water	1 litre
B	Ammonium chloride	0.5 gm

Boiling immersion (A)   (Fifteen minutes)
Boiling immersion (A+B)   (Ten minutes)

A reddish-purple colour is produced on rough or as-cast surfaces, which tend also to retain a bloom of greyish-green patina.

The article is immersed in the boiling copper sulphate solution for about fifteen minutes or until the colour is well developed. It is then removed to a bath of hot water, while the ammonium chloride is added to the colouring solution, and then re-immersed. The colour is brightened and tends to become more red. After about ten minutes it is removed and washed in hot water. If the immersion is prolonged a bloom of greyish-green patina will tend to form, which may need to be removed with a soft bristle-brush. The article is dried in sawdust and finally wax finished.

1.75★   Rich brown   Satin/semi-gloss   Pl. I

Antimony trisulphide	30 gm
Ferric oxide	10 gm
Ammonium sulphide (16% solution) to form a paste	

Applied paste (sulphide paste)

A variegated surface can be obtained by brushing the paste onto the whole surface and allowing it to dry for about three hours, until it is superficially dry and cracking, and then burnishing it away with a nylon brush or a brass scratch-brush. This may be repeated, as necessary, to darken the colour.

▲ Ammonium sulphide solution is very corrosive and precautions should be taken to protect the eyes and skin while the paste is being prepared. It also liberates harmful fumes of ammonia and of hydrogen sulphide, and the preparation should therefore be carried out in a fume cupboard or well ventilated area. The paste once prepared is relatively safe to use but should be kept away from the skin and eyes.

▲ A nose and mouth face mask fitted with a fine dust filter should be worn when the residues are brushed away. The dust produced in this case contains antimony compounds, which are toxic.

The ingredients are mixed to a thick paste by the very gradual addition of a small quantity of ammonium sulphide solution, using a pestle and mortar. The paste is applied to the whole surface of the article to be coloured, with a bristle-brush. A stiff nylon brush is then used to work the surface gradually in small areas. The nylon brush should be intermittently dipped in a mixture of the dry ingredients (in the same proportion as above), and the surface worked in this way until the paste becomes dry and falls from the surface. When the whole surface is dry and completely free of any residue or dust, it may be burnished with a brass scratch-brush and wax finished.

1.76   Dark reddish-brown (slight grain-enhancement)   Semi-matt   Pl. I

Sodium thiosulphate	6.25 gm
Ferric nitrate	50 gm
Water	1 litre

Hot immersion (One minute)

During preparation, the nitrate solution is darkened by the addition of the thiosulphate, but slowly clears to give a straw-coloured solution.

An even dark brown/black is produced on rough or as-cast surfaces. On polished faces the dark reddish-brown colour is slightly translucent revealing a darker mottling of the grain structure.

Higher temperatures should not be used. Tests at 70–80°C produced a stripping of the surface and etching effects.

A succession of lustre colours is rapidly produced when the article is immersed in the hot solution (50–60°C), changing to a more opaque purplish colour after about forty-five seconds. The colour darkens quickly and the article is removed after about one minute. After thorough washing it is allowed to dry in air, and then wax finished.

1.77*   Reddish-brown (*or* black on reddish-brown) Semi-matt   *Pl. I, Pl. XII*

Copper sulphate	50 gm
Potassium chlorate	20 gm
Potassium permanganate	5 gm
Water	1 litre

Boiling immersion   (Fifteen minutes)

The object is immersed in the boiling solution, and rapidly becomes coated with a black/brown layer. Immersion is continued for fifteen minutes, after which the object is removed and gently rinsed in a bath of warm water. It is then allowed to dry in air, leaving a dark powdery layer on the surface. When dry this is gently brushed with a soft brush to remove any dry residue. This treatment will remove some of the top layer, partly revealing the underlying colour, and should be continued until the desired surface effect is obtained. A stiffer bristle-brush may be required. After any residual dust has been removed with a soft brush, the object is wax finished.

On polished surfaces, the bulk of the dark dry residue is removed by gentle brushing with a bristle-brush. Some of the residue may be more adherent and brushes into the surface to produce a darker mottling on the reddish-brown. On rough or as-cast surfaces the dark layer is much more tenacious, and an even black finish is generally produced.

▲ Potassium chlorate is a powerful oxidising agent, and must not be allowed to come into contact with combustible material. The solid or solutions should be prevented from coming into contact with the eyes, skin or clothing.

▲ The potassium chlorate and the potassium permanganate must not be mixed together in the solid state. They should each be separately dissolved in a portion of the water, and then mixed together.

1.78   Variegated reddish-brown Semi-matt

A	Copper sulphate	125 gm
	Ferrous sulphate	100 gm
	Water	1 litre
B	Ammonium chloride	1 gm

Boiling immersion (A)   (Fifteen minutes)
Boiling immersion (A+B) (Five minutes)

The object is immersed in the boiling solution of copper sulphate and ferrous sulphate, and immersion continued for about fifteen minutes, until a brown colour has developed. The object is removed and rinsed in a tank of hot water, while the ammonium chloride is added to the colouring solution. The object is re-immersed in the colouring solution for about five minutes, causing a brightening and development of the surface to a red colour. It is then removed and washed very thoroughly in hot water. After drying in sawdust, it may be wax finished.

An even reddish-brown colour is produced on rough or as-cast surfaces.

The object must be thoroughly washed after immersion, to prevent the formation of a greenish bloom or patches of green. If there is a build-up of bloom during immersion, then this should be cleared by agitating the object, or by removing and rinsing in hot water if necessary.

If shorter immersion times are used before the addition of the ammonium chloride, then the results tend to be lighter and more glossy, but are more prone to unevenness.

1.79*   Dark purplish-brown Gloss

Ammonium sulphide (16% solution)	100 cm³
Water	1 litre

Hot bristle-brushing   (Several minutes)

The article is directly bristle-brushed with the hot solution (60°C) for several minutes, alternating with bristle-brushing under hot water. The procedure is repeated three or four times until the colour has developed and the article is then washed in warm water. After thorough drying in air or sawdust, it is wax finished.

An even dark brown or black colour is produced on as-cast surfaces.

Scratch-brushing with the hot solution produces lighter 'bronzing' colours, which are better if a cold solution is used, see recipe 1.68.

▲ Ammonium sulphide solution liberates a vapour containing ammonia and hydrogen sulphide gas, which should not be inhaled. Preparation and colouring must be carried out in a well ventilated area. The solution must be prevented from coming into contact with the eyes or skin, as it causes burns or severe irritation.

1.80   Black/dark purplish-brown Gloss/semi-gloss

Barium sulphide	10 gm
Water	1 litre

Hot immersion   (Forty minutes)

The object is immersed in the hot solution (60°C) and the surface is gradually tinted by a succession of lustre colours, which develop and recede. These slowly darken during the latter half of the period of immersion until the final black colour develops, after about forty minutes. The object is removed and washed carefully in warm water, and allowed to dry in air. When the surface is thoroughly dry, the object may be wax finished.

An even dark purplish-brown or black is also produced on rough or as-cast surfaces.

For results obtained using a cold solution, see recipe 1.166.

The surface tends to be fragile when removed from the hot solution, and should be treated with care until dry.

▲ Barium sulphide is poisonous and very harmful if taken internally. Some hydrogen sulphide gas will be liberated as the solution is heated, and colouring should be carried out in a well ventilated area.

1.81* Dark purplish-brown Satin/semi-gloss

Antimony trisulphide	20 gm
Ferric oxide	20 gm
Potassium polysulphide	2 gm
Ammonia (.880 solution)	2 cm³
Water to form a paste	(approx 4 cm³)

Applied paste (suphide paste)

The potassium polysulphide is ground using a pestle and mortar and the other solid ingredients added and mixed. These are made up to a thick paste by the gradual addition of the ammonia and a small quantity of water. The paste is applied to the whole surface of the object to be coloured, using a soft bristle-brush. A stiff nylon brush is then used to work the surface gradually in small areas. The nylon should be intermittently dipped in a mixture of the dry ingredients of the paste (in the same proportion as above), and the surface worked in this way until the paste becomes dry and falls from the surface. When the whole surface is dry and completely free of any residue or dust, it may be burnished with a brass scratch-brush and wax finished.

A variegated surface can be obtained by brushing the paste on to the whole surface and allowing it to dry for about three hours, until it is superficially dry and cracking, and then burnishing it away with a nylon brush or a brass scratch-brush. This may be repeated as necessary to darken the colour.

▲ A nose and mouth face mask should be worn when the residues are brushed away. The dust produced in this case contains antimony compounds, which are toxic.

1.82 Variegated dark reddish-brown/black (as-cast surfaces only) Semi-matt

A	Potassium polysulphide	4 gm
	Water	1 litre
B	Ferric nitrate	10 gm
	Water	1 litre
A	Torch technique	
B	Torch technique	

The metal is heated with a blow torch and the polysulphide solution applied sparingly with a brass scratch-brush using small circular motions and working across the surface until a dark brown/black colour is established. The metal is then gently heated as the ferric nitrate solution is stippled onto the surface using a bristle-brush or a soft cloth, giving a reddish-brown tone. When treatment is complete, the object is allowed to cool and dry thoroughly, and may then be wax finished.

▲ Ferric nitrate is corrosive and should not be allowed to come into contact with the eyes or skin. A face shield or goggles should be worn to protect the eyes from hot splashes, as either of these solutions are applied to the heated metal.

▲ The fumes evolved as either of these solutions are applied to the heated metal must not be inhaled. Adequate ventilation should be provided.

1.83 Dark purple-brown Semi-gloss

Potassium sulphide	5 gm
Barium sulphide	10 gm
Ammonia (.880 solution)	20 cm³
Water	1 litre

Hot immersion and scratch-brushing (Twenty-five minutes)

The article is immersed in the hot solution (80–90°C), and a grey/black layer forms on the surface after about three minutes. This should be brass-brushed to clear any non-adherent material. Immersion is continued and the surface brass-brushed every few minutes for the first ten minutes. After twenty-five minutes the article is removed, washed in hot water and allowed to dry in air. When dry, it may be wax finished.

The purple-brown colour obtained on polished surfaces may be fragile. It should not be overworked with the scratch-brush while in solution, and should be allowed to dry thoroughly without handling before finishing. A dark brown colour is obtained on rough or as-cast surfaces.

▲ Barium sulphide is poisonous and may be very harmful if taken internally.

▲ Ammonia vapour will be liberated during preparation and colouring, which must be carried out in a well ventilated area. Some hydrogen sulphide gas may also be liberated. The ammonia solution must be prevented from coming into contact with the eyes and skin, as it will cause burns or severe irritation.

1.84 Dark brown Semi-matt

Ammonium chloride	5 gm
Sodium chloride	5 gm
Water	1 litre

Torch technique

The solid ingredients are mixed with the water to form a solution, which is left to stand for several hours. The object is heated with a blow torch and the solution dabbed onto the surface using a cloth or bristle-brush. This gradually darkens the surface. When treatment is complete, the object should be allowed to cool and dry thoroughly before wax finishing.

▲ The fumes evolved as the solution is applied to the heated metal should not be be inhaled. Adequate ventilation must be provided.

▲ A face mask or goggles should be worn to protect the eyes from hot splashes as the solution is applied.

1.85*   **Dark brown/black   Satin**

Potassium sulphide	20 gm
Ammonium chloride	30 gm
Water	1 litre

Cold scratch-brushing   (Several minutes)

A dark brown/black colour is produced on as-cast surfaces.

▲ The procedure should be carried out in a well ventilated area, as hydrogen sulphide gas is liberated by the solution.

The object is scratch-brushed with the cold solution, which is worked hard into the surface. No intermittent washing with water should be used. When the colour has developed, the object is rinsed in cold water and allowed to dry in air. When the surface is thoroughly dry, it is lightly burnished with a dry scratch-brush, and finally wax finished.

1.86*   **Dark brown/black   Semi-matt**   *Pl. I*

Ferric nitrate	50 gm
Sodium thiosulphate	6 gm
Water	1 litre

Torch technique

▲ Ferric nitrate is corrosive and should be prevented from coming into contact with the eyes and skin. A face shield or goggles should be worn to protect the eyes from hot splashes as the solution is applied.

▲ The fumes evolved as the solution is applied to the heated metal should not be inhaled. Adequate ventilation should be provided and a nose and face mask fitted with the correct filter should be worn.

The object is heated with a blow torch and the solution dabbed onto the surface with a cloth. The surface quickly darkens to a dark brown/black and the variegation that tends to occur can be evenly controlled by careful dabbing. When treatment is complete, the object is allowed to cool and dry thoroughly, and may then be wax finished.

1.87   **Dark brown (slight lustre)   Gloss**

Potassium sulphate	100 gm
Potassium sulphide	8 gm
Ammonium chloride	8 gm
Water	1 litre

Boiling immersion and scratch-brushing   (Several minutes)

On rough or as-cast surfaces, an even dark purplish-brown is produced. This surface will burnish to a dull shine when waxed.

▲ Colouring should be carried out in a well ventilated area, as some hydrogen sulphide gas will be liberated from the hot solution.

Immersion in the boiling solution produces an immediate darkening of the surface with lustre colours, followed by the gradual development of more opaque colour. This tends to occur unevenly. After a few minutes immersion, the object is removed and scratch-brushed with the hot solution. When the surface has evened and the colour developed, the object is washed in hot water, and allowed to dry in air. When dry, it is wax finished.

1.88   **Dark slate on brown   Semi-gloss**

Sodium thiosulphate	65 gm
Copper sulphate	12 gm
Copper acetate	10 gm
Arsenic trioxide	5 gm
Sodium chloride	5 gm
Water	1 litre

Hot immersion   (Thirty minutes)

The surface tends to be very uneven. The procedure is difficult to control and cannot generally be recommended, for polished surfaces.

An even dark slate/black colour is produced on rough or as-cast surfaces.

▲ Arsenic trioxide is very toxic and will give rise to very harmful effects if taken internally. The dust must not be inhaled and neither must the fumes evolved from the heated solution. Localised extraction should be used.

Immersion in the hot solution (50–60°C) produces patchy lustre colours during the first fifteen minutes. Continued immersion produces a progressively more opaque and darker colour. After about thirty minutes, the article is removed and washed in warm water. It is allowed to dry in air, and may be wax finished when dry.

1.89   **Variegated dark brown/black on dark brown ground   Semi-matt**

Potassium polysulphide	4 gm
Water	1 litre

Torch technique

▲ The fumes evolved as the solution is applied to the heated metal should not be inhaled. Adequate ventilation should be provided.

▲ A face mask or goggles should be worn to protect the eyes from hot splashes as the solution is applied.

The metal is heated with a blow torch and the solution applied sparingly with a scratch-brush using small circular motions and working across the surface until a matt black colour appears. The amount of heat and solution used are critical. If too much solution is applied then the surface tends to strip, and if too little heat is used then the solution does not act on the metal. When the matt black colour has been obtained, it should be lightly brushed with a dry scratch-brush. This brings up the surface and highlights any faults, that can then be worked on with further heating and applications of solution. When treatment is complete, the object is allowed to cool and dry thoroughly, and may then be wax finished.

1.90*   Dark reddish-brown (variegated) Semi-gloss

Ferric chloride	10 gm
Water	1 litre

Torch technique

The object is heated with a blow torch and the solution applied sparingly with a scratch-brush using small circular motions and working across the surface. The metal gradually darkens to a reddish-brown tone. The variegated surface produced can be made more even by lightly dabbing with a cloth dampened with the solution. When treatment is complete, the object is allowed to cool and dry thoroughly, and may then be wax finished.

▲ The fumes evolved as the solution is applied to the heated metal should not be inhaled. Adequate ventilation must be provided and a nose and mouth face mask fitted with the correct filter should be worn.

▲ Ferric chloride is corrosive and should not be allowed to come into contact with the eyes and skin. A face mask or goggles should be worn to protect the eyes from hot splashes as the solution is applied.

1.91   Orange-brown/dark brown (variegated) (as-cast surfaces only)

A	Potassium polysulphide	4 gm
	Water	1 litre
B	Ferric chloride	10 gm
	Water	1 litre

A Torch technique
B Torch technique

The metal is heated with a blow torch and the polysulphide solution applied sparingly with a brass scratch-brush, using small circular motions and working across the surface until a dark brown or black colour is established. The metal is then gently heated as the ferric chloride solution is stippled onto the surface using a bristle-brush or cloth, to give an orange-brown tone. When treatment is complete, the object is allowed to cool and dry thoroughly, and is then wax finished.

▲ The fumes evolved as either solution is applied to the heated metal should not be inhaled. Adequate ventilation must be provided.

▲ A face mask or goggles should be worn to protect the eyes from hot splashes as the solutions are applied.

▲ Ferric chloride is corrosive and should be prevented from coming into contact with the eyes and skin.

1.92   Light bronzing and bare metal Semi-matt

A	Copper sulphate	50 gm
	Water	1 litre
B	Potassium sulphide	12.5 gm
	Water	1 litre

Hot immersion. Scratch-brushing (A)
Cold immersion. Scratch-brushing (B)

The article is immersed in the copper sulphate solution for about thirty seconds and transferred to the cold potassium sulphide solution, and immersed for about one minute. It is then removed and scratch-brushed with each solution in turn. The surface is finished with a light scratch-brushing in the potassium sulphide solution. The article is washed thoroughly in warm water and allowed to dry before wax finishing.

An even but thin black colour is produced on rough or as-cast surfaces. Polished faces require delicate working and produce only a light bronzing effect similar to a relieved surface. An adequate finish is difficult to obtain and the recipe cannot be recommended.

▲ Colouring should be carried out in a well ventilated area, as some hydrogen sulphide gas will be liberated.

1.93   Dark greenish-brown with dull lustre Semi-matt/semi-gloss

Copper carbonate	100 gm
Ammonia (.880 solution)	500 cm³
Water	500 cm³

Warm immersion (One hour)

The ammonia is added to the water, followed by the carbonate which will partially dissolve. Some excess carbonate should remain in suspension. The solution is heated to about 50°C, and the article is immersed. The colour develops slowly during the period of immersion. After about one hour, when the colour is fully developed, the article is removed and washed thoroughly in warm water. It is then carefully dried in sawdust, and wax finished when dry.

On polished surfaces a dark, purplish-brown colour, with a green tinge and some lustre, is produced. A dark purplish grey is produced on rough or as-cast surfaces.

▲ This procedure should only be carried out where there is adequate ventilation. Ammonia vapour will be freely liberated when the solution is prepared, and during colouring, which can severely irritate the eyes and respiratory system. The skin, and particularly the eyes, must be protected against accidental splashes of this corrosive alkali.

1.94★ Very dark purplish-brown Gloss   *Pl. II*

Sodium hydroxide	25 gm
Potassium persulphate	10 gm
Water	1 litre

Hot immersion (Thirty minutes)

The potassium persulphate is dissolved in the water, and the sodium hydroxide then added in small quantities while stirring. The solution is then brought up to the correct temperature (70–80°C) and the article immersed. A lustrous reddish-gold colour is produced which gradually darkens as immersion continues. The full dark colour develops in about thirty minutes. When the required colour has been obtained, the object is removed and washed in warm water. It is dried in sawdust, and may be wax finished when dry.

The lustrous red colours produced during the course of immersion tend to be tonally variegated, although the final colour is even.

A dark purplish-brown or black colour is produced on as-cast surfaces.

▲ Potassium persulphate is a powerful oxidant and should not be allowed to come into contact with combustible material. The sodium hydroxide and the persulphate must not be allowed to come into contact in the solid state.

▲ Sodium hydroxide is a powerful caustic alkali which can cause severe burns. Both the solid and solutions should be prevented from coming into contact with the eyes and skin.

1.95 Dull brown Semi-matt/matt

Copper nitrate	35 gm
Ammonium chloride	35 gm
Calcium chloride	35 gm
Water	1 litre

Applied liquid to heated metal (Repeat several times)

The article is either heated on a hotplate or with a torch to a temperature such that the solution sizzles gently when it is applied. The solution is applied with a soft brush, and dries out to produce a superficial powdery green layer. The procedure is repeated several times, each fresh application tending to remove the green powdery layer, until an underlying colour has developed. Any residual green is then removed by washing and the article is allowed to dry in air. When dry it may be wax finished.

A greenish patina on a brown ground is produced on an as-cast surface.

If the solution is persistently applied as the surface is heated with a torch, a more adherent pale blue-green patina tends to form. However, solutions containing little or no chlorides are preferable for this effect.

1.96 Purple on brown (as-cast surfaces only) Matt

Copper sulphate	8 gm
Copper acetate	3.25 gm
Potassium aluminium sulphate	2 gm
Water	1 litre
Acetic acid (6% solution)	20 cm³

Boiling immersion (About one hour)

The article is immersed in the boiling solution and after a few minutes the surface darkens and becomes more opaque as the colour develops. The surface develops gradually as immersion continues, and after about one hour the article is removed and washed in hot water. It is dried in sawdust and may be wax finished when completely dry.

A dull bronzing is produced on polished surfaces.

1.97★ Blue-black Semi-matt   *Pl. II*

Potassium sulphide	10 gm
Water	1 litre
Ammonia (.880 solution)	1 cm³

Warm immersion. Bristle/scratch-brushing (About ten minutes)

The object is immersed in the warm solution (40°C) and becomes coated with a black film within one or two minutes. The surface is evened by bristle-brushing when this has developed, and immersion continued. After about six minutes it is removed and brushed gently with a brass-brush to even the surface. The solution is then brushed on to the surface liberally with a bristle-brush and lightly worked as the colour develops. The object is then washed thoroughly in cold water and allowed to dry in air. When dry, it is wax finished.

Immersion without bristle-brushing and scratch-brushing produces a matt black surface which tends to be non-adherent on polished surfaces, when dry. An adherent matt black is produced on rough or as-cast surfaces by simple immersion.

As an alternative to immersion, the solution can be applied liberally to the surface using a soft brush, and without 'working' the surface.

▲ Some hydrogen sulphide gas will be liberated from the solution. Colouring should be carried out in a well ventilated area.

1.98★ Black Semi-gloss

Sodium thiosulphate	50 gm
Ferric nitrate	12.5 gm
Water	1 litre

Hot immersion (Twenty minutes)

The article is immersed in the hot solution (60–70°C) and after about one minute the surface is coloured with a purple-blue lustre, which gradually recedes. After five minutes a brown colour slowly appears which changes to a slate grey with continued immersion. After about twenty minutes the article is removed, washed in hot water and allowed to dry in air. The surface may be fragile at this stage and should be handled as little as possible. When completely dry the article is wax finished.

If the temperature is too high the colour layer will be patchy and more fragile when removed from solution, revealing a tan colour beneath the black.

More even results were obtained in tests in which the surface was cleared with a soft bristle-brush, under hot water, after five minutes and after ten minutes.

1.99   Black/dark purplish-grey   Gloss/semi-gloss

Sodium thiosulphate	250 gm
Water	1 litre
Copper sulphate	80 gm
Water	1 litre

Cold immersion   (Two hours)

A dark purplish-grey/black is also produced on rough or as-cast surfaces.

The lustre colours produced tend to be less clear than those obtained using some other thiosulphate solutions.

The chemicals should be separately dissolved, and the two solutions then mixed together. The article is suspended in the cold solution, and after ten minutes immersion shows signs of darkening. After twenty minutes a reddish-purple lustre appears and gradually changes to a blue-green lustre after thirty minutes. The sequence of lustre colours is repeated, becoming more opaque and producing a maroon colour after one hour. With continued immersion the colour slowly darkens to black. After about two hours the article is removed, washed thoroughly and allowed to dry in air. When dry it may be wax finished.

1.100   Very dark brown/black   Semi-gloss/gloss

Ammonium persulphate	10 gm
Sodium hydroxide	30 gm
Water	1 litre

Boiling immersion   (Twenty minutes)

A very dark brown or black colour is also produced on rough or as-cast surfaces.

▲ Ammonium persulphate (peroxodisulphate) is a powerful oxidant, and must be kept out of contact with all combustible materials. It should be prevented from coming into contact with the eyes, skin or clothing.

▲ Sodium hydroxide is a powerful caustic alkali and must be prevented from coming into contact with the eyes and skin.

The ammonium persulphate is dissolved in half the water, and the sodium hydroxide separately dissolved in the remaining water. The two solutions are then mixed together and the mixture heated to boiling. The article to be coloured is immersed in the boiling solution, which produces golden and then brown lustre colours. Continued immersion darkens these colours to an opaque but glossy finish. When the colour has developed, after about twenty minutes, the article is removed, washed in hot water and dried in sawdust. When dry, it may be wax finished.

1.101   Black   Matt

Potassium permanganate	5 gm
Copper sulphate	50 gm
Ferric sulphate	5 gm
Water	1 litre

Boiling immersion   (Twenty minutes)

The surface tended to be finely spotted with uncoloured metal in tests.

A variegated black on a pinkish-brown ground is produced on rough or as-cast surfaces.

The article is immersed in the boiling solution. A black layer forms on the surface, which should not be disturbed. Immersion is continued for about twenty minutes, and the article is then removed and allowed to dry in air, without washing. The surface should not be handled at this stage. After several hours drying, the article may be wax finished.

1.102   Black/grey (variegated)   Semi-gloss

Sodium thiosulphate	65 gm
Water	1 litre
Antimony trichloride	until cloudy

Hot immersion   (Thirty minutes)

The results produced on polished surfaces tend to be lustrous, but clouded with grey in places. The surface also tends to be fragile on polished faces, and cannot be generally recommended. On rough or as-cast surfaces an even and adherent slate grey/black is produced.

▲ Antimony trichloride is poisonous if taken by mouth. It is also irritating to the skin, eyes and respiratory system, and contact must be avoided. The vapour evolved from the hot solution must not be inhaled.

The sodium thiosulphate is dissolved in the water and the antimony trichloride added by the drop, to form a white precipitate. The solution begins to become orange and cloudy at the top, and then gradually throughout. The solution is heated to 50°C, and when that temperature is reached, the object is immersed. The surface initially passes through the lustre cycle—golden after three minutes, purple after four minutes, pale metallic blue after ten minutes. After about fifteen minutes grey/white streaks appear on the surface, which darkens to produce an appearance like a black powdery layer. The article is removed after thirty minutes and washed gently in warm water, removing the powdery black deposit, to reveal a rather fragile variegated grey/black surface. The article is finally wax finished.

1.103   Dark brown/black Matt

Sodium thiosulphate	60 gm
Copper sulphate	42 gm
Potassium hydrogen tartrate	22 gm
Water	1 litre

Hot immersion   (Fifteen minutes)

An even black colour is produced on rough or as-cast surfaces.

The green layer is superficial, and easily removed, but may leave a slight green tinge to the underlying colour. On polished surfaces there is some variation in the underlying dark colour, some areas having a reddish tinge.

After one minute of immersion in the solution (at 50–60°C), a yellow-orange colour forms on the surface of the article and gradually turns to bright green. The article is removed from the solution after fifteen minutes and allowed to dry in air, without washing in water. When the article is completely dry it is gently rubbed with a soft cloth, which removes the green layer, revealing the underlying dark brown/black. It is then wax finished.

1.104   Dark grey-brown (as-cast only) Matt

Ammonium sulphate	30 gm
Sodium hydroxide	40 gm
Water	1 litre

Boiling immersion (Thirty minutes)

A dark grey-brown/black layer is also produced on polished surfaces, but this tends to be non-adherent and can be partially removed by bristle-brushing when dry. Some reddish-brown colour tends to remain. The results obtained in tests suggest that the procedure can be used to produce 'antique' or relieved effects.

▲ Sodium hydroxide is a powerful caustic alkali, the solid and strong solutions causing severe burns to the eyes and skin. When preparing the solution, small quantities of the sodium hydroxide should be added to large quantities of water and *not* vice-versa. The reaction is exothermic and will warm the solution. When the sodium hydroxide is completely dissolved, the ammonium sulphate is added and the solution may be heated.

The object is immersed in the boiling solution and the surface gradually darkens to produce a black colour, and acquires a slightly crystalline appearance. After thirty minutes the object is removed, washed in warm water and allowed to dry in air. Relieving is then carried out, and the object wax finished.

1.105   Dark slate with reddish tinge Semi-matt

A	Copper nitrate	100 gm
	Water	1 litre
B	Potassium sulphide	5 gm
	Water	1 litre
	Hydrochloric acid (35%)	5 cm³

Boiling immersion (A)   (Fifteen seconds)
Cold immersion (B)   (Thirty minutes)

This recipe is suitable for producing antiquing and relieved effects only, on polished surfaces. On rough and as-cast surfaces it yields an even dark grey, which may also be relieved. The polished surfaces are prone to spotting, and the recipe cannot be generally recommended.

▲ Solution B should only be prepared and used in a fume cupboard or a very well ventilated area. When the acid is added to the potassium sulphide solution, quantities of hydrogen sulphide gas are evolved which are toxic and can be dangerous even in moderate concentrations.

The object is immersed in the boiling solution A for fifteen seconds and then transferred to solution B. The surface darkens initially to produce a series of uneven petrol blue/purple interference colours, and then gradually changes to a slate grey. After thirty minutes the object is removed and rinsed, relieved by gentle scratch-brushing or pumice if desired, and finally washed and dried in sawdust. When completely dry it is wax finished.

1.106   Mid-dark grey Semi-gloss/semi-matt   *Pl. II*

Copper sulphate	6.25 gm
Copper acetate	1.25 gm
Potassium nitrate	1.25 gm
Sodium chloride	2 gm
Sulphur (flowers of sulphur)	3.5 gm
Acetic acid (6% solution)	1 litre

Boiling immersion (About one hour)

A mid- to dark grey colour, with a slight green patina, is also produced on as-cast surfaces.

▲ Potassium nitrate is a powerful oxidising agent and must not be allowed to come into contact with combustible material. It should not be mixed with the sulphur in the solid state.

The article is immersed in the boiling solution and after a few minutes the surface darkens and becomes more opaque as the colour develops. The surface develops gradually as immersion continues, and after about one hour the article is removed and washed in hot water. It is dried in sawdust and may be wax finished when completely dry.

1.107   Dark slate Semi-matt/semi-gloss

Selenous acid	6 cm³
Water	1 litre
Sodium hydroxide solution (250 gm/litre)	50 drops

Warm immersion (About ten minutes)

The acid is dissolved in the water, and the sodium hydroxide added by the drop, while stirring; the solution is then heated to 25–30°C. The article to be coloured is immersed in the solution, producing a purplish-brown lustre after about fifteen seconds, which darkens unevenly and then fades. The surface begins to darken again after two or three minutes, slowly developing to a slate grey colour after five minutes. When this has developed fully, after about ten minutes, the article is removed, washed in cold water and allowed to dry in air. Any relieving should then be carried out with a bristle-brush, and the article finally wax finished.

The early lustre colours produced are generally of poor quality. The dark slate layer that develops on both polished and as-cast surfaces tends to be only partially adherent, and can be bristle-brushed, when dry, to produce variegated or relieved effects. The relieved grey layer reveals the bronze surface beneath. This solution can only be recommended for variegated and relieved effects and not for producing even-coloured surfaces.

▲ Selenous acid is poisonous and very harmful if swallowed or inhaled. Colouring must be carried out in a well ventilated area. Contact with the skin or eyes must be avoided to prevent severe irritation.

▲ Sodium hydroxide is a powerful caustic alkali. Contact with the eyes and skin must be prevented.

1.108*   Black Satin   *Pl. I*

Heating in a kiln (About one hour)

The kiln is pre-heated to a temperature of 600°C and the object placed inside in a position that will favour even heating. The size of the object will determine the length of time needed for the colour to form. The object should be inspected from time to time to monitor progress. When the colour has fully developed, the object is removed and allowed to cool. When it is cool it may be wax finished.

The object must be perfectly clean and grease-free to obtain an even colour.

The tests relating to this recipe were carried out using an electric kiln. The effects produced with a gas kiln are not known.

1.109   Black Matt

A	Copper sulphate	50 gm
	Water	1 litre
B	Sodium thiosulphate	50 gm
	Lead acetate	12.5 gm
	Water	1 litre

Hot immersion (A)   (One minute)
Hot immersion (B)   (Two minutes)

The object is immersed in the hot copper sulphate solution (80°C) for one minute, producing a darkening of the surface. It is then transferred to solution B (80°C) for two minutes immersion. These alternating immersions are repeated a number of times until an even black layer is produced on the surface. The object is then washed in warm water and allowed to dry in air. When dry, it is rubbed gently with a soft cloth or bristle-brush, and wax finished.

An even matt black is produced on rough or as-cast surfaces. Tests showed that on polished surfaces the results tend to be non-adherent. Careful drying and brushing can produce a variegated black and light brown surface, but this recipe cannot be recommended for producing an even black colour on polished surfaces.

▲ Lead acetate is a highly toxic substance, and every precaution should be taken to prevent inhalation of the dust during preparation. The solution is very harmful if taken internally. Colouring should be carried out in a well ventilated area.

1.110   Black Semi-matt

Potassium polysulphide	10 gm
Sodium chloride	10 gm
Water	1 litre

Hot immersion (A few minutes)

The object is immersed in the hot solution (60°C), which immediately blackens the surface. Light bristle-brushing can be used to help in producing an even surface finish. After a few minutes, the object is removed and rinsed in warm water. It is allowed to dry in air, and wax finished when dry.

The black colour is also obtained on as-cast surfaces, but tends to be tenuous on both as-cast and polished surfaces.

▲ Colouring should be carried out in a well ventilated area, as hydrogen sulphide gas is liberated from the hot solution.

1.111   Black (as-cast surfaces only)

A	Barium sulphide	2 gm
	Potassium sulphide	2 gm
	Ammonium sulphide (16% solution)	2 cm³
	Water	1 litre
B	Copper nitrate	20 gm
	Ammonium chloride	20 gm
	Calcium hypochlorite	20 gm
	Water	1 litre

Cold immersion (A) (Thirty seconds)
Cold immersion (B) (A few minutes)

The article is immersed in the mixed sulphide solution, which immediately blackens the surface, and removed after about thirty seconds. If the surface is uneven then it can be lightly brushed with a scratch-brush and re-immersed for a similar length of time. The article is washed and transferred to solution B, which changes the colour to a dark slate or grey-brown. After a few minutes the article is removed and allowed to dry in air. When dry it may be wax finished.

The colour is not adherent on a polished surface, which is slightly dulled by the treatment. The procedure may be used for obtaining relieved effects.

The secondary treatment in solution B tended to produce results that were only tenuously adherent, and cannot be generally recommended.

Similar poor results were also obtained with an after treatment of a solution where the calcium hypochlorite is replaced by an equal quantity of calcium chloride.

▲ Ammonia vapour and hydrogen sulphide gas are liberated from the mixed sulphide solution. Colouring should be carried out in a fume cupboard or very well ventilated area. It should be prevented from coming into contact with the eyes and skin.

▲ Barium compounds are toxic, and harmful if taken internally.

▲ The fine dust from the calcium hypochlorite (bleaching powder) must not be inhaled.

1.112*   Pale greenish-grey on buff ground Semi-matt   *Pl. II*

Copper carbonate	100 gm
Ammonium chloride	100 gm
Ammonia (.880 solution)	20 cm³
Water	1 litre

Hot immersion (Fifteen minutes)

The article is immersed in the hot solution (50°C) producing immediate etching effects, typical with chloride-containing solutions. After about fifteen minutes, the article is removed and washed thoroughly in warm water. It is allowed to dry in air, and may be wax finished when dry.

A pale greenish-grey on a brown ground is obtained on as-cast surfaces.

▲ When preparing the solution, care should be taken to avoid inhaling the ammonia vapour. The eyes and skin should be protected from accidental splashes of the concentrated ammonia solution.

1.113*   Dull metallic grey (variegated) on brown Semi-matt   *Pl. I, Pl. XIII*

Nickel ammonium sulphate	50 gm
Sodium thiosulphate	50 gm
Water	1 litre

Hot immersion (Fifteen minutes)

A purplish-grey colour quickly forms on the surface of the object on immersion, becoming patchy and metallic grey after one minute. After two minutes the object should be removed from the solution and lightly scratch-brushed with water, to remove the non-adherent deposits that build up on the surface. The object is then re-immersed and scratch-brushing repeated twice at intervals of five minutes. After fifteen minutes the object is again lightly scratch-brushed, briefly re-immersed and then washed in hot water. After thorough drying in sawdust, it may be wax finished.

On rough or as-cast surfaces, an even slate grey is produced. On polished surfaces a very uneven but coherent surface is produced.

In preparing the solution, the two chemicals should be separately dissolved in two portions of the water. The thiosulphate solution is warmed, and the nickel ammonium sulphate solution added gradually. The light green solution is used at a temperature of 60–80°C.

▲ Nickel salts are a common cause of sensitive skin reactions. The vapour evolved from the heated solution should not be inhaled.

1.114*   Grey lustre on mid-brown Gloss

Potassium hydrogen tartrate	20 gm
Sodium tartrate	10 gm
Tartaric acid	10 gm
Copper acetate	10 gm
Stannic chloride (hydrated)	3 gm
Sodium metabisulphite	19 gm
Water	1 litre

Hot immersion (Several minutes)

The object is immersed in the hot solution (80°C), which produces brown or purplish-brown lustre colours after two or three minutes. These tend to be uneven, but become more even after about five minutes as the colour changes to grey. When the greyish colour has just developed, the object is removed, washed in hot water and allowed to dry in air. When dry, it may be wax finished.

An even dark brown colour is produced on rough or as-cast surfaces.

The timing is difficult. If immersion is carried on too long, the colour is lost. If the object is removed too soon, then the surface tends to be uneven.

▲ Stannic chloride is corrosive, and should be prevented from coming into contact with the eyes and skin.

1.115* Dark green patina Semi-matt   *Pl. I, Pl. XII*

Copper nitrate	110 gm
Water	110 cm³
Ammonia (.880 solution)	440 cm³
Acetic acid (6% solution)	440 cm³
Ammonium chloride	110 gm

Applied liquid (Five days)

The solution is applied to the surface by dabbing with a soft cloth, and the article then left to dry in air. The metal quickly darkens and a whitish/blue-green powdery patina forms gradually on the surface. When dry, this is rubbed with a soft dry cloth to remove any loose material and to smooth the surface, prior to the next application of solution. This procedure is repeated over a period of five days, during which time the patina darkens and becomes integral with the surface. When treatment is complete and the surface thoroughly dry, the article may be wax finished.

On rough or as-cast rather than polished surfaces, this treatment results in an even blue-green patina on a very dark olive ground.

▲ Colouring must be carried out in a well ventilated area. Ammonia vapour will be liberated from the strong solution, which will cause irritation of the eyes and respiratory system. The solution must be prevented from coming into contact with the eyes or skin, as it can cause burns or severe irritation.

1.116* Blue-green patina on variegated brown ground Semi-matt   *Pl. II*

Copper sulphate	20 gm
Copper acetate	20 gm
Ammonium chloride	10 gm
Acetic acid (6% solution)	to form a paste

Applied paste (Several days)

The ingredients are ground to a creamy paste with a little acetic acid, using a pestle and mortar. The paste is applied to the object with a soft brush, giving quite a thick coating which is then allowed to dry for a day. The dry residue is then washed away under cold water, using a soft brush. A thin layer of paste is then wiped onto the object with a cloth, and the object left to dry for a day. The residue is again washed off. This procedure of applying thin layers and drying is repeated in the same way until a good variegated patina is produced. The object should be allowed to dry thoroughly when treatment is complete, and may then be wax finished.

1.117* Grey-green patina on reddish-brown ground Matt/semi-matt   *Pl. II*

Sodium chlorate	100 gm
Copper nitrate	25 gm
Water	1 litre

Boiling immersion (Two hours)

The object is immersed in the boiling solution, and the surface is initially coloured yellow-orange. Continued immersion produces a gradual development of the ground colour during the first hour. A greenish-grey bloom slowly builds up on this ground, in the form of a light dusting at first and as an even coherent layer after about two hours. When the required finish has been obtained, the object is removed and washed in hot water. It is allowed to dry in air, and may be wax finished when dry.

An even grey-green patina on a reddish-purple ground is produced on both polished and rough or as-cast surfaces.

▲ Sodium chlorate is a powerful oxidant and must not be allowed to come into contact with combustible material or acids. It should also be prevented from coming into contact with the eyes, skin or clothing.

1.118* Dark green/dark bluish-green Semi-matt   *Pl. XIII*

Ammonia (.880 solution)	
Sodium chloride	20 gm/litre

Vapour technique (About two days)

The object is placed or suspended in a container, into the bottom of which the solution of sodium chloride in ammonia is introduced. The object must be placed well clear of the surface of the liquid, and in a position that will favour an even distribution of vapour around it. The object should be well clear of the walls and lid of the container. The container is sealed and the vapour allowed to act on the object for about two days. The object is then removed and washed in cold water. It is allowed to dry thoroughly for at least a day, and may then be wax finished.

A more intense blue-green colour tends to develop on as-cast surfaces, with some darker grey variegation.

The exact method used will depend on the size of the object being coloured. It is essential to plan the procedure carefully. Ammonia vapour will concentrate in the container and should not simply be released. See 'Metal colouring techniques', 9.

After-treatments involving the use of the vapour from concentrated acetic acid, suggested by some sources, tended to make the colour less adherent.

▲ Ammonia solution is highly corrosive and must be prevented from coming into contact with the eyes or skin. The vapour is extremely irritating to the eyes and respiratory system, and will irritate the skin. Adequate ventilation must be provided.

1.119*   **Light green patina on reddish-brown ground Matt**   *Pl. II*

Copper nitrate	200 gm
Sodium chloride	200 gm
Water	1 litre

Applied liquid   (Twice a day for five days)

The solution is dabbed on to the object with a soft cloth and left to dry in air. This process is repeated twice a day for five days. The object is then allowed to dry in air, without treatment, for a further five days. The ground colour develops gradually throughout this period and after two or three days some blue-green patina forms on the surface. This continues to build up as long as the surface is moist. Finishing should only be carried out when it is certain that the surface is dry and patina development complete.

The patina is more dense and the ground colour less evident on rough or as-cast surfaces.

A dabbing technique was found to be better than wiping which tends to inhibit the formation of the patina.

If there is an initial surface resistance to 'wetting', the liquid can be applied by scratch-brushing in the early stages.

In damp weather conditions the surface may take several weeks to dry completely. Although apparently dry, the surface may 'sweat' intermittently. The drying process should not be rushed and wax finishing only carried out when it is certain that chemical action has ceased.

1.120*   **Blue-green patina on brown/black ground Semi-matt**   *Pl. I, Pl. XIV*

Copper nitrate	100 gm
Nitric acid (70% solution)	40 cm³
Water	1 litre

Torch technique

The object is heated with a blow torch and the solution applied sparingly with a bristle-brush or paint brush. The liquid quickly turns dark brown, and becomes yellowish where it is not being directly heated. Continued heating makes the surface blacken and areas of blue-green patina begin to form. The blue-green patina tends to be superficial at this stage and the surface should be further heated and stippled with solution until a good dark brown or black ground has been established. The object is then gently heated and dabbed with an almost dry brush or cloth, barely damp with the solution, until an evenly distributed blue-green patina is obtained. When treatment is complete, the object is allowed to cool and dry thoroughly, and may then be wax finished.

▲ Nitric acid is very corrosive and must not be allowed to come into contact with the eyes, skin or clothing. A face shield or goggles should be worn to protect the eyes from hot splashes of solution as it is applied to the heated metal.

▲ The fumes evolved as the solution is applied to the heated metal should not be inhaled. Adequate ventilation must be provided, and a nose and mouth face mask fitted with the correct filter should be worn.

1.121   **Blue-green patina on mid-brown ground Semi-matt/matt**

Ammonium chloride	35 gm
Copper acetate	20 gm
Water	1 litre

Applied liquid   (Several days)

The ingredients are ground with a little of the water using a pestle and mortar. They are then added to the remaining water. The solution is applied to the object by dabbing and wiping, using a soft cloth. The solution should be applied sparingly, to leave an evenly moist surface. The object is then allowed to dry in air. This procedure is repeated once a day for several days, producing a gradual development of the ground colour and blue-green patina. When treatment is complete, the object should be left to dry for several days, during which time there is further patina development. When the object is completely dry, and there is no further surface change, the object is wax finished.

There is a greater development of a more intense blue-green patina on rough or as-cast surfaces.

It is essential to ensure that all patina development is complete, and the surface is completely dry, before wax finishing is carried out. The final drying period may have to be extended to a matter of weeks in damp or humid conditions.

1.122   **Blue-green and yellow-green patina on brown ground Semi-matt**

Copper nitrate	20 gm
Ammonium chloride	20 gm
Calcium hypochlorite	20 gm
Water	1 litre

Cold immersion   (Twenty hours)

The article is suspended in the cold solution. After an initial 'bright dip' effect which enhances the metal surface to a reddish-gold colour, there is a gradual build-up of colour and an etching effect that is typical with chloride solutions. The article is left in the solution for twenty hours and then removed, rinsed in cold water and allowed to dry in air. Drying in air produces a green powdery layer which is only partially adherent on polished surfaces. When the green layer has developed and the surface is completely dry, the object is wax finished.

A variegated green and blue green patina on a darker ground is produced on rough or as-cast surfaces. Patina development on polished faces is slow, resulting in thin variegated patches on a light brown ground.

It is essential to ensure that patina development is complete and the surface dry, before wax finishing. Drying time may need to be extended in damp or humid conditions.

▲ Calcium hypochlorite (bleaching powder) can cause irritation of the eyes, skin and respiratory system. It is essential to avoid inhaling the fine dust. Both the powder and solutions should be prevented from coming into contact with the eyes and skin.

1.123* Blue-green patina (blue-green patina on black ground)

Copper nitrate	200 gm
Water	1 litre

Torch technique

The metal surface is heated with a blow-torch, and the solution applied with a soft brush until it is covered with an even blue-green patina. If the surface is then heated, without further application of the solution, it will become black. The solution may then be applied to this black ground, heating the surface as necessary, to form a blue-green patina on the ground colour. The surface quality obtained can be varied by the precise method of application—a stippled surface is produced by stippling with a relatively dry brush, or a more mottled or marbled effect obtained using the same technique with a brush that is more 'loaded' with the solution. When the required finish is achieved, the object is allowed to dry out, and may then be wax finished.

These colours are also produced on as-cast surfaces.

▲ Colouring should be carried out in a well ventilated area, so that inhalation of the vapours is avoided.

1.124* Olive-green on black ground Matt

A	Copper nitrate	200 gm
	Water	1 litre
B	Potassium polysulphide	50 gm
	Water	1 litre

Torch technique (A)
Torch technique (B)

The article is heated with a blow torch and the copper nitrate solution applied with a soft brush until an even thin layer of blue-green patina covers the surface. This is again heated, without further application of the solution, until it turns black. Any loose residue should be brushed off with a soft dry brush. The copper nitrate is then again applied, heating as necessary, until the blue-green colour re-appears and is evenly distributed. While the surface is still hot, the polysulphide solution is applied until the required depth of colour is achieved. When treatment is complete, the article is allowed to dry for some time, and may then be wax finished.

These colours are also produced on as-cast surfaces.

The concentrations of both solutions can be varied within quite wide limits.

▲ Colouring should be carried out in a well ventilated area, so that inhalation of the vapours and of the hydrogen sulphide gas that are evolved is avoided.

1.125 Blue-green patina on red-brown/maroon ground Semi-matt

Copper nitrate	200 gm
Water	1 litre

Applied liquid (Five days)

The article is dipped in the cold solution for a few seconds, drained and allowed to dry in air. This procedure is repeated twice a day for about five days, during which time the surface darkens gradually to a red-brown. After about two days, patches of powdery blue-green patina also begin to appear on the surface. When treatment is complete, the article should be left to dry in air for a period of at least three days to allow the patina to develop fully and dry out. When dry, the article is wax finished.

On rough and as-cast rather than polished surfaces, the ground colour tends to be very dark brown.

Some sources suggest a scratch-brushed application of the solution. Tests carried out using this technique produced similar but slightly lighter results over the same period of time.

Tests carried out in which the metal was briefly 'pickled' in a 10% solution of nitric acid prior to the initial dip, as suggested by some sources, produced very similar results with somewhat duller surfaces.

The patina tends to develop in streaks and patches corresponding to the draining of the solution from the surface. This can be minimised by 'brushing out' with a soft brush after dipping, to prevent runs and pooling.

If the ground colour alone is required, then after two days of dipping and drying, any patina should be brushed off scrupulously and application continued by wiping-on very sparingly. A completely patina-free ground is difficult to achieve.

1.126 Greenish patina on light brown ground Semi-matt   *Pl. I*

Copper nitrate	30 gm
Zinc chloride	30 gm
Water	to form a paste

Applied paste (Two or three hours)

The ingredients are mixed by grinding together using a pestle and mortar. Water should be added by the drop, to form a creamy paste. An excess of water should be avoided, as the paste tends to thin rather suddenly. The paste is applied to the object with a soft brush, and allowed to dry for two or three hours. The residual paste is gently washed away with cold water, and the object then allowed to dry in air for several days. Exposure to daylight will cause a variegated darkening of the ground colours. When the surface is completely dry, and colour change has ceased, it may be wax finished.

Both the ground colour and the patina tend to be very variegated on polished surfaces. A greenish patina on a more even mid-brown ground is produced on as-cast surfaces.

Repeated applications do not improve the results, and tend to require far longer final drying periods.

It is essential to ensure that the surface is completely dry, and that all colour change has ceased, before wax finishing. This may take some weeks, as the finish is prone to intermittent 'sweating'.

1.127   Pale green patina on black/brown ground Semi-matt   *Pl. II, Pl. XIV*

Copper nitrate	15 gm
Zinc nitrate	15 gm
Ferric chloride	0.5 gm
Hydrogen peroxide (100 vols.)	to form a paste

Applied paste (Four hours)

The ingredients are ground using a mortar and pestle, and the hydrogen peroxide added in small quantities to form a thin creamy paste. The paste is applied to the object with a soft brush, and left to dry in air. After four hours, the residue is washed away with cold water and the object is allowed to dry in air for several hours. When completely dry, it may be wax finished.

A variegated black/brown colour is produced on rough and as-cast surfaces.

The surface of the object should be handled as little as possible until dry, and should not be brushed during washing.

▲ Hydrogen peroxide should not be allowed to come into contact with combustible material. It should be prevented from coming into contact with the eyes and skin, as it may cause severe irritation. It is very harmful if taken internally. The vapour evolved as it is added to the solid ingredients of the paste must not be inhaled.

1.128   Watery pale blue and green Semi-matt

Ammonium carbonate	180 gm
Copper sulphate	60 gm
Copper acetate	20 gm
Oxalic acid	1.5 gm
Ammonium chloride	0.5 gm
Acetic acid (10% solution)	1 litre

Boiling immersion (Twenty-five minutes)

The object is immersed in the boiling solution, which produces some etching effects and light colouring after a few minutes. As the immersion continues streaky greenish films gradually develop, beginning to become dark after about ten minutes. The object is removed after twenty-five minutes and the hot solution applied with a soft brush, to leave the surface evenly moist. The object should then be allowed to dry thoroughly in air for several hours, without washing. When dry, it is wax finished.

On drying, patches of watery pale blue develop on a light green surface. On rough and as-cast surfaces, the pale blue patina develops on a light brown ground.

▲ The ammonium carbonate should be added to the acetic acid in small quantities. The reaction is effervescent, and this may be very vigorous if large amounts are added.

▲ Oxalic acid is harmful if taken internally and irritating to the eyes and skin.

1.129   Blue-green patina Matt   *Pl. I*

Ammonium carbonate	24 gm
Potassium hydrogen tartrate	6 gm
Sodium chloride	6 gm
Copper sulphate	6 gm
Acetic acid (6% solution)	to form a paste

Applied paste (Several days)

The solid ingredients are made to a paste with a little of the acetic acid, using a pestle and mortar. The paste is applied with a soft brush. The object is left to dry for about two days, and the application is then repeated. It is then left to dry thoroughly in air, which generally takes several days. When completely dry, the surface is brushed gently with a bristle-brush to remove any loose material, rubbed with a soft cloth to smooth the surface, and wax finished.

On polished surfaces the patina tends to build up as an even incrustation. On rough or as-cast surfaces the underlying dark ground is more evident giving a greener tinge.

If the paste is made up with stronger solutions of acetic acid (15–30%) then there is a greater tendency for the blue-green patina to be non-adherent, partially revealing orange-brown ground colours.

▲ A nose and mouth face mask fitted with a fine dust filter should be worn when brushing away the dry residue.

1.130   Pale green on reddish-brown Semi-matt

Copper sulphate	125 gm
Sodium chlorate	50 gm
Ferrous sulphate	25 gm
Water	1 litre

Boiling immersion (Thirty minutes)

The article is immersed in the boiling solution, which produces an ochre colour after about five minutes. Continued immersion produces a slow change to a reddish-brown colour, and the gradual development of a pale green patina which is deposited on this ground. This initially takes the form of a greenish-grey bloom but later becomes more substantial and forms variegated areas of green or blue-green on the brown ground. When the colour has developed, after about thirty minutes, the article is removed and washed in hot water. It is dried in sawdust, and may be wax finished when dry.

Similar results are also produced on rough or as-cast surfaces.

▲ Sodium chlorate is a powerful oxidising agent and must not be allowed to come into contact with combustible materials or acids. Inhalation of the dust must be avoided. The solid and its solutions must be prevented from coming into contact with the eyes, skin and clothing.

1.131   Grey-green patina on brownish ground Matt/semi-matt

Copper chloride	50 gm
Water	1 litre

Hot immersion (Twenty minutes)

A greenish patina on an olive-brown ground is produced on rough or as-cast surfaces.

The object is immersed in the hot solution (80°C), which produces etching effects and pale grey colouration. With continued immersion, a denser colour is produced. After twenty minutes the object is removed and washed in hot water, which tends to remove a superficial grey layer, and leach some yellowish-green colour from the surface. It is then allowed to dry in air for several hours, and wax finished when dry.

1.132   Green patina on brownish ground Semi-matt

Copper nitrate	35 gm
Ammonium chloride	35 gm
Calcium chloride	35 gm
Water	1 litre

Cold immersion (Fifteen minutes)

On polished surfaces the result takes the form of a brownish enhanced grain with little patina development. On rough and as-cast surfaces the pale green patina develops fully.

▲ A nose and mouth face mask with a fine dust filter should be worn when the surface is brushed.

Immersing the object in the cold solution produces an etching effect which brings out the grain structure of a polished surface. Continued immersion darkens the colour to a pale olive. After fifteen minutes the object is removed and washed in hot water, which brings out traces of green patina. It is then allowed to dry in air, without washing, and further patina development occurs. When completely dry the surface is gently brushed to remove any loose dust and wax finished.

1.133   Pale green patina on olive ground Semi-matt/semi-gloss

Copper nitrate	200 gm
Sodium chloride	200 gm
Water	1 litre

Boiling immersion (Thirty minutes)

The patina and ground colour develop better on a rough or as-cast surface. On a highly polished surface a thin veil of patina is produced on a slightly olive grain-enhanced ground.

Tests were carried out using lower concentrations of the ingredients, and additions of ammonium chloride and potassium hydrogen tartrate, as suggested by some sources. The results were poor, producing drab grey-brown grounds with thin traces of dull patina.

An immediate etching of the surface occurs when the article is immersed in the boiling solution, followed by the gradual development of a mottled greyish-white colour. This darkens appreciably during the later stages of immersion. After thirty minutes the article is removed and washed in hot water, which produces a yellowish-green colour. The article is then left to dry in air. When completely dry the article is wax finished. The yellow-green colouration is much less apparent when the surface is dry, particularly after waxing, and becomes a more whitish-green after a short time.

1.134   Blue-green patina Matt

Copper nitrate	80 gm
Water	1 litre
Ammonia (.880 solution)	3 cm³

Applied liquid (Three days)

Prolonging the treatment beyond three days does not appear to produce any further development. In tests extending to twenty days no notable changes could be detected after three days.

▲ Inhalation of the fine spray is harmful. It is essential to wear a suitable nose and mouth face mask provided with the correct filter, and to ensure adequate ventilation.

The solution is applied to the object by spraying with a moderately fine atomising spray to produce a misty coating. It is essential to apply only a fine misty coating and to avoid any pooling of the solution or runs on the surface. The object is then left to dry. This procedure is repeated twice a day for three days, after which the surface should be allowed to dry out thoroughly for several days before finishing with wax.

1.135   Light blue-green patina on variegated buff/brown ground Semi-matt

Copper sulphate	20 gm
Zinc chloride	20 gm
Water	to form a paste

Applied paste (Twenty-four hours)

A thin greyish-green patina develops on a light brown ground on as-cast surfaces. Some patina may also occur on polished surfaces, particularly if the procedure is repeated.

The paste should be applied sparingly, and should not be too thick, or incrustations tend to occur that 'sweat' intermittently during the drying period.

The ingredients are ground to a thin creamy paste with a little water, using a pestle and mortar. The paste is applied sparingly to the surface of the object with a soft brush. It is then allowed to dry completely, and the residual dry paste washed off with cold water. The object is then allowed to dry in air for about twelve hours, and the procedure repeated if required. When treatment is complete, the object should be allowed to dry for several days at least. Exposure to daylight causes a variegated darkening of the ground colours. When the surface is completely dry and colour change has ceased, the object may be wax finished.

1.136   Dark green on mid-brown ground Matt

Ammonium sulphate	105 gm
Copper sulphate	3.5 gm
Ammonia (.880 solution)	2 cm³
Water	1 litre

Torch technique

The metal surface is heated with a blow-torch, and the solution applied with a soft brush, until a dark green colour on a mid-brown ground is obtained. If the metal is too hot when the solution is applied then impermanent light grey or pinkish-red colours are produced. Continued application of the solution as the metal cools will remove these colours and encourage the brown ground and the dark green patina. If the green patina is heated once it has developed, by playing the torch across the surface, then it will tend to become black. When the desired surface finish has been achieved, the object is left to cool and dry for some time, and may then be wax finished.

A wide variety of surface effects are obtained, which depend on the temperature of the metal, and the manner in which the solution is applied (eg. liberally or sparingly; by 'stippling' or by 'painting' etc).

▲ Colouring should be carried out in a well ventilated area, to avoid inhalation of the vapours.

1.137   Green patina on brown ground Semi-matt

Copper nitrate	100 gm
Hydrochloric acid (35% solution)	10 cm³
Water	1 litre

Torch technique

The object is heated with a blow-torch and the solution sparingly applied with a bristle-brush or paint brush. A heavy deposit of yellowish-green patina quickly forms, but this is rather non-adherent at this stage. Continued heating and the application of small amounts of solution by brushing gradually darkens the surface. When the brown ground colour has been established, the green patina is built up either by stippling with a nearly dry brush, or by using a cloth which is barely damp with the solution, heat being frequently applied to the area being worked. When a good patina has been obtained, the surface can be burnished with a dry cloth. Wax finishing is carried out when the object is cool and thoroughly dry.

In some cases the patina may have a tendency to 'sweat' shortly after treatment. If this occurs then the object should be left in a warm dry place for some time, and the wax only applied when the surface is thoroughly dry.

▲ Hydrochloric acid is corrosive and should not be allowed to come into contact with the eyes and skin. A face shield or goggles should be worn to protect the eyes from hot splashes as the solution is applied.

▲ The fumes evolved as the solution is applied to the heated metal should not be inhaled. Adequate ventilation must be provided and a nose and mouth face mask fitted with the correct filter should be worn.

1.138   Light olive-green Matt

Copper sulphate	125 gm
Sodium chlorate	50 gm
Water	1 litre

Boiling immersion (Ten minutes)

The article is immersed in the boiling solution, and develops lustre colours in about one minute. Continued immersion produces a rapid development of more opaque colours. After ten minutes the article is removed and washed thoroughly in hot water. After careful drying in sawdust, it is wax finished.

A more grey-green and patina-like finish is produced on rough or as-cast rather than polished surfaces.

Prolonging the immersion produces a darker colour, but this is increasingly accompanied by a grey-green patina like layer being built up. This tends to be uneven on polished surfaces, and although it may be made more even by intermittent bristle-brushing during immersion, it is difficult to control. On rough or as-cast surfaces the results are more even.

▲ Sodium chlorate is a powerful oxidising agent and must not be allowed to come into contact with combustible material or acids. Inhaling the dust must be avoided, and the eyes and skin should be protected against the corrosive and irritating effects of the solid, or splashes of solution.

1.139   Green/blue-green patina on light brown ground Semi-matt

Ammonium carbonate	150 gm
Copper acetate	60 gm
Sodium chloride	50 gm
Potassium hydrogen tartrate	50 gm
Acetic acid (10% solution)	1 litre

Applied liquid (Twice a day for several days)

The solution is dabbed on to the object with a soft cloth, and then allowed to dry thoroughly in air. This process is repeated twice a day for several days until the desired colour is achieved. Any powder or other loose material that forms during drying periods should be gently brushed away with a soft dry cloth, prior to the next application of solution. When treatment is complete the surface should be allowed to dry in air for several days. When completely dry, it may be wax finished.

Sparing use of the solution produces a thin glaze of green patina on a light brown ground, on polished surfaces. On rough or as-cast surfaces a more developed blue-green patina on a darker ground is obtained. Less sparing use of the solution in the latter stages of treatment encourages a more dense patina.

For results obtained using a more concentrated form of this solution, see recipe 1.141.

▲ The ammonium carbonate should be added to the acetic acid in small quantities. The reaction is effervescent and if large quantities are added, then the effervescence is very vigorous.

1.140   Variegated (greenish patina on brown ground)   Semi-matt/semi-gloss

Copper nitrate	20 gm
Sodium chloride	16 gm
Potassium hydrogen tartrate	12 gm
Ammonium chloride	4 gm
Water	to form a paste

Applied paste   (Four hours)

On rough and as-cast surfaces a dull grey-brown colour is produced, with some slight pale patina.

▲ A nose and mouth face mask fitted with a fine dust filter should be worn when brushing the dry surface.

The ingredients are ground to a paste with a pestle and mortar by the very gradual addition of a small amount of water. The creamy paste is brushed onto the surface with a soft brush, and is left to dry. After four hours the dry residue is removed with a bristle-brush. The object is washed in cold water and allowed to dry in air. Exposure to daylight will gradually darken the colour from a variegated buff orange and green to produce a thin greenish patina on a brown ground. When the colour has stabilised, the object may be wax finished.

1.141   Slight blue-green patina on brownish etched ground   Matt

Ammonium carbonate	600 gm
Sodium chloride	200 gm
Copper acetate	200 gm
Potassium hydrogen tartrate	200 gm
Acetic acid (10% solution)	1 litre

Applied liquid   (Twice a day for several days)

Polished surfaces are unevenly etched by the solution to produce dull patches. A darker brown but patchy finish was produced on as-cast surfaces. The surface finish is difficult to control and adequate results could not be produced in tests. This procedure cannot therefore be recommended.

For results obtained with a more dilute form of this solution, see recipe 1.139.

It is essential to ensure that the surface is dry and patina development complete before wax finishing. In damp or humid conditions the drying time may need to be extended.

▲ The ammonium carbonate should be added to the acetic acid in small quantities. The reaction is effervescent and if large amounts are added the effervescence is very vigorous.

The solution is applied to the article to be coloured, by wiping sparingly with a soft cloth. It should then be allowed to dry thoroughly in air. This procedure should be repeated twice a day for several days, until the ground colour has developed and the patina has formed. The article should then be left to dry in air for several days, during which time the patina will continue to develop. When patina development has ceased and the surface is completely dry, the article may be wax finished.

1.142   Blue-green patina on red-brown ground   Semi-matt

Ammonium chloride	40 gm
Ammonium carbonate	120 gm
Sodium chloride	40 gm
Water	1 litre

Cold immersion. Applied liquid   (One hour)

On polished surfaces patina development tends to be slight. On rough or as-cast surfaces the patina develops more easily, and the ground colour tends to be darker.

It is essential to ensure that patina development is complete, and the surface dry, before wax finishing. In damp or humid conditions the drying time may need to be extended.

The object to be coloured is immersed in the cold solution for about one hour, producing a slight etching effect and a gradual development of the ground colour. It is then removed and allowed to dry thoroughly in air. Any excessive surface moisture should be gently brushed out with a soft brush, to prevent 'runs' or 'pooling' of the solution and to leave an evenly moistened surface. The patina develops slowly as the surface dries out. If further patina development is desired, then the solution should be applied sparingly with a soft cloth and the object again left to dry thoroughly. When patina development is complete and the surface dry, the object may be wax finished.

1.143   Green patina on light brown ground   Semi-matt

Ammonium carbonate	150 gm
Sodium chloride	50 gm
Copper acetate	60 gm
Potassium hydrogen tartrate	50 gm
Water	1 litre

Applied liquid   (Several days)

A dull glaze of green patina on a light brown ground is produced on polished surfaces if the minimum of residual moisture is left on the surface after the initial application. This can be achieved by brushing out any excess with a soft brush. A wetter surface will encourage a more opaque blue-green patina. On rough or as-cast surfaces a well developed blue-green patina is produced, as the moisture tends to be retained by the surface texture.

The solution is applied generously to the surface of the article using a soft cloth. It is allowed to dry in air. This procedure is repeated twice a day for several days, during which time the colour gradually develops. It is then left to dry in air, without further treatment, to allow the patina to develop. When patina development is complete and the surface is completely dry, the article is wax finished.

1.144   Pale blue (*or* green) patina on light brown/ochre ground Semi-matt

Ammonium carbonate	180 gm
Copper sulphate	60 gm
Copper acetate	20 gm
Oxalic acid	1.5 gm
Ammonium chloride	0.5 gm
Acetic acid (10% solution)	1 litre

Torch technique

The ingredients are dissolved and the solution boiled for about ten minutes. It is then left to stand for several days. The mixture should be shaken immediately prior to use. The surface of the object to be coloured is heated with a blow torch; and the solution applied with a soft brush. This process is continued until a pale blue patina is produced, on a light brown or ochre ground. If the torch is then gently played across the surface, the pale blue will change to green. Varying effects can be obtained by 'stippling' or 'dragging' the brush on the surface. When the required finish is obtained, the object is left to cool and dry for some time and may then be wax finished.

Continuous application of liberal quantities of the solution will tend to inhibit patina formation, but encourages the development of the ground colour.

▲ The ammonium carbonate should be added to the acetic acid in small quantities. The reaction is effervescent, and if large amounts are added, this may be violent.

▲ Colouring should be carried out in a well ventilated area, so that inhalation of the vapours is avoided.

1.145   Blue-green patina on pale brown ground Semi-matt/matt

Ammonium carbonate	150 gm
Ammonium chloride	100 gm
Water	1 litre

Applied liquid (Several days)

The solution is applied to the object by wiping it on to the surface with a soft cloth. The solution should be used sparingly, leaving the surface evenly moist. The object is then allowed to dry in air. This procedure is repeated once a day for several days, until the grey-green patina is sufficiently developed. When treatment is complete, the object should be allowed to dry out thoroughly for several days, and is wax finished when completely dry.

On a highly polished surface, the patina develops rather slowly, tending to produce only a very thin but adherent dusting. On rough or as-cast surfaces a well developed patina is produced.

It is essential to ensure that the surface is completely dry and patina development complete, before wax finishing. The drying time may need to be extended to a matter of weeks in damp or humid atmospheric conditions.

1.146   Blue-green patina on light brown ground Semi-matt/matt

Copper acetate	30 gm
Copper carbonate	15 gm
Ammonium chloride	30 gm
Hydrochloric acid (18% solution)	2 cm³
Water	to form a paste

Applied paste (Two or three days)

The ingredients are ground to a paste with a little water and the hydrochloric acid, using a pestle and mortar. The paste is applied to the object with a soft brush and then left for several hours to dry out. The dry residue is removed with a stiff bristle-brush. The procedure is repeated until a brown ground has developed with a thin bloom of patina. The paste is then thinned out to a liquid and applied using the same procedure, but with a soft brush to remove residue. When the surface is sufficiently developed and well dried out, the object may be wax finished.

A blue-green patina on a greenish-brown ground is produced on rough or as-cast surfaces.

A thick incrustation tends to form if the paste is made too thick. This is difficult to remove by brushing, when dry, but tends to flake eventually.

▲ A nose and mouth face mask should be worn when brushing away the dry residue.

1.147   Thin blue-green patina (grain-enhanced metal) Semi-matt

Ammonium carbonate	300 gm
Acetic acid (10% solution)	35 cm³
Water	1 litre

Warm scratch-brushing (Two or three minutes)

The ammonium carbonate is added to the water, while stirring, and the acetic acid then added very gradually. The solution is then warmed to 30–40°C. The solution is applied to the object liberally with a soft brush, and then immediately scratch-brushed. The surface is worked for a short time, while applying more solution, and then allowed to dry in air. Any excess moisture on the object should be brushed out with a soft bristle-brush, to prevent 'pooling' or 'running' and to leave an evenly moist surface. When the surface is completely dry, which may take several days, it is wax finished.

A more developed patina on a light brown ground is produced on rough or as-cast surfaces. On polished faces, the patina is thin and patchy and the ground colour undeveloped. Some etching, giving rise to grain enhancement, occurs on polished surfaces.

The results obtained in tests with this solution were poor. Although the colours obtained are interesting, the surfaces were very prone to streaking and patchiness, and little control over this could be gained. The technique cannot therefore be generally recommended.

1.148   Thin blue-green patina on dull metal ground Semi-matt

Ammonium sulphate	105 gm
Copper sulphate	3.5 gm
Ammonia (.880 solution)	2 cm³
Water	1 litre

Applied liquid   (Several days)

An olive-brown ground with slight blue-green patination is produced on rough or as-cast surfaces. Polished surfaces tend to remain undeveloped in both ground colour and patination.

This recipe, which has been used commercially to induce patination on copper roofing, requires weathering in the open air to achieve the best results. In tests carried out in workshop conditions, only slight patination and patchy ground colours could be obtained.

The solution is applied to the surface of the object with a soft cloth. It is then allowed to dry in air. This procedure is repeated twice a day for several days until the ground colour has developed and traces of patina begin to appear. The solution should then be applied more sparingly by dabbing with a soft cloth. The object is allowed to dry slowly in a damp atmosphere to encourage the development of patina. It is finally allowed to dry in a warm dry atmosphere, and wax finished when completely dry.

1.149   Slight blue-green patina on light ochre/dull metal ground Matt

A	Ammonium sulphate	80 gm
	Water	1 litre
B	Copper sulphate	100 gm
	Sodium hydroxide	10 gm
	Water	1 litre

Applied liquid (A)   (Five to ten days)
Applied liquid (B)   (Two or three days)

A blue-green patina on a light brown ground is produced on rough or as-cast surfaces.

In tests, the solution tended to retract into pools and failed to 'wet' the surface. The addition of wetting agents produced no improvement, and the surface developed in patches rather than evenly.

▲ Sodium hydroxide is corrosive and should be prevented from coming into contact with the eyes and skin.

The ammonium sulphate solution is applied to the surface of the object by dabbing vigorously with a soft cloth, leaving it sparingly moist. It is then allowed to dry completely. This procedure is repeated twice a day for up to ten days. Solution B is then applied in the same way for two or three days. When treatment is complete, the object is allowed to dry in air for several days. When completely dry, it may be wax finished.

1.150   Slight blue-green patina on dull metal Semi-matt/matt

Ammonium chloride	10 gm
Ammonium acetate	10 gm
Water	1 litre

Applied liquid   (Twice a day for ten days)

A blue-green patina on a light brown ground is produced on rough or as-cast surfaces.

Colour development is generally slow, the ground developing unevenly at first, and the patina appearing later during the periods of air-drying. Patina development is more pronounced in surface features such as punched marks or textured areas.

The solution is applied to the surface by wiping with a soft cloth or by brushing with a soft brush. It is then allowed to dry in air. This procedure is repeated twice a day for about ten days. The article is then allowed to dry in air, without treatment, for a further five days. When completely dry, and when patina development has ceased, it may be wax finished.

1.151   Pale greenish patina on grey-brown Matt

Copper sulphate	200 gm
Ferric chloride	25 gm
Water	1 litre

Hot immersion   (Thirty minutes)

An incrustation of patina tends to develop on a very uneven ground. Considerable growth of this patina occurred some time after treatment, and there is some doubt as to the stability of the finish. The procedure cannot therefore be recommended.

▲ Ferric chloride is corrosive and should be prevented from coming into contact with the eyes and skin.

The object is immersed in the hot solution (60°C), which etches the surface and produces a pale greyish colour, typical of chloride-containing solutions. Continued immersion produces a gradual change to a more brown colour. After thirty minutes the object is removed and washed in hot water, which tends to leach a slight yellowish-green colour from the surface. After thorough washing, the object is allowed to dry in air and exposed to daylight, which tends to cause an uneven darkening of the colour. When the surface development is complete and the object thoroughly dried out, which may take several days, it can be wax finished.

1.152   Grey-green (as-cast surfaces only) Matt

A	Ammonium chloride	10 gm
	Copper sulphate	10 gm
	Water	1 litre
B	Hydrogen peroxide (100 vols.)	200 cm³
	Sodium chloride	5 gm
	Acetic acid (glacial)	5 cm³
	Water	1 litre

Cold immersion (A) (Five minutes)
Hot immersion (B) (A few seconds)

The object is immersed in solution A for about five minutes, producing a slight etching effect, and a pale greyish colour typical of chloride-containing solutions. The object is then briefly immersed in solution B (at 60–70°C). The reaction tends to be violent, causing the solution to 'boil', and immersion should not be prolonged. A greenish-brown layer is produced. If this is uneven, the surface should be gently scratch-brushed and the object briefly re-immersed. Prolonged or repeated immersion in solution B will tend to strip the greenish-brown colour. When treatment is complete, the object is washed in warm water and allowed to dry in air, and may be wax finished when dry.

The greenish-brown colour obtained on a polished surface tended to be quickly removed in solution B leaving an even, lightly etched pink/buff finish.

The surface finish is difficult to control, and the procedure cannot be generally recommended.

▲ Hydrogen peroxide can cause severe burns, and must be prevented from coming into contact with the eyes and skin. When preparing the solution, it should be added to the water prior to the addition of the other ingredients.

▲ Glacial acetic acid is very corrosive and must be prevented from coming into contact with the eyes or skin.

1.153   Dark slate with slight blue-green patina (as-cast surfaces only) Matt

Copper nitrate	20 gm
Zinc sulphate	30 gm
Mercuric chloride	30 gm
Water	1 litre

Applied liquid (Several minutes)

The solution is wiped on to the surface of the object with a soft cloth, producing a slightly adherent grey-black layer. Continued rubbing with the cloth clears this layer and produces a surface that is evenly coated with mercury. This tends to be non-adherent when dry. The solution is then applied to the surface by dabbing with a soft cloth, and allowed to dry in air. The procedure may be repeated if necessary, omitting the initial wiping. When the colour has developed, the object is washed in cold water and allowed to dry in air. When dry it may be wax finished.

The colour noted could only be obtained on as-cast surfaces. On polished faces, only a dull bare metal surface variegated with brown patches could be produced.

▲ Mercuric chloride is very toxic and must be prevented from coming into contact with the skin and eyes. The dust must not be inhaled. If taken internally, very serious effects are produced.

▲ In view of the high risks involved, and the very limited colouring potential, this solution cannot be generally recommended.

1.154* Pink lustre on green-brown ground Gloss

Potassium hydroxide	100 gm
Copper sulphate	30 gm
Sodium tartrate	30 gm
Water	1 litre

Warm immersion (Twenty to thirty minutes)

The potassium hydroxide is added gradually to the water and stirred. Heat is evolved as it dissolves, raising the temperature to about 35°C. The other ingredients are added and allowed to dissolve. There is no need to heat the solution. The object is immersed in the warm solution, and immersion continued until the lustre colour develops. The object is then removed and washed in warm water. It is allowed to dry in air, after any excess moisture has been removed by dabbing with absorbent tissue. When dry, it is wax finished.

The lustre colour is only obtained on polished surfaces. An interim lustre colour, a golden brown, is produced after ten or fifteen minutes.

On as-cast or rough rather than polished surfaces, a variegated pink and greenish-brown colour is produced.

▲ Potassium hydroxide is a powerful corrosive alkali and contact of the solid or solutions with the eyes or skin must be prevented.

1.155* Dark brown with slight blue lustre Gloss

Sodium thiosulphate	20 gm
Antimony potassium tartrate	10 gm
Water	1 litre

Hot immersion (Forty-five minutes)

The article is immersed in the hot solution (60°C) for ten minutes, producing a slight darkening of the surface and an orange tinge. The temperature of the solution is then gradually raised to 80°C and immersion continued. After forty-five minutes the article is removed, washed in warm water and allowed to dry in air. When dry, it is wax finished.

An even brown colour is produced on rough or as-cast surfaces.

Timings for the immersion may be variable. If the desired colour is produced in a shorter time, the object should be removed at that stage. The immersion should not be continued longer than necessary.

▲ Antimony potassium tartrate is poisonous and may be harmful if taken internally. The vapour evolved from the heated solution must not be inhaled.

1.156* Golden lustre/mid-brown bronzing (lustre) Gloss

Copper nitrate	115 gm
Tartaric acid (crystals)	50 gm
Sodium hydroxide	65 gm
Ammonia (.880 solution)	15 cm³
Water	1 litre

Boiling immersion (Fifteen minutes)

Immersion in the boiling solution produces a series of lustre colours, which begin with a blue that develops after about one minute to a golden lustre after about two minutes. If the golden colour is required then the object should be immediately removed and washed in hot water, and allowed to dry in air. It is wax finished when dry. Continuing the immersion will cause a gradual pink tinge to appear which subsequently darkens, to produce the final colour after about fifteen minutes. The object is removed and finished as noted above.

The lustrous bronzing is only produced on polished surfaces. On rough and as-cast surfaces a matt golden shade is produced.

Although a number of colour changes occur, the two colours noted are the only ones in this lustre series which are both rich and even.

▲ Sodium hydroxide is a powerful corrosive alkali, and the solid or solutions must be prevented from coming into contact with the eyes and skin. When preparing the solution, the sodium hydroxide should be added to the water in small amounts, and the other ingredients added after it has dissolved.

▲ Preparation and colouring should be carried out in a well ventilated area. Ammonia vapour will be liberated, which if inhaled causes irritation of the respiratory system. A combination of caustic vapours will be liberated from the solution as it boils.

▲ A dense precipitate is formed which may cause severe bumping. Glass vessels should not be used.

1.157 Lustre series Gloss

Sodium thiosulphate	125 gm
Lead acetate	35 gm
Water	1 litre

Hot immersion (Various)

The article is immersed in the hot solution (50–60°C), which produces a series of lustre colours in the following sequence: golden-yellow/orange; purple; blue; pale blue; pale grey. The timing tends to be variable, but it takes roughly five minutes for the purple stage to be reached, and about fifteen minutes for pale blue. (Tests failed to produce a further series of colours, the surface tending to remain pale grey after twenty minutes.) When the desired colour has been produced, the article is removed and immediately washed in cold water. It is allowed to dry in air, any excess moisture being removed with an absorbent tissue. When dry it may be wax finished or coated with a fine lacquer.

A brown colour, tinged with the lustre series colours, is produced on as-cast surfaces.

More dilute solutions produce the lustre colour sequence, but take a correspondingly longer time to produce.

▲ Lead acetate is very toxic and will give rise to serious conditions if taken internally. The vapour evolved from the heated solution must not be inhaled.

1.158   Dark golden lustre (reddish tinge) Gloss

Potassium permanganate	10 gm
Water	1 litre

Hot immersion (Three to five minutes)

The article is immersed in the hot solution (90°C). A golden lustre colour develops within one minute, gradually becoming more intense. When the lustre colour is fully developed, which may take from three to five minutes, the article is removed and washed in hot water which is gradually cooled during washing. The article is finally washed in cold water before being carefully dried either in sawdust, or by gently blotting the surface with absorbent tissue paper to remove excess moisture and allowing it to dry in air. When dry, it is wax finished.

The effect is only obtained on polished or directionally grained satin surfaces. On rough, as-cast and matt surfaces a dull orange or orange-brown is produced.

Extending the immersion time produces a slight darkening initially to a more pink colour, which rapidly fades to a pale silvery grey. These latter colours are usually uneven. Colours produced at longer immersion times tend to be more fragile when wet, and should not be handled at that stage.

1.159   Mid-brown (slight lustre) Semi-matt

Ammonium carbonate	50 gm
Copper acetate	25 gm
Acetic acid (6% solution)	1 litre

Boiling immersion (Thirty minutes)

The object is immersed in the boiling solution, and the surface gradually darkens to a reddish-brown colour. After thirty minutes, when the colour has fully developed, the object is removed and washed in warm water. After careful drying in sawdust, it may be wax finished.

A dull greenish-brown colour is produced on rough or as-cast surfaces.

Tests involving the addition of 5 gm of ammonium chloride and 2 gm of oxalic acid per litre, to the solution when the colours have developed, produced poor results. The colour initially becomes lighter but is then subject to streaking and etching effects, producing very patchy matt yellow surfaces. The variation cannot be recommended for polished surfaces. On rough or as-cast surfaces an even light orange/sand colour is produced. Very similar results were obtained with small additions of zinc chloride or copper chloride.

▲ The ammonium carbonate should be added to the acetic acid in small quantities. The reaction is effervescent, and if large amounts are added, the effervescence is very vigorous.

1.160   Lustre series Gloss

Sodium thiosulphate	280 gm
Lead acetate	25 gm
Water	1 litre
Potassium hydrogen tartrate	30 gm

Warm immersion (Various)

The sodium thiosulphate and the lead acetate are dissolved in the water, and the temperature is adjusted to 30–40°C. The potassium hydrogen tartrate is added immediately prior to use. The article to be coloured is immersed in the warm solution and a series of lustre colours are produced in the following sequence: golden-yellow/orange; purple; blue; pale blue/pale grey. The colour sequence takes about ten minutes to complete at 40°C. (Hotter solutions are faster, cooler solutions slower.) A second series of colours follows in the same sequence, but these tend to be less distinct. When the desired colour has been reached, the article is removed and immediately washed in cold water. It is allowed to dry in air, any excess moisture being removed by dabbing with an absorbent tissue. When dry it may be wax finished, or coated with a fine lacquer.

Similar colours are produced on as-cast surfaces, but these are tinged with the brown colour of the underlying metal.

Citric acid may be used instead of potassium hydrogen tartrate, and in the same quantity. Tests carried out with both these alternatives produced identical results

▲ Lead acetate is very toxic and will give rise to serious conditions if taken internally. The vapour evolved from the heated solution must not be inhaled.

1.161   Warm brown lustre Gloss

Potassium permanganate	10 gm
Sodium hydroxide	25 gm
Water	1 litre

Boiling immersion (Thirty minutes)

The object is immersed in the boiling solution, and after two or three minutes the surface shows signs of darkening. After ten minutes this develops to an uneven dark lustrous surface. Continued immersion causes some further darkening. After thirty minutes the object is removed, washed in warm water and allowed to dry in air. When dry, it may be wax finished.

An even mid-brown colour is obtained on rough or as-cast surfaces.

▲ Sodium hydroxide is a powerful caustic alkali which can cause severe burns. Both the solid and solutions must be prevented from coming into contact with the eyes and skin. When preparing the solution, small quantities of sodium hydroxide should be added to large amounts of water and *not* vice-versa.

1.162 Greenish-brown (slight lustre) Semi-gloss

Ammonium carbonate	60 gm
Copper sulphate	30 gm
Oxalic acid (crystals)	1 gm
Acetic acid (6% solution)	1 litre

Hot immersion (Thirty minutes)

A mid-brown or grey-brown colour is produced on rough or as-cast surfaces.

▲ Oxalic acid is poisonous and may be harmful if taken internally.

▲ The ammonium carbonate should be added to the acetic acid in small quantities. The reaction is effervescent, and if large amounts are added the effervescence becomes very vigorous.

The object is immersed in the hot solution (80°C) producing light brownish colours after two or three minutes, which may be uneven initially. With continued immersion the colours become even and darken gradually. When the colour has fully developed, after about thirty minutes, the object is removed and washed in hot water. After thorough drying in sawdust, it may be wax finished.

1.163 Dull pink lustre Gloss

Antimony trichloride	50 gm
Olive oil	to form a paste

Applied paste (Five minutes)

The dull lustre is only produced on polished surfaces. On as-cast or rough surfaces a dark slate colour is produced.

If the paste is left to dry on the surface, it becomes difficult to remove without damaging the colour finish.

Tests carried out using aqueous solutions produced similar results, but these tended to be patchy.

▲ Antimony trichloride is highly toxic and every precaution must be taken to avoid contact with the eyes and skin. It is irritating to the respiratory system and eyes, and can cause skin irritation and dermatitis.

The antimony trichloride is ground to a creamy paste with a little olive oil using a pestle and mortar. The paste is brushed on to the surface of the object to be coloured. A purple lustre colour develops after about two minutes, changing to a pink lustre after about five minutes. A steel-blue appears if the paste is left for longer. When the desired colour has developed, the residual paste is washed from the surface with pure distilled turpentine. The object is finally rubbed with a soft cloth, and wax finished.

1.164 Greyish-brown lustre Gloss

Sodium sulphide	50 gm
Antimony trisulphide	50 gm
Water	1 litre
Acetic acid (10% solution)	100 cm³

Hot immersion (Twelve minutes)

The lustre colour is only produced on polished surfaces. On rough and as-cast surfaces a purplish-brown is produced.

▲ Sodium sulphide is corrosive, particularly in solid form, and should be prevented from coming into contact with the eyes or skin.

▲ Addition of the acetic acid, and colouring, should be carried out in a fume cupboard or well ventilated area as toxic hydrogen sulphide gas will be evolved, which can be harmful even in moderate concentrations.

The solid ingredients are dissolved in the water by gradually raising the temperature to boiling. The solution is allowed to cool to 50°C and then acidulated gradually with the acetic acid, producing a dense brown precipitate. The object is immersed in the solution and examined occasionally for colour development. After approximately twelve minutes, when the lustre has developed, the object is removed and washed. It is allowed to dry in air and wax finished when completely dry.

1.165 Uneven dark lustre Semi-gloss

Butyric acid	20 cm³
Sodium chloride	17 gm
Sodium hydroxide	7 gm
Sodium sulphide	5 gm
Water	1 litre

Boiling immersion (One hour)

A variegated purplish-brown and light brown colour is produced on rough or as-cast surfaces.

The results produced tend to be uneven or patchy, and the procedure cannot be generally recommended.

▲ Butyric acid should be prevented from coming into contact with the eyes and skin, as it can cause severe irritation. It also has a foul odour which tends to cling to the hair and clothing. Preparation and colouring should be carried out in a well ventilated area.

▲ Both sodium hydroxide and sodium sulphide are corrosive, and must be prevented from coming into contact with the eyes and skin.

The solid ingredients are added to the water and allowed to dissolve, and the butyric acid then added, and the solution heated to boiling. The object is immersed in the boiling solution, which rapidly produces a series of patchy lustre colours. Continued immersion gradually produces a more opaque dark purplish-green colour, which tends to be uneven. After about one hour the object is removed and washed in hot water. It is allowed to dry in air, and may be wax finished when dry.

1.166 Purplish-brown Semi-gloss

Barium sulphide	10 gm
Water	1 litre

Cold immersion (Two hours)

A mottled purple/purple-brown colour is obtained on rough or as-cast surfaces.

For results obtained with hot solutions, see recipe 1.80.

▲ Barium sulphide is poisonous and may be very harmful if taken internally.

After a few minutes immersion in the cold solution, the surface of the object is coloured a lustrous brown. Continued immersion produces a series of gradually changing lustre colours, and more opaque colours during the latter stages. When the full colours have developed, after about two hours, the object is removed and washed in cold water and allowed to dry in air. When dry, the object is wax finished.

1.167   Lustre sequence   Gloss

Antimony trisulphide	50 gm
Ammonium sulphide (16% solution)	100 cm³
Water	1 litre

Cold immersion   (Two to ten minutes)

The article is immersed in the cold solution, producing a rapid sequence of lustre colours initially. After about two minutes the colour obtained is predominantly blue with some local red or golden surface variation. This changes gradually to a greenish lustre after four minutes, which slowly fades. The final colours will have developed after about six to eight minutes, and immersion beyond ten minutes produces no detectable change. When the desired colour has been reached, at any stage in the immersion, the article is removed, washed in cold water and allowed to dry in air. When dry, it is wax finished.

The colours produced at an early stage in the immersion are the most intense, but they are also subject to unpredictable variegated effects.

The best results obtained in tests occurred after about two minutes and after ten minutes immersion. After two minutes a dark straw/brown lustre is produced, with a light brown on an as-cast surface. After ten minutes a dark purplish lustre occurs, with an even light grey on an as-cast surface. Intermediate colours tended to be variegated.

▲ The preparation and colouring should be carried out in a well ventilated area. Ammonium sulphide solution liberates a mixed vapour of ammonia and hydrogen sulphide, which must not be inhaled, and which will irritate the eyes. The solution must be prevented from coming into contact with the eyes and skin, as it will cause burns or severe irritation.

1.168*   Pink/green lustre on orange ground   Gloss

Copper sulphate	25 gm
Sodium hydroxide	5 gm
Ferric oxide	25 gm
Water	1 litre

Hot immersion. Applied liquid. Bristle-brushing   (A few minutes)

The copper sulphate is dissolved in the water and the sodium hydroxide added. The ferric oxide is added, and the mixture heated and boiled for about ten minutes. When it has cooled to about 70°C, the object is immersed in it for about ten seconds and becomes coated with a brown layer. The object is transferred to a hotplate and gently heated until the brown layer has dried. The dry residue is brushed off with a soft bristle-brush. The turbid solution is then applied to the metal with a brush, as the object continues to be heated on the hotplate. The residue is brushed off, when dry. The application is repeated two or three times, until a green/pink lustre is produced. After a final brushing with the bristle-brush, the object is wax finished.

The results tend to be variegated rather than even, and may appear patchy when viewed in some lights.

▲ Contact with the sodium hydroxide in the solid state must be avoided when preparing the solution. It is very corrosive to the skin and in particular to the eyes.

▲ A nose and mouth face mask with a fine dust filter should be worn when brushing off the dry residue.

1.169   Lustre series   Gloss

Sodium thiosulphate	240 gm
Copper acetate	25 gm
Water	1 litre
Citric acid (crystals)	30 gm

Cold immersion   (Various)

The sodium thiosulphate and the copper acetate are dissolved in the water, and the citric acid added immediately prior to use. The article is immersed in the cold solution and a series of lustre colours are produced in the following sequence: golden-yellow/orange; brown; purple; blue; pale grey. This sequence takes about thirty minutes to complete, the brown changing to purple after about fifteen minutes. It is followed by a second series of colours, in the same sequence, which takes a further thirty minutes to complete. The latter colours in the second series tend to be slightly more opaque. A third series of colours follows, if immersion is continued, but these tend to be indistinct and murky. (If the temperature of the solution is raised then the sequences are produced more rapidly. However, it was found in tests that the results tended to be less even. Temperatures in excess of about 35–40°C tend to cause the solution to decompose, with a concomitant loss of effect.) When the required colour has been produced, the article is removed and washed in cold water. It is allowed to dry in air, any excess moisture being removed by gentle blotting with a tissue. When dry it is wax finished or coated with a fine lacquer.

Similar colours are produced on rough or as-cast surfaces, although these are more dull and affected by the colour of the underlying surface.

1.170   Brown lustre   Gloss

Sodium thiosulphate	90 gm
Potassium hydrogen tartrate	30 gm
Zinc chloride	15 gm
Water	1 litre

Hot immersion   (Thirty minutes)

The article is immersed in the hot solution (50–60°C) which darkens the surface and gradually produces a bloom, which should be removed by bristle-brushing. Continued immersion gives rise to a brown colour, which slowly darkens. After about thirty minutes, when the colour has developed, the article is removed and washed in warm water. It is either dried in sawdust or allowed to dry in air, and may be wax finished when dry.

An even dark brown colour is produced on as-cast surfaces.

1.171   Grey on dark ground   Semi-gloss/semi-matt

Sodium thiosulphate	300 gm
Antimony potassium tartrate	15 gm
Ammonium chloride	20 gm
Water	1 litre

Hot immersion   (Twenty minutes)

An even matt slate-grey is produced on rough or as-cast surfaces. On polished surfaces the results tend to be uneven, producing areas of 'frosted' metallic grey on a dark ground. Slight local lustre patches are visible in some lights.

▲ Antimony potassium tartrate is poisonous and may be harmful if taken internally.

▲ The vapour evolved from the hot solution should not be inhaled.

Immersion in the hot solution (60°C) rapidly produces a series of lustre colours on the surface of the object. These are pale and pass from purple to brown and blue, becoming a very pale blue grey after about two minutes. This is followed by a gradual darkening of the surface, producing dark greys and browns. After about twenty minutes the object is removed, washed in warm water and dried in sawdust. It may be wax finished when dry.

1.172   Bright golden lustre   Gloss

Copper sulphate	125 gm
Sodium chlorate	50 gm
Water	1 litre

Boiling immersion   (One minute)

The lustre colour is only produced on polished surfaces.

▲ Sodium chlorate is a powerful oxidising agent and must not be allowed to come into contact with combustible material or acids. Inhaling the dust must be avoided, and the eyes and skin should be protected from contact with either the solid or solutions.

The object is immersed in the boiling solution and removed after about one minute, when a bright golden lustre colour has developed. It should be washed immediately in hot water, and dried in sawdust. When dry, it is wax finished.

1.173   Pink/green lustre on mottled brown ground   Gloss

Copper sulphate	25 gm
Nickel ammonium sulphate	25 gm
Potassium chlorate	25 gm
Water	1 litre

Boiling immersion   (One or two minutes)

This effect is produced only on polished surfaces.

▲ Nickel salts are a common cause of sensitive skin reactions.

▲ Potassium chlorate is a powerful oxidising agent and should not be allowed to come into contact with combustible material. The solid, and solutions, should be prevented from coming into contact with the eyes, skin or clothing.

The object is immersed in the boiling solution and rapidly removed when the mottled surface and lustre colour have developed. It should be washed immediately in hot water. After drying in sawdust, it is wax finished.

1.174   Pale brown lustre

Potassium sulphide	2 gm
Ammonium sulphate	0.5 gm
Water	1 litre

Cold immersion   (One or two minutes)

The technique is difficult to control. The colour produced tends to be variegated rather than even.

The lustre colour is only produced on highly polished surfaces. Matt or other unpolished surfaces yield only patchy dull brownish colours.

Immerse the article in the cold solution. A golden lustre colour develops fairly quickly, and in roughly one to two minutes will begin to change to a mustard colour. As soon as this occurs the object should be transferred quickly to a container of cold water and thoroughly rinsed. It should then be washed in clean cold water and dried in sawdust. When completely dry it is wax finished.

1.175   Light brown lustre   Semi-gloss

A	Copper sulphate	50 gm
	Water	1 litre
B	Potassium sulphide	12.5 gm
	Water	1 litre

Hot immersion (A)   (Thirty seconds)
Cold immersion (B)   (One minute)

An even thin black colour is produced on rough or as-cast surfaces.

If the results produced in solution A are streaky, then these will tend to be visible in some lights, after treatment with solution B. Application with the bristle-brush darkens and improves the results. Producing an adequate finish evenly is difficult, and the recipe cannot be generally recommended.

For results obtained using a scratch-brushed technique see recipe 1.92.

▲ Colouring should be carried out in a well ventilated area as some hydrogen sulphide gas will be liberated.

The article is immersed in the hot solution A (80°C) for about thirty seconds to produce a green/pink lustre. It is then transferred to the cold solution B and immersed for about one minute, producing a clouding of the surface and a smokey greenish-pink. If this is uneven the solutions may be applied alternately with a bristle-brush. When treatment is complete, the article is washed in warm water, allowed to dry in air and then wax finished.

1.176★ Blue-green patina/brown and etched metal (stippled) Matt *Pl. XIV*

Ammonium carbonate	120 gm
Ammonium chloride	40 gm
Sodium chloride	40 gm
Water	1 litre

Sawdust technique (Twenty to thirty hours)

The polished surface is etched by the solution, producing a textured surface of etched areas and incrustations of the patina.

On as-cast surfaces, patina development tended to be light and even on a brown ground, where the medium is sparingly moist. Patches of a greener patina occurred where the medium was more moist.

The article to be coloured is laid-up in sawdust which has been evenly moistened with the solution, and is then left for a period of about twenty or thirty hours. The sawdust should be kept moist throughout the period of treatment. When treatment is complete, the article should be washed in cold water and allowed to dry in air. When it is completely dry, it may be wax finished.

1.177★ Blue-green patina on brown ground (stipple) Semi-matt *Pl. I*

Ammonium chloride	15 gm
Potassium aluminium sulphate	7 gm
Copper nitrate	30 gm
Sodium chloride	30 gm
Acetic acid (6% solution)	1 litre

Sawdust technique (Twenty to thirty hours)

Patina development is greater on rough or as-cast surfaces, giving an even blue/blue-green. The ground colour is less apparent.

The object to be coloured is laid-up in sawdust which has been evenly moistened with the solution, and left for about twenty or thirty hours. The object is then removed and allowed to dry in air for several hours, without washing. When dry, the surface is brushed with a stiff bristle-brush to remove any particles of sawdust or loose material. The object is then left for a further period of several hours, after which it may be wax finished.

1.178★ Bright metal stipple on buff/pale greenish-blue, mottled ground Semi-matt *Pl. II*

Ammonium chloride	10 gm
Ammonia (.880 solution)	20 cm³
Acetic acid (30% solution)	80 cm³
Water	1 litre

Sawdust technique (One or two days)

The surface is selectively etched by the solution, to produce a buff ground mottled with a greenish-blue or blue-grey colour. An even scattering of unetched bright metal remains, giving the appearance of a metallic stipple.

On rough or as-cast rather than polished surfaces, a stippled blue-green patina on a buff or light brown ground is produced.

▲ Preparation and colouring should be carried out in a well ventilated area. Ammonia vapour and some acetic acid vapour will be liberated, which can irritate the eyes and respiratory system. These solutions should be prevented from coming into contact with the eyes and skin, or severe irritation or burns may result.

The object to be coloured is laid-up in sawdust which has been evenly moistened with the solution, and left for one or two days. Colour development should be monitored using a sample, so that the object itself remains undisturbed. When the colour has developed satisfactorily, the object is removed, washed in cold water and allowed to dry in air for several hours. When dry, it may be wax finished.

1.179★ Blue-green patina/orange-brown ground Semi-matt *Pl. I*

Copper nitrate	200 gm
Water	1 litre

Sawdust technique (Two days)

The mottled orange-brown ground develops first, and darkens as the blue-green patina appears. The patina tends to develop further as the moist medium dries out. The ground colour can be darkened by removing the object before the medium becomes dry, and re-packing in freshly moistened sawdust. On rough or as-cast surfaces a stipple of blue-green patina develops on a dark brown ground.

The article to be coloured is laid-up in sawdust which has been evenly moistened with the solution, and left for about two days. Colour development may be monitored using a sample, but no deleterious effects were produced, in tests, by unpacking and re-packing the object itself. When the colour has developed satisfactorily, the article is removed and washed in cold water. After several hours drying in air, it may be wax finished.

1.180★ Greenish-grey patina Matt *Pl. II*

Ammonium chloride	100 gm
Ammonium carbonate	150 gm
Water	1 litre

Sawdust technique (To twenty or thirty hours)

The polished surface is etched by the solution, producing an evenly textured surface. The greenish-grey patina develops over the whole surface, but is slightly darker on the least etched parts. Some spots of a more blue patina may also form as small incrustations.

Rough or as-cast surfaces, tend to develop incrustations of the blue patina on a brown and grey ground.

The article to be coloured is laid-up in sawdust which has been moistened evenly with the solution. It may then be left for periods of up to twenty or thirty hours. When treatment is complete, the object is washed in cold water and allowed to dry in air. When it is completely dry, it may be wax finished.

1.181★ Bronze and greyish-blue patina (stipple) Semi-matt   *Pl. I*

Ammonium carbonate	150 gm
Copper acetate	60 gm
Sodium chloride	50 gm
Potassium hydrogen tartrate	50 gm
Water	1 litre

Sawdust technique (Ten to twenty hours)

The bronze surface is selectively etched where the granular sawdust is in close contact with the metal, producing a stipple of etched areas that acquire a blue-grey patina.

A mottled blue-green patina on a light brown ground is produced on rough or as-cast surfaces. This may develop in the form of a light incrustation.

The object to be coloured is laid-up in sawdust which has been evenly moistened with the solution. It is left for a period of from ten to twenty hours, colour development being monitored with a sample, so that the object remains undisturbed. When treatment is complete, the object is removed, washed in cold water and allowed to dry in air for several hours. When dry, it may be wax finished.

1.182★ Green/blue-green patina on grey-brown (mottled) Matt

Copper nitrate	100 gm
Water	200 cm³
Ammonia (.880 solution)	400 cm³
Acetic acid (6% solution)	400 cm³
Ammonium chloride	100 gm

Sawdust technique (Twenty to thirty hours)

The surface is etched to a matt finish and a green/blue-green patina is produced on a grey-brown ground, to give an overall mottled appearance. Patina development may be considerable, particularly if a more moist medium is used, and may form substantial adherent incrustations.

A variegated blue-green patina develops on a darker grey-brown ground on rough or as-cast surfaces.

▲ Preparation and colouring should be carried out in a well ventilated area. Ammonia vapour and some acetic acid vapour will be liberated, which cause irritation of the eyes and respiratory system.

▲ The strong ammonia solution must be prevented from coming into contact with the eyes and skin, as it will cause burns or severe irritation.

The copper nitrate is dissolved in the water and the ammonia gradually added. The acetic acid is then added, followed by the ammonium chloride. The article to be coloured is laid-up in sawdust which has been evenly moistened with this solution, and is then left for a period of twenty to thirty hours. When treatment is complete, the article is washed in cold water and allowed to dry in air. When completely dry, after several hours, it may be wax finished.

1.183★ Green/bluish-green patina on brown ground Semi-matt

Ammonium chloride	350 gm
Copper acetate	200 gm
Water	1 litre

Sawdust technique (Ten to twenty hours)

A stipple of green/bluish-green patina develops on a brown ground. If the medium is too moist, patina development tends to be inhibited and the brown ground is dominant.

A variegated or stippled green patina on a light brown ground is produced on as-cast surfaces.

The object to be coloured is laid-up in sawdust which has been moistened evenly with the solution, and left for a period of from ten to twenty hours. After ten hours it should be examined and repacked if further treatment is required. It should subsequently be examined every few hours until the required surface finish is obtained. The sawdust should be kept moist during the full period of treatment. The object should then be removed, washed in cold water and allowed to dry in air. When completely dry, it may be wax finished.

1.184★ Dark grey/brown and greenish-blue patina (stipple) Semi-matt

Copper acetate	60 gm
Copper carbonate	30 gm
Ammonium chloride	60 gm
Water	1 litre
Hydrochloric acid (15% solution)	5 cm³

Sawdust technique (Ten to twenty hours)

The polished surface is coloured a dark grey/brown, and selectively etched where the granular sawdust is in close contact with the metal, producing a stipple of etched areas that acquire a greenish-blue patina. The etched and patinated areas become more extensive as treatment is prolonged.

An uneven bluish-green patina on a light brown ground, tends to develop on rough or as-cast surfaces.

▲ Hydrochloric acid can cause burns or severe irritation, and must be prevented from coming into contact with the eyes and skin.

The object to be coloured is laid-up in sawdust which has been evenly moistened with the solution. It is left for a period of about ten hours and then examined. If further treatment is required, it should be re-packed and examined at intervals of a few hours. When treatment is complete, the object is removed, washed thoroughly in cold water and allowed to dry in air. When completely dry, it may be wax finished.

1.185 Frosted metal and grey-green patina (stipple) Semi-matt

Ammonium carbonate	300 gm
Acetic acid (10% solution)	35 cm³
Water	1 litre

Sawdust technique (To twenty hours)

The surface is selectively etched, where the granular sawdust is in close contact with the metal, producing a stipple of etched areas that acquire a grey-green patina. The remaining metal surface takes on a frosted finish.

A pale blue-green patina develops on a light brown ground, on rough or as-cast surfaces. The patina may take the form of an incrustation.

The ammonium carbonate is added to the water and the acetic acid then added gradually, while stirring. The object to be coloured is laid-up in sawdust which has been evenly moistened with this solution. It is then left for a period of up to twenty hours, ensuring that the sawdust remains moist throughout the period of treatment. After about ten hours, and at intervals of several hours subsequently, the object should be examined to check its progress. When the desired surface finish has been obtained, the object is removed, washed in cold water and allowed to dry in air. When thoroughly dry, it may be wax finished.

1.186* Pale blue-green patina on yellow-brown ground Semi-matt

Copper sulphate	100 gm
Lead acetate	100 gm
Ammonium chloride	10 gm
Water	1 litre

Sawdust technique (To twenty hours)

The object to be coloured is laid-up in sawdust which has been evenly moistened with the solution. It is then left to develop for a period of up to about twenty hours. When treatment is complete, the object is removed, washed in cold water and allowed to dry in air. It may be wax finished when completely dry.

The polished surface is coloured a reddish-orange, and selectively etched where the granular sawdust is in close contact with the metal, producing a stipple of finely etched yellow areas. As treatment progresses the yellow colouration tends to predominate, and a stipple of blue-green patina develops.

On rough or as-cast surfaces, a blue-green patina tends to develop on a mottled reddish and light brown ground.

▲ Lead acetate is a toxic substance, and every precaution should be taken to prevent inhalation of the dust, when preparing the solution. It is very harmful if taken internally.

1.187* Metallic green stipple on buff ground Semi-matt

Copper carbonate	300 gm
Ammonia (.880 solution)	200 cm³
Water	1 litre

Sawdust technique (To twenty or thirty hours)

The article to be coloured is laid-up in sawdust which has been evenly moistened with the above solution, and left for a period of up to twenty or thirty hours. The sawdust should remain moist throughout the period of treatment. When treatment is complete, the article is removed and washed in cold water. It is then allowed to dry in air for some time before wax finishing.

The surface is coloured brown and then green by the solution, and selectively etched where the granular sawdust is in close contact with the metal, producing a stipple of finely etched buff ground.

As-cast surfaces were coloured a mottled dull brown by the treatment. Slight traces of green were also obtained, in tests.

▲ The strong ammonia solution will liberate ammonia vapour, which irritates the eyes and respiratory system. Preparation and colouring must therefore be carried out in a well ventilated area.

▲ Ammonia solution causes burns and severe irritation, and must be prevented from coming into contact with the eyes and skin.

1.188* Orange-brown with dark and light grey (stipple) Semi-matt

Copper nitrate	280 gm
Water	850 cm³
Silver nitrate	15 gm
Water	150 cm³

Sawdust technique (Twenty hours)

The copper nitrate and the silver nitrate are dissolved in separate portions of water, and the two solutions then mixed. The article to be coloured is laid-up in sawdust which has been evenly moistened with this mixed solution. It is left for a period of about twenty hours, ensuring that the sawdust remains moist throughout this time. When treatment is complete, the article is removed and washed in cold water. After allowing it to dry thoroughly in air for some time, it may be wax finished.

The surface is coloured an orange-brown by the treatment, and is also etched where the granular sawdust is in close contact with the metal, producing a stipple of light and dark grey etched areas.

▲ Silver nitrate must be prevented from coming into contact with the skin, and more particularly the eyes, as it may cause burns or severe irritation.

1.189 Dark green/black stipple on light brown bronze Gloss

Potassium hydroxide	100 gm
Copper sulphate	30 gm
Sodium tartrate	30 gm
Water	1 litre

Sawdust technique (Twenty hours)

The object is laid-up in sawdust which has been evenly moistened with the solution, and is then left undisturbed for a period of about twenty hours. The sawdust should remain moist throughout this period. When treatment is complete, the object is removed and washed thoroughly in cold water. After it has been allowed to dry in air, it may be wax finished.

The stipple and the ground colours tend to be lustrous.

A stippled dark brown and light brown colour is produced on as-cast surfaces. Some green patina may also occur.

▲ Potassium hydroxide is a strong caustic alkali, and precautions should be taken to prevent it from coming into contact with the eyes and skin.

▲ When preparing the solution, the potassium hydroxide should be added to the water in small quantities, and *not* vice-versa. When it has dissolved completely, the other ingredients may be added.

1.190* Dark green/black stipple on bronze Semi-matt

Ammonium chloride	16 gm
Sodium chloride	16 gm
Ammonia (.880 solution)	30 cm³
Water	1 litre

Sawdust technique (Twenty-four hours)

The object to be coloured is laid-up in sawdust which has been evenly moistened with the solution. It is then left for about twenty-four hours. The object should not be disturbed while colouring is in progress, as re-packing tends to cause a loss of surface quality and definition. Progress should be monitored by including a small sample of the object metal, when laying-up, which can be examined as necessary. When treatment is complete, the object is washed thoroughly in cold water and allowed to dry in air. It may be wax finished, when completely dry.

On polished surfaces a sharply defined dark stipple is produced, which may include slight traces of blue-green patina. This will tend to be more marked in any surface features such as punched or engraved marks. On rough or as-cast surfaces a blue/blue-green patina, with some grey, on a light brown ground is obtained.

▲ Ammonia solution causes burns or severe irritation, and must be prevented from coming into contact with the eyes or skin. The vapour should not be inhaled as it irritates the respiratory system. Colouring should be carried out in a well ventilated area.

1.191   Black mottling on metal ground Semi-gloss

Butyric acid	20 cm³
Sodium chloride	17 gm
Sodium hydroxide	7 gm
Sodium sulphide	5 gm
Water	1 litre

Sawdust technique (Twenty hours)

The object to be coloured is laid-up in sawdust which has been evenly moistened with the solution, and left for a period of up to twenty hours. The timing of colour development tends to be unpredictable, and it is essential to include a sample of the same material as the object, when laying-up, so that progress can be monitored without disturbing the object. When treatment is complete, the object is removed and washed thoroughly with cold water. After a period of several hours air drying, it may be wax finished.

On polished surfaces, the black colour tends to occur as a sparse mottle on a predominantly bare metal ground. On rough or as-cast surfaces, a variegated purple-grey/black colour is produced covering the whole surface.

▲ Butyric acid causes burns, and contact with the skin, eyes and clothing must be prevented. It also has a foul pungent odour, which tends to cling to the clothing and hair, and is difficult to eliminate.

▲ Sodium hydroxide is a powerful caustic alkali causing burns, and must be prevented from coming into contact with the eyes or skin.

▲ Sodium sulphide is corrosive and causes burns. It must be prevented from coming into contact with the eyes or skin. When preparing the solution it must not be mixed directly with the acid, or toxic hydrogen sulphide gas will be evolved. The acid should be added to the water and then neutralised with the sodium hydroxide, before the remaining ingredients are added.

1.192   Slight grey-green patina/light bronze (stipple) Semi-matt

Ammonium chloride	10 gm
Copper sulphate	4 gm
Potassium binoxalate	20 gm
Water	1 litre

Sawdust technique (Several days)

The object to be coloured is laid up in sawdust which has been moistened with the solution, and left for several days. Colour development is very slow, and should be monitored using a sample so that the object remains undisturbed. When the colour has developed satisfactorily, the object is removed, washed in cold water and allowed to dry in air for several hours. It may be wax finished when dry.

On polished surfaces a light bronze stipple on an etched buff ground is produced, on which a slight grey-green patina may occur. The etched buff areas tend to become more dominant as treatment is prolonged. A stippled or variegated green/olive on a mid-brown ground is obtained on rough or as-cast surfaces.

▲ Potassium binoxalate is harmful if taken internally. It can also cause irritation, and should be prevented from coming into contact with the eyes or skin.

1.193   Stippled blue-green patina on bronze ground Semi-matt

Copper nitrate	200 gm
Sodium chloride	100 gm
Water	1 litre

Sawdust technique (Twenty or thirty hours)

The article to be coloured is laid-up in sawdust which has been moistened evenly with the solution. It should then be left for a period of twenty or thirty hours. After twenty hours the object should be examined, and repacked for shorter periods of time, as necessary. The sawdust should be kept moist throughout the period of treatment. When treatment is complete, the article should be washed thoroughly in cold water and allowed to dry in air. When completely dry, after several hours, it may be wax finished.

The surface is selectively etched by the solution, where the granular sawdust is in close contact with the metal, causing an even stipple of fine pits in the metal. The blue-green patina develops locally in the region of the pits. The remainder of the surface is not etched but acquires lustre colours initially, which recede to leave a dull metal surface if treatment is prolonged.

A stippled green/blue-green patina on a mid-brown ground is produced on as-cast surfaces.

1.194   Light golden bronzing (fine stipple) Gloss

'Rokusho'	30 gm
Copper sulphate	5 gm
Water	1 litre

Sawdust technique (Several days)

The 'Rokusho' and the copper sulphate are added to the water, and the mixture boiled for a short time and then allowed to cool to room temperature. The object to be coloured is laid-up in sawdust which has been moistened with the solution. It is then left for several days. Colour development is very slow, and should be monitored using a sample so that the object remains undisturbed. When the colour has satisfactorily developed, the object is removed, washed in cold water and allowed to dry in air. It may be wax finished when dry.

On polished surfaces, a very fine stipple with an orange tint is produced, giving an overall pale golden bronzing effect. On rough or as-cast surfaces, an even mid-brown colour is produced.

For description of 'Rokusho' and its method of preparation, see appendix 1.

1.195   Buff etched stipple on dull yellow-bronze Semi-matt

Copper nitrate	20 gm
Ammonium chloride	20 gm
Calcium hypochlorite	20 gm
Water	1 litre

Sawdust technique (Twenty or thirty hours)

The object to be coloured is laid-up in sawdust which has been moistened evenly with the solution, and left for a period of twenty to thirty hours. The sawdust should be kept moist throughout this time. After treatment, the object is washed in cold water and allowed to dry in air. When thoroughly dry, it may be wax finished.

A lightly etched stipple with a buff colour on a dull yellow-bronze, is produced on polished surfaces.

A mottled light brown colour is produced on as-cast surfaces.

▲ The corrosive fine dust of the calcium hypochlorite (bleaching powder) must not be inhaled, as it can severely irritate the respiratory system. The solution must be prevented from coming into contact with the eyes, which would be severely irritated.

1.196   Light ochre stipple Semi-matt

Ammonium sulphate	85 gm
Copper nitrate	85 gm
Ammonia (.880 solution)	3 cm³
Water	1 litre

Sawdust technique (Thirty hours)

The article to be coloured is laid-up in sawdust which has been moistened evenly with the solution. It should then be left for a period of twenty or thirty hours. The sawdust should be kept moist throughout the period of treatment. After thirty hours the object should be examined, and re-packed for shorter periods of time as necessary. When treatment is complete, the object should be washed thoroughly in cold water and allowed to dry in air. When dry it may be wax finished.

A light ochre stipple occurs where the granular sawdust is in close contact with the metal surface. The remainder of the surface remains polished. If the medium is too moist then uneven patches of dull grain-enhancement tend to occur.

A grey-brown stipple on a lighter mid-brown ground is produced on as-cast surfaces.

1.197   Very slight grey stipple Semi-matt

Di-ammonium hydrogen orthophosphate	100 gm
Water	1 litre

Sawdust technique (Thirty hours)

The object to be coloured is laid-up in sawdust which has been evenly moistened with the solution. It is left for about thirty hours and then removed and washed in cold water. If further colour development is required the treatment should be repeated for additional shorter periods. When treatment is complete, the object is finally washed and allowed to dry in air, and wax finished when completely dry.

Colour development on polished surfaces is rather slight. On rough or as-cast surfaces an even pale buff/grey colour is produced. The procedure cannot generally be recommended.

Tests carried out with this solution, using various other techniques, including cold and hot direct application, and hot and cold immersion, produced little or no effect.

1.198*   Bluish-green patina on brown ground Matt

Ammonium carbonate	120 gm
Ammonium chloride	40 gm
Sodium chloride	40 gm
Water	1 litre

Cotton-wool technique (Twenty to thirty hours)

Cotton-wool, moistened with the solution, is applied to the surface of the object to be coloured. It is important to ensure that the moist cotton-wool is in full contact with the surface of the object. It is left for a period of about twenty to thirty hours, ensuring that the cotton-wool remains moist throughout the period of treatment. When treatment is complete, the object should be washed in cold water, and allowed to dry in air. When completely dry, it is wax finished.

The surface is etched by the solution to a brown matt finish. The bluish-green patina develops in patches and has a powdery appearance. It tends to develop more locally at a later stage to form more substantial incrustations.

1.199*   Grey-brown Matt

Ammonium chloride	350 gm
Copper acetate	200 gm
Water	1 litre

Cotton-wool technique (Ten to twenty hours)

Cotton-wool, moistened with the solution, is applied to the surface of the object. This is then left for a period of ten or twenty hours, ensuring that the cotton-wool remains moist. After the first few hours, the surface should be periodically examined by exposing a small portion, to check the progress. When the desired surface finish has been reached, the object is removed, washed in cold water and allowed to dry in air. When completely dry, it may be wax finished.

The surface is etched to a matt finish, and takes on a grey-brown colour. Patina development tends to occur only on rough or as-cast surfaces.

1.200* Green and greyish-green patina (variegated) Matt   *Pl. II*

Ammonium chloride	100 gm
Ammonium carbonate	150 gm
Water	1 litre

Cotton-wool technique (To twenty or thirty hours)

The polished surface is etched to a matt finish by the solution, and takes on a greyish colour. The patina develops to produce a variegated surface, which has a 'powdery' appearance.

On rough or as-cast surfaces the patina tends to develop more in the form of an irregular incrustation on a grey-brown ground.

Cotton-wool, moistened with the solution, is applied to the surface of the object to be coloured. It is then left for periods of up to twenty or thirty hours. The cotton-wool should be kept moist throughout the period of treatment. After about ten or fifteen hours the surface should be examined, by carefully exposing a small portion, to determine the extent of colour development. Subsequently a check should be made every few hours until the desired surface finish is obtained. When treatment is complete, the object is washed in cold water and allowed to dry in air. When completely dry, it may be wax finished.

1.201* Variegated blue-green and green patina Semi-matt   *Pl. I*

Ammonium carbonate	150 gm
Copper acetate	60 gm
Sodium chloride	50 gm
Potassium hydrogen tartrate	50 gm
Water	1 litre

Cotton-wool technique (Ten to twenty hours)

The polished surface is more or less evenly etched back by the solution, and acquires a variegated blue-green and green patina.

A blue-green patina on a light brown ground tends to be produced on as-cast surfaces.

Cotton-wool which has been moistened with the solution is applied to the surface of the article to be coloured. It is important to ensure that the cotton-wool is in full contact with the surface. It is left for a period of from ten to twenty hours, during which time the cotton-wool should remain moist. Colour development should be monitored with a sample, so that the article remains undisturbed. When treatment is complete, the article is removed and washed in cold water. After being allowed to dry in air for several hours, it may be wax finished.

1.202* Green/blue-green patina on grey brown (variegated) Matt

Copper nitrate	100 gm
Water	200 cm³
Ammonia (.880 solution)	400 cm³
Acetic acid (6% solution)	400 cm³
Ammonium chloride	100 gm

Cotton-wool technique (Twenty to thirty hours)

The polished surface is etched to a matt finish. An uneven variegated patina forms on a brown ground, to give an overall cloudy appearance. Patina development may be considerable, giving rise to substantial incrustations if a more moist medium is used

A variegated blue/blue-green patina on a dark ground is produced on as-cast surfaces.

▲ Preparation and colouring should be carried out in a well ventilated area. Ammonia vapour and some acetic acid vapour will be liberated, which cause irritation of the eyes and respiratory system.

▲ The strong ammonia solution must be prevented from coming into contact with the eyes and skin, as it will cause burns or severe irritation.

The copper nitrate is dissolved in the water and the ammonia gradually added. The acetic acid is then added, followed by the ammonium chloride. Cotton-wool is wetted with the solution and applied to the surface of the object to be coloured. It is then left for a period of about twenty to thirty hours. The cotton-wool should be kept moist throughout the period of treatment. When treatment is complete, the object is washed in cold water and allowed to dry in air. When dry, it is wax finished.

1.203 Greyish-brown on brown Matt

Copper acetate	60 gm
Copper carbonate	30 gm
Ammonium chloride	60 gm
Water	1 litre
Hydrochloric acid (15% solution)	5 cm³

Cotton-wool technique (Ten to twenty hours)

The results obtained in tests on polished surfaces were generally poor.

A bluish patina on an uneven brown ground tends to develop on rough or as-cast surfaces. This may take the form of an incrustation.

▲ Hydrochloric acid can cause burns or severe irritation, and must be prevented from coming into contact with the eyes and skin.

Cotton-wool, moistened with the above solution, is applied to the surface of the article, and left for a period of about ten hours. The surface is then examined, by carefully lifting a small portion, to determine the stage it has reached. When the desired surface finish has been obtained, the article is washed thoroughly in cold water and allowed to dry in air. When completely dry, it may be wax finished.

1.204* Variegated grey-green patina Matt

Ammonium carbonate	300 gm
Acetic acid (10% solution)	35 cm³
Water	1 litre

Cotton-wool technique (Twenty hours)

A blue-green patina on a light buff ground tends to be produced on as-cast surfaces.

The ammonium carbonate is added to the water and the acetic acid then added gradually, while stirring. Cotton-wool, moistened with this solution, is applied to the surface of the object, and left for a period of about twenty hours. It is important to ensure that the cotton-wool is in full contact with the surface, and that it remains moist throughout. When treatment is complete, the object is washed in cold water and allowed to dry in air. When completely dry, it may be wax finished.

1.205 Pale blue-green patina on ochre ground Semi-matt

Copper sulphate	100 gm
Lead acetate	100 gm
Ammonium chloride	10 gm
Water	1 litre

Cotton-wool technique (Ten to twenty hours)

A variegated ochre ground is produced on rough or as-cast surfaces, which tend to develop a more substantial green/blue-green patina.

▲ Lead acetate is a toxic substance, and every precaution should be taken to prevent inhalation of the dust when preparing the solution. It is very harmful if taken internally.

Cotton-wool, moistened with the solution, is applied to the surface of the object to be coloured, and is left in place for a period of up to twenty hours. The cotton-wool should remain moist throughout the period of treatment. When treatment is complete, the object is removed and washed in cold water. It is then dried in air, and gradually darkens on exposure to daylight. When colour change has ceased, the object may be wax finished.

1.206 Mid-brown with slight petrol lustre Semi-matt

Copper carbonate	300 gm
Ammonia (.880 solution)	200 cm³
Water	1 litre

Cotton-wool technique (About twenty hours)

The polished surface is etched to a matt finish, and coloured a dull brown by the solution. A sheen of petrol lustre tends to occur.

A variegated dark brown is produced on as-cast surfaces.

▲ The strong ammonia solution will liberate ammonia vapour, which irritates the eyes and respiratory system. Preparation and colouring must therefore be carried out in a well ventilated area.

▲ Ammonia solution causes burns and severe irritation, and must be prevented from coming into contact with the eyes and skin.

Cotton-wool, moistened with the solution, is applied to the surface of the object to be coloured, which is then left for a period of twenty hours. It is important to ensure that the moist cotton-wool is in full contact with the surface, and that it remains moist throughout the period of treatment. When treatment is complete, the article is removed and washed in cold water. It is then allowed to dry in air for some time before wax finishing.

1.207* Cloudy grey on light brown Matt

Copper nitrate	280 gm
Water	850 cm³
Silver nitrate	15 gm
Water	150 cm³

Cotton-wool technique (Twenty hours)

The surface is etched to a light brown, and acquires a thin cloudy grey layer. A variegated purple-brown on light brown tends to be produced on as-cast surfaces. Results obtained in tests with this technique tended to be very variable, and it is suggested that in practice a sample should be used to monitor colour development.

▲ Silver nitrate must be prevented from coming into contact with the skin, and more particularly the eyes, as it may cause burns or severe irritation.

The copper nitrate and the silver nitrate are dissolved in separate portions of water, and the two solutions then mixed. Cotton-wool, moistened with this mixed solution, is applied to the surface of the object to be coloured. It is important to ensure that the cotton-wool is in full contact with the surface, and that it remains moist during the period of treatment. The object is removed after about twenty hours and washed in cold water. After thorough drying in air, it may be wax finished.

1.208 Purple/green petrol lustre Gloss

Potassium hydroxide	100 gm
Copper sulphate	30 gm
Sodium tartrate	30 gm
Water	1 litre

Cotton-wool technique (Twenty hours)

The actual colour obtained depends upon the length of treatment, and colour development should be monitored with a sample. The lustre colour is only produced on polished surfaces. A variegated dark brown tends to be produced on rough or as-cast surfaces.

▲ Potassium hydroxide is a strong caustic alkali, and should be prevented from coming into contact with the eyes or skin.

▲ When preparing the solution, the potassium hydroxide should be added to the water in small quantities, and not vice-versa.

Cotton-wool, moistened with the solution, is applied to the surface of the object, which is then left for a period about twenty hours. It is important to ensure that the moist cotton-wool is in full contact with the surface, and that it remains moist throughout the period of treatment. When treatment is complete, the object is washed in cold water and allowed to dry in air. When dry, it may be wax finished.

1.209   Pale blue-green patina on etched ground Matt

Copper nitrate	200 gm
Sodium chloride	200 gm
Water	1 litre

Cotton-wool technique (Twenty hours)

The polished surface is etched by the solution, and tends to become dull. The blue-green patina develops during drying, and has a powdery appearance. On rough or as-cast surfaces, traces of blue-green patina form on a mid-brown ground.

The results tend to be dull and uneven, and the technique cannot be generally recommended.

Cotton-wool, moistened with the solution, is applied to the surface of the object to be coloured. It is then left for a period of about twenty hours. It is essential to ensure that the cotton-wool is in full contact with the surface and is kept moist throughout the period of treatment. Breaks in the colour and patches of patina may result if the cotton-wool is allowed to dry out or lift from the surface. When treatment is complete, the object should be washed thoroughly in cold water, and allowed to dry in air. When dry, it may be wax finished.

1.210   Mottled buff etch Matt

Copper nitrate	20 gm
Ammonium chloride	20 gm
Calcium hypochlorite	20 gm
Water	1 litre

Cotton-wool technique (Twenty hours)

Polished surfaces are etched to a matt finish by this treatment, and acquire a slightly mottled buff colour.

On rough or as-cast surfaces, a bluish patina tends to develop on a light brown ground. Results tend to be uneven.

▲ The corrosive fine dust of the calcium hypochlorite (bleaching powder) must not be inhaled, as it can severely irritate the respiratory system. The solution must be prevented from coming into contact with the eyes, which would be severely irritated.

Cotton-wool, moistened with the solution, is applied to the surface of the object to be coloured, and left for a period of about twenty hours. It is important to ensure that the cotton-wool is in close contact with the surface, and that it remains moist throughout the period of treatment. When treatment is complete, the object is removed and washed thoroughly in cold water. After a period of drying in air, it may be wax finished.

1.211*   Variegated lustre Gloss

Ammonium sulphate	85 gm
Copper nitrate	85 gm
Ammonia (.880 solution)	3 cm^3
Water	1 litre

Cotton-wool technique (Thirty hours)

The technique is only suitable for use on polished surfaces. A variegated golden lustre is produced, fringed with reddish and greenish bands.

Cotton-wool, moistened with the solution, is applied to the surface of the object to be coloured. It is then left for a period of about thirty hours. It is essential to ensure that the moist cotton-wool is in full contact with the surface of the object, or breaks in the colour will occur. The cotton-wool should be kept moist throughout the period of treatment. When treatment is complete, the object should be washed thoroughly in cold water and allowed to dry in air. When dry, it is wax finished.

1.212   Thin blue-green patina on light ground Semi-matt

Ammonium carbonate	20 gm
Oxalic acid (crystals)	10 gm
Acetic acid (6% solution)	1 litre

Cloth technique (Twenty-four hours)

A blue-green patina on an uneven reddish-brown ground is obtained on rough or as-cast surfaces. On polished surfaces the ground colour remains light and tends to take on the texture of the applied cloth. Patina development is patchy in both cases.

If the cloth dries completely on the surface, it adheres, damaging the surface when removed.

Other tests were carried out. Temperatures from cold to boiling produced no effect, even after prolonged immersion. Hot or cold application by dabbing and wiping also produced no effect, as the surface resists wetting.

▲ The carbonate must be added to the acetic acid in small quantities, to avoid violent effervescence.

▲ Oxalic acid is harmful if taken internally. Contact with the eyes and skin should be avoided.

Strips of soft cotton cloth, or other absorbent material, are soaked with the solution. These are applied to the surface of the object and stippled into place using the end of a stiff brush which has been moistened with the solution. They are left on the surface until very nearly dry, which may take up to twenty-four hours, and then removed. If patina development is to be encouraged, the surface is then moistened by wiping the solution on using a soft cloth. If the ground colour and a minimum of patina is required, then the surface is allowed to dry without applying any more solution. When the surface is completely dry, it may be wax finished.

1.213   Blue-green patina on mottled light brown Semi-matt

Ammonium sulphate	90 gm
Copper nitrate	90 gm
Ammonia (.880 solution)	1 cm^3
Water	1 litre

Cloth technique (Twenty hours)

The disposition of colour on polished surfaces follows the texture of the applied cloth.

An even blue-green patina on a brown ground is produced on as-cast surfaces.

If a patina-free ground colour is required, the cloth should be removed at an earlier stage when it is still wet, and the object thoroughly washed. The cloth application may be repeated, for a more developed ground colour.

Tests carried out involving direct application of this solution were not successful. The surface stubbornly resists 'wetting', even with additions of wetting agents to the solution.

Soft cotton cloth which has been soaked with the solution is applied to the surface of the object, and stippled into place with a stiff bristle-brush. The object is then left for a period of about twenty hours. The cloth should be removed when it is very nearly dry, and the object then left to dry in air without washing. The blue-green patina tends to develop during the drying period. When treatment is complete and the surface thoroughly dried out, it may be wax finished.

2 Cast yellow brass

The particular brass to which the recipes and results refer is SCB 3 (copper 65%, zinc 35%), which is commonly used as a general purpose casting brass. Brasses with a significantly different composition will in some cases yield variant colourings.

The main colour heading for each recipe refers to the results obtained on polished surfaces, unless otherwise stated. Reference is made to the colour and surface obtained, and to grain-enhancement where this has occurred. The results obtained on rough or as-cast surfaces are indicated in the accompanying marginal notes.

	Recipe numbers
Purple and orange Red-purple/purple-brown/orange-brown	2.1 – 2.14
Brown Light bronzing/brown/dark brown	2.15 – 2.85
Black Black/dark slate/grey	2.86 – 2.108
Green patina Blue-green/green/yellow-green	2.109 – 2.145
Lustre colours	2.146 – 2.170
Special techniques	2.171 – 2.208

2.1★   Purple Satin/semi-matt

Copper nitrate	80 gm
Water	1 litre
Ammonia (.880 solution)	3–5 cm³

Boiling immersion (Thirty minutes)

The article is immersed in the boiling solution. After five minutes a pink tinge is apparent, followed by a gradual and even build-up of colour. After thirty minutes the article is removed and washed thoroughly in hot water before drying in sawdust. When dry the surface has a characteristic greyish bloom which becomes nearly invisible after wax finishing, and only apparent when viewed from oblique angles.

Rough and as-cast surfaces yield a deeper, less ruddy purple.

Increasing the amount of ammonia added, up to about 10 cm³ produces a deeper colour, but this is accompanied by a decrease in adherence. As-cast/sandblasted surfaces respond well to the increase producing a fine variegated red and purple colour.

Decreasing the overall concentration of ingredients or decreasing the ammonia content produces a more grey-brown colour.

2.2★   Purplish-red Satin   *Pl. IV, Pl. XII*

Copper sulphate	120 gm
Ammonia (.880 solution)	30 cm³
Water	1 litre

Boiling immersion (Twenty minutes)

Immersion in the boiling solution produces a vivid green lustre colour after one minute, followed by the gradual change to a more opaque orange-pink which is well developed after seven or eight minutes. Continued immersion produces a gradual and even deepening of the colour. After twenty minutes the object is removed, washed thoroughly in hot water and dried in sawdust. When completely dry it is wax finished.

A slightly more purple colour is produced on rough or as-cast surfaces.

▲ A dense precipitate is formed which tends to sediment in the immersion vessel, causing 'bumping'. Glass vessels should not be used.

▲ Preparation of the solution, and colouring, should be carried out in a well ventilated area to prevent irritation of the eyes and respiratory system by the ammonia vapour. The skin, and particularly the eyes, should be protected from accidental splashes of this corrosive alkali.

2.3★   Dark purple/purple-brown Semi-matt/semi-gloss   *Pl. III, Pl. XI*

Copper nitrate	200 gm
Water	1 litre

Boiling immersion (Twenty minutes)

The article is immersed in the boiling solution. After five minutes the surface darkens and acquires a pink tinge, gradually becoming brown as immersion continues. After twenty minutes a drab grey layer forms. The article is removed and washed thoroughly in hot water to remove the grey layer and reveal the purplish-brown colour beneath. After thorough washing and drying in sawdust, the surface exhibits a characteristic grey bloom, which becomes almost invisible when the article is wax finished.

Rough and as-cast surfaces show a more distinctly purple colour, with less tendency to brown.

Tests carried out using a range of lower concentrations of copper nitrate produced colours that were less distinctly purple and more grey-brown. An example is shown in plate XI.

Concentrated hot solutions (200 gm/litre at 70°C) produced mid-tan colourations, with slight silvery streaking.

Prolonged immersion in cold solutions (eg. 200 gm/litre at 15–20°C for 36 hours) also produced mid-tan colours with silvery streaks.

2.4★   Purplish-brown/reddish-brown Semi-gloss

Copper sulphate	25 gm
Ammonia (.880 solution)	3 cm³
Water	1 litre

Boiling immersion (Thirty minutes)

Immersion in the boiling solution rapidly produces lustre colours within a minute, which quickly change to a more opaque orange colour after three minutes. Continued immersion produces a gradual darkening of the surface to reddish-brown. After about thirty minutes, when the colour is fully developed, the object is removed, washed in hot water and dried in sawdust. When dry it is wax finished.

A more purple colour is obtained on rough or as-cast than on polished surfaces.

Increasing the ammonia content tends to inhibit colour development. At a concentration of 10 cm³/litre of ammonia, with the stated concentration of copper sulphate, no colour is produced.

2.5   Purplish-brown/reddish-brown Semi-matt

Copper acetate	80 gm
Water	1 litre

Boiling immersion (Fifty minutes)

The article is immersed in the boiling solution, which produces dull and uneven lustre colours initially, and immersion is continued until the brown colour is produced. After about fifty minutes the article is removed and washed in hot water. It is dried in sawdust and may subsequently be wax finished when dry.

The results obtained in tests tended to be variegated on both polished and rough or as-cast surfaces.

2.6   Watery purple mottle on brass Semi-matt

Copper nitrate	40 gm
Ammonia (.880 solution)	to form a paste

Applied paste (Several hours)

The copper nitrate is ground using a pestle and mortar, and a very small amount of ammonia added by the drop to form a crystalline 'paste'. This is applied to the object using a soft brush, and the surface then allowed to dry for several hours. When dry, the residue is washed away with cold water. The process may be repeated if necessary. After a final washing and air-drying, the object may be wax finished.

A mottled variegation of watery purple on brass was produced on polished surfaces. On rough or as-cast surfaces a variegated reddish-brown is obtained. In tests, the procedure did not require repetition.

A true paste will not form, but the moist crystalline mass that is obtained can be applied with a brush. This tends to become watery and may not dry out completely.

2.7   Purplish-brown Semi-gloss

Copper sulphate	25 gm
Water	1 litre

Boiling immersion (Ten to thirty minutes)

The object is immersed in the boiling solution. After one or two minutes the surface takes on a drab orange colour which gradually darkens with continued immersion. When the surface has attained the purple-brown colour, which may take from ten to thirty minutes, it is removed and washed thoroughly in hot water. After drying in sawdust it is wax finished.

A purple colour is produced on rough or as-cast surfaces.

Tests have shown that the time taken for the full colour to develop is very variable, and in some cases may be longer than thirty minutes.

Tests with more concentrated solutions produced similar results for both polished and rough or as-cast surfaces.

Scratch-brushing with the hot solution may be used to even the surface during immersion. It may also be used as an after treatment, for 'relieved' effects, producing a warm brownish bronzing.

2.8*   Purple-brown Gloss

Copper acetate	70 gm
Copper sulphate	40 gm
Potassium aluminium sulphate	6 gm
Acetic acid (10% solution)	10 cm³
Water	1 litre

Boiling immersion (Fifteen minutes)

The object is immersed in the boiling solution. The colour development is gradual and even throughout the period of immersion, producing purple colours after about five minutes. The object is removed after fifteen minutes, washed thoroughly in hot water and dried in sawdust. When completely dry it is wax finished.

The purple-brown colour is obtained on both polished and rough or as-cast surfaces.

2.9*   Dark purplish-brown Satin/semi-matt

Copper sulphate	125 gm
Ferrous sulphate	100 gm
Acetic acid (glacial)	6.5 cm³
Water	1 litre

Boiling immersion (Thirty to forty minutes)

The object is immersed in the boiling solution. The surface darkens gradually during the course of immersion, initially tending to be drab and rather uneven, but later becoming more even and generally darker and richer in colour. When the colour has developed, after thirty to forty minutes, the object is removed and washed in hot water. It is dried in sawdust, and wax finished when dry.

An even reddish or purple-brown is produced on rough or as-cast surfaces

▲ Acetic acid in the highly concentrated 'glacial' form is corrosive, and must not be allowed to come into contact with the skin, and more particularly the eyes. It liberates a strong vapour which is irritating to the respiratory system and eyes. Preparation of the solution should be carried out in a fume cupboard or well ventilated area.

2.10*   Orange-brown/reddish-brown Gloss   *Pl. III*

Copper acetate	25 gm
Copper sulphate	19 gm
Water	1 litre

Hot immersion (Fifteen minutes)

The article is immersed in the hot solution (80°C). The colour develops gradually to produce a light terracotta after about five minutes, which darkens evenly as immersion continues. After ten minutes the surface is orange-brown in colour. The article is removed after fifteen minutes immersion, washed thoroughly and dried in sawdust. When dry, the surface has a characteristic grey bloom which is invisible after wax finishing.

The colour obtained on rough and as-cast surfaces is more red-brown in colour.

2.11   Terracotta Satin

Copper acetate	50 gm
Copper nitrate	50 gm
Water	1 litre

Boiling immersion (Thirty minutes)

On rough and as-cast surfaces the colour produced is reddish-purple.

Immersing the object in the boiling solution produces lustrous golden colours after one minute, which gradually change to more opaque matt pink/orange tints after five minutes. Continued immersion produces a gradual darkening to a terracotta. After thirty minutes the object is removed, washed in hot water and dried in sawdust. When completely dry it is wax finished.

2.12   Slightly variegated orange-brown Gloss

Sodium hydroxide	100 gm
Sodium tartrate	60 gm
Copper sulphate	60 gm
Water	1 litre

Hot immersion (Forty minutes)

A darker colour tended to be produced at the edges of polished surfaces.

A dark brown with a reddish tinge is produced on as-cast surfaces.

▲ Sodium hydroxide is a powerful caustic alkali. The solid or solutions can cause severe burns and must be prevented from coming into contact with the eyes or skin. When preparing the solution, small quantities of the sodium hydroxide should be added to large quantities of water and *not* vice versa. The solution will become warm as the sodium hydroxide is dissolved. When it is completely dissolved, the other ingredients are added.

The object is immersed in the hot solution (60°C) producing dull lustre colours after about one minute. The surface gradually becomes more opaque and colour slowly turns brown as immersion continues. After about forty minutes, when the colour has developed, the object is removed and washed in hot water and either dried in sawdust or allowed to dry in air. It may be wax finished when dry.

2.13   Reddish-brown Semi-matt

A	Copper sulphate	25 gm
	Water	1 litre
B	Ammonium chloride	0.5 gm

Boiling immersion (A)   (Fifteen minutes)
Boiling immersion (A+B) (Ten minutes)

If the immersion is prolonged a bloom of greyish-green patina will tend to form, which may need to be removed with a soft bristle brush.

A purple colour is produced on rough or as-cast surfaces, which tend also to retain a bloom of greenish-grey patina.

The article is immersed in the boiling copper sulphate solution for about fifteen minutes or until the colour is well developed. It is then removed to a bath of hot water, while the ammonium chloride is added to the colouring solution, and then re-immersed. The colour is brightened and tends to become more red. After about ten minutes it is removed and washed in hot water. The article is dried in sawdust and finally wax finished.

2.14*   Orange-brown Semi-matt

Copper sulphate	25 gm
Nickel ammonium sulphate	25 gm
Potassium chlorate	25 gm
Water	1 litre

Boiling immersion (Twenty minutes)

A yellow-green patina on a brown ground is produced on rough or as-cast rather than polished surfaces.

▲ Nickel salts are a common cause of sensitive skin reactions.

▲ Potassium chlorate is a powerful oxidising agent and should not be allowed to come into contact with combustible material. The solid, and solutions, should not be allowed to come into contact with the eyes, skin or clothing.

Immersing the object in the boiling solution produces an immediate lustrous mottling of the surface. Continued immersion induces a change to a more even orange colour which darkens gradually. After twenty minutes the object is removed and washed in hot water. After thorough drying in sawdust it is wax finished.

2.15   Dull golden yellow (slight grain-enhancement) Semi-matt/satin

Sodium dichromate	150 gm
Nitric acid (10% solution)	20 cm³
Hydrochloric acid (15% solution)	6 cm³
Ethanol	1 cm³
Water	1 litre

Cold immersion (Five minutes)

A variegated light greenish-brown/grey colour is produced on as-cast surfaces.

▲ Sodium dichromate is a powerful oxidant, and must not be allowed to come into contact with combustible material. The dust must not be inhaled. The solid and solutions should be prevented from coming into contact with the eyes and skin, or clothing. Frequent exposure can cause skin ulceration, and more serious effects through absorption.

▲ Both hydrochloric and nitric acids are corrosive and should be prevented from coming into contact with the eyes, skin or clothing.

The sodium dichromate is dissolved in the water, and the other ingredients are then added separately while stirring the solution. The article is immersed in the cold solution for a period of about five minutes, and then removed. It is washed in cold water, to remove any greenish layer that may have been deposited, and allowed to dry in air. When dry it may be wax finished.

2.16   Light grey-green (as-cast surface only) Matt

A	Ammonium chloride	10 gm
	Copper sulphate	10 gm
	Water	1 litre
B	Hydrogen peroxide (100 vols)	200 cm³
	Sodium chloride	5 gm
	Acetic acid (glacial)	5 cms³
	Water	1 litre

Cold immersion (A) (Five minutes)
Hot immersion (B) (A few seconds)

The greenish-brown colour obtained on a polished surface is easily removed, leaving an even, lightly etched, dull gold finish.

The surface finish is difficult to control, and the procedure cannot be generally recommended.

▲ Hydrogen peroxide can cause severe burns, and must be prevented from coming into contact with the eyes and skin. When preparing the solution, it should be added to the water prior to the addition of the other ingredients.

▲ Glacial acetic acid is very corrosive and must be prevented from coming in contact with the eyes or skin.

The object is immersed in solution A for about five minutes, which produces a slight etching effect and a pale greyish colour typical of chloride-containing solutions. The object is then briefly immersed in solution B (at 60–70°C). The reaction tends to be violent, causing the solution to 'boil', and immersion should not be prolonged. A greenish-brown layer is produced. If this is uneven, the surface should be gently scratch-brushed and the object briefly re-immersed. Prolonged or repeated immersion in solution B will tend to strip the greenish-brown colour. When treatment is complete, the object is washed in warm water and allowed to dry in air, and may be wax finished when dry.

2.17   Slight yellow-green tint to brass Semi-gloss/semi matt

Copper acetate	20 gm
Sodium tetraborate (Borax)	5 gm
Potassium nitrate	5 gm
Mercuric chloride	2.5 gm
Olive oil	to form a paste

Applied paste (Two or three days)

A variegated grey and dull brown colour is produced on rough and as-cast surfaces. The colouring effect on polished surfaces is very slight.

▲ Mercuric chloride is a highly toxic substance. It is essential to prevent any contact with the skin, and to prevent any chance of inhaling the dust. Gloves and masks must be worn. A nose and mouth face mask fitted with a fine dust filter must be worn when brushing away the dry residue.

▲ Potassium nitrate is a powerful oxidising agent and should not be allowed to come into contact with combustible material.

▲ In view of the limited colouring potential, and the high risks involved, this method cannot be recommended.

The ingredients are ground to a thin paste with a little olive oil, using a pestle and mortar. The paste is applied to the object, either with a brush or a soft cloth, and is allowed to dry for several hours. When dry, the object is brushed with a bristle-brush to remove the dry residue. This procedure is repeated as necessary. When treatment is complete, the surface is again brushed, and then cleaned with pure distilled turpentine. When this has dried, it is wax finished.

2.18   Pale golden bronzing Semi-gloss

Sodium sulphide	50 gm
Antimony trisulphide	50 gm
Water	1 litre
Acetic acid (10% solution)	100 cm³

Hot scratch-brushing (Several minutes)

The pale golden bronzing effect occurs only on polished surfaces. On rough and as-cast surfaces a purplish-brown colour is produced.

▲ Sodium sulphide is corrosive and should be prevented from coming into contact with the eyes and skin.

▲ Addition of the acetic acid and colouring should be carried out in a well ventilated area, as toxic hydrogen sulphide gas will be evolved.

The solid ingredients are added to the water and dissolved by gradually raising the temperature to boiling. The solution is allowed to cool to 60°C and then acidulated gradually with the acetic acid, producing a dense brown precipitate. The object is scratch-brushed with this turbid solution, alternating with hot and then cold water. When the colour has developed it is washed in cold water, allowed to dry in air, and wax finished when dry.

2.19   Very pale bronzing Semi-matt

Ammonium sulphide (16% solution)	100 cm³
Water	1 litre

Hot bristle-brushing (Several minutes)

An even dark brown or black colour is produced on rough or as-cast surfaces.

▲ Ammonium sulphide solution liberates a vapour containing ammonia and hydrogen sulphide gas, which should not be inhaled. Preparation and colouring must be carried out in a well-ventilated area. The solution must be prevented from coming into contact with the eyes or skin, as it causes burns or severe irritation.

The article is directly bristle-brushed with the hot solution (60°C) for several minutes, alternating with bristle-brushing under hot water. The procedure is repeated three or four times until the colour has developed and the article is then washed in warm water. After thorough drying in air or sawdust, it is wax finished.

2.20   Light grey bronzing Semi-matt

A	Copper sulphate	50 gm
	Water	1 litre
B	Potassium sulphide	12.5 gm
	Water	1 litre

Hot immersion. Scratch-brushing (A)
Cold immersion. Scratch-brushing (B)

An even but light grey-brown colour is produced on rough or as-cast surfaces. On polished faces, the surface tends to be uneven and the grey bronzing light and difficult to obtain evenly. The recipe cannot be generally recommended.

▲ Colouring should be carried out in a well ventilated area, as some hydrogen sulphide gas will be liberated.

The article is immersed in the copper sulphate solution for about thirty seconds and transferred to the cold potassium sulphide solution, and immersed for about one minute. It is then removed and scratch-brushed with each solution in turn. The surface is finished with a light scratch-brushing in the potassium sulphide solution. The article is washed thoroughly in warm water and allowed to dry in air before wax finishing.

2.21   Dull golden bronzing Semi-matt

Potassium sulphide	2 gm
Ammonium sulphate	0.5 gm
Water	1 litre

Cold scratch-brushing (Several minutes)

A mid-brown colour is produced on rough or as-cast surfaces.

This recipe has the disadvantage with this metal that unevenness, once produced, is very difficult to correct.

▲ Colouring should be carried out in a well ventilated area, as some hydrogen sulphide gas will be liberated.

The object is scratch-brushed with the cold solution, alternating with hot and then cold water. This process is continued until the colour has developed. Any relieving that is required should then be carried out. The object is washed thoroughly in cold water and allowed to dry in air. When completely dry it is wax finished.

2.22   Pale golden bronzing Semi-gloss

Potassium sulphide	1 gm
Ammonium chloride	4 gm
Water	1 litre

Cold scratch-brushing (Five minutes)

A grey-brown/pale olive colour is produced on rough and as-cast surfaces.

Tests carried out using stronger solutions, and also hot solutions, produced only very pale results.

▲ The procedure should be carried out in a well ventilated area, as some hydrogen sulphide gas is liberated from the solution.

The object is scratch-brushed with the cold solution for a short time and then washed in hot water. This alternation is repeated several times until the colour develops. The surface is then gently burnished with the scratch-brush in hot water, and finally with cold water. After thorough rinsing and drying in air the object may be wax finished.

2.23   Light brown bronzing Gloss

Antimony trisulphide	12.5 gm
Sodium hydroxide	40 gm
Water	1 litre

Hot scratch-brushing (Several minutes)

A variegated light brown is produced on rough or as-cast surfaces.

▲ Sodium hydroxide is a powerful caustic alkali which can cause severe burns to the eyes and skin, in the solid form or in solution. When preparing the solution, small quantities of the sodium hydroxide should be added to a large quantity of water and *not* vice-versa. When the sodium hydroxide has completely dissolved, the antimony trisulphide is added.

The hot solution (60°C) is repeatedly applied liberally to the surface of the object with a soft brush, allowing time for the colour to develop. The surface is then worked lightly with the scratch-brush, while applying the hot solution and hot water alternately. The treatment is concluded by lightly burnishing the surface with the scratch-brush and hot water. After rinsing it is dried in sawdust, and wax finished when dry.

2.24   Light bronzing Semi-gloss/gloss

Sodium sulphide	30 gm
Sulphur (Flowers of sulphur)	4 gm
Water	1 litre

Hot immersion and scratch-brushing (Fifteen minutes)

A grey-brown bronzing is produced on rough or as-cast surfaces.

▲ Sodium sulphide in solid form and in solution must be prevented from coming into contact with the eyes and skin, as it can cause severe burns. Some hydrogen sulphide gas may be evolved when the solution is heated. Colouring should be carried out in a well ventilated area.

The solids are added to about 200 cm³ of the water, and boiled until they dissolve and yield a clear deep orange solution. This is added to the remaining water and the temperature adjusted to 60–70°C. The object is immersed in the hot solution and the colour allowed to develop for up to ten minutes. It is then removed and scratch-brushed with the hot solution, until the surface is even and the colour developed. The object should be washed thoroughly in hot water and dried in sawdust. When dry, it may be wax finished.

2.25 Light bronzing (slightly lustrous) Gloss

Potassium sulphate	100 gm
Potassium sulphide	8 gm
Ammonium chloride	8 gm
Water	1 litre

Boiling immersion and scratch-brushing (Several minutes)

Immersion in the boiling solution produces an immediate darkening of the surface with lustre colours, followed by the gradual development of more opaque colour. This tends to occur unevenly. After a few minutes' immersion, the object is removed and scratch-brushed with the hot solution. When the surface has evened and the colour developed, the object is washed in hot water, and allowed to dry in air. When dry, it is wax finished.

On rough or as-cast surfaces, an even grey-brown is produced, which will burnish to a dull shine when waxed.

▲ Colouring should be carried out in a well ventilated area, as some hydrogen sulphide gas will be liberated from the hot solution.

2.26 Greenish-yellow grain enhancement (slight spotting) Semi-gloss

A	Copper nitrate	20 gm
	Ammonium chloride	20 gm
	Calcium hypochlorite	20 gm
	Water	1 litre
B	Water	

Cold immersion (A)   (Ten minutes)
Cold immersion (B)   (Twenty hours)

The article is immersed in solution A for about ten minutes, which produces a brightening of the surface to a slightly golden colour. The article is rinsed and transferred to a container of cold water, in which it is left for a period of about twenty hours. Small spots or patches of a darker colour develop slowly during the period of cold water immersion. When the surface is sufficiently developed, the article is removed and either dried in sawdust or allowed to dry in air. When dry, it is wax finished. The darker areas tend to darken slightly on exposure to daylight.

The spotted surface effect tends to be rather variable, and did not always appear.

A greenish-brown colour is produced on as-cast surfaces.

▲ Care should be taken to avoid inhaling the fine dust that may be raised from the calcium hypochlorite (bleaching powder). It should also be prevented from coming into contact with the eyes or skin.

2.27 Greenish-yellow Semi-matt

Copper nitrate	80 gm
Water	1 litre
Ammonia (.880 solution)	3 cm³

Applied liquid (Twice a day for ten days)

The solution is applied to the metal surface by dabbing with a soft cloth. The solution should be applied sparingly and may require vigorous dabbing in the initial stages to ensure adequate wetting. After each application the surface is allowed to dry completely in air. This procedure is repeated twice a day for ten days. The surface colouration develops gradually throughout the period of treatment. The article should be allowed to dry for several days before wax finishing.

Rough or as-cast surfaces yield a mid olive-green colour with this treatment.

On polished surfaces the greenish-yellow colour development is accompanied by a slight etching which reveals the grain structure of the casting.

If the solution is used less sparingly particularly in the later stages of treatment then traces of blue-green patina may form.

2.28* Warm mid-brown Semi-matt/matt

Copper sulphate	120 gm
Potassium chlorate	60 gm
Water	1 litre

Hot immersion (Ten or fifteen minutes)

The object is immersed in the hot solution (70–80°C). A rapid series of lustre colours is produced which become more opaque and darken to an orange-buff within two or three minutes. Continued immersion produces a gradual darkening to a dark brown colour. After about ten minutes a greenish layer begins to be deposited on the object, as a dense precipitate is formed in the solution. If this is not required then the surface should be gently bristle-brushed in warm water, and the object re-immersed. After about fifteen minutes the object is removed and washed. It should be bristle-brushed under warm water if the greenish layer has re-formed. After thorough washing it is dried in sawdust and wax finished when dry. A greyish or greenish-grey bloom tends to remain.

Results produced in tests were variable. A good yellowish-brown grain-enhanced surface is produced after about five minutes, which gradually becomes brown and more opaque. After about ten or fifteen minutes this is well developed, but may be fragile, revealing a more red and less matt layer beneath. As-cast surfaces were mid-brown with a 'dusting' of adherent grey-green patina.

▲ Potassium chlorate is a powerful oxidising agent and must not be allowed to come into contact with combustible material. Contact with the eyes, skin or clothing must be avoided.

2.29   Greenish-yellow grain enhancement Semi-matt   *Pl. IV*

Copper nitrate	80 gm
Nitric acid (10% solution)	100 cm³
Water	1 litre

Hot immersion (Five minutes)

A light greenish-yellow colour is produced on as-cast surfaces.

▲ Nitric acid is corrosive and may cause severe irritation or burns. It should be prevented from coming into contact with the eyes or skin.

The article is immersed in the hot solution (60–70°C), which causes etching of the surface and a gradual darkening. After five minutes it is removed and washed in warm water, and allowed to dry in air. When dry, it may be wax finished.
(Some sources suggest using this as a ground for a greenish patina, produced by applying a solution of ammonium carbonate (100 gm/litre) by wiping with a soft cloth and allowing to dry. Tests for extended periods produced only the slightest traces of a green tinge.)

2.30★   Light brown (grain enhanced) Gloss

Ammonium acetate	50 gm
Copper acetate	30 gm
Copper sulphate	4 gm
Ammonium chloride	0.5 gm
Water	1 litre

Boiling immersion (Seven to ten minutes)

A reddish-brown is produced on rough or as-cast surfaces.

After one or two minutes of immersion in the boiling solution, the object takes on an olive colouration, which gradually darkens to a more ochre colour as immersion continues. The full colour tends to develop in about seven minutes, although it may take a little longer. Prolonging the immersion tends to cause the colour to become more pallid, and does not improve surface consistency. The object is removed and washed in hot water. After thorough drying in sawdust, it is wax finished.

2.31   Reddish-brown (variegated) Semi-gloss   *Pl. IV*

Ferric nitrate	10 gm
Water	1 litre

Torch technique

A darker variegation can be obtained by finally dabbing with a stronger solution. (Ferric nitrate 40 gm/litre.)

▲ The fumes evolved as the solution is applied to the heated metal should not be inhaled. Adequate ventilation must be provided and a nose and mouth face mask worn, fitted with the correct filter.

▲ Ferric nitrate is corrosive and should not be allowed to come into contact with the eyes or skin. A face mask or goggles should be worn to protect the eyes from hot splashes as the solution is applied.

The object is heated with a blow torch and the solution applied sparingly with a scratch-brush, using small circular motions and working across the surface. The metal gradually darkens to a reddish-brown and the surface can be finished by lightly dabbing with a cloth dampened with the solution. When treatment is complete and the object has been allowed to cool and dry thoroughly, it may be wax finished.

2.32★   Orange-brown with grain-enhancement Semi-gloss   *Pl. III*

Copper acetate	7 gm
Ammonium chloride	3.5 gm
Acetic acid (6% solution)	100 cm³
Water	1 litre

Boiling immersion (Twenty minutes)

On rough or as-cast surfaces, an even warm brown colour is produced. The orange-brown colour on polished surfaces is translucent in quality, the slightly lighter enhanced grain giving a fine mottled appearance.

The object is immersed in the boiling solution. The surface gradually darkens and a slight etching effect slowly reveals the grain structure of the casting, on polished surfaces. When the surface colour is orange-brown, after about twenty minutes, the object is removed, washed in hot water, and dried in sawdust. When dry, it is wax finished.

2.33★   Orange-brown grain-enhanced surface Semi-matt   *Pl. IV*

Potassium chlorate	50 gm
Copper sulphate	25 gm
Ferrous sulphate	25 gm
Water	1 litre

Boiling immersion (Twenty minutes)

An orange-brown colour is produced on as-cast surfaces.

▲ Potassium chlorate is a powerful oxidant and must not be allowed to come into contact with combustible materials. It should also be prevented from coming into contact with the eyes, skin or clothing.

The object is immersed in the boiling solution. The colour develops gradually during immersion, becoming darker and tending to acquire a bloom of greenish-grey patina during the latter stages. If the bloom becomes excessive it should be removed by gently bristle-brushing the surface. When the required colour has developed, after about twenty minutes, the object is removed and washed in hot water. It is dried in sawdust, and may be wax finished when dry.

2.34*   Mid-brown (grain-enhanced) Gloss

'Rokusho'	5 gm
Copper sulphate	5 gm
Water	1 litre

Boiling immersion (About one hour)

The article is immersed in the boiling solution and after a few minutes the surface darkens and becomes more opaque as the colour develops. The surface develops gradually as immersion continues, and after about one hour the article is removed and washed in hot water. It is dried in sawdust and may be wax finished when completely dry.

A dark brown colour is produced on rough or as-cast surfaces.

For description of 'Rokusho' and its method of preparation, see appendix 1.

A darker and slightly more reddish-brown colour is produced, if the copper sulphate content is increased to 20–30 gm/litre. The results tend to be less even.

2.35   Brownish-yellow (grain-enhanced) Semi-matt

Copper nitrate	85 gm
Zinc nitrate	85 gm
Ferric chloride	3 gm
Hydrogen peroxide (3% solution)	1 litre

Applied liquid (Twice a day for five days)

The cold solution is dabbed on to the surface of the object with a soft cloth, and the object left to dry thoroughly in air. This procedure is repeated twice a day for a period of about five days. An initial lustrous colouration is gradually replaced by a more opaque colour. Traces of blue-green patina may also appear during the later stages of treatment, if the application is not sufficiently sparing, and these should be brushed away when dry if not required. The object is wax finished when completely dry.

A dark olive-green colour is produced on rough or as-cast surfaces.

Patina development can be encouraged, particularly on rough or as-cast surfaces, by a less sparing use of the solution in the later stages of treatment.

▲ Hydrogen peroxide must be kept from contact with combustible material. In the concentrated form it can cause burns to the eyes and skin, and when diluted will at least cause severe irritation.

2.36   Light brown (grain-enhanced) Semi-matt

Copper nitrate	15 gm
Zinc nitrate	15 gm
Ferric chloride	0.5 gm
Hydrogen peroxide (100 vols)	to form a paste

Applied paste (Four hours)

The ingredients are ground using a mortar and pestle, and the hydrogen peroxide added in small quantities to form a thin creamy paste. The paste is applied to the object with a soft brush, and left to dry in air. After four hours, the residue is washed away with cold water and the object is allowed to dry in air for several hours. When completely dry, it may be wax finished.

The grain-enhanced polished surface is only very lightly coloured. A variegated dark brown and orange-brown colour is produced on rough or as-cast surfaces. Some pale patina may also be produced.

The surface of the object should be handled as little as possible until dry, and should not be brushed during washing.

▲ Hydrogen peroxide should not be allowed to come into contact with combustible material. It should be prevented from coming into contact with the eyes and skin, as it may cause severe irritation. It is very harmful if taken internally. The vapour evolved as it is added to the solid ingredients of the paste must not be inhaled.

2.37   Greenish-yellow (variegated and grain enhanced) Matt

Copper sulphate	10 gm
Lead acetate	10 gm
Ammonium chloride	5 gm
Water	to form a paste

Applied paste (Several days)

The ingredients are ground to a creamy paste with a little water, using a pestle and mortar. The paste is applied to the object with a soft brush, and is left to dry out. When it is completely dry, after several hours, the dry residue is brushed away with a bristle brush. This process is repeated three or four times. When treatment has finished, the object should be allowed to stand in a dry place for several days to ensure thorough drying. After a final brushing with the bristle brush, it is wax finished.

On rough or as-cast surfaces an even warm olive-green colour is produced. A similar but lighter colour on a grain enhanced ground is produced on polished surfaces.

▲ Lead acetate is highly toxic, and every precaution should be taken to prevent the inhalation of the dust during weighing, and preparation of the paste. A mask fitted with a fine dust filter must be worn when brushing away the dry residue. A clean working method is essential throughout.

2.38   Greenish-yellow (grain enhanced) Semi-gloss

Copper sulphate	100 gm
Ammonium chloride	0.5 gm
Water	1 litre

Applied liquid on heated metal (Several minutes)

The object is gently heated on a hotplate and the solution applied either by brushing or with a soft cloth. The object should be heated to a temperature such that when the solution is applied, it sizzles gently. Application causes a slight enhancement of the grain structure of the surface, and some slight colouring. When the colour has developed, the object is wax finished.

A mid-brown colour is produced on rough or as-cast surfaces. The colouring effect on polished faces is rather slight. Some traces of a bloom of patina may occur as the surface dries out during heating.

2.39   Greenish-yellow (spangled enhanced grain) Semi-matt/matt

Copper sulphate	25 gm
Ammonium chloride	25 gm
Lead acetate	25 gm
Water	1 litre

Hot immersion (Fifteen minutes)

The spangled grain caused by etching is only produced on polished surfaces. On rough or as-cast surfaces, a slightly variegated dark grey-brown colour is obtained.

▲ Lead acetate is very toxic and every precaution should be taken to prevent inhalation of the dust during weighing and preparation of the solution. The vapour evolved from the heated solution must not be inhaled.

The object is immersed in the hot solution (80°C) producing slight etching effects and pale orange/pink colours. After fifteen minutes, the object is removed washed in hot water and allowed to dry in air. Exposure to daylight causes a darkening of the colour. When the colour has stabilised and the object completely dry, it is wax finished.

2.40*   Dull olive with bright spangled grain Matt   *Pl. III*

Copper nitrate	30 gm
Sodium chloride	30 gm
Ammonium chloride	15 gm
Potassium aluminium sulphate	7 gm
Acetic acid (6% solution)	1 litre

Boiling immersion (Ten minutes)

On rough and as-cast surfaces a drab olive colour without grain enhancement is produced.

The article is immersed in the boiling solution. After two or three minutes the etching effect of the solution brings out the grain structure of the cast surface. Continued immersion gradually darkens the object, which is removed after ten minutes, washed, and allowed to dry in air. When dry it is wax finished.

2.41   Yellowish-green and grey (grain enhanced) Semi-matt

Copper acetate	40 gm
Copper sulphate	200 gm
Potassium nitrate	40 gm
Sodium chloride	65 gm
Acetic acid (6% solution)	1 litre

Cold immersion (Ten minutes)

An olive-green colour is produced on rough or as-cast surfaces. Tests on polished surfaces tended to produce uneven finishes, and little control could be gained over the distribution of the colours.

▲ Potassium nitrate is a powerful oxidising agent and should not be allowed to come into contact with combustible material.

The object is suspended in the cold solution, which produces an immediate etching effect. After five minutes a pale cream or pink colour develops unevenly. After ten minutes the object is removed and scratch-brushed with the cold solution to even the surface. It should then be allowed to dry in air without washing. Any wet areas on the surface should be very gently brushed out with a soft brush, to prevent streaking or pooling. When dry, the object is exposed evenly to daylight, causing a darkening of the colours. When the colour has stabilised the object is wax finished.

2.42   Dull greenish-yellow (grain enhanced) Matt

Copper acetate	40 gm
Copper sulphate	200 gm
Potassium nitrate	40 gm
Sodium chloride	65 gm
Flowers of sulphur	110 gm
Acetic acid (6% solution)	1 litre

Hot immersion (Five minutes)

On rough or as-cast surfaces an even yellowish-grey colour is produced. The colouring effect on polished surfaces is rather slight, consisting of a slight darkening of a dull etched and grain-enhanced surface. The recipe cannot be generally recommended.

▲ Potassium nitrate is a powerful oxidising agent and should not be allowed to come into contact with combustible material. It must not be brought into contact with the sulphur in a dry state, but should be completely dissolved before the flowers of sulphur are added.

The article is immediately etched on immersion in the hot solution (70–80°C). A greenish colour, typical with chloride-containing solutions, develops within two or three minutes. After five minutes the article is removed, and the uneven surface is scratch-brushed with the hot solution to improve the surface consistency. The article is then washed in hot water and allowed to dry in air. When dry the surface is exposed evenly to daylight, causing a darkening of the colour. When the colour has stabilised and no further signs of change can be detected, it is wax finished.

2.43 Dull greenish-yellow (grain enhanced)

A	Copper sulphate	25 gm
	Ammonia (.880 solution)	30 cm³
	Water	1 litre

B	Potassium hydroxide	16 gm
	Water	1 litre

Hot immersion (A)   (Ten minutes)
Hot immersion (B)   (One minute)

An even greenish-brown colour is produced on rough or as-cast surfaces.

▲ Preparation of solution A, and colouring should be carried out in a well ventilated area, to prevent irritation of the eyes and respiratory system by the ammonia vapour liberated.

▲ Potassium hydroxide, in solid form and in solution, is corrosive to the skin and particularly the eyes. Adequate precautions should be taken.

The article is immersed in the hot solution A (60°C) for about ten minutes, producing a greenish-brown, drab, grain-enhanced surface. The article is then transferred to the hot solution B (60°C) for about one minute. It is removed, washed in warm water and allowed to dry in air. When dry it is wax finished.

2.44 Olive-tinged grain enhanced surface Semi-matt

Copper nitrate	35 gm
Ammonium chloride	35 gm
Calcium chloride	35 gm
Water	1 litre

Torch technique

A slight greenish patina on an olive-brown ground tends to be obtained on an as-cast surface.

If the solution is persistently applied as the surface is heated with a torch, a more adherent pale blue-green patina tends to form. However, solutions containing little or no chlorides are preferable for this effect.

The article is heated either on a hotplate or with a torch to a temperature such that the solution sizzles gently when it is applied. The solution is applied with a soft brush, and dries out to produce a superficial powdery green layer. The procedure is repeated several times, each fresh application tending to remove the green powdery layer, until an underlying colour has developed. Any residual green is then removed by washing and the article is allowed to dry in air. When dry it may be wax finished.

2.45 Olive-tinged grain-enhanced surface Matt

Ferric oxide	25 gm
Nitric acid (70% solution)	100 cm³
Iron filings	5 gm
Water	1 litre

Cold immersion (Fifteen minutes)

An olive-green etched finish is produced on as-cast surfaces.

▲ Preparation and colouring should be carried out in a fume cupboard. Toxic and corrosive brown nitrous fumes are evolved which must not be inhaled.

▲ Concentrated nitric acid is extremely corrosive, causing burns, and must be prevented from coming into contact with the eyes and skin.

The ferric oxide is dissolved in the nitric acid (producing toxic fumes), and the mixture is added to the water. The iron filings are finally added. The article is immersed in the solution, which produces an etched surface, and removed after about fifteen minutes. It is washed thoroughly in cold water and allowed to dry in air, and may be wax finished when dry.

2.46 Greenish-brown Semi-gloss   *Pl. III*

Ferric chloride	10 gm
Water	1 litre

Torch technique

A reddish-brown colour is produced on as-cast surfaces.

If a stronger solution of ferric chloride (40 gm/litre) is used for the final dabbing application, a darker variegated surface can be obtained.

▲ Ferric chloride is corrosive and should not be allowed to come into contact with the eyes or skin. A face shield or goggles should be worn to protect the eyes from hot splashes as the solution is applied.

▲ The fumes evolved as the solution is applied to the heated metal must not be inhaled. Adequate ventilation should be provided and a nose and mouth face mask fitted with the correct filter should be worn.

The object is heated with a blow-torch and the solution applied sparingly with a brass scratch-brush using small circular motions and working across the surface. The metal gradually darkens, and the variegation that tends to occur can be evened by lightly dabbing the surface with a cloth that is dampened with the solution. When treatment is complete, the object is left to cool and dry thoroughly, and may then be wax finished.

2.47 Dull greenish-brown grain enhanced surface Semi-matt

Copper carbonate	100 gm
Ammonium chloride	100 gm
Ammonia (.880 solution)	20 cm³
Water	1 litre

Hot immersion (Fifteen minutes)

A greenish-brown/olive colour is obtained on rough or as-cast surfaces.

▲ When preparing the solution, care should be taken to avoid inhaling the ammonia vapour. The eyes and skin should be protected from accidental splashes of the concentrated ammonia solution.

The article is immersed in the hot solution (50°C), which produces immediate etching effects, typical with chloride-containing solutions. After about fifteen minutes, the article is removed and washed thoroughly in warm water. It is allowed to dry in air, and may be wax finished when dry.

2.48   Grey-brown grain enhancement Matt

Copper sulphate	200 gm
Ferric chloride	25 gm
Water	1 litre

Hot immersion   (Thirty minutes)

In tests, a slight greenish patina on a brown or greyish-brown ground occurred on as-cast surfaces.

▲ Ferric chloride is corrosive and should be prevented from coming into contact with the eyes and skin.

The object is immersed in the hot solution (60°C), which etches the surface and produces a pale greyish colour, typical of chloride-containing solutions. Continued immersion produces a gradual change to a more brown colour. After thirty minutes the object is removed and washed in hot water, which tends to leach a slight yellowish-green colour from the surface. After thorough washing, the object is allowed to dry in air and exposed to daylight, which tends to cause an uneven darkening of the colour. When the surface development is complete and the object thoroughly dried out, which may take several days, it can be wax finished.

2.49   Thin reddish-brown on bare metal Semi-matt

Copper nitrate	40 gm
Water	to form a paste

Applied paste   (Four to six hours)

Colouring tends to be slight on polished surfaces and requires repetition. A variegated dull brown is produced on as-cast surfaces. The procedure cannot generally be recommended.

A true paste will not form, but the damp crystalline mass that is obtained can be applied with a brush. This tends to become watery and may not dry out.

The copper nitrate is ground using a pestle and mortar, and a very small amount of water added by the drop to form a crystalline 'paste'. The paste is applied using a soft brush, and the surface then allowed to dry for several hours. When dry, the residue is washed away with cold water. The process is repeated if necessary. After washing and air-drying, the object may be wax finished.

2.50   Mid-brown (orange-brown areas) Semi-matt

Potassium chlorate	50 gm
Copper sulphate	25 gm
Water	1 litre

Hot immersion   (Fifteen or twenty minutes)

An even brownish ochre is produced on rough or as-cast surfaces.

▲ Potassium chlorate is a powerful oxidising agent and should not be allowed to come into contact with combustible material. Contact with the eyes, skin and clothing must be avoided.

▲ A nose and mouth face mask fitted with a fine dust filter should be worn when brushing away any dry powder from the surface.

The object is immersed in the hot solution (80°C) and is initially coloured with an uneven golden lustre finish which changes to a more opaque yellow-green. Continued immersion darkens the colour to orange and then finally to a more brown colour. After fifteen or twenty minutes, when the full colour has developed, the object is removed and washed in hot water. When it has been thoroughly dried in sawdust, it is wax finished. A green powdery film may develop when the sample is dried, which can be brushed away prior to waxing. The wax finish tends to make any residual green bloom invisible.

2.51   Purple on brown (as-cast surfaces only) Matt

Copper sulphate	8 gm
Copper acetate	3.25 gm
Potassium aluminium sulphate	2 gm
Water	1 litre
Acetic acid (6% solution)	20 cm³

Boiling immersion   (About one hour)

Slight uneven lustre colours are produced on polished surfaces.

The article is immersed in the boiling solution and after a few minutes the surface darkens and becomes more opaque as the colour develops. The surface develops gradually as immersion continues, and after about one hour the article is removed and washed in hot water. It is dried in sawdust and may be wax finished when completely dry.

2.52★   Bloom of green patina on reddish-brown Matt   *Pl. III, Pl. XIV*

Potassium chlorate	50 gm
Copper sulphate	25 gm
Nickel sulphate	25 gm
Water	1 litre

Boiling immersion   (Thirty to forty minutes)

Very similar colours are produced on rough or as-cast and polished surfaces.

If the green colour is to be kept to a minimum, the object should be removed after thirty minutes.

▲ Potassium chlorate is a powerful oxidising agent and must not be allowed to come into contact with combustible material. The solid or solutions must be prevented from coming into contact with the eyes, skin or clothing.

▲ Nickel compounds are a common cause of sensitive skin reactions. Precautions should be taken to prevent inhalation of the dust, during preparation of the solution.

After two or three minutes' immersion in the boiling solution an orange colour develops on the surface of the object, which gradually darkens. The full colour develops in about thirty minutes, together with a green tinge which appears after about twenty minutes. Colour development and surface quality are generally better with a longer immersion, the green tending to enhance the surface. The object is removed after about forty minutes, washed in warm water, and either dried in sawdust or allowed to dry in air. When dry, it is wax finished.

2.53   Light purplish-grey bloom on light orange Semi-matt   *Pl. III*

	A	Potassium nitrate	100 gm
		Water	1 litre
	B	Copper sulphate	25 gm
		Ferrous sulphate	20 gm
		Acetic acid (Glacial)	1 cm^3

Boiling immersion (Thirty minutes)

The object is immersed in the potassium nitrate solution for a period of about fifteen minutes, which produces a slight brightening of the surface. The object is removed to a bath of hot water, while the copper sulphate, ferrous sulphate and acetic acid are separately added to the colouring solution. The object is re-immersed in the boiling solution, which gradually produces a greyish-brown colour, and removed after about fifteen minutes. It is washed in hot water and allowed to dry in air. When dry it is wax finished.

A light purplish-grey colour is produced on rough or as-cast surfaces.

▲ Potassium nitrate is a powerful oxidant and must be prevented from coming into contact with combustible materials.

▲ Glacial acetic acid is very corrosive and must be prevented from coming into contact with the eyes and skin.

2.54   Mid-brown bronze Semi-gloss

Barium chlorate	52 gm
Copper sulphate	40 gm
Water	1 litre

Hot immersion (Fifteen minutes)

Immersion in the hot solution (70°C) produces an immediate golden lustre colour, which changes gradually to pink/green and then to a brownish lustre after about three or four minutes. Continued immersion produces a greater opacity of colour and a gradual development to a yellow-brown or orange-brown colour. When the colour has fully developed, after fifteen minutes, the object is removed and washed in hot water. It is wax finished, after thorough drying in sawdust.

An even mid-brown/yellowish-brown is produced on rough or as-cast surfaces.

The lustre colours produced in the early stages of immersion are generally even, but are less bright than the equivalent colours produced by other methods.

▲ Barium compounds are poisonous and harmful if taken internally. Barium chlorate is a strong oxidising agent and should not be allowed to come into contact with combustible material. It should be prevented from coming into contact with the eyes, skin or clothing.

2.55   Light greenish-brown (grain enhanced) Semi-matt

Copper sulphate	50 gm
Ferrous sulphate	5 gm
Zinc sulphate	5 gm
Potassium permanganate	2.5 gm
Water	1 litre

Boiling immersion (Ten to fifteen minutes)

The object is immersed in the boiling solution, and becomes coated with a dark brown layer. After one or two minutes the object is removed and the brown layer brushed away under hot water, using a bristle-brush. It is then re-immersed for two minutes, and the bristle-brushing again carried out when the dark brown layer has reformed. After two or three such immersions and bristle-brushing, the brown layer will tend not to be adherent when it re-forms, and can be removed by agitating the object while in the solution. Immersion is continued to a total of ten or fifteen minutes, after which the object is removed and given a final bristle-brushing under hot water. It is thoroughly rinsed in warm water, dried in sawdust and wax finished when completely dry.

A slightly darker greenish-brown is obtained on rough or as-cast surfaces.

Darker colours with a reddish bias can be produced using this solution, if the technique is modified. See recipe 2.77.

2.56   Orange-brown/yellow-brown, with greenish bloom Matt

Copper sulphate	125 gm
Sodium chlorate	50 gm
Water	1 litre

Boiling immersion (Ten minutes)

The article is immersed in the boiling solution, and develops lustre colours in about one minute. Continued immersion produces a rapid development of more opaque colours. After ten minutes the article is removed and washed thoroughly in hot water. After careful drying in sawdust, it is wax finished.

A more grey-green and patina-like finish is produced on rough or as-cast rather than polished surfaces.

Prolonging the immersion produces a darker colour, but this is increasingly accompanied by a grey-green patina-like layer being built up. This tends to be uneven on polished surfaces, and although it may be made more even by intermittent bristle-brushing during immersion, it is difficult to control. On rough or as-cast surfaces the results are more even.

▲ Sodium chlorate is a powerful oxidising agent and must not be allowed to come into contact with combustible material or acids. Inhaling the dust must be avoided, and the eyes and skin should be protected against the corrosive and irritating effects of the solid, or splashes of solution.

2.57   Greenish-brown tint to brass Semi-matt

Copper nitrate	20 gm
Zinc sulphate	30 gm
Mercurous chloride	30 gm
Water	1 litre

Applied liquid (Twice a day for ten days)

The solution is initially applied by vigorous dabbing with a soft cloth, to overcome resistance to wetting. Once some surface effect has been produced, the solution may be applied by wiping. The surface is allowed to dry completely after each application, which should be carried out twice a day for at least ten days. If the solution is used very sparingly then a patina-free ground colour tends to be produced. A more liberal use of the solution during the later stages of treatment will tend to encourage the formation of a blue-green patina. When treatment is complete, the article should be allowed to dry in air for several days before wax finishing.

A variegated greenish-brown colour was obtained on an as-cast surface.

In tests, the blue-green patina was found to be difficult to produce.

▲ Mercurous chloride may be harmful if taken internally. There are dangers of cumulative effects.

2.58   Pale olive Matt

Copper acetate	30 gm
Ammonium chloride	15 gm
Ferric oxide	30 gm
Water	1 litre

Hot immersion, hot scratch-brushing (Ten to fifteen minutes)

The solution is brought to the boil and then removed from the source of heat. The object is then immersed, and becomes coated with a thick brown layer. After ten minutes immersion the object is removed and gently heated on a hotplate, the surface being worked with the scratch-brush. More solution is applied and worked in with the scratch-brush, occasionally allowing the surface to dry off without scratch-brushing. When the colour has developed it is allowed to dry on the hotplate for a few minutes, removed and the dry residue removed with a stiff bristle-brush. When the surface is free of residue it is wax finished.

A reddish-brown variegated with dark brown is produced on rough or as-cast surfaces.

An alternative technique which is more suited to larger objects is to heat locally with a torch and apply the pasty liquid with the scratch-brush, without prior immersion.

▲ A nose and mouth face mask fitted with a fine dust filter should be worn when brushing off the dry residue.

2.59   Greenish-brown (slightly variegated) Semi-gloss/semi-matt

Copper sulphate	25 gm
Lead acetate	25 gm
Water	1 litre

Hot immersion (Thirty minutes)

The object is immersed in the hot solution (80°C) which produces lustre colours after about one minute. Continued immersion produces a gradual change to opaque colours that slowly become more brown. After thirty minutes the object is removed and washed in hot water. After thorough drying in sawdust, it is wax finished.

A mid-brown colour is produced on rough or as-cast surfaces.

Tests showed that a significant brightening of the colour to a more reddish-brown could be obtained by the addition of a small quantity of ammonium chloride. This should be added at a rate of 2.5 gm/litre at the end of immersion, which is then continued for an extra three or four minutes.

▲ Lead acetate is highly toxic, and every precaution should be taken to prevent inhalation of the dust during weighing and preparation of the solution. Colouring should be carried out in a well ventilated area. The vapour evolved from the hot solution must not be inhaled.

2.60   Variegated brown (slight petrol lustre) Semi-gloss

Copper sulphate	60 gm
Copper acetate	10 gm
Potassium aluminium sulphate	25 gm
Water	1 litre

Boiling immersion (Forty minutes)

The object is immersed in the boiling solution, which produces a gradual darkening of the surface to a drab ochre/brown colour after ten minutes. A slight lustre may also be apparent. Continued immersion produces further darkening, although colour development is slow. After forty minutes the object is removed, washed in hot water and dried in sawdust. When dry, it is wax finished.

An even purplish-brown colour is produced on rough or as-cast surfaces.

2.61   Orange-brown Semi-gloss

Ammonium carbonate	50 gm
Copper acetate	25 gm
Ammonium chloride	5 gm
Oxalic acid (crystals)	1 gm
Acetic acid (6% solution)	1 litre

Boiling immersion (One hour)

The article is immersed in the boiling solution. After about five minutes a patchy orange-yellow colouration with traces of green appears. Continued immersion in the boiling solution darkens the colour, which appears to be best after about one hour. The article should then be removed, washed in warm water and dried in sawdust. When dry, it may be wax finished.

On rough or as-cast rather than polished surfaces, a yellow-brown or greenish-brown colour is produced. Some slight unevenness and streaking may occur on polished surfaces.

Prolonging the immersion time tends to increase the tendency for streaks to occur.

▲ When preparing the solution, the ammonium carbonate should be added in very small quantities. It reacts effervescently with the acid, and this reaction is very vigorous if large quantities are added.

▲ Oxalic acid is poisonous and may be harmful if taken internally.

2.62   Greenish-brown (slight petrol lustre) Semi-gloss

Copper sulphate	100 gm
Copper acetate	40 gm
Potassium nitrate	40 gm
Water	1 litre

Boiling immersion (Ten minutes)

The article is immersed in the boiling solution, which produces lustrous green colours after one minute. These gradually darken and become more opaque with continued immersion. The article is removed when the colour has developed, after about ten minutes, and washed in hot water. After thorough drying in sawdust, it is wax finished.

An even reddish-brown colour is produced on rough or as-cast surfaces.

▲ Potassium nitrate is a powerful oxidising agent and must not be allowed to come into contact with combustible material.

2.63*   Mid-brown (*or variegated black on brown*) Semi-matt

Potassium permanganate	2.5 gm
Copper sulphate	50 gm
Ferric sulphate	5 gm
Water	1 litre

Boiling immersion (Fifteen minutes)

The article is immersed in the boiling solution, and the surface becomes coated with a superficial dark brown film, which should be removed by bristle-brushing in a bath of hot water. The article is re-immersed in the boiling solution and again becomes coated with dark brown. (If a plain brown surface is required, the brown film should be removed every few minutes as immersion continues. After about fifteen minutes the article is removed and washed thoroughly in hot water. After drying in air it may be wax finished.) If a variegated surface is required, then the brown film should be left on the surface as immersion continues. After fifteen minutes, the article is removed, and allowed to dry in air for a few minutes. The surface is then gently brushed with a dry bristle-brush, which will partially remove the brown layer, leaving a variegated surface. The article should be wax finished.

The variegated black effect is difficult to obtain.

Similar colours are produced on rough or as-cast surfaces.

The solution should not be boiled for any length of time prior to the immersion of the object, or the dark layer tends not to be deposited, and the variegated effect will then not be produced.

2.64   Mid-brown Gloss

Copper sulphate	6.25 gm
Copper acetate	1.25 gm
Potassium nitrate	1.25 gm
Water	1 litre

Boiling immersion (About one hour)

The article is immersed in the boiling solution and after a few minutes the surface darkens and becomes more opaque as the colour develops. The surface develops gradually as immersion continues, and after about one hour the article is removed and washed in hot water. It is dried in sawdust and may be wax finished when completely dry.

The colour produced on polished surfaces tends to have a slightly thin and translucent quality.

Variegated and more opaque brown colours are produced on as-cast surfaces.

▲ Potassium nitrate is a powerful oxidising agent and must not be allowed to come into contact with combustible material.

2.65   Orange-brown (slight grain enhancement) Semi-matt

Copper nitrate	200 gm
Water	1 litre

Applied liquid (Twice a day for five days)

A darker grey-brown was produced in tests carried out on as-cast surfaces, with no development of patina.

The solution is initially dabbed on vigorously with a soft cloth, to ensure that the surface is evenly 'wetted'. Subsequent applications should be wiped on, and used very sparingly, to inhibit the formation of green patina. After each application the article is left to dry in air. This procedure is repeated twice a day for about five days. The article is then allowed to dry out thoroughly in air, and then wax finished when completely dry.

2.66   Orange-brown Gloss

Copper sulphate	100 gm
Ammonium chloride	0.5 gm
Water	1 litre

Hot immersion (Fifteen minutes)

On rough or as-cast surfaces a dark grey-brown colour is produced.

The object is immersed in the hot solution (70–80°C) producing a pale orange colouration, which darkens gradually with continued immersion. This is accompanied by a slight grain enhancement due to etching. After fifteen minutes the object is removed, washed thoroughly in warm water and dried in sawdust. Exposure to daylight will cause a darkening of the colour initially. When no further darkening is taking place the object is wax finished.

2.67   Reddish-brown Semi-gloss

Copper sulphate	125 gm
Sodium acetate	12.5 gm
Water	1 litre

Boiling immersion (Ten to fifteen minutes)

An even dark brown is produced on rough or as-cast surfaces. On polished surfaces, the reddish-brown colour tends to be slightly variegated with a lighter brown.

After one minute of immersion in the boiling solution a slightly lustrous orange/pink colour develops on the surface of the object. This gradually darkens tending to become brown as immersion continues. When the colour has developed, after ten or fifteen minutes, the object is removed and washed in hot water. It is wax finished after thorough drying in sawdust.

2.68   Purplish-brown (slight grain enhancement) Gloss

Copper acetate	6 gm
Copper sulphate	1.5 gm
Sodium chloride	1.5 gm
Water	1 litre

Boiling immersion (About one hour)

A variegated purple and brown colour is produced on as-cast surfaces.

The article is immersed in the boiling solution and after a few minutes the surface darkens and becomes more opaque as the colour develops. The surface develops gradually as immersion continues, and after about one hour the article is removed and washed in hot water. It is dried in sawdust and may be wax finished when completely dry.

2.69*   Mid to dark brown (grain enhanced) Gloss   *Pl. III, Pl. XII*

'Rokusho'	3.25 gm
Copper sulphate	3.25 gm
Potassium aluminium sulphate	2 gm
Water	1 litre

Boiling immersion (About one hour)

A greenish-brown colour is produced on rough or as-cast surfaces.

For description of 'Rokusho' and its method of preparation, see appendix 1.

The article is immersed in the boiling solution and after a few minutes the surface darkens and becomes more opaque as the colour develops. The surface develops gradually as immersion continues, and after about one hour the article is removed and washed in hot water. It is dried in sawdust and may be wax finished when completely dry.

2.70   Reddish-brown (variegated with pinkish-brown) Semi-gloss

A	Copper sulphate	125 gm
	Ferrous sulphate	100 gm
	Water	1 litre
B	Ammonium chloride	1 gm

Boiling immersion (A)   (Fifteen minutes)
Boiling immersion (A+B)   (Five minutes)

The object is immersed in the boiling solution of copper sulphate and ferrous sulphate, and immersion continued for about fifteen minutes, until a brown colour has developed. The object is removed and rinsed in a tank of hot water, while the ammonium chloride is added to the colouring solution. The object is re-immersed in the colouring solution for about five minutes, causing a brightening and development of the surface to a red colour. It is then removed and washed very thoroughly in hot water. After drying in sawdust, it may be wax finished.

A slight greenish patina on a reddish-brown ground is produced on rough or as-cast surfaces.

The object must be thoroughly washed after immersion, to prevent the formation of a greenish bloom or patches of green. If there is a build-up of bloom during immersion, then this should be cleared by agitating the object, or by removing and rinsing in hot water if necessary.

If shorter immersion times are used before the addition of the ammonium chloride, then the results tend to be lighter and more glossy, but are more prone to unevenness.

2.71   Reddish-brown (slight grain enhancement) Semi-matt

Copper sulphate	6.25 gm
Copper acetate	1.25 gm
Sodium chloride	2 gm
Potassium nitrate	1.25 gm
Water	1 litre

Boiling immersion (About one hour)

The article is immersed in the boiling solution and after a few minutes the surface darkens and becomes more opaque as the colour develops. The surface develops gradually as immersion continues, and after about one hour the article is removed and washed in hot water. It is dried in sawdust and may be wax finished when completely dry.

Variegated brown colours are produced on as-cast surfaces.

▲Potassium nitrate is a powerful oxidising agent, and should not be allowed to come into contact with combustible materials.

2.72   Dark brown Matt

Copper sulphate	30 gm
Potassium permanganate	6 gm
Water	1 litre

Boiling immersion and scratch-brushing (Several minutes)

On immersion in the boiling solution, the object becomes coated with a brownish film, which develops to a dark purplish-brown after one minute. This brown colour film is loosely adherent and sloughs off gradually to reveal uneven underlying grey-brown colours. After three or four minutes the object is removed and gently scratch-brushed with the hot solution. Overworking will cut through to the underlying metal. If the surface is unsatisfactory then the process is repeated. When the desired colour has been obtained, the object is washed in hot water, dried either in sawdust or in air, and wax finished when dry.

An even dark brown/black colour is produced on rough or as-cast surfaces. The results obtained in tests on polished surfaces were somewhat variable in both depth of colour and surface quality.

The development of the loose brown layer appears to be essential to successful colouring. If the solution is boiled for too long before the object is immersed, then a more dense precipitate is formed in the solution, and coating does not take place.

2.73*   Dark brown Gloss

Ammonium carbonate	60 gm
Copper sulphate	30 gm
Oxalic acid (crystals)	1 gm
Acetic acid (6% solution)	1 litre

Hot immersion (Thirty-five minutes)

The object is immersed in the hot solution (80°C) producing light brownish colours after two or three minutes, which may be uneven initially. With continued immersion the colours become even and darken gradually. When the colour has fully developed, after about thirty-five minutes, the object is removed and washed in hot water. After thorough drying in sawdust, it may be wax finished.

A dark brown/black colour is produced on rough or as-cast surfaces.

▲ Oxalic acid is poisonous and may be harmful if taken internally.

▲ The ammonium carbonate should be added to the acetic acid in small quantities. The reaction is effervescent, and if large amounts are added the effervescence becomes very vigorous.

2.74   Ochre/brown   Satin/semi-gloss

Antimony trisulphide	20 gm
Ferric oxide	20 gm
Potassium polysulphide	2 gm
Ammonia (.880 solution)	2 cm³
Water	to form a paste (approx 4 cm³)

Applied paste (sulphide paste) (Several minutes)

The potassium polysulphide is ground using a pestle and mortar and the other solid ingredients added and mixed. These are made up to a thick paste by the gradual addition of the ammonia and a small quantity of water. The paste is applied to the whole surface of the object to be coloured, using a soft bristle-brush. A stiff nylon brush is then used to work the surface gradually in small areas. The nylon brush should be intermittently dipped in a mixture of the dry ingredients of the paste (in the same proportion as above), and the surface worked in this way until the paste becomes dry and falls from the surface. When the whole surface is dry and completely free of any residue or dust, it may be burnished with a brass scratch-brush and then wax finished.

A variegated surface can be obtained by brushing the paste on to the whole surface and allowing it to dry for about three hours, until it is superficially dry and cracking, and then burnishing it away with a nylon brush or a brass scratch-brush. This may be repeated as necessary to darken the colour.

▲ A nose and mouth face mask should be worn when the residues are brushed away. The dust produced in this case contains antimony compounds, which are toxic.

2.75   Orange brown/dark brown (variegated) (as-cast surfaces only)

A	Potassium polysulphide	4 gm
	Water	1 litre
B	Ferric chloride	10 gm
	Water	1 litre

Torch technique (A and B)

The metal is heated with a blow torch and the polysulphide solution applied sparingly with a brass scratch-brush, using small circular motions and working across the surface until a dark brown colour is established. The metal is then gently heated as the ferric chloride solution is stippled on to the surface using a bristle-brush or cloth, to give an orange-brown tone. When treatment is complete, the object is allowed to cool and dry thoroughly, and is then wax finished.

▲ The fumes evolved as either solution is applied to the heated metal should not be inhaled. Adequate ventilation must be provided.

▲ A face mask or goggles should be worn to protect the eyes from hot splashes as the solutions are applied.

▲ Ferric chloride is corrosive and should be prevented from coming into contact with the eyes and skin.

2.76   Brownish-purple   Semi-gloss/gloss

Copper carbonate	160 gm
Sodium hydroxide	80 gm
Water	1 litre

Boiling immersion (Thirty minutes)

The sodium hydroxide is added to the water in small quantities, while stirring. Heat is evolved as it dissolves, warming the solution. When it has completely dissolved, the copper carbonate is added and the solution heated to boiling. The object is immersed in the boiling solution, and gradually acquires a pale grey colour which darkens to a greyish-mauve as immersion proceeds. After about thirty minutes the object is removed and dried in sawdust. When completely dry, it may be wax finished.

An even purple colour is produced on rough or as-cast surfaces.

Similar results can be obtained using less concentrated solutions for correspondingly longer times, but these tend to be slightly less even.

▲ Sodium hydroxide is a powerful caustic alkali, and the solid or solutions must be prevented from coming into contact with the eyes and skin. The sodium hydroxide, in small quantities, should be added to the water and *not* vice versa. Colouring should be carried out in a well ventilated area, as inhalation of the caustic vapours should be avoided.

2.77*   Dark brown   Satin/semi-matt   *Pl. IV*

Copper sulphate	50 gm
Ferrous sulphate	5 gm
Zinc sulphate	5 gm
Potassium permanganate	2.5 gm
Water	1 litre

Boiling immersion (Thirty minutes)

When the object is immersed in the gently boiling solution, a dark brown layer forms on the surface, which should not be disturbed. Immersion is continued for fifteen minutes, after which time the object is removed and the dark layer washed away in hot water using the bristle-brush. The object is then re-immersed for a further fifteen minutes and then removed and bristle-brushed gently with hot water. After thorough rinsing it is dried in sawdust, and wax finished when completely dry.

An even and darker reddish-brown is produced on rough or as-cast surfaces.

Lighter colours with a more green bias can be produced using this solution. See recipe 2.55

2.78   Dark greenish-brown (variegated) Semi-matt

Copper sulphate	25 gm
Sodium chloride	100 gm
Potassium polysulphide	10 gm
Water	1 litre

Hot immersion (Fifteen minutes)

The object is immersed in the hot solution (60°C), producing a slight matting of the surface due to etching, and the gradual development of dull brown colours. After about fifteen minutes, when the brown has developed, the object is removed. Relieving may be carried out at this stage, using the hot solution with the scratch-brush. The object is then removed washed in warm water and allowed to dry in air. When dry, it is wax finished.

A more even dark greenish-brown is produced on rough or as-cast surfaces than on polished surfaces. Gentle relieving with the scratch-brush reveals a lighter grain enhanced surface.

▲ The colouring should be carried out in a well ventilated area, as some hydrogen sulphide gas will be liberated.

2.79   Dark brown/black (stipple) Semi-matt

Copper sulphate	50 gm
Potassium chlorate	20 gm
Potassium permanganate	5 gm
Water	1 litre

Boiling immersion (Fifteen minutes)

The object is immersed in the boiling solution, and rapidly becomes coated with a chocolate-brown layer. Immersion is continued for fifteen minutes, after which the object is removed and gently rinsed in a bath of warm water. It is then allowed to dry in air, leaving a dark powdery layer on the surface. When dry this is gently brushed with a soft brush to remove any dry residue. This treatment will remove some of the top layer, partly revealing the underlying colour, and should be continued until the desired surface effect is obtained. A stiffer bristle-brush may be required. After any residual dust has been removed with a soft brush, the object is wax finished.

When the dark layer on a polished surface is lightly bristle-brushed, it tends to take the underlying layer with it. However, it also tends to leave an even fine stipple of dark brown on the metal surface, which is adherent. If the brushing is carefully carried out, a fine, even, stippled effect can be obtained. On rough or as-cast surfaces the dark layer is more tenacious, producing an even matt black, after light brushing.

▲ Potassium chlorate is a powerful oxidising agent, and must not be allowed to come into contact with combustible material. The solid or solutions should be prevented from coming into contact with the eyes, skin or clothing.

▲ The potassium chlorate and the potassium permanganate must not be mixed together in the solid state. They should each be separately dissolved in a portion of the water, and then mixed together.

2.80★   Very dark brown Gloss   *Pl. IV, Pl. XIII*

Ammonium carbonate	50 gm
Copper acetate	25 gm
Acetic acid (6% solution)	1 litre

Boiling immersion (Fifteen minutes)

The object is immersed in the boiling solution, and the surface gradually darkens to a reddish-brown colour. After fifteen minutes, when the colour has fully developed, the object is removed and washed in warm water. After careful drying in sawdust, it may be wax finished.

An even dark brown/black colour is produced on rough or as-cast surfaces.

Tests involving the addition of 5 gm of ammonium chloride and 2 gm of oxalic acid per litre, to the solution when the colours have developed, produced poor results. The colour initially becomes lighter but is then subject to streaking and etching effects, producing very patchy matt yellow surfaces. The variation cannot be recommended for polished surfaces. On rough or as-cast surfaces, an even light orange/sand colour is produced. Very similar results were obtained with small additions of zinc chloride or copper chloride.

▲ The ammonium carbonate should be added to the acetic acid in small quantities. The reaction is effervescent, and if large amounts are added, the effervescence is very vigorous.

2.81   Light reddish-brown Satin/semi-gloss   *Pl. IV*

Antimony trisulphide	30 gm
Ferric oxide	10 gm
Ammonium sulphide (16% solution) to form a paste	

Applied paste (sulphide paste)

The ingredients are mixed to a thick paste, using a pestle and mortar, by the very gradual addition of a small quantity of ammonium sulphide solution. The paste is applied to the whole surface of the article to be coloured, using a soft bristle-brush. A stiff nylon brush is then used to work the surface gradually in small areas. The nylon brush should be intermittently dipped in a mixture of the dry ingredients of the paste (in the same proportion as above), and the surface worked in this way until the paste becomes dry and falls from the surface. When the whole surface is dry and completely free of any residue or dust, it may be burnished with a brass scratch-brush and wax finished.

A variegated surface can be obtained by brushing the paste on to the whole surface and allowing it to dry for about three hours, until it is superficially dry and cracking, and then burnishing it away with a nylon brush or a brass scratch-brush. This may be repeated, as necessary, to darken the colour.

▲ Ammonium sulphide solution is very corrosive and precautions should be taken to protect the eyes and skin while the paste is being prepared. It also liberates harmful fumes of ammonia and of hydrogen sulphide, and the preparation should therefore be carried out in a fume cupboard or well ventilated area. The paste once prepared is relatively safe to use but should be kept away from the skin and eyes.

▲ A nose and mouth face mask fitted with a fine dust filter should be worn when the residues are brushed away. The dust produced in this case contains antimony compounds, which are toxic.

2.82*   Brown/dark brown (slightly grain-enhanced) Gloss

Copper carbonate	200 gm
Ammonia (.880 solution)	250 cm³
Water	750 cm³

Cold immersion (Fifteen to thirty minutes)

The ammonia is added to the water, followed by the carbonate, which will partially dissolve. Some excess carbonate should remain in suspension. The article is suspended in the cold solution. The colour develops slowly, and the surface will show slight signs of grain enhancement. When the desired depth of colour has been reached, the article is removed and washed in cold water. It should be dried carefully in fine sawdust, and wax finished when dry.

The colour will only develop on polished surfaces; rough or as-cast surfaces remain unaffected.

To obtain a black colour using this solution, see recipe 2.86.

▲ This procedure should only be carried out where there is adequate ventilation. Ammonia vapour will be freely liberated when the solution is prepared, and in colouring, which can severely irritate the eyes and respiratory system. The skin, and particularly the eyes, must be protected against accidental splashes from this corrosive alkali.

2.83*   Dark brown (slightly grain-enhanced) Gloss   *Pl. IV*

A	Ammonium carbonate	180 gm
	Copper sulphate	60 gm
	Acetic acid (6% solution)	1 litre
B	Oxalic acid	1 gm
	Ammonium chloride	0.25 gm
	Acetic acid (6% solution)	500 cm³

Hot immersion . (Fifteen minutes)

Solution A is boiled and allowed to gradually evaporate until its volume is reduced by half. Solution B is then added and the mixture heated to bring the temperature to 80°C. The article is immersed in the hot solution, the colour gradually darkening until it becomes brown after ten minutes. After fifteen minutes, when the colour has fully developed, the article is removed and rinsed carefully in warm water. The surface should not be disturbed at this stage. It is allowed to dry in air, and wax finished when completely dry.

A slightly darker and more grey-brown is produced on rough or as-cast surfaces. The slight bloom that may appear on polished faces during the latter stages of immersion is virtually invisible after wax finishing.

▲ The ammonium carbonate should be added to the acetic acid in small quantities. The reaction is effervescent, and if large amounts are added the effervescence is very vigorous.

▲ Preparation of the solution, and colouring, should be carried out in a well ventilated area. Acetic acid vapour will be liberated when the solutions are boiled.

▲ Oxalic acid is harmful if taken internally, and irritating to the eyes and skin.

2.84   Variegated dark reddish-brown/black (as-cast surfaces only) Semi-matt

A	Potassium polysulphide	4 gm
	Water	1 litre
B	Ferric nitrate	10 gm
	Water	1 litre

Torch technique (A and B)

The metal is heated with a blow torch and the polysulphide solution applied sparingly with a brass scratch-brush using small circular motions and working across the surface until a dark brown/black colour is established. The metal is then gently heated as the ferric nitrate solution is stippled onto the surface using a bristle-brush or a soft cloth, giving a reddish-brown tone. When treatment is complete, the object is allowed to cool and dry thoroughly, and may then be wax finished.

▲ Ferric nitrate is corrosive and should not be allowed to come into contact with the eyes or skin. A face shield or goggles should be worn to protect the eyes from hot splashes, as either of these solutions are applied to the heated metal.

▲ The fumes evolved as either of these solutions are applied to the heated metal must not be inhaled. Adequate ventilation should be provided.

2.85*   Dark brown/black Matt/semi-matt

Nickel sulphate	35 gm
Copper sulphate	25 gm
Potassium permanganate	5 gm
Water	1 litre

Hot immersion (Five to thirty minutes)

The object is immersed in the hot solution (80–90°C) and the surface immediately becomes covered with a dark brown layer. After four or five minutes it is removed and gently bristle-brushed with hot water. The dark layer is removed revealing a brown 'bronzing' beneath. If darker colours are required then the process should be repeated. If, on the other hand a full very dark brown/black colour with a more matt finish is required, the object should be re-immersed, and the dark brown layer will re-form and should be left on the surface as immersion is continued. After thirty minutes the object is removed, and gently bristle-brushed under hot water. After thorough rinsing, and drying in sawdust, it is wax finished.

Similar colours are produced using either technique. If the repeated bristle-brush method is used then a more glossy surface is produced. On rough or as-cast rather than polished surfaces, an even dark brown/black can be obtained with either method.

▲ Nickel compounds are a common cause of sensitive skin reactions. Precautions should also be taken to avoid inhalation of the dust when preparing the solution.

2.86*   Blue-black/black   Semi-gloss

Copper carbonate	125 gm
Ammonia (.880 solution)	250 cm³
Water	750 cm³

Warm immersion   (One hour)

The ammonia is added to the water, followed by the carbonate, which will partially dissolve. Some excess carbonate should remain in suspension. The solution is heated to about 50°C, and the object is immersed. The colour develops slowly during the period of immersion. After about one hour, when the colour has fully developed, the object is removed and washed thoroughly in warm water. It should be carefully dried in sawdust, and wax finished when dry.

An even black colour is also produced on rough and as-cast surfaces.

▲ This procedure should only be carried out where there is adequate ventilation. Ammonia vapour will be freely liberated when the solution is prepared, and during colouring, which can severely irritate the eyes and respiratory system. The skin, and particularly the eyes, must be protected against accidental splashes of this corrosive alkali.

2.87*   Black   Semi-gloss

Sodium thiosulphate	6.25 gm
Ferric nitrate	50 gm
Water	1 litre

Hot immersion   (One minute)

A succession of lustre colours is rapidly produced when the article is immersed in the hot solution (50–60°C), changing to a more opaque purplish colour after about forty-five seconds. The colour darkens quickly and the article is removed after about one minute. After thorough washing it is allowed to dry in air, and then wax finished.

During preparation, the nitrate solution is darkened by the addition of the thiosulphate, but slowly clears to give a straw coloured solution.

An even black colour is also produced on rough or as-cast surfaces.

Higher temperatures should not be used. Tests at 70–80°C produced a stripping of the surface colour and strong etching effects.

2.88   Black   Semi-gloss

Sodium thiosulphate	50 gm
Ferric nitrate	12.5 gm
Water	1 litre

Hot immersion   (Twenty minutes)

The article is immersed in the hot solution (60–70°C) and after about one minute the surface is coloured with a purple-blue lustre, which gradually recedes. After five minutes a brown colour slowly appears which changes to a slate grey with continued immersion. After about twenty minutes the article is removed, washed in hot water and allowed to dry in air. The surface may be fragile at this stage and should be handled as little as possible. When completely dry the article is wax finished.

If the temperature is too high the colour layer will be patchy and more fragile when removed from solution, revealing a tan colour beneath the black.

2.89   Mottled dark slate-grey   Semi-gloss

Selenous acid	6 cm³
Water	1 litre
Sodium hydroxide solution (250 gm/litre)	50 drops

Warm immersion   (About five minutes)

The acid is dissolved in the water, and the sodium hydroxide added by the drop, while stirring. The solution is then heated to 25–30°C. The article to be coloured is immersed in the solution, producing a purplish-brown lustre after about fifteen seconds, which darkens unevenly and then fades. Continued immersion produces further darkening after about two minutes, and the surface gradually acquires a grey colour, mottled with bluish lustrous patches. After about five minutes the article is removed, washed in cold water and allowed to dry in air. When dry it is wax finished.

Prolonging the immersion beyond about five minutes tends to cause a loss of surface quality, and of the blue lustre. An even but thin black colour is produced on rough or as-cast surfaces.

▲ Selenous acid is poisonous and very harmful if swallowed or inhaled. Colouring must be carried out in a well ventilated area. Contact with the skin or eyes must be avoided to prevent severe irritation.

▲ Sodium hydroxide is a powerful caustic alkali. Contact with the eyes and skin must be prevented.

2.90*   Black   Matt   *Pl. III*

Potassium permanganate	5 gm
Copper sulphate	50 gm
Ferric sulphate	5 gm
Water	1 litre

Boiling immersion   (Twenty minutes)

The article is immersed in the boiling solution. A black layer forms on the surface, which should not be disturbed. Immersion is continued for about twenty minutes, and the article is then removed and allowed to dry in air, without washing. The surface should not be handled at this stage. After several hours drying, the article may be wax finished.

An even matt black colour is also produced on a rough or as-cast surface.

2.91   Black Matt

Sodium thiosulphate	60 gm
Copper sulphate	42 gm
Potassium hydrogen tartrate	22 gm
Water	1 litre

Hot immersion (Fifteen minutes)

After one minute of immersion in the solution (at 50–60°C), a yellow-orange colour forms on the surface of the article and gradually turns to bright green. The article is removed from the solution after fifteen minutes and allowed to dry in air, without washing in water. When the article is completely dry it is gently rubbed with a soft cloth, which removes the green layer, revealing the underlying black colour. It is then wax finished.

The black colour is produced both on rough or as-cast and polished surfaces.

The green layer is superficial, and easily removed, but may leave a slight green tinge to the underlying colour.

2.92   Dark slate-grey Semi-matt/semi-gloss

A	Copper nitrate	100 gm
	Water	1 litre
B	Potassium sulphide	5 gm
	Water	1 litre
	Hydrochloric acid (35%)	5 cm³

Boiling immersion (A) (Fifteen seconds)
Cold immersion (B) (Thirty minutes)

The object is immersed in the boiling solution A for fifteen seconds and then transferred to solution B. The surface darkens initially to produce a series of uneven petrol blue/purple interference colours, and then gradually changes to a slate grey. After thirty minutes the object is removed and rinsed, relieved by gentle scratch-brushing or pumice if desired, and finally washed and dried in sawdust. When completely dry it is wax finished.

This recipe is suitable for producing antiquing and relieving effects only, on polished surfaces. On rough and as-cast surfaces it yields an even dark grey, which may also be relieved. The polished surfaces are prone to spotting.

▲ Solution B should only be prepared and used in a fume cupboard or a very well ventilated area. When the acid is added to the potassium sulphide solution, quantities of hydrogen sulphide gas are evolved, which is toxic and can be dangerous even in moderate concentrations.

▲ Hydrochloric acid is corrosive, and should be prevented from coming into contact with the eyes and skin.

2.93   Greenish-grey grain-enhanced surface Semi-gloss/semi-matt   *Pl. IV*

Copper sulphate	6.25 gm
Copper acetate	1.25 gm
Potassium nitrate	1.25 gm
Sodium chloride	2 gm
Sulphur (flowers of sulphur)	3.5 gm
Acetic acid (6% solution)	1 litre

Boiling immersion (About one hour)

The article is immersed in the boiling solution and after a few minutes the surface darkens and is slightly etched as the colour develops. The surface develops gradually as immersion continues, and after about one hour the article is removed and washed in hot water. It is dried in sawdust and may be wax finished when completely dry.

A greenish-grey colour is also produced on as-cast surfaces.

The colour appears to darken slightly on exposure to daylight.

▲ Potassium nitrate is a powerful oxidising agent and should not be allowed to come into contact with combustible materials. It must not be mixed with the sulphur in the solid state.

2.94   Dark grey-brown (as-cast surfaces only) Matt

Ammonium sulphate	30 gm
Sodium hydroxide	40 gm
Water	1 litre

Boiling immersion (Thirty minutes)

The object is immersed in the boiling solution and the surface gradually darkens to produce a black colour, and acquires a slightly crystalline appearance. After thirty minutes the object is removed, washed in warm water and allowed to dry in air. Relieving is then carried out, and the object wax finished.

A dark grey-brown/black layer is also produced on polished surfaces, but this tends to be non-adherent and can be partially removed by bristle-brushing when dry. Some reddish-brown colour tends to remain. The results obtained in tests suggest that the procedure can be used to produce 'antique' or 'relieved' effects.

▲ Sodium hydroxide is a powerful caustic alkali, the solid and strong solutions causing severe burns to the eyes and skin. When preparing the solution, small quantities of the sodium hydroxide should be added to large quantities of water and *not* vice-versa. The reaction is exothermic and will warm the solution. When the sodium hydroxide is completely dissolved, the ammonium sulphate is added and the solution may be heated.

2.95   Black on brown ground   Semi-gloss

Sodium thiosulphate	65 gm
Copper sulphate	12 gm
Copper acetate	10 gm
Arsenic trioxide	5 gm
Sodium chloride	5 gm
Water	1 litre

Hot immersion   (Thirty minutes)

Immersion in the hot solution (50–60°C) produces patchy lustre colours during the first fifteen minutes. Continued immersion produces a progressively more opaque and darker colour. After about thirty minutes, the article is removed and washed in warm water. It is allowed to dry in air, and may be wax finished when dry.

The surface tends to be very uneven. The procedure is difficult to control and cannot generally be recommended for polished surfaces.

An even dark slate/black colour is produced on rough and as-cast surfaces.

▲ Arsenic trioxide is very toxic and will give rise to very harmful effects if taken internally. The dust must not be inhaled. The vapour evolved from the solution must not be inhaled.

2.96   Black   Matt

A Copper sulphate	50 gm
Water	1 litre
B Sodium thiosulphate	50 gm
Lead acetate	12.5 gm
Water	1 litre

Hot immersion (A)   (One minute)
Hot immersion (B)   (Two minutes)

The object is immersed in the hot copper sulphate solution (80°C) for one minute, producing a darkening of the surface. It is then transferred to solution B (80°C) for two minutes immersion. These alternating immersions are repeated a number of times until an even black layer is produced on the surface. The object is then washed in warm water and allowed to dry in air. When dry, it is either rubbed gently with a soft cloth or bristle-brushed, and wax finished.

An even matt black is produced on rough or as-cast surfaces. Tests showed that on polished surfaces the results tend to be non-adherent. Careful drying and brushing can produce a variegated black and dull bare metal surface, but this recipe cannot be recommended for producing an even black on polished surfaces.

▲ Lead acetate is a highly toxic substance, and every precaution should be taken to prevent inhalation of the dust during preparation. The solution is very harmful if taken internally. Colouring should be carried out in a well ventilated area, as the vapour from the solution must not be inhaled.

2.97   Light grey on black (as-cast surfaces only)   Matt

Potassium sulphide	5 gm
Barium sulphide	10 gm
Ammonia (.880 solution)	20 cm³
Water	1 litre

Hot immersion and scratch-brushing   (One hour)

The article is immersed in the hot solution (80–90°C), and a grey/black layer forms on the surface after about three minutes. This should be brass-brushed to clear any non-adherent material. Immersion is continued and the surface brass-brushed every few minutes for the first ten minutes. After one hour the article is removed, washed in hot water and allowed to dry in air. When dry, it may be wax finished.

On rough or as-cast surfaces a black surface is produced, which is covered with an adherent 'dusting' of light grey. The black layer that develops on polished faces is not adherent, and when removed reveals a dulled brass surface.

▲ Barium sulphide is poisonous and may be very harmful if taken internally.

▲ Ammonia vapour will be liberated during preparation and colouring, which must be carried out in a well ventilated area. Some hydrogen sulphide gas may also be liberated. The ammonia solution must be prevented from coming into contact with the eyes and skin, as it will cause burns or severe irritation.

2.98   Black (as-cast surfaces only)   Matt

Sodium thiosulphate	250 gm
Water	1 litre
Copper sulphate	80 gm
Water	1 litre

Cold immersion   (Two hours)

The chemicals should be separately dissolved, and the two solutions then mixed together. The article is suspended in the cold solution, and after ten minutes immersion shows signs of darkening. After twenty minutes a reddish-purple lustre appears, which gradually changes to a blue-green lustre after thirty minutes. The sequence of lustre colours is repeated, becoming more opaque and producing a maroon colour after about one hour. With continued immersion the colour slowly darkens to black. After about two hours the article is removed, washed thoroughly and allowed to dry in air. When dry it may be wax finished.

Although a black layer is also produced on polished surfaces, this tends to be non-adherent if worked with a bristle-brush when dry. Removal of the black layer reveals a darkened 'antique' brass colour. The results obtained in tests suggest that the procedure can be used to obtain good 'antique' or relieved effects.

The lustre colours produced during immersion tend to be less clear than those obtained with some other thiosulphate solutions.

2.99   Thin dark grey (as-cast surfaces only)

A	Barium sulphide	2 gm
	Potassium sulphide	2 gm
	Ammonium sulphide (16% solution)	2 cm³
	Water	1 litre
B	Copper nitrate	20 gm
	Ammonium chloride	20 gm
	Calcium hypochlorite	20 gm
	Water	1 litre

Cold immersion (A)   (Thirty seconds)
Cold immersion (B)   (A few minutes)

The article is immersed in the mixed sulphide solution, which immediately blackens the surface, and removed after about thirty seconds. If the surface is uneven then it can be lightly brushed with a scratch-brush and re-immersed for a similar length of time. The article is washed and transferred to solution B, which changes the colour to a dark slate or grey-brown. After a few minutes the article is removed and allowed to dry in air. When dry it may be wax finished.

The dark colour is not adherent on a polished surface, but tends to leave a grey-brown underlying tone. The procedure may be used for relieved or 'antique' effects.

The secondary treatment in solution B tended to produce results that were only tenuously adherent, and cannot be generally recommended. Similar poor results were also obtained with an after treatment of a solution where the calcium hypochlorite is replaced by an equal quantity of calcium chloride.

▲ Ammonia vapour and hydrogen sulphide gas are liberated from the mixed sulphide solution. Colouring should be carried out in a fume cupboard or very well ventilated area. It should be prevented from coming into contact with the eyes and skin.

▲ Barium compounds are toxic, and harmful if taken internally.

▲ The fine dust from the calcium hypochlorite (bleaching powder) must not be inhaled.

2.100   Dark grey-brown (as-cast surfaces only) Matt

Copper nitrate	20 gm
Zinc sulphate	30 gm
Mercuric chloride	30 gm
Water	1 litre

Cold application (Several minutes)

The solution is wiped onto the surface of the object with a soft cloth, producing a slightly adherent grey-black layer. Continued rubbing with the cloth clears this layer and produces a surface that is evenly coated with mercury. This tends to be non-adherent when dry. The solution is then applied to the surface by dabbing with a soft cloth, and allowed to dry in air. The procedure may be repeated if necessary, omitting the initial wiping. When the colour has developed, the object is washed in cold water and allowed to dry in air. When dry it may be wax finished.

The colour noted could only be obtained on as-cast surfaces. On polished faces, a dull bare metal remains once the non-adherent coating of mercury has been removed.

▲ Mercuric chloride is very toxic and must be prevented from coming into contact with the skin and eyes. The dust must not be inhaled. If taken internally, very serious effects are produced.

▲ In view of the high risks involved, and the very limited colouring potential, this solution cannot be generally recommended.

2.101   Reddish-brown (as-cast surfaces only)

Ferric oxide	30 gm
Lead dioxide	15 gm
Water	to form a paste

Applied paste and heat (Repeat several times)

The ingredients are mixed to a paste with a little water. The paste is applied to the surface with a soft brush, while the object is gently heated on a hotplate or with a torch. When the paste is completely dry, it is brushed off with a stiff bristle-brush. The procedure is repeated a number of times, until the colour has developed. After the final brushing, the object may be wax finished.

A slight brownish tinge is produced on polished surfaces, and an orange-brown colour in features such as punched marks. The procedure is suitable for producing a contrast between as-cast and polished surfaces.

▲ Lead dioxide is toxic and very harmful if taken internally. The dust must not be inhaled. A nose and mouth face mask with a fine dust filter must be worn when brushing away the dry residue.

2.102   Grey lustre on greenish-brown Gloss

Potassium hydrogen tartrate	20 gm
Sodium tartrate	10 gm
Tartaric acid	10 gm
Copper acetate	10 gm
Stannic chloride (hydrated)	3 gm
Sodium metabisulphite	19 gm
Water	1 litre

Hot immersion (Several minutes)

The object is immersed in the hot solution (80°C), which produces brown or purplish-brown lustre colours after two or three minutes. These tend to be uneven, but become more even after about five minutes as the colour changes to grey. When the greyish colour has just developed, the object is removed, washed in hot water and allowed to dry in air. When dry, it may be wax finished.

A mid-brown/greenish-brown colour is produced on rough or as-cast surfaces.

The timing is difficult. If immersion is carried on too long, the colour is lost. If the object is removed too soon, then the surface tends to be uneven.

▲ Stannic chloride is corrosive, and should be prevented from coming into contact with the eyes and skin.

2.103   Cloudy black on dull brass Semi-gloss

Copper nitrate	300 gm
Water	100 cm³
Silver nitrate	2 gm
Water	10 cm³

Applied liquid (Several minutes)

The copper nitrate and the silver nitrate are separately dissolved in the two portions of water, and then mixed together. The solution is warmed to about 40°C. The article to be coloured is immersed for a few minutes in a cold solution of sulphuric acid (50% solution), and then washed. The warm colouring solution is applied to the article by wiping with a soft cloth, producing a superficial grey layer which is washed away with warm water, revealing the underlying surface effect. When the article is completely dry, it may be wax finished.

The underlying colour tends not to be adherent, but leaves a cloudy black colour on a brass ground. A variegated brown and black finish is produced on as-cast surfaces. The procedure might be used for producing 'antique' effects, but is not generally recommended.

▲ Silver nitrate should be prevented from coming into contact with the skin, and particularly the eyes which may be severely irritated.

▲ Sulphuric acid is very corrosive, causing severe burns, and must be prevented from coming into contact with the eyes and skin.

2.104*   Light grey on ochre Matt   *Pl. IV*

Bismuth nitrate	50 gm
Nitric acid (25% solution)	300 cm³
Water	700 cm³

Cold immersion (Twenty minutes)

The object to be coloured is briefly pickled in sulphuric acid (50% solution), and rinsed in cold water. It is then immersed in the bismuth nitrate solution, which initially darkens the surface and subsequently produces a grey colouration. After about twenty minutes the object is removed and rinsed in a bath of cold water and allowed to dry in air. When dry, it may be wax finished.

The grey layer is only partially adherent on polished surfaces, leaving an even thin grey colour on the ochre ground. A light grey or white colour is retained in features such as punched marks or textured areas.

A more developed grey is produced on rough or as-cast surfaces, which also tends to retain some near white colour in surface features.

▲ Nitric acid and sulphuric acid are corrosive and must be prevented from coming into contact with the eyes and skin.

2.105   Dull metallic grey (frosted) on black Semi-matt/matt

Nickel ammonium sulphate	50 gm
Sodium thiosulphate	50 gm
Water	1 litre

Hot immersion (Fifteen minutes)

A purplish-grey colour quickly forms on the surface of the object on immersion, becoming patchy and metallic grey after one minute. After two minutes the object should be removed from the solution and lightly scratch-brushed with water, to remove the non-adherent deposits that build up on the surface. The object is then re-immersed and the scratch-brushing repeated twice at intervals of five minutes. After fifteen minutes the object is again lightly scratch-brushed, briefly re-immersed and then washed in hot water. After thorough drying in sawdust, it may be wax finished.

An even slate grey is produced on rough or as-cast surfaces. Results obtained with polished surfaces were variable, and the black ground may not be revealed beneath the grey.

In preparing the solution, the two chemicals should be separately dissolved in two portions of the water. The thiosulphate solution is warmed, and the nickel ammonium sulphate solution added gradually. The light green solution is used at a temperature of 60–80°C.

▲ Nickel salts are a common cause of sensitive skin reactions.

2.106   Metallic grey Gloss

Sodium thiosulphate	300 gm
Antimony potassium tartrate	15 gm
Ammonium chloride	20 gm
Water	1 litre

Hot immersion (Twenty minutes)

Immersion in the hot solution (60°C) rapidly produces a series of lustre colours on the surface of the object. These are pale and pass from purple to brown and blue, becoming a very pale blue-grey after about two minutes. This is followed by a gradual darkening of the surface, producing dark greys and browns. After about twenty minutes the object is removed, washed in warm water and dried in sawdust. It may be wax finished when dry.

On rough or as-cast surfaces, a variegated dull brown and slate grey surface is produced. On polished surfaces, a glossy metallic grey finish is produced. This is generally even, but some areas of a slightly darker tone may be visible in some lights.

▲ Antimony potassium tartrate is poisonous and may be harmful if taken internally.

2.107   Greenish-grey (variegated) Semi-matt

Object heated and plunged into boiling water

The object is heated evenly to a full red/orange colour using a blow torch or preferably a kiln or muffle furnace. It is then immediately plunged into a bath of turbulently boiling water. When it has cooled to the temperature of the water, it is removed from the bath and washed in cold water. It may then be dried in sawdust and wax finished when dry.

The temperature of the object as it enters the water appears to be critical. The equipment should be arranged so that as little heat as possible is lost during the transfer from heat source to immersion in the boiling water.

Although good results are readily achieved with small scale items and thin materials, as the scale and wall thickness increases so it becomes more difficult to obtain an even colour.

2.108   Buff/grey (variegated)   Satin/semi-matt   *Pl. IV*
Heating in a kiln   (About one hour)

The kiln is preheated to a temperature of 600°C and the object placed inside in a position that will favour even heating. The kiln temperature is then allowed to rise slowly to 800°C. The size of the object will determine the length of the time needed for the colour to form, and the object should be inspected from time to time to check the progress. When the colour has developed, the object is removed and allowed to cool. When cool it may be wax finished.

The object must be perfectly clean and grease-free to obtain an unblemished surface.

The tests relating to this recipe were carried out using an electric kiln. The effects produced with a gas kiln are not known.

2.109*   Blue-green patina on olive ground   Semi-matt

Copper nitrate	110 gm
Water	110 cm³
Ammonia (.880 solution)	440 cm³
Acetic acid (6% solution)	440 cm³
Ammonium chloride	110 gm

Applied liquid   (Five days)

The solution is applied to the surface by dabbing with a soft cloth, and the article then left to dry in air. The metal quickly darkens and a whiteish blue-green powdery patina forms gradually on the surface. When dry, this is rubbed with a soft dry cloth to remove any loose material and to smooth the surface, prior to the next application of solution. This procedure is repeated over a period of five days, during which time the patina darkens and becomes integral with the surface. When treatment is complete and the surface thoroughly dry, the article may be wax finished.

On rough or as-cast rather than polished surfaces, this treatment results in an even blue-green patina on a very dark olive ground.

▲ Colouring must be carried out in a well ventilated area. Ammonia vapour will be liberated from the strong solution, which will cause irritation of the eyes and respiratory system. The solution must be prevented from coming into contact with the eyes or skin, as it can cause burns or severe irritation.

2.110*   Blue-green patina on variegated brown ground   Semi-matt   *Pl. IV*

Copper sulphate	20 gm
Copper acetate	20 gm
Ammonium chloride	10 gm
Acetic acid (6% solution)	to form a paste

Applied paste   (Several days)

The ingredients are ground to a creamy paste with a little acetic acid, using a pestle and mortar. The paste is applied to the object with a soft brush, giving quite a thick coating which is then allowed to dry for a day. The dry residue is then washed away under cold water, using a soft brush. A thin layer of paste is then wiped on to the object with a cloth, and the object left to dry for a day. The residue is again washed off. This procedure of applying thin layers and drying is repeated in the same way until a good variegated patina is produced. The object should be allowed to dry thoroughly when treatment is complete, and may then be wax finished.

2.111*   Blue-green patina on pale brown ground   Semi-matt/matt   *Pl. IV*

Ammonium chloride	35 gm
Copper acetate	20 gm
Water	1 litre

Cold application   (Several days)

The ingredients are ground with a little of the water using a pestle and mortar, and then added to the remaining water. The solution is applied to the object by dabbing and wiping, using a soft cloth. Application should be sparing, to leave an evenly moist surface. The object is then allowed to dry in air. This procedure is repeated once a day for several days, producing a gradual development of the ground colour and blue-green patina. When treatment is complete, the object should be left to dry for several days, during which time there is further patina development. When completely dry, and surface change has ceased, the object is wax finished.

There is greater development of a more intense blue-green patina on rough and as-cast surfaces.

It is essential to ensure that all patina development is complete, and the surface is completely dry, before wax finishing is carried out. The final drying period may have to be extended to a matter of weeks in damp or humid conditions.

2.112* Blue-green patina on brown/black ground Semi-matt   *Pl. III*

Copper nitrate	100 gm
Nitric acid (70% solution)	40 cm³
Water	1 litre

Torch technique

The object is heated with a blow torch and the solution applied sparingly with a bristle-brush or paint brush. The liquid quickly turns dark brown, and becomes yellowish where it is not being directly heated. Continued heating makes the surface blacken and areas of blue-green patina begin to form. The blue-green patina tends to be superficial at this stage and the surface should be further heated and stippled with solution until a good dark brown or black ground has been established. The object is then gently heated and dabbed with an almost dry brush or cloth, barely damp with the solution, until an evenly distributed blue-green patina is obtained. When treatment is complete, the object is allowed to cool and dry thoroughly, and may then be wax finished.

▲ Nitric acid is very corrosive and must not be allowed to come into contact with the eyes, skin or clothing. A face shield or goggles should be worn to protect the eyes from hot splashes of solution as it is applied to the heated metal.

▲ The fumes evolved as the solution is applied to the heated metal should not be inhaled. Adequate ventilation must be provided, and a nose and mouth face mask fitted with the correct filter should be worn.

2.113* Blue-green patina (blue-green patina on black ground)

| Copper nitrate | 200 gm |
| Water | 1 litre |

Torch technique

The metal surface is heated with a blow torch, and the solution applied with a soft brush until it is covered with an even blue-green patina. If the surface is then heated, without further application of the solution, it will become black. The solution may then be applied to this black ground, heating the surface as necessary, to form a blue-green patina on the ground colour. The surface quality obtained can be varied by the method of application—a stippled surface may be produced by stippling with a relatively dry brush, or a more mottled or marbled effect obtained using the same technique with a brush that is more 'loaded' with solution. When the required finish has been achieved, the object is allowed to dry out, and may then be wax finished.

These colours are also produced on as-cast surfaces.

▲ Colouring should be carried out in a well ventilated area, so that inhalation of the vapours is avoided.

2.114* Olive-green on black ground Matt

A	Copper nitrate	200 gm
	Water	1 litre
B	Potassium polysulphide	50 gm
	Water	1 litre

Torch technique (A)
Torch technique (B)

The article is heated with a blow torch and the copper nitrate solution applied with a soft brush until an even thin layer of blue-green patina covers the surface. This is again heated, without further application of the solution, until it turns black. Any loose residue should be brushed off with a soft dry brush. The copper nitrate is then again applied, heating as necessary, until the blue-green colour re-appears and is evenly distributed. While the surface is still hot, the polysulphide solution is applied until the required depth of colour is achieved. When treatment is complete, the article is allowed to dry for some time, and may then be wax finished.

These colours are also produced on as-cast surfaces.

The concentrations of both solutions can be varied within quite wide limits.

▲ Colouring should be carried out in a well ventilated area, so that inhalation of the vapours and the hydrogen sulphide gas that are evolved is avoided.

2.115* Blue-green patina on light brown ground Semi-matt/matt

Copper nitrate	200 gm
Sodium chloride	200 gm
Water	1 litre

Applied liquid (Twice a day for five days)

The solution is dabbed on to the object with a soft cloth and left to dry in air. This process is repeated twice a day for five days. The object is then allowed to dry in air, without treatment, for a further five days. The ground colour develops gradually throughout this period and after two or three days some blue-green patina forms on the surface. This continues to build up as long as the surface is moist. Finishing should only be carried out when it is certain that the surface is dry and patina development complete.

The patina is more dense and the ground colour darker on rough or as-cast surfaces.

A dabbing technique was found to be better than wiping which tends to inhibit the formation of the patina.

If there is an initial resistance to 'wetting' the liquid can be applied by scratch-brushing in the early stages.

In damp weather conditions the surface may take several weeks to dry completely. Although apparently dry, the surface may 'sweat' intermittently. The drying process should not be rushed and wax finishing only carried out when it is certain that chemical action has ceased.

2.116 Blue-green patina Matt

Ammonium carbonate	24 gm
Potassium hydrogen tartrate	6 gm
Sodium chloride	6 gm
Copper sulphate	6 gm
Acetic acid (6% solution)	to form a paste

Applied paste (Several days)

The solid ingredients are made to a paste with a little of the acetic acid, using a pestle and mortar. The paste is applied with a soft brush. The object is then left to dry for about two days, and the application is then repeated. It is then left to dry thoroughly in air, generally taking several days to dry. When completely dry, the surface is brushed gently with a bristle-brush to remove any loose material, rubbed with a soft cloth to smooth the surface, and wax finished.

The colour produced on as-cast or rough rather than polished surfaces, tends to be darker and slightly more green as the underlying ground colour is more evident.

If the paste is made up with stronger solutions of acetic acid (15–30%) then there is a greater tendency for the blue-green patina to be non-adherent, partially revealing orange-brown ground colours.

▲ A nose and mouth face mask fitted with a fine dust filter should be worn when brushing away the dry residue.

2.117* Blue-green patina on black/dark brown ground *Pl. III*
Semi-matt/matt

Copper nitrate	80 gm
Water	1 litre
Ammonia (.880 solution)	3 cm³

Applied liquid (Four to five days)

The solution is applied to the object by spraying with a moderately fine atomising spray to produce a misty coating. It is essential to apply only a fine misty coating and to avoid any pooling of the solution or runs on the surface. The object is then left to dry. This procedure is repeated twice a day for four or five days, after which the surface should be allowed to dry out thoroughly for several days before wax finishing.

Prolonging the treatment beyond five days produces little further development. In tests extending to twenty days no notable change could be detected after five days.

▲ Inhalation of the fine spray is harmful. It is essential to wear a suitable nose and mouth face mask fitted with the correct filter, and to ensure adequate ventilation.

2.118 Dark green on mid-brown ground Matt

Ammonium sulphate	105 gm
Copper sulphate	3.5 gm
Ammonia (.880 solution)	2 cm³
Water	1 litre

Torch technique

The metal surface is heated with a blow torch, and the solution applied with a soft brush, until a dark green colour on a mid-brown ground is obtained. If the metal is too hot when the solution is applied then impermanent light grey or pinkish-red colours are produced. Continued application of the solution as the metal cools will remove these colours and encourage the brown ground and the dark green patina. If the green patina is heated once it has developed, by playing the torch across the surface, then it will tend to become black. When the desired surface finish has been achieved, the object is left to cool and dry for some time, and may then be wax finished.

A wide variety of surface effects are obtained, which depend on the temperature of the metal, and the manner in which the solution is applied (eg. liberally or sparingly; by 'stippling' or by 'painting' etc).

▲ Colouring should be carried out in a well ventilated area, to avoid inhalation of the vapours.

2.119 Green patina on brown ground Semi-matt

Copper nitrate	100 gm
Hydrochloric acid (35% solution)	10 cm³
Water	1 litre

Torch technique

The object is heated with a blow torch and the solution sparingly applied with a bristle-brush or paint brush. A heavy deposit of yellowish-green patina quickly forms, but this is rather non-adherent at this stage. Continued heating and the application of small amounts of solution by brushing gradually darkens the surface. When the brown ground colour has been established, the green patina is built up either by stippling with a nearly dry brush, or by using a cloth which is barely damp with the solution, heat being frequently applied to the area being worked. When a good patina has been obtained, the surface can be burnished with a dry cloth. Wax finishing is carried out when the object is cool and thoroughly dry.

In some cases the patina may have a tendency to 'sweat' shortly after treatment. If this occurs then the object should be left in a warm dry place for some time, and the wax only applied when the surface is thoroughly dry.

▲ Hydrochloric acid is corrosive and should not be allowed to come into contact with the eyes and skin. A face shield or goggles should be worn to protect the eyes from hot splashes as the solution is applied.

▲ The fumes evolved as the solution is applied to the heated metal should not be inhaled. Adequate ventilation must be provided and a nose and mouth face mask fitted with the correct filter should be worn.

2.120* Grey-green patina on ochre ground Matt

Sodium chlorate	100 gm
Copper nitrate	25 gm
Water	1 litre

Boiling immersion (Two hours)

The object is immersed in the boiling solution, and the surface is initially coloured yellow-orange. Continued immersion produces a gradual development of the ground colour during the first hour. A greenish-grey bloom slowly builds up on this ground, in the form of a light dusting at first and as an even coherent layer after about two hours. When the required finish has been obtained, the object is removed and washed in hot water. It is allowed to dry in air, and may be wax finished when dry.

An even grey-green patina is produced on both polished and rough or as-cast surfaces. The ground colour on as-cast surfaces is a more reddish-brown.

An interesting surface effect is produced after about one hour. A thin even dusting of the patina is produced, on a dull gold ground. At this stage, as-cast surfaces tend to be a light reddish-brown.

▲ Sodium chlorate is a powerful oxidant and must not be allowed to come into contact with combustible material or acids. It should also be prevented from coming into contact with the eyes, skin or clothing.

2.121 Light green patina on yellowish ground Matt

Copper nitrate	30 gm
Zinc chloride	30 gm
Water	to form a paste

Applied paste (Two or three hours)

The ingredients are mixed by grinding together using a pestle and mortar. Water should be added by the drop, to form a creamy paste. Excess water should be avoided, as the paste tends to thin rather suddenly. The paste is applied to the object with a soft brush, and allowed to dry for two or three hours. The residual paste is gently washed away with cold water, and the object then allowed to dry in air for several days. Exposure to daylight will cause a variegated darkening of the ground colours. When the surface is completely dry, and colour change has ceased, it may be wax finished.

The light green patina tends to be very variegated, on polished surfaces. A greenish patina on a light brown ground is produced on as-cast surfaces.

Repeated applications do not improve the results, and tend to require far longer final drying periods.

It is essential to ensure that the surface is completely dry before wax finishing. This may take some weeks, as the surface is prone to intermittent 'sweating'.

2.122 Pale green patina on light olive ground Matt

Copper nitrate	35 gm
Ammonium chloride	35 gm
Calcium chloride	35 gm
Water	1 litre

Cold immersion (Fifteen minutes)

Immersing the object in the cold solution produces an etching effect which brings out the grain structure of a polished surface. Continued immersion darkens the colour slightly to a pale olive. After fifteen minutes the object is removed and washed in hot water. It is then allowed to dry in air *without* washing, and some patina development occurs. When completely dry, the surface is brushed gently to remove any loose dust and wax finished.

This recipe produces a thin pale green patina on a light olive grain-enhanced ground on polished surfaces. On rough and as-cast surfaces both the ground colour and the patina develop fully.

▲ A nose and mouth face mask with a fine dust filter should be worn when the dry surface is brushed.

2.123 Yellowish-green patina on grey-brown ground Matt

Copper nitrate	80 gm
Sodium chloride	40 gm
Ammonium chloride	10 gm
Potassium hydrogen tartrate	10 gm
Water	1 litre

Boiling immersion (Twenty minutes)

Immersing the object in the boiling solution produces an immediate fine etching of the surface. A grey colouration with a greenish tinge develops gradually as immersion continues. After twenty minutes the object is removed and washed in hot water, then allowed to dry in air. Exposure to daylight will produce a slight darkening of the ground colour initially. Wax-finishing may be carried out when the object is completely dry, and all colour change has ceased.

On polished surfaces the result takes the form of a thin yellowish-green patina on a grey-brown grain-enhanced ground. On rough and as-cast surfaces the appearance is of a variegated yellowish-green and grey-brown mixture.

Tests carried out using stronger solutions of these ingredients, as suggested by some sources, produced little ground colour development and a rather dull patina.

2.124 Variegated green and greyish patina on grain-enhanced ground Semi-matt

Copper nitrate	20 gm
Sodium chloride	16 gm
Potassium hydrogen tartrate	12 gm
Ammonium chloride	4 gm
Water	to form a paste

Applied paste (Four hours)

The ingredients are ground to a paste with a pestle and mortar by the very gradual addition of cold water. The creamy paste is applied with a soft brush, and left to dry. After four hours the dry residue is removed with a bristle-brush. The object is washed in cold water and allowed to dry in air. Exposure to daylight darkens the initially brighter colours to a thin patina of mixed colours on a pale grain-enhanced ground. When all colour change has ceased, the object may be wax finished.

On rough and as-cast surfaces a dull brown colour is produced, with very slight patina development.

▲ A nose and mouth face mask fitted with a fine dust filter should be worn when brushing away the dry residue.

2.125 Watery pale blue on dark olive-brown ground Semi-matt

Ammonium carbonate	180 gm
Copper sulphate	60 gm
Copper acetate	20 gm
Oxalic acid	1.5 gm
Ammonium chloride	0.5 gm
Acetic acid (10% solution)	1 litre

Boiling immersion (Twenty-five minutes)

The object is immersed in the boiling solution, which produces some etching effects and light colouring after a few minutes. As the immersion continues streaky greenish films gradually develop, beginning to become dark after about ten minutes. The object is removed after twenty-five minutes and the hot solution applied with a soft brush, to leave the surface evenly moist. The object should then be allowed to dry thoroughly in air for several hours, without washing. When dry, it is wax finished.

On drying, patches of watery pale blue develop on the olive-brown ground. On rough and as-cast surfaces a slight bluish patina develops on the more olive-green ground.

▲ The ammonium carbonate should be added to the acetic acid in small quantities. The reaction is effervescent, and this may be very vigorous if large amounts are added.

▲ Oxalic acid is harmful if taken internally and irritating to the eyes and skin.

2.126 Yellow-green patina on terracotta ground Semi-matt

Copper sulphate	125 gm
Sodium chlorate	50 gm
Ferrous sulphate	25 gm
Water	1 litre

Boiling immersion (Thirty minutes)

The object is immersed in the boiling solution. Orange colours develop after a few minutes and gradually darken with continued immersion. After about fifteen minutes a greenish layer begins to be deposited on the now brown ground colour. Agitating the object within the solution tends to inhibit the deposition of the green layer. After thirty minutes the object is removed and washed thoroughly in hot water. After drying in sawdust, it is wax finished.

If the ground colour only is required, then the object should be intermittently bristle-brushed in hot water, during immersion. A residual greenish-grey bloom tends to remain.

▲ Sodium chlorate is a powerful oxidising agent and must not be allowed to come into contact with combustible material or acids. Inhalation of the dust must be avoided. The eyes and skin should be protected from contact with either the solid or solutions.

2.127★ Blue-green patina on red-brown/maroon ground Semi-matt

Copper nitrate	200 gm
Water	1 litre

Dipping in cold solution (Twice a day for five days)

The article is dipped in the cold solution for a few seconds, drained and allowed to dry in air. This procedure is repeated twice a day for about five days, during which time the surface darkens gradually to a red-brown. After about two days, patches of powdery blue-green patina also begin to appear on the surface. When treatment is complete, the article should be left to dry in air for a period of at least three days to allow the patina to develop fully and dry out. When dry, the article is wax finished.

On rough and as-cast rather than polished surfaces, the ground colour tends to be a very dark brown.

Some sources suggest a scratch-brushed application of the solution. Tests carried out using this technique produced similar but slightly lighter results over the same period of time.

Tests carried out in which the metal was briefly 'pickled' in a 10% solution of nitric acid prior to the initial dip, as suggested by some sources, produced very similar results with somewhat duller surfaces.

The patina tends to develop in streaks and patches corresponding to the draining of the solution from the surface. This can be minimised by 'brushing out' with a soft brush after dipping, to prevent runs and pooling.

If the ground colour alone is required, then after two days of dipping and drying, any patina should be brushed off scrupulously and application continued by wiping-on very sparingly. A completely patina-free ground is difficult to achieve.

2.128 Green patina on grey-brown ground Semi-matt   *Pl. III*

Copper nitrate	200 gm
Sodium chloride	200 gm
Water	1 litre

Boiling immersion (Thirty minutes)

The article is immersed in the boiling solution. An immediate etching of the surface takes place, followed by the gradual development of a mottled greyish-white colour which darkens appreciably during the later stages of immersion. After thirty minutes the article is removed and washed in hot water. A yellow-green colour will develop as the hot water is applied. The article is dried in air and wax finished when completely dry. The yellow-green colouration is less pronounced when the article is dry, particularly after waxing.

On polished surfaces the result takes the form of a yellowish-green tint on a grey-brown grain-enhanced ground. On rough and as-cast surfaces the ground is more olive in colour and patina development more pronounced. The colour obtained after drying and waxing, on these surfaces, is a whiter and less transparent green.

2.129*   Thin grey-green patina on mottled buff ground Semi-matt   *Pl. III*

Copper sulphate	20 gm
Zinc chloride	20 gm
Water	to form a paste

Applied paste (Twenty-four hours)

A greenish patina on a light brown ground is produced on as-cast surfaces. Patina development is encouraged on both polished and as-cast surfaces, by repeated applications.

The paste should be applied sparingly and should not be used too thick, or incrustations will form that tend to 'sweat' intermittently during the drying period.

The ingredients are ground to a thin creamy paste with a little water, using a pestle and mortar. The paste is applied sparingly to the surface of the object with a soft brush. It is then allowed to dry completely, and the residual dry paste washed off with cold water. The object is then allowed to dry in air for about twelve hours, and the procedure repeated if required. When treatment is complete, the object should be allowed to dry for several days at least. Exposure to daylight causes a variegated darkening of the ground colours. When the surface is completely dry and colour change has ceased, the object may be wax finished.

2.130   Greenish-yellow (variegated) Semi-matt

Copper nitrate	200 gm
Zinc chloride	200 gm
Water	1 litre

Applied liquid (Twice a day for five days)

An even bloom of greyish patina on an olive-brown ground was produced in tests on rough and as-cast surfaces. On polished faces the greenish-yellow was variegated with some darker areas and traces of pale blue-green patina.

The solution is wiped onto the article with a soft cloth and allowed to dry in air. When it is completely dry, the surface is gently rubbed smooth with a soft dry cloth to remove any loose dust, and a further application of solution wiped on. This procedure is repeated twice a day for about five days. The solution takes effect immediately, giving rise to an etching effect and pale cream and pink colours typical with chloride solutions. These colours darken on exposure to daylight. When treatment is complete, the article is left to dry thoroughly in air, and when completely dry, it is wax finished.

2.131*   Green and blue-green patina on brown ground Semi-matt

Ammonium carbonate	150 gm
Sodium chloride	50 gm
Copper acetate	60 gm
Potassium hydrogen tartrate	50 gm
Water	1 litre

Applied liquid (Twice a day for several days)

A dull glaze of green and blue-green patina on a brown ground is produced on polished surfaces, if excess moisture is brushed out before allowing the surface to dry. A wetter surface tends to produce a more opaque blue-green patina. A dense blue-green patina on a darker ground tends to be produced on rough or as-cast surfaces, where the moisture is retained by the surface texture.

The solution is applied generously to the surface of the article using a soft cloth. It is then allowed to dry in air. This procedure is repeated twice a day for several days, during which time the colour gradually develops. It is then left to dry in air, without further treatment, to allow the patina to develop. When patina development is complete and the surface is completely dry, the article is wax finished.

2.132*   Green/blue-green patina on orange-brown ground Semi-matt

Ammonium carbonate	150 gm
Copper acetate	60 gm
Sodium chloride	50 gm
Potassium hydrogen tartrate	50 gm
Acetic acid (10% solution)	1 litre

Applied liquid (Twice a day for several days)

Less sparing use of the solution in the latter stages of treatment gives a more dense blue-green patina. On rough or as-cast faces the ground colour is a darker reddish-brown and the patina better developed and more blue-green in colour.

For results obtained using a more concentrated form of this solution, see recipe 2.135.

It is essential to ensure that patina development is complete, and the surface completely dry, before wax finishing. In damp or humid conditions, the drying time may need to be extended.

▲ The ammonium carbonate should be added to the acetic acid in small quantities. The reaction is effervescent and if large quantities are added, then the effervescence is very vigorous.

The solution is dabbed onto the object with a soft cloth, and then allowed to dry thoroughly in air. This process is repeated twice a day for several days until the desired colour is achieved. Any powder or other loose material that forms during drying periods should be gently brushed away with a soft dry cloth, prior to the next application of solution. When treatment is complete the surface should be allowed to dry in air for several days. When completely dry, it may be wax finished.

2.133   Blue-green patina on variegated reddish-brown ground Semi-matt

Ammonium sulphate	105 gm
Copper sulphate	3.5 gm
Ammonia (.880 solution)	2 cm³
Water	1 litre

Applied liquid (Twice a day for several days)

A well-developed blue-green patina on a darker ground is produced on rough or as-cast surfaces. The red-brown ground obtained on polished faces tends to be thin, giving localised lighter variation. The blue-green patina appears on polished faces in the form of cloudy areas on the dark ground.

This recipe, which has been used commercially to induce patination on copper roofing, requires weathering in the open air to achieve the best results. In tests carried out in workshop conditions, only slight patination and patchy ground colours could be obtained.

The solution is applied to the surface of the object with a soft cloth. It is then allowed to dry in air. This procedure is repeated twice a day for several days until the ground colour has developed and traces of patina begin to appear. The solution should then be applied more sparingly by dabbing with a soft cloth. The object is allowed to dry slowly in a damp atmosphere to encourage the development of patina. It is finally allowed to dry in a warm dry atmosphere, and wax finished when completely dry.

2.134   **Blue-green patina on variegated ochre ground**   **Semi-matt**

Ammonium chloride	40 gm
Ammonium carbonate	120 gm
Sodium chloride	40 gm
Water	1 litre

Cold immersion, applied liquid   (One hour)

The object to be coloured is immersed in the cold solution for about one hour, producing a slight etching effect and a gradual development of the ground colour. It is then removed and allowed to dry thoroughly in air. Any excessive surface moisture should be gently brushed out with a soft brush, to prevent 'runs' 'or 'pooling' of the solution and to leave an evenly moistened surface. The patina develops slowly as the surface dries out. If further patina development is desired, then the solution should be applied sparingly with a soft cloth and the object again left to dry thoroughly. When patina development is complete and the surface dry, the object may be wax finished.

On polished surfaces the ground produced is an ochre colour, variegated with dull bare metal. Patina development tends to be slight. On rough or as-cast surfaces an even brown ground is obtained, and the blue-green patina develops more readily.

It is essential to ensure that patina development is complete, and the surface dry, before wax finishing. In damp or humid conditions the drying time may need to be extended.

2.135   **Slight blue-green patina (dull olive and etched ground)**   **Matt**

Ammonium carbonate	600 gm
Sodium chloride	200 gm
Copper acetate	200 gm
Potassium hydrogen tartrate	200 gm
Acetic acid (10% solution)	1 litre

Applied liquid   (Twice a day for several days)

The solution is applied to the article to be coloured, by wiping sparingly with a soft cloth. It should then be allowed to dry thoroughly in air. This procedure should be repeated twice a day for several days, until the ground colour has developed and the patina has formed. The article should then be left to dry in air for several days, during which time the patina will continue to develop. When patina development has ceased and the surface is completely dry, the article may be wax finished.

Polished surfaces are unevenly etched by the solution, producing some even dull olive areas and some tinted etched areas. The overall appearance tends to be patchy. Patina development on polished faces is slight. A well developed blue-green patina on a darker and more even ground is produced on rough or as-cast surfaces.

For results obtained with a more dilute form of this solution, see recipe 2.132

It is essential to ensure that the surface is dry and patina development complete before wax finishing. In damp or humid conditions the drying time may need to be extended.

▲ The ammonium carbonate should be added to the acetic acid in small quantities. The reaction is effervescent and if large amounts are added the effervescence is very vigorous.

2.136   **Pale blue (*or* green) patina on light brown/ochre ground**   **Semi-matt**

Ammonium carbonate	180 gm
Copper sulphate	60 gm
Copper acetate	20 gm
Oxalic acid	1.5 gm
Ammonium chloride	0.5 gm
Acetic acid (10% solution)	1 litre

Torch technique

The ingredients are dissolved and the solution boiled for about ten minutes. It is then left to stand for several days. The mixture should be shaken immediately prior to use. The surface of the object to be coloured is heated with a blow torch, and the solution applied with a soft brush. This process is continued until a pale blue-patina is produced, on a light brown or ochre ground. If the torch is then gently played across the surface, the pale blue will change to green. Varying effects can be obtained by 'stippling' or 'dragging' the brush on the surface. When the required finish is obtained, the object is left to cool and dry for some time and may then be wax finished.

Continuous application of liberal quantities of the solution will tend to inhibit patina formation, but encourages the development of the ground colour.

▲ The ammonium carbonate should be added to the acetic acid in small quantities. The reaction is effervescent, and if large amounts are added, this may be very vigorous.

▲ Colouring should be carried out in a well ventilated area, so that inhalation of the vapours is avoided.

▲ Oxalic acid is poisonous, and may be harmful if taken internally.

2.137*   **Dull olive green**   **Matt/semi-matt**   *Pl. III*

Copper chloride	50 gm
Water	1 litre

Hot immersion   (Twenty minutes)

The object is immersed in the hot solution (80°C), which produces etching effects and pale grey colouration. With continued immersion, a denser colour is produced. After twenty minutes the object is removed and washed in hot water, which tends to remove a superficial grey layer, and leach some yellowish-green colour from the surface. It is then allowed to dry in air for several hours, and wax finished when dry.

A greenish patina on an olive-brown ground is produced on rough or as-cast surfaces.

2.138   Blue-green patina Semi-matt/matt

Ammonium carbonate	150 gm
Ammonium chloride	100 gm
Water	1 litre

Applied liquid   (Several days)

The solution is applied to the object, by wiping it on to the surface with a soft cloth. The solution should be used sparingly, leaving the surface evenly moist. The object is then allowed to dry in air. This procedure is repeated once a day for several days, until the blue-green patina is sufficiently developed. When treatment is complete, the object should be allowed to dry out thoroughly for several days, and is wax finished when completely dry.

On a highly polished surface, the patina develops slowly, tending to leave areas of dulled bare metal, producing a variegated surface. On rough or as-cast surfaces a well developed patina is produced.

It is essential to ensure that the surface is completely dry and patina development complete before wax finishing. The drying time may need to be extended to a matter of weeks in damp or humid atmospheric conditions.

2.139   Blue-green patina on brown ground Semi-matt/matt

Copper acetate	30 gm
Copper carbonate	15 gm
Ammonium chloride	30 gm
Hydrochloric acid (18% solution)	2 cm³
Water	to form a paste

Applied paste   (Two or three days)

The ingredients are ground to a paste with a little water and the hydrochloric acid, using a pestle and mortar. The paste is applied to the object with a soft brush and then left for several hours to dry out. The dry residue is removed with a stiff bristle-brush. The procedure is repeated until a brown ground has developed with a thin bloom of patina. The paste is then thinned out to a liquid and applied using the same procedure, but with a soft brush to remove residue. When the surface is sufficiently developed and well dried out, the object may be wax finished.

A blue-green patina on a greenish-brown ground is produced on rough or as-cast surfaces.

A thick incrustation tends to form if the paste is made too thick. This is difficult to remove by brushing, when dry, but tends to flake eventually.

▲ A nose and mouth face mask should be worn when brushing away the dry residue.

2.140   Blue-green patina on grain-enhanced metal Semi-matt/matt

Copper nitrate	20 gm
Ammonium chloride	20 gm
Calcium hypochlorite	20 gm
Water	1 litre

Cold immersion   (Twenty hours)

The article is suspended in the cold solution. After an initial 'bright dip' effect which enhances the metal surface to a reddish-gold colour, there is a gradual build-up of colour and an etching effect that is typical with chloride solutions. The article is left in the solution for twenty hours and then removed, rinsed in cold water and allowed to dry in air. Drying in air produces a green powdery layer which is only partially adherent on polished surfaces. When the green layer has developed and the surface is completely dry, the object is wax finished.

An even blue-green patina on a dark ground is produced on rough or as-cast surfaces. Patina development on polished faces is partial, resulting in patches on an etched (grain-enhanced) ground.

It is essential to ensure that patina development is complete and the surface dry, before wax finishing. The drying time may need to be extended in damp or humid conditions.

▲ Calcium hypochlorite (bleaching powder) can cause irritation of the eyes, skin and respiratory system. It is essential to avoid inhaling the fine dust. Both the powder and solutions should be prevented from coming into contact with the eyes and skin.

2.141   Blue-green patina on dark brown ground (as-cast surfaces only) Matt

A	Ammonium sulphate	80 gm
	Water	1 litre
B	Copper sulphate	100 gm
	Sodium hydroxide	10 gm
	Water	1 litre

Applied liquid (A) (Five to ten days)
Applied liquid (B) (Two or three days)

The ammonium sulphate solution is applied to the surface of the object by dabbing vigorously with a soft cloth, leaving it sparingly moist. It is then allowed to dry completely. This procedure is repeated twice a day for up to ten days. Solution B is then applied in the same way for two or three days. When treatment is complete, the object is allowed to dry in air for several days. When completely dry, it may be wax finished.

A blue-green patina on a dark brown ground is produced on rough or as-cast surfaces.

In tests on polished surfaces, the solution tended to retract into pools and failed to 'wet' the surface. The addition of wetting agents produced no improvement, and the surface developed in patches rather than evenly.

2.142 Slight patina on grain-enhanced ground Semi-matt

Ammonium carbonate	300 gm
Acetic acid (10% solution)	35 cm³
Water	1 litre

Warm scratch-brushing (Two or three minutes)

The ammonium carbonate is added to the water, while stirring, and the acetic acid then added very gradually. The solution is then warmed to 30–40°C. The solution is applied to the object liberally with a soft brush, and then immediately scratch-brushed. The surface is worked for a short time, while applying more solution, and then allowed to dry in air. Any excess moisture on the object should be brushed out with a soft bristle-brush, to prevent 'pooling' or 'running' and to leave an evenly moist surface. When the surface is completely dry, which may take several days, it is wax finished.

On polished surfaces there is little development of either the ground colour or the patina. On rough or as-cast surfaces a yellowish/light brown variegated ground is produced, with patches of more developed blue-green patina.

The results obtained in tests with this solution were poor. Although the colours obtained are interesting, the surfaces were very prone to streaking and patchiness, and little control over this could be gained. The technique cannot therefore be generally recommended.

2.143 Pale yellow-green grain-enhancement Semi-matt/matt

Copper acetate	120 gm
Ammonium chloride	60 gm
Water	1 litre

Boiling immersion, scratch-brushing (Ten minutes)

The object is immersed in the boiling solution, rapidly producing an etching effect typical of chloride solutions. Immersion is continued for ten minutes, and the object is then removed, scratch-brushed with the hot solution and washed in hot water. It is then allowed to dry in air, and wax finished when dry.

The pale yellow-green colour develops when the surface is dried. On rough and as-cast, surfaces a grey-green and yellow-brown mixed colour is produced.

Tests carried out produced only patchy results, and the technique cannot therefore be recommended.

2.144 Blue-green patina on dark brown (as-cast surfaces only) Semi-matt

Ammonium chloride	10 gm
Ammonium acetate	10 gm
Water	1 litre

Applied liquid (Twice a day for ten days)

The solution is applied to the surface by wiping with a soft cloth or by brushing with a soft brush. It is then allowed to dry in air. This procedure is repeated twice a day for about ten days. The article is then allowed to dry in air, without treatment, for a further five days. When completely dry, and when patina development has ceased, it may be wax finished.

A blue-green patina on a dark brown ground is produced on rough or as-cast surfaces.

Colour development is generally slow, the ground developing unevenly at first, and the patina appearing later during the periods of air-drying. Patina development is more pronounced in surface features such as punched marks or textured areas.

2.145 Dark green Semi-matt   *Pl. IV*

Ammonium chloride	5 gm
Copper acetate	5 gm
Water	1 litre

Torch technique

The solid ingredients are mixed with the water to form a solution, which is left to stand for several hours. The object is heated with a blow torch and the solution dabbed on to the surface using a cloth or bristle-brush. This gradually darkens the surface. When treatment is complete, the object should be allowed to cool and dry thoroughly before it is wax finished.

▲ The fumes evolved as the solution is applied to the heated metal should not be inhaled. Adequate ventilation must be provided.

▲ A face mask or goggles should be worn to protect the eyes from hot splashes as the solution is applied.

2.146★ Brown lustre with pale blue sheen Gloss   *Pl. III*

Sodium thiosulphate	20 gm
Antimony potassium tartrate	10 gm
Water	1 litre

Hot immersion (Forty-five minutes)

The article is immersed in the hot solution (60°C) for ten minutes, producing a slight darkening of the surface and an orange tinge. The temperature of the solution is then gradually raised to 80°C and immersion continued. After forty-five minutes the article is removed, washed in warm water and allowed to dry in air. When dry, it is wax finished.

An even grey-brown colour is produced on rough or as-cast surfaces.

Timings for the immersion may be variable. If the desired colour is produced in a shorter time, the object should be removed at that stage. The immersion should not be continued longer than necessary.

▲ Antimony potassium tartrate is poisonous and may be harmful if taken internally.

▲ The vapour evolved from the hot solution should not be inhaled.

2.147★ Green lustre on brown ground Gloss   *Pl. III, Pl. XIII*

Potassium hydroxide	100 gm
Copper sulphate	30 gm
Sodium tartrate	30 gm
Water	1 litre

Warm immersion (Five to ten minutes)

The potassium hydroxide is added gradually to the water and stirred. Heat is evolved as it dissolves, raising the temperature to about 35°C. The other ingredients are added and allowed to dissolve. There is no need to heat the solution. The object is immersed in the warm solution, and immersion continued until the lustre colour develops. The object is then removed and washed in warm water. It is allowed to dry in air, after any excess moisture has been removed by dabbing with absorbent tissue. When dry, it is wax finished.

The lustre colour is only obtained on polished surfaces. On rough or as-cast surfaces, variegated pink and green colours are produced.

Prolonging the immersion or raising the temperature of the solution tends to produce a duller and more opaque colour.

▲ Potassium hydroxide is a powerful corrosive alkali and contact of the solid or solutions with the eyes or skin must be prevented.

2.148★ Vivid green lustre Gloss   *Pl. IV*

Copper sulphate	120 gm
Ammonia (.880 solution)	30 cm³
Water	1 litre

Boiling immersion (One minute)

The object is immersed in the boiling solution, and removed when the vivid green lustre has developed, after about one minute. It is washed thoroughly in hot water and dried in sawdust. When dry it is wax finished.

On rough or as-cast surfaces, variegated green and red-brown tints are produced.

For results obtained with prolonged immersion in this solution see recipe 2.2.

▲ Preparation of the solution and colouring should be carried out in a well ventilated area to prevent irritation of the eyes and respiratory system by the ammonia vapour. The skin, and particularly the eyes, should be protected from accidental splashes of this corrosive alkali.

2.149★ Greenish-yellow (slight petrol lustre) Satin/semi-gloss   *Pl. III*

Copper sulphate	25 gm
Ammonia (.880 solution)	3–5 cm³
Water	1 litre

Boiling immersion (Two to three minutes)

Immersion in the boiling solution rapidly produces lustre colours within a minute. These quickly change to a more even and opaque orange colour. The object is removed when this colour has developed evenly, after two or three minutes. It is thoroughly washed in hot water, dried in sawdust, and wax finished when dry.

A warm brown colour is produced on rough or as-cast surfaces.

Increasing the ammonia content beyond the stated amount tends to inhibit colour development. At a concentration of 10 cm³/litre of ammonia, with this concentration of copper sulphate, no colour is produced.

2.150 Lustre series Gloss

Sodium thiosulphate	125 gm
Lead acetate	35 gm
Water	1 litre

Hot immersion (Various)

The article is immersed in the hot solution (50–60°C), which produces a series of lustre colours in the following sequence: golden-yellow/orange; purple; blue; pale blue; pale grey. The timing varies, but it takes roughly five minutes for the purple stage to be reached, and about fifteen minutes for pale blue. (Tests failed to produce a further series of colours, the surface tending to remain pale grey after twenty minutes.) When the desired colour has been produced, the article is removed and immediately washed in cold water. It is left to dry in air, any excess moisture being removed with an absorbent tissue. When dry it may be wax finished or coated with a fine lacquer.

A brown colour, tinged with the lustre series colours, is produced on as-cast surfaces.

More dilute solutions produce the lustre colour sequence, but take a correspondingly longer time to produce.

▲ Lead acetate is very toxic and will give rise to serious conditions if taken internally.

▲ The vapour evolved from the hot solution must not be inhaled.

2.151* Golden lustre Gloss

Sodium hydroxide	25 gm
Potassium persulphate	10 gm
Water	1 litre

Hot immersion (One minute)

The potassium persulphate is dissolved in the water, and the sodium hydroxide then added in small quantities while stirring. The solution is then brought up to the correct temperature (70–80°C) and the article immersed. A golden lustre colour is produced within one minute. (Continued immersion will tend to produce some reddish lustre which tinges the gold unevenly. This eventually recedes, leaving a somewhat paler golden lustre.) When the required colour is obtained, the article is removed and washed in warm water. It is dried in sawdust, and may be wax finished when dry.

A brown colour is produced on as-cast surfaces. which darkens gradually as immersion proceeds.

▲ Potassium persulphate is a powerful oxidant and should not be allowed to come into contact with combustible material. The sodium hydroxide and the persulphate must not be allowed to come into contact in the solid state.

▲ Sodium hydroxide is a powerful caustic alkali which can cause severe burns. Both the solid and solutions should be prevented from coming into contact with the eyes and skin.

2.152* Brown lustre with pink tinge Gloss

Copper nitrate	115 gm
Tartaric acid (crystals)	50 gm
Sodium hydroxide	65 gm
Ammonia (.880 solution)	15 cm³
Water	1 litre

Boiling immersion (Fifteen minutes)

Immersion in the boiling solution produces a series of lustre colours, which begins with a blue that develops after about one minute, and continues to an even green lustre after about two minutes. If the green colour is required then the object should be immediately removed and washed in hot water, and allowed to dry in air. It is wax finished when dry. Continuing the immersion will cause a gradual pink tinge to appear which subsequently darkens, to produce the final colour after about fifteen minutes. The object is removed and finished as noted above.

The brown lustre is only produced on polished surfaces. On rough or as-cast surfaces a golden matt is produced.

Although a number of colour changes occur, the two colours noted are the only ones in this lustre series which are both rich and even.

▲ Sodium hydroxide is a powerful corrosive alkali, and the solid or solutions must be prevented from coming into contact with the eyes and skin. It should be added to the solution in small quantities.

▲ Preparation and colouring should be carried out in a well ventilated area. Ammonia vapour will be liberated, which if inhaled causes irritation of the respiratory system. A combination of caustic vapours will be liberated from the solution as it boils.

▲ A dense precipitate is formed which may cause severe bumping. Glass vessels should not be used.

2.153 Variegated golden-brown lustre Gloss

Potassium permanganate	10 gm
Water	1 litre

Hot immersion (Three to five minutes)

The article is immersed in the hot solution (90°C). A golden lustre colour develops within one minute, gradually becoming more intense. When the lustre colour is fully developed, which may take from three to five minutes, the article is removed and washed in hot water which is gradually cooled during washing. The article is finally washed in cold water before being carefully dried either in sawdust, or by gently blotting the surface with absorbent tissue paper to remove excess moisture and allowing it to dry in air. When dry, it is wax finished.

The effect is only obtained on polished, or directionally grained satin surfaces. On rough, as-cast and matt surfaces a dull orange or orange-brown is produced.

Extending the immersion time produces a slight darkening initially to a more pink colour, which rapidly fades to a pale silvery grey. These latter colours are usually uneven. Colours produced at longer immersion times tend to be more fragile when wet, and should not be handled at that stage.

2.154* Pink/green lustre on orange ground Gloss *Pl. IV*

Copper sulphate	25 gm
Sodium hydroxide	5 gm
Ferric oxide	25 gm
Water	1 litre

Hot immersion. Applied liquid and bristle-brushing (A few minutes)

The copper sulphate is dissolved in the water and the sodium hydroxide added. The ferric oxide is added, and the mixture heated and boiled for about ten minutes. When it has cooled to about 70°C, the object is immersed in it for about ten seconds and becomes coated with a brown layer. The object is transferred to a hotplate and gently heated until the brown layer has dried. The dry residue is brushed off with a soft bristle-brush. The turbid solution is then applied to the metal with a brush, as the object continues to be heated on the hotplate. The residue is brushed off, when dry. The application is repeated two or three times, until a green/pink lustre is produced. After a final brushing with the bristle-brush, the object is wax finished.

The results tend to be variegated rather than even, and may appear patchy when viewed in some lights.

▲ Contact with the sodium hydroxide in the solid state must be avoided when preparing the solution. It is very corrosive to the skin and in particular to the eyes.

2.155* Lustrous bronze-brown Gloss

Potassium permanganate	10 gm
Sodium hydroxide	25 gm
Water	1 litre

Boiling immersion (Thirty minutes)

The object is immersed in the boiling solution, and after two or three minutes the surface shows signs of darkening. After ten minutes this develops to an uneven dark lustrous surface. Continued immersion causes some further darkening. After thirty minutes the object is removed, washed in warm water and allowed to dry in air. When dry, it may be wax finished.

▲ Sodium hydroxide is a powerful caustic alkali which can cause severe burns. Both the solid and solutions must be prevented from coming into contact with the eyes and skin. When preparing the solution, small quantities of sodium hydroxide should be added to large amounts of water and *not* vice-versa.

2.156 Brown lustre Gloss

Sodium thiosulphate	90 gm
Potassium hydrogen tartrate	30 gm
Zinc chloride	15 gm
Water	1 litre

Hot immersion (Thirty minutes)

The article is immersed in the hot solution (50–60°C) which darkens the surface and gradually produces a bloom, which should be removed by bristle-brushing. Continued immersion gives rise to a brown colour, which slowly darkens. After about thirty minutes, when the colour has developed, the article is removed and washed in warm water. It is either dried in sawdust or allowed to dry in air, and may be wax finished when dry.

An even dark grey/black colour is obtained on rough or as-cast surfaces.

2.157* Pale pink/brown lustre Gloss

Sodium thiosulphate	65 gm
Water	1 litre
Antimony trichloride	until cloudy

Hot immersion (Thirty minutes)

The sodium thiosulphate is dissolved in the water and the antimony trichloride added by the drop, to form a white precipitate. The solution begins to become orange and cloudy at the top, and then gradually throughout. The solution is heated to 50°C, and when this temperature is reached, the object is immersed. The surface initially passes through the lustre cycle—golden after three minutes, purple after four minutes, pale metallic blue after ten minutes. After about fifteen minutes grey/white streaks appear on the surface, which darkens to produce an appearance like a black powdery layer. The article is removed after thirty minutes and washed gently in warm water, removing the powdery black deposit to reveal the pale pink/brown lustre. After drying in sawdust or in air, the article is wax finished.

Some grey clouding or spotting tends to occur on the lustre produced on polished surfaces. On rough or as-cast surfaces an even slate-grey/black is produced.

▲ Antimony trichloride is poisonous if taken by mouth. It is also irritating to the skin, eyes and respiratory system, and contact must be avoided.

▲ The vapour evolved from the hot solution must not be inhaled.

2.158* Light bronzing (slight petrol lustre) Gloss

A	Copper sulphate	50 gm
	Water	1 litre
B	Potassium sulphide	12.5 gm
	Water	1 litre

Hot immersion (A) (Thirty seconds)
Cold immersion (B) (One minute)

The article is immersed in the hot copper sulphate solution (80°C) for about thirty seconds to produce a green/pink lustre colour. It is then transferred to the cold potassium sulphide solution and immersed for about one minute, producing a clouding of the surface and a smokey greenish-pink colour. This tends to be uneven and can be improved by applying the solutions alternately using a bristle-brush. When treatment is complete, the article is washed in warm water and allowed to dry in air. When dry, it is wax finished.

An even but thin dark brown colour is produced on rough or as-cast surfaces.

If the results produced in solution A are streaky, then the streaks will tend to be visible in some lights, after treatment with solution B. Application with the bristle-brush darkens and improves the results. Producing an adequate finish evenly is difficult, and the recipe cannot be generally recommended.

For results obtained using a scratch-brushed technique see recipe 2.20.

▲ Colouring should be carried out in a well ventilated area as some hydrogen sulphide gas will be liberated.

2.159   Lustre sequence Gloss

Antimony trisulphide	50 gm
Ammonium sulphide (16% solution)	100 cm³
Water	1 litre

Cold immersion (Two to ten minutes)

The article is immersed in the cold solution, producing a rapid sequence of lustre colours initially. After about two minutes the colour obtained is predominantly blue with some local red or golden surface variation. This changes gradually to a greenish lustre after four minutes, which slowly fades. The final colours will have developed after about six to eight minutes, and immersion beyond ten minutes produces no detectable change. When the desired colour has been reached, at any stage in the immersion, the article is removed, washed in cold water and allowed to dry in air. When dry, it is wax finished.

The colours produced at an early stage in the immersion are the most intense, but they are also subject to unpredictable variegated effects.

The best colours obtained in tests occurred after about two or three minutes' immersion. A dark brown lustre is produced, tinged with pale blue, with an even grey-brown on as-cast surfaces.

▲ The preparation and colouring should be carried out in a well ventilated area. Ammonium sulphide solution liberates a mixed vapour of ammonia and hydrogen sulphide, which must not be inhaled, and which will irritate the eyes. The solution must be prevented from coming into contact with the eyes and skin, as it will cause burns or severe irritation.

2.160   Dark straw lustre Gloss

Barium sulphide	10 gm
Water	1 litre

Hot immersion (Two or three minutes)

The object is immersed in the hot solution (50°C), and the surface quickly takes on a golden colour, which darkens to a straw-coloured or brown lustrous finish. The object should be quickly removed and rinsed thoroughly in warm water. It is then allowed to dry in air, and wax finished when completely dry.

If immersion is continued for too long, the colour will quickly become very pale and may acquire a blue tinge.

Cold solutions will produce similar results, given very extended immersion times. In tests, it was found that prolonged immersion in a cold solution tends to produce less even results.

▲ Barium sulphide is poisonous and may be very harmful if taken internally. Some hydrogen sulphide gas will be liberated as the solution is heated. Colouring should be carried out in a well ventilated area.

2.161   Lustre series Gloss

Sodium thiosulphate	240 gm
Copper acetate	25 gm
Water	1 litre
Citric acid (crystals)	30 gm

Cold immersion (Various)

The sodium thiosulphate and the copper acetate are dissolved in the water, and the citric acid added immediately prior to use. The article is immersed in the cold solution and a series of lustre colours are produced in the following sequence: golden-yellow/orange; brown; purple; pale grey. This sequence takes about thirty minutes to complete, the brown changing to purple after about fifteen minutes. It is followed by a second series of colours, in the same sequence, which take a further thirty minutes to complete. The latter colours in the second series tend to be slightly more opaque. A third series of colours follows, if immersion is continued, but these tend to be indistinct and murky. (If the temperature of the solution is raised then the sequences are produced more rapidly. However, it was found in tests that the results tended to be less even. Temperatures in excess of about 35–40°C tend to cause the solution to decompose, with a concomittant loss of effect). When the required colour has been produced, the article is removed and washed in cold water. It is allowed to dry in air, any excess moisture being removed by gentle blotting with an absorbent tissue. When dry it may be wax finished or coated with a fine lacquer.

Similar colours are produced on rough or as-cast surfaces, although these are modified by the colour of the underlying surface and tend to be more dull.

2.162   Lustre series Gloss

Sodium thiosulphate	280 gm
Lead acetate	25 gm
Water	1 litre
Potassium hydrogen tartrate	30 gm

Warm immersion (Various)

The sodium thiosulphate and the lead acetate are dissolved in the water, and the temperature is adjusted to 30–40°C. The potassium hydrogen tartrate is added immediately prior to use. The article to be coloured is immersed in the warm solution and a series of lustre colours are produced in the following sequence: golden-yellow/orange; purple; blue; pale blue/pale grey. The colour sequence takes about ten minutes to complete at 40°C. (Hotter solutions are faster, cooler solutions slower.) A second series of colours follows in the same sequence, but these tend to be less distinct. When the desired colour has been reached, the article is removed and immediately washed in cold water. It is allowed to dry in air, any excess moisture being removed by dabbing with an absorbent tissue. When dry it may be wax finished, or coated with a fine lacquer.

Similar colours are produced on rough or as-cast surfaces, but these are more affected by the colour of the underlying metal, producing duller brown tinted results.

Citric acid may be used instead of potassium hydrogen tartrate, and in the same quantity. Tests carried out with both these alternatives produced identical results.

▲ Lead acetate is very toxic and will give rise to serious conditions if taken internally.

▲ The vapour evolved from the hot solution must not be inhaled.

2.163   Green lustre (pink tinge) Gloss

Potassium sulphide	10 gm
Water	1 litre
Ammonia (.880 solution)	1 cm³

Hot immersion (Ten minutes)

The object is immersed in the hot solution (50°C) and the surface gradually takes on a lustrous quality, which develops to a green lustre in about ten minutes. The object is then removed, washed in cold water, and allowed to dry in air. When dry, it may be wax finished.

The green lustre tends to be unevenly tinged with pink.

A greenish-grey/black colour tends to be produced on rough or as-cast surfaces.

As an alternative to immersion, the solution can be applied liberally to the surface using a soft brush, and without 'working' the surface.

▲ Some hydrogen sulphide gas will be liberated from the solution. Colouring should be carried out in a well ventilated area.

2.164   Variegated light grey/brown lustre Gloss

Potassium polysulphide	10 gm
Sodium chloride	10 gm
Water	1 litre

Hot immersion and scratch-brushing (Several minutes)

The article is immersed in the hot solution (60–70°C) for one or two minutes, until a golden-brown lustre colour is produced. This tends to be uneven. The article is removed and scratch-brushed with the hot solution and then re-immersed for a few minutes until the colour has developed. It is then removed and washed in warm water, allowed to dry in air, and wax finished when dry.

A cloudy grey colour on a light greenish-brown ground is produced on as-cast surfaces.

▲ Colouring should be carried out in a well ventilated area, as hydrogen sulphide gas is evolved from the hot solution.

2.165   Pale brown/pale grey lustre Gloss

Potassium sulphide	2 gm
Ammonium sulphate	0.5 gm
Water	1 litre

Cold immersion (One or two minutes)

Immerse the article in the cold solution. A golden lustre colour develops fairly quickly, and in roughly one to two minutes will begin to change to a mustard colour. As soon as this occurs the object should be transferred quickly to a container of cold water and thoroughly rinsed. It should then be washed in clean cold water and dried in sawdust. When completely dry it is wax finished.

The technique is difficult to control. The colour produced tends to be variegated rather than even.

The lustre colour is only produced on highly polished surfaces. Matt or other unpolished surfaces yield only patchy dull brownish colours.

2.166   Dull pink lustre Gloss

Antimony trichloride	50 gm
Olive oil	to form a paste

Applied paste (Five minutes)

The antimony trichloride is ground to a creamy paste with a little olive oil using a pestle and mortar. The paste is brushed onto the surface of the object to be coloured. A purple lustre colour develops after about two minutes, changing to a pink lustre after about five minutes. A steel blue appears if the paste is left for longer. When the desired colour has developed, the residual paste is washed from the surface with pure distilled turpentine. The object is finally rubbed with a soft cloth, and wax finished.

The dull lustre only occurs on polished surfaces. Localised effects can be produced by selective application of brush strokes. On rough or as-cast surfaces a dark slate colour with bare metal highlights is produced.

If the paste is left to dry on the surface, it becomes difficult to remove without damaging the colour finish.

Tests carried out using aqueous solutions produced similar results, but these tended to be patchy.

▲ Antimony trichloride is highly toxic and every precaution must be taken to avoid contact with the eyes and skin. It is irritating to the respiratory system and eyes, and can cause skin irritation and dermatitis.

2.167   Mottled reddish-brown (slightly lustrous) Gloss

Ammonium persulphate	10 gm
Sodium hydroxide	30 gm
Water	1 litre

Boiling immersion (Thirty minutes)

The ammonium persulphate is dissolved in half the water, and the sodium hydroxide separately dissolved in the remaining water. The two solutions are then mixed together and the mixture heated to boiling. The article to be coloured is immersed in the boiling solution, which produces golden and then brown lustre colours. Continued immersion darkens these colours to an opaque but glossy finish. When the colour has developed, after about thirty minutes, the article is removed, washed in hot water and dried in sawdust. When dry, it may be wax finished.

A very dark brown or black colour is produced on as-cast surfaces.

▲ Ammonium persulphate (peroxodisulphate) is a powerful oxidant, and must be kept out of contact with all combustible materials. It should be prevented from coming into contact with the eyes, skin or clothing.

▲ Sodium hydroxide is a powerful caustic alkali and must be prevented from coming into contact with the eyes and skin.

2.168   Green and pink lustre Gloss

Copper sulphate	120 gm
Potassium chlorate	60 gm
Water	1 litre

Hot immersion (Thirty seconds)

The article is immersed in the hot solution (70–80°C) and quickly removed after thirty seconds immersion, when the green lustre has developed. It is immediately washed thoroughly in hot water. After drying in sawdust, it is wax finished. The colour appears to change on drying, and again on waxing, due to the change in the interference colour caused by the presence of the layer of water, and then the layer of wax on the surface. The final colour is a green lustre variegated with pink.

The lustre effect is only obtained on polished surfaces.

Beyond the green lustre phase, the surface rapidly becomes orange-buff coloured. See also recipe 2.28.

▲ Potassium chlorate is a powerful oxidising agent and must not be allowed to come into contact with combustible material. Contact with the eyes, skin or clothing must be avoided.

2.169   Pink/green lustre on mottled brown ground Gloss

Copper sulphate	25 gm
Nickel ammonium sulphate	25 gm
Potassium chlorate	25 gm
Water	1 litre

Boiling immersion (One to two minutes)

The object is immersed in the boiling solution and rapidly removed when the mottled surface and lustre colour has developed. It should be washed immediately in hot water. After drying in sawdust, it is wax finished.

This effect is only produced on polished surfaces.

▲ Nickel salts are a common cause of sensitive skin reactions.

▲ Potassium chlorate is a powerful oxidising agent and should not be allowed to come into contact with combustible material. The solid, and solutions, should be prevented from coming into contact with the eyes, skin or clothing.

2.170   Uneven purplish lustre Semi-matt

Butyric acid	20 cm³
Sodium chloride	17 gm
Sodium hydroxide	7 gm
Sodium sulphide	5 gm
Water	1 litre

Boiling immersion (One hour)

The solid ingredients are added to the water and allowed to dissolve, and the butyric acid then added, and the solution heated to boiling. The object is immersed in the boiling solution, which rapidly produces a series of patchy lustre colours. Continued immersion gradually produces a more opaque dark purplish-green colour, which tends to be uneven. After about one hour the object is removed and washed in hot water. It is allowed to dry in air, and may be wax finished when dry.

A fairly even black colour was obtained on an as-cast surface.

The results produced tend to be uneven or patchy, and the procedure cannot be generally recommended.

▲ Butyric acid should be prevented from coming into contact with the eyes and skin, as it can cause severe irritation. It also has a foul odour which tends to cling to the hair and clothing. Preparation and colouring should be carried out in a well ventilated area.

▲ Both sodium hydroxide and sodium sulphide are corrosive, and must be prevented from coming into contact with the eyes and skin.

2.171* Blue-green patina/orange-brown ground Semi-matt   *Pl. IV*

Copper nitrate	200 gm
Water	1 litre

Sawdust technique (Two days)

The article to be coloured is laid-up in sawdust which has been evenly moistened with the solution, and left for about two days. Colour development may be monitored using a sample, but no deleterious effects were produced, in tests, by unpacking and re-packing the object itself. When the colour has developed satisfactorily, the article is removed and washed in cold water. After several hours drying in air, it may be wax finished.

The mottled orange-brown ground develops first. The blue-green patina which also occurs in the form of a mottle, develops most markedly as the medium dries out. The ground colour can be enriched and the patina inhibited, by removing the object before the medium dries out, and repacking if necessary with freshly moistened sawdust.

On rough or as-cast surfaces stippled patina develops on a mottled dark reddish-brown ground.

2.172* Blue patina (stipple) on grey-brown ground Matt

Ammonium carbonate	120 gm
Ammonium chloride	40 gm
Sodium chloride	40 gm
Water	1 litre

Sawdust technique (Twenty to thirty hours)

The article to be coloured is laid-up in sawdust which has been evenly moistened with the solution, and is then left for a period of about twenty or thirty hours. The sawdust should be kept moist throughout the period of treatment. When treatment is complete, the article should be washed in cold water and allowed to dry in air. When it is completely dry, it may be wax finished.

The surface is selectively etched by the solution, producing a stippled texture which is coloured grey-brown. The stippled blue patina occurs evenly, in the form of a fine incrustation.

On rough or as-cast surfaces, patches of blue-green patina develop on a mixed brown and grey ground.

2.173* Pale blue patina on dark grey-brown ground Matt   *Pl. IV, Pl. XIV*

Ammonium chloride	100 gm
Ammonium carbonate	150 gm
Water	1 litre

Sawdust technique (Twenty to thirty hours)

The article to be coloured is laid-up in sawdust which has been moistened evenly with the solution. It may then be left for periods of up to twenty or thirty hours. When treatment is complete, the object is washed in cold water and allowed to dry in air. When it is completely dry, it may be wax finished.

The surface is selectively etched by the solution where the granular sawdust is in close contact with the metal, producing a textured finish which is coloured grey-brown. The pale blue patina occurs as a thin stippled layer on the least etched parts of the surface.

A stippled blue patina on a dark grey ground is produced on as-cast surfaces. The patina takes the form of small spots of incrustation.

2.174* Orange-brown with light and dark grey (stipple) Semi-matt

Copper nitrate	280 gm
Water	850 cm³
Silver nitrate	15 gm
Water	150 cm³

Sawdust technique (Twenty hours)

The copper nitrate and the silver nitrate are dissolved in separate portions of water, and the two solutions then mixed. The article to be coloured is laid-up in sawdust which has been evenly moistened with this mixed solution. It is left for a period of about twenty hours, ensuring that the sawdust remains moist throughout this time. When treatment is complete, the article is removed and washed in cold water. After allowing it to dry thoroughly in air for some time, it may be wax finished.

The surface is coloured orange-brown by the treatment, and is also selectively etched where the granular sawdust is in close contact with the metal, producing a stipple of light grey etched areas. The dark grey colour develops as a fringe around these etched spots.

A grey stipple on a light brown ground is produced on as-cast surfaces.

▲ Silver nitrate must be prevented from coming into contact with the skin, and more particularly the eyes, as it may cause burns or severe irritation.

2.175* Purplish-grey stipple (with blue patina) on brass Semi-matt   *Pl. III*

Copper nitrate	100 gm
Water	200 cm³
Ammonia (.880 solution)	400 cm³
Acetic acid (6% solution)	400 cm³
Ammonium chloride	100 gm

Sawdust technique (Twenty to thirty hours)

The copper nitrate is dissolved in the water and the ammonia gradually added. The acetic acid is then added, followed by the ammonium chloride. The article to be coloured is laid-up in sawdust which has been evenly moistened with this solution, and is then left for a period of twenty to thirty hours. When treatment is complete, the article is washed in cold water and allowed to dry in air. When completely dry, after several hours, it may be wax finished.

The surface is selectively etched, where the granular sawdust is in close contact with the metal, producing a stipple of purplish-grey etched spots. The blue patina also develops in spots in the region of the etched areas. The remainder of the surface remains brass-coloured, but takes on a slightly frosted finish. Etching is more pronounced and extensive if a wetter medium is used.

A blue patina on a brown and black ground is produced on as-cast surfaces.

▲ Preparation and colouring should be carried out in a well ventilated area. Ammonia vapour and some acetic acid vapour will be liberated, which cause irritation of the eyes and respiratory system.

▲ The strong ammonia solution must be prevented from coming into contact with the eyes and skin, as it will cause burns or severe irritation.

2.176   Metallic stipple on greyish etched ground Semi-matt

Ammonium chloride	10 gm
Ammonia (.880 solution)	20 cm³
Acetic acid (30% solution)	80 cm³
Water	1 litre

Sawdust technique   (One or two days)

The object to be coloured is laid-up in sawdust which has been evenly moistened with the solution, and left for one or two days. Colour development should be monitored using a sample, so that the object itself remains undisturbed. When the colour has developed satisfactorily, the object is removed, washed in cold water and allowed to dry in air for several hours. When dry, it may be wax finished.

The surface is selectively etched by the solution, producing a greyish etched ground, and leaving an evenly distributed stipple of unetched metal with a slightly lustrous finish.

On rough or as-cast surfaces, a light dusting of pale blue-green patina develops on a variegated brown ground.

▲ Preparation and colouring should be carried out in a well ventilated area. Ammonia vapour and some acetic acid vapour will be liberated, which can irritate the eyes and respiratory system. These solutions should be prevented from coming into contact with the eyes and skin, or severe irritation or burns may result.

2.177   Slight grey-green patina/greenish-yellow metal (stipple) Semi-matt

Ammonium chloride	10 gm
Copper sulphate	4 gm
Potassium binoxalate	20 gm
Water	1 litre

Sawdust technique   (Several days)

The object to be coloured is laid-up in sawdust which has been moistened with the solution, and left for several days. Colour development is very slow, and should be monitored using a sample so that the object remains undisturbed. When the colour has developed satisfactorily, the object is removed, washed in cold water and allowed to dry in air for several hours. It may be wax finished when dry.

Patches of a dull etched metal tend to develop if treatment is too long, or if the sawdust is unevenly moist. On rough or as-cast surfaces, a dull variegated brown colour is produced, with some slight grey-green patina development.

▲ Potassium binoxalate is harmful if taken internally. It can also cause irritation, and should be prevented from coming into contact with the eyes or skin.

2.178   Dark brown and black stipple on golden bronze Gloss

Potassium hydroxide	100 gm
Copper sulphate	30 gm
Sodium tartrate	30 gm
Water	1 litre

Sawdust technique   (Twenty hours)

The object is laid-up in sawdust which has been evenly moistened with the solution, and is then left undisturbed for a period of about twenty hours. The sawdust should remain moist throughout this period. When treatment is complete, the object is removed and washed thoroughly in cold water. After it has been allowed to dry in air, it may be wax finished.

Both the stipple and the ground colours tend to be lustrous.

A stippled dark brown colour on an ochre ground is produced on rough or as-cast surfaces. Some green colouration may also occur.

▲ Potassium hydroxide is a strong caustic alkali, and precautions should be taken to prevent it from coming into contact with the eyes and skin.

▲ When preparing the solution, the potassium hydroxide should be added to the water in small quantities, and not vice-versa. When it has dissolved completely, the other ingredients may be added.

2.179   Blue-green patina and dark grey stipple on greenish brown Semi-matt

Ammonium carbonate	150 gm
Copper acetate	60 gm
Sodium chloride	50 gm
Potassium hydrogen tartrate	50 gm
Water	1 litre

Sawdust technique   (Ten to twenty hours)

The object to be coloured is laid-up in sawdust which has been evenly moistened with the solution. It is left for a period of from ten to twenty hours, colour development being monitored with a sample, so that the object remains undisturbed. When treatment is complete, the object is removed, washed in cold water and allowed to dry in air for several hours. When dry, it may be wax finished.

The polished surface is bronzed by the solution, and selectively etched where the granular sawdust is in close contact with the metal, producing a stipple of etched areas. These are coloured dark grey and greenish-brown. The unetched metal surface tends to acquire a stipple of blue-green patina.

A mottled bluish-grey on a greenish-brown ground is produced on rough or as-cast surfaces.

2.180   Stippled blue-green patina on metal ground Semi-matt

Copper nitrate	200 gm
Sodium chloride	100 gm
Water	1 litre

Sawdust technique   (Twenty or thirty hours)

The article to be coloured is laid-up in sawdust which has been moistened evenly with the solution. It should then be left for a period of twenty or thirty hours. After twenty hours the object should be examined, and repacked for shorter periods of time, as necessary. The sawdust should be kept moist throughout the period of treatment. When treatment is complete, the article should be washed thoroughly in cold water and allowed to dry in air. When completely dry, after several hours, it may be wax finished.

The surface is selectively etched by the solution, where the granular sawdust is in close contact with the metal, causing an even stipple of fine pits in the surface. The blue-green patina develops locally in the region of the pits. The remainder of the surface is not etched, but acquires lustre colours initially, which recede to leave a dull metal surface if treatment is prolonged.

Patches of green/blue-green patina on a mid-brown ground are produced on as-cast surfaces.

2.181 Dark brown stipple on etched brass ground Semi-matt

Copper carbonate	300 gm
Ammonia (.880 solution)	200 cm³
Water	1 litre

Sawdust technique (Twenty to thirty hours)

The article to be coloured is laid-up in sawdust which has been evenly moistened with the above solution, and left for a period of up to twenty or thirty hours. The sawdust should remain moist throughout the period of treatment. When treatment is complete, the article is removed and washed in cold water. It is then allowed to dry in air for some time before wax finishing.

The surface is coloured dark brown by the treatment, and selectively etched where the granular sawdust is in close contact with the metal, producing a stipple of finely etched brass ground. As treatment is prolonged, the dark brown areas recede to leave an etched brass surface. Traces of green patina tend to appear around the dark brown areas.

A stipple of blue-green patina on a mottled dark ground is produced on as-cast surfaces.

▲ The strong ammonia solution will liberate ammonia vapour, which irritates the eyes and respiratory system. Preparation and colouring must therefore be carried out in a well ventilated area.

▲ Ammonia solution causes burns and severe irritation, and must be prevented from coming into contact with the eyes and skin.

2.182 Light golden bronzing (fine stipple) Gloss

'Rokusho'	30 gm
Copper sulphate	5 gm
Water	1 litre

Sawdust technique (Several days)

The 'Rokusho' and the copper sulphate are added to the water, and the mixture boiled for a short time and then allowed to cool to room temperature. The object to be coloured is laid-up in sawdust which has been moistened with the solution. It is then left for several days. Colour development is very slow, and should be monitored using a sample so that the object remains undisturbed. When the colour has satisfactorily developed, the object is removed, washed in cold water and allowed to dry in air. It may be wax finished when dry.

A very fine stipple is produced on polished surfaces, which develops evenly to give an overall yellow or light golden bronzing effect. On rough or as-cast surfaces an even mid-brown colour is obtained.

For description of 'Rokusho' and its method of preparation, see appendix 1.

2.183 Light brown stipple Semi-matt

Ammonium sulphate	85 gm
Copper nitrate	85 gm
Ammonia (.880 solution)	3 cm³
Water	1 litre

Sawdust technique (Thirty hours)

The article to be coloured is laid-up in sawdust which has been moistened evenly with the solution. It should then be left for a period of twenty or thirty hours. The sawdust should be kept moist throughout the period of treatment. After thirty hours the object should be examined, and repacked for shorter periods of time as necessary. When treatment is complete, the object should be washed thoroughly in cold water and allowed to dry in air. When dry it may be wax finished.

A light brown stipple occurs where the granular sawdust is in close contact with the metal surface, caused by slight etching. The remainder of the surface remains polished. If the medium is too moist, uneven patches of grain enhancement tend to occur.

A grey stipple on a light brown ground is produced on as-cast surfaces.

2.184 Yellowish-brown etched stipple Semi-matt

Ammonium chloride	350 gm
Copper acetate	200 gm
Water	1 litre

Sawdust technique (Ten to twenty hours)

The object to be coloured is laid-up in sawdust which has been moistened evenly with the solution, and left for a period of from ten to twenty hours. After ten hours it should be examined and repacked if further treatment is required. It should subsequently be examined every few hours until the the required surface finish is obtained. The sawdust should be kept moist during the full period of treatment. The object should then be removed, washed in cold water and allowed to dry in air. When completely dry, it may be wax finished.

The surface is selectively etched by this treatment, producing an even distribution of fine 'pits', which take on a brown colour. The remainder of the surface retains its brass colour, and is lustrous initially, becoming a dull yellow as treatment proceeds. Some greenish patina may develop, particularly in any surface features such as punched marks.

On rough or as-cast surfaces a greenish patina tends to develop unevenly on a variegated yellow-brown and brown ground.

2.185 Blue-green patina on buff ground Semi-matt

Ammonium chloride	15 gm
Potassium aluminium sulphate	7 gm
Copper nitrate	30 gm
Sodium chloride	30 gm
Acetic acid (6% solution)	1 litre

Sawdust technique (Twenty to thirty hours)

The object to be coloured is laid-up in sawdust which has been evenly moistened with the solution, and left for about twenty or thirty hours. The object is then removed and allowed to dry in air for several hours, without washing. When dry, the surface is brushed with a stiff bristle-brush to remove any particles of sawdust or loose material. The object is then left for a further period of several hours, after which, it may be wax finished.

In tests, the patina produced on polished surfaces consisted of a thin glaze of blue-green, stippled with a darker blue-green. On rough or as-cast surfaces, patina development is greater and slightly darker in appearance.

2.186   Slight stipple of grey-green/light buff on brass Semi-matt

Di-ammonium hydrogen orthophosphate	100 gm
Water	1 litre

Sawdust technique (Thirty hours)

The object to be coloured is laid-up in sawdust which has been evenly moistened with the solution. It is left for about thirty hours and then removed and washed in cold water. If further colour development is required the treatment should be repeated for additional shorter periods. When treatment is complete, the object is finally washed and allowed to dry in air, and wax finished when completely dry.

Colour development on polished surfaces is rather slight, and extended times may be required. An even but dull variegated grey/pale brown is produced on rough or as-cast surfaces.

Tests carried out with this solution, using various other techniques, including cold and hot direct application, and hot and cold immersion, produced little or no effect.

2.187   Light brown bronzing (slight darker stipple) Gloss

Butyric acid	20 cm³
Sodium chloride	17 gm
Sodium hydroxide	7 gm
Sodium sulphide	5 gm
Water	1 litre

Sawdust technique (Twenty hours)

The object to be coloured is laid-up in sawdust which has been evenly moistened with the solution, and left for a period of up to twenty hours. The timing of colour development tends to be unpredictable, and it is essential to include a sample of the same material as the object, when laying-up, so that progress can be monitored without disturbing the object. When treatment is complete, the object is removed and washed thoroughly with cold water. After a period of several hours air drying, it may be wax finished.

A thin black covering is produced on as-cast surfaces.

▲ Butyric acid causes burns, and contact with the skin, eyes and clothing must be prevented. It also has a foul pungent odour, which tends to cling to the clothing and hair, and is difficult to eliminate.

▲ Sodium hydroxide is a powerful caustic alkali causing burns, and must be prevented from coming into contact with the eyes or skin.

▲ Sodium sulphide is corrosive and causes burns. It must be prevented from coming into contact with the eyes or skin. When preparing the solution it must not be mixed directly with the acid, or toxic hydrogen sulphide gas will be evolved. The acid should be added to the water and then neutralised with the sodium hydroxide, before the remaining ingredients are added.

2.188   Dark green/blue-green patina (stipple) on light brown ground (as-cast surfaces only)

Ammonium chloride	16 gm
Sodium chloride	16 gm
Ammonia (.880 solution)	30 cm³
Water	1 litre

Sawdust technique (Twenty-four hours)

The object to be coloured is laid-up in sawdust which has been evenly moistened with the solution. It is then left for about twenty-four hours. The object should not be disturbed while colouring is in progress, as re-packing tends to cause a loss of surface quality and definition. Progress should be monitored by including a small sample of the object metal, when laying-up, which can be examined as necessary. When treatment is complete, the object is washed thoroughly in cold water and allowed to dry in air. It may be wax finished, when completely dry.

Colour development on polished surfaces tends to be slight, taking the form of a slight greenish tint to the brass surface, with a very light stipple of pale green or grey.

▲ Ammonia solution causes burns or severe irritation, and must be prevented from coming into contact with the eyes or skin. The vapour should not be inhaled as it irritates the respiratory system. Colouring should be carried out in a well ventilated area.

2.189   Orange-brown stipple on etched yellow-brown Semi-matt

Copper sulphate	100 gm
Lead acetate	100 gm
Ammonium chloride	10 gm
Water	1 litre

Sawdust technique (Ten to twenty hours)

The object to be coloured is laid-up in sawdust which has been evenly moistened with the solution. It is then left to develop for a period of up to about twenty hours. When treatment is complete, the object is removed, washed in cold water and allowed to dry in air. It may be wax finished when completely dry.

The surface is initially coloured a slightly lustrous orange-brown, and selectively etched by the medium, producing a stipple of etched yellow ground. The orange-brown surface is gradually lost as treatment is prolonged, leaving a textured yellow-brown surface.

On rough or as-cast surfaces, a pale blue-green patina tends to develop on a mid-brown or orange-brown ground.

▲ Lead acetate is a toxic substance, and every precaution should be taken to prevent inhalation of the dust, when preparing the solution. It is very harmful if taken internally. A clean working method is essential throughout.

2.190   Slight brown stipple on coarsely etched brass Matt

Copper acetate	60 gm
Copper carbonate	30 gm
Ammonium chloride	60 gm
Water	1 litre
Hydrochloric acid (15% solution)	5 cm³

Sawdust technique   (Ten to twenty hours)

The object to be coloured is laid-up in sawdust which has been evenly moistened with the solution. It is left for a period of about ten hours and then examined. If further treatment is required, it should be re-packed and examined at intervals of a few hours. When treatment is complete, the object is removed, washed thoroughly in cold water and allowed to dry in air. When completely dry, it may be wax finished.

The surface is coloured brown initially, but is rapidly etched back to produce a coarse grain-enhanced surface with a dull brown tinge.

Variegated brown colours, with some blue-green patina development, are produced on rough or as-cast surfaces.

▲ Hydrochloric acid can cause burns or severe irritation, and must be prevented from coming into contact with the eyes and skin.

2.191   Grey and brown stipple on frosted yellow Matt

Ammonium carbonate	300 gm
Acetic acid (10% solution)	35 cm³
Water	1 litre

Sawdust technique   (To twenty hours)

The ammonium carbonate is added to the water and the acetic acid then added gradually, while stirring. The object to be coloured is laid-up in sawdust which has been evenly moistened with this solution. It is then left for a period of up to twenty hours, ensuring that the sawdust remains moist throughout the period of treatment. After about ten hours, and at intervals of several hours subsequently, the object should be examined to check its progress. When the desired surface finish has been obtained, the object is removed, washed in cold water and allowed to dry in air. When thoroughly dry, it may be wax finished.

The surface is strongly etched to produce a textured surface that acquires a mixed grey and brown colour on a frosted yellow ground. The coloured areas gradually give way to the frosted ground as treatment continues. Timing appears to be unpredictable, and surface development is best monitored by using a sample.

2.192   Pale grey stipple on yellow lustre Semi-gloss

Copper nitrate	20 gm
Ammonium chloride	20 gm
Calcium hypochlorite	20 gm
Water	1 litre

Sawdust technique   (Twenty to thirty hours)

The object to be coloured is laid-up in sawdust which has been moistened evenly with the solution, and left for a period of twenty to thirty hours. The sawdust should remain moist throughout this time. After treatment, the object is washed in cold water and allowed to dry in air. When thoroughly dry, it may be wax finished.

A lightly etched pale grey stipple on a slight lustrous yellow, is produced on polished surfaces.

A mid-brown colour with a grey stipple is produced on rough or as-cast surfaces.

▲ The corrosive fine dust of the calcium hypochlorite (bleaching powder) must not be inhaled, as it can severely irritate the respiratory system. The solution must be prevented from coming into contact with the eyes, which would be severely irritated.

2.193*   Grey/brown mottle with traces of blue patina Matt

Ammonium carbonate	120 gm
Ammonium chloride	40 gm
Sodium chloride	40 gm
Water	1 litre

Cotton-wool technique   (Twenty to thirty hours)

Cotton-wool, moistened with the solution, is applied to the surface of the object to be coloured. It is important to ensure that the moist cotton-wool is in full contact with the surface of the object. It is then left for a period of about twenty to thirty hours, ensuring that the cotton-wool remains moist throughout the period of treatment. When treatment is complete, the object should be washed in cold water, and allowed to dry in air. When completely dry, it is wax finished.

The surface is etched by the solution, and a smooth mottled grey/brown layer forms, tinged with blue. Some traces of a more substantial blue patina develop locally, particularly where the cotton-wool is not in close contact with the surface.

A blue-green patina on a grey or brown ground, tends to develop on rough or as-cast surfaces.

2.194*   Variegated grey tinged with greenish-blue patina Matt   *Pl. IV*

Ammonium chloride	100 gm
Ammonium carbonate	150 gm
Water	1 litre

Cotton-wool technique   (To twenty or thirty hours)

Cotton-wool, moistened with the solution, is applied to the surface of the object to be coloured. It is then left for periods of up to twenty or thirty hours. The cotton-wool should be kept moist throughout the period of treatment. After about ten or fifteen hours the surface should be examined, by carefully exposing a small portion, to determine the extent of colour development. Subsequently a check should be made every few hours until the desired surface finish is obtained. When treatment is complete, the object is washed in cold water and allowed to dry in air. When completely dry, it may be wax finished.

The metal is etched by the solution, producing a randomly textured surface of a variegated grey colour, which takes on a blue tint from the thin film of patina that develops.

More substantial development of a blue-green patina on a grey-brown ground tends to occur on as-cast surfaces.

2.195* Dull brown mottle on etched surface Matt

Copper nitrate	280 gm
Water	850 cm³
Silver nitrate	15 gm
Water	150 cm³

Cotton-wool technique (Twenty hours)

The copper nitrate and the silver nitrate are dissolved in separate portions of water, and the two solutions then mixed. Cotton-wool, moistened with this mixed solution, is applied to the surface of the object to be coloured. It is important to ensure that the cotton-wool is in full contact with the surface, and that it remains moist during the period of treatment. The object is removed after about twenty hours and washed in cold water. After thorough drying in air, it may be wax finished.

The polished surface is coarsely etched by this treatment, partly revealing the grain structure, and producing a dull brownish mottled colouration. A variegated dull brown is produced on as-cast surfaces. Tests involving this procedure produced very poor results, and the technique cannot be recommended.

▲ Silver nitrate must be prevented from coming into contact with the skin, and more particularly the eyes, as it may cause burns or severe irritation.

2.196* Variegated grey Matt   Pl. III

Copper nitrate	100 gm
Water	200 cm³
Ammonia (.880 solution)	400 cm³
Acetic acid (6% solution)	400 cm³
Ammonium chloride	100 gm

Cotton-wool technique (Twenty to thirty hours)

The copper nitrate is dissolved in the water and the ammonia gradually added. The acetic acid, is then added, followed by the ammonium chloride. Cotton-wool is wetted with the solution and applied to the surface of the object to be coloured. It is then left for a period of about twenty to thirty hours. The cotton-wool should be kept moist throughout the period of treatment. When treatment is complete, the object is washed in cold water and allowed to dry in air. When dry, it is wax finished.

The polished surface is etched unevenly by the solution, producing a variegated grey and unevenly textured surface.

A variegated blue/blue-green patina and a dark grey/brown ground is produced on as-cast surfaces.

▲ Preparation and colouring should be carried out in a well ventilated area. Ammonia vapour and some acetic acid vapour will be liberated, which cause irritation of the eyes and respiratory system.

▲ The strong ammonia solution must be prevented from coming into contact with the eyes and skin, as it will cause burns or severe irritation.

2.197 Dark brown with lustrous patches Gloss

Potassium hydroxide	100 gm
Copper sulphate	30 gm
Sodium tartrate	30 gm
Water	1 litre

Cotton-wool technique (Twenty to thirty hours)

Cotton-wool, moistened with the solution, is applied to the surface of the object, which is then left for a period of twenty to thirty hours. It is important to ensure that the moist cotton-wool is in full contact with the surface, and that it remains moist throughout the period of treatment. When treatment is complete, the object is washed in cold water and allowed to dry in air. When dry, it may be wax finished.

The surface goes through a series of lustre colour changes during treatment, but tends to an even dark brown. Some lustrous patches may remain.

A dark brown or black colour is produced on as-cast surfaces.

▲ Potassium hydroxide is a strong caustic alkali, and should be prevented from coming into contact with the eyes or skin.

▲ When preparing the solution, the potassium hydroxide should be added to the water in small quantities, and *not* vice-versa.

2.198* Variegated greenish-brown Semi-matt

Ammonium carbonate	150 gm
Copper acetate	60 gm
Sodium chloride	50 gm
Potassium hydrogen tartrate	50 gm
Water	1 litre

Cotton-wool technique (Ten to twenty hours)

Cotton-wool moistened with the solution is applied to the surface of the article. It is important to ensure that the cotton-wool is in full contact with the surface. It is left for ten to twenty hours, during which time the cotton-wool should remain moist. Colour development should be monitored with a sample, so that the article remains undisturbed. When treatment is complete, the article is removed and washed in cold water. After drying in air for several hours, it may be wax finished.

The polished surface is more or less evenly etched back by the solution, and acquires a variegated greenish-brown colour.

A bluish patina on a greenish-brown ground tends to be produced on as-cast surfaces.

2.199 Coarse grain-enhanced surface (slight patina) Matt

Copper nitrate	200 gm
Sodium chloride	200 gm
Water	1 litre

Cotton-wool technique (Twenty hours)

Cotton-wool, moistened with the solution, is applied to the surface of the object to be coloured. It is then left for a period of about twenty hours. It is essential to ensure that the cotton-wool is in full contact with the surface and is kept moist throughout the period of treatment. Breaks in the colour and patches of patina may result if the cotton-wool is allowed to dry out or lift from the surface. When treatment is complete, the object should be washed thoroughly in cold water, and allowed to dry in air. When dry, it may be wax finished.

The surface is strongly etched by the solution, revealing the grain structure of the cast surface, but giving a coarse appearance. In tests, traces of blue-green patina also developed, particularly if the surface was allowed to dry without washing.

Patches of blue-green patina on a mid-brown were produced on as-cast surfaces.

The technique cannot be generally recommended, particularly with polished surfaces.

2.200* Mottled dark brown Gloss

Copper carbonate	300 gm
Ammonia (.880 solution)	200 cm³
Water	1 litre

Cotton-wool technique (Twenty hours)

Cotton-wool, moistened with the solution, is applied to the surface of the object to be coloured, when is then left for a period of twenty hours. It is important to ensure that the moist cotton-wool is in full contact with the surface, and that it remains moist throughout the period of treatment. When treatment is complete, the article is removed and washed in cold water. It is then allowed to dry in air for some time before wax finishing.

On rough or as-cast surfaces a bluish patina tends to develop, on a variegated brown ground.

▲ The strong ammonia solution will liberate ammonia vapour, which irritates the eyes and respiratory system. Preparation and colouring must therefore be carried out in a well ventilated area.

▲ Ammonia solution causes burns and severe irritation, and must be prevented from coming into contact with the eyes and skin.

2.201* Variegated and mottled lustre Gloss

Ammonium sulphate	85 gm
Copper nitrate	85 gm
Ammonia (.880 solution)	3 cm³
Water	1 litre

Cotton-wool technique (Thirty hours)

Cotton-wool, moistened with the solution, is applied to the surface of the object to be coloured. It is then left for a period of about thirty hours. It is essential to ensure that the moist cotton-wool is in full contact with the surface of the object, or breaks in the colour will occur. The cotton-wool should be kept moist throughout the period of treatment. When treatment is complete, the object should be washed thoroughly in cold water and allowed to dry in air. When dry, it is wax finished.

The technique is only suitable for use on polished surfaces. A slightly variegated golden lustre is produced, which tends to become mottled with a purplish-brown.

2.202 Greenish patina on coarse grain-enhanced metal Matt

Ammonium chloride	350 gm
Copper acetate	200 gm
Water	1 litre

Cotton-wool technique (Ten to twenty hours)

Cotton-wool, moistened with the solution, is applied to the surface of the object. This is then left for a period of ten to twenty hours, ensuring that the cotton-wool remains moist. After the first few hours, the surface should be periodically examined by exposing a small portion, to check its progress. When the desired surface finish has been reached, the object is removed, washed in cold water and allowed to dry in air. When completely dry, it may be wax finished.

The polished surface is coarsely etched by the solution, producing a dull enhancement of the grain structure. Patches of green patina develop on this ground. A more developed blue-green patina on a brown ground tends to occur on as-cast surfaces.

2.203 Yellow grain-enhanced surface with greenish tinge Semi-matt

Copper sulphate	100 gm
Lead acetate	100 gm
Ammonium chloride	10 gm
Water	1 litre

Cotton-wool technique (Ten to twenty hours)

Cotton-wool, moistened with the solution, is applied to the surface of the object to be coloured, and is left in place for a period of up to twenty hours. The cotton-wool should remain moist throughout the period of treatment. When treatment is complete, the object is removed and washed in cold water. It is then dried in air, and gradually darkens on exposure to daylight. When colour change has ceased, the object may be wax finished.

The polished surface is etched by the solution, revealing the grain structure of the cast surface, which becomes tinged with a pale greenish film.

On rough or as-cast surfaces, a variable patina develops on a yellowish-brown ground.

▲ Lead acetate is a toxic substance, and every precaution should be taken to prevent inhalation of the dust when preparing the solution. It is very harmful if taken internally. A clean working method is essential throughout.

2.204 Dull brown grain enhancement Matt

Copper acetate	60 gm
Copper carbonate	30 gm
Ammonium chloride	60 gm
Water	1 litre
Hydrochloric acid (15% solution)	5 cm³

Cotton-wool technique (Ten to twenty hours)

Cotton-wool, moistened with the above solution, is applied to the surface of the article, and left for a period of about ten hours. The surface is then examined, by carefully lifting a small portion, to determine the stage it has reached. When the desired surface finish has been obtained, the article is washed thoroughly in cold water and allowed to dry in air. When completely dry, it may be wax finished.

The surface is coarsely etched by this treatment, producing a dull brown grain-enhanced surface. The technique cannot be recommended for polished surfaces.

An uneven bluish patina on a brown ground tends to occur on rough or as-cast surfaces.

▲ Hydrochloric acid can cause burns or severe irritation, and must be prevented from coming into contact with the eyes and skin.

2.205   Variegated black and dull grey on frosted brownish metal Matt

Ammonium carbonate	300 gm
Acetic acid (10% solution)	35 cm³
Water	1 litre

Cotton-wool technique (Twenty hours)

The surface is strongly etched by this treatment, producing an uneven surface which is partly black or dull grey, but predominantly a dull frosted brown.

A bluish patina on a grey-brown ground tends to be produced on as-cast surfaces.

The ammonium carbonate is added to the water and the acetic acid then added gradually, while stirring. Cotton-wool, moistened with this solution, is applied to the surface of the object, and left for a period of about twenty hours. It is important to ensure that the cotton-wool is in full contact with the surface, and that it remains moist throughout. When treatment is complete, the object is washed in cold water and allowed to dry in air. When completely dry, it may be wax finished.

2.206   Yellowish grain-enhanced surface Matt

Copper nitrate	20 gm
Ammonium chloride	20 gm
Calcium hypochlorite	20 gm
Water	1 litre

Cotton-wool technique (Twenty hours)

Polished surfaces are etched by this treatment, revealing the grain structure of the metal and producing a slightly yellow tinge.

A bluish patina, on a brown ground, tends to develop on rough or as-cast surfaces. The result is usually uneven.

▲ The corrosive fine dust of the calcium hypochlorite (bleaching powder) must not be inhaled, as it can severely irritate the respiratory system. The solution must be prevented from coming into contact with the eyes, which would be severely irritated.

Cotton-wool, moistened with the solution, is applied to the surface of the object to be coloured, and left for a period of about twenty hours. It is important to ensure that the cotton-wool is in close contact with the surface, and that it remains moist throughout the period of treatment. When treatment is complete, the object is removed and washed thoroughly in cold water. After a period of drying in air, it may be wax finished.

2.207   Thin blue-green patina on black/brown ground Semi-matt

Ammonium carbonate	20 gm
Oxalic acid (crystals)	10 gm
Acetic acid (6% solution)	1 litre

Cloth technique (Twenty-four hours)

A more developed patina on a reddish-brown ground is produced on rough or as-cast surfaces. On polished faces, there is limited patina development on a black and light brown ground, which takes on the texture of the applied cloth.

If the cloth is allowed to dry completely on the surface, it becomes very adherent and tends to damage the surface when removed.

A variety of tests were carried out using this solution. Immersion at temperatures from cold to boiling, produced no effect, even after prolonged immersion. Direct application of the cold or hot solution by dabbing and wiping also produced no effect, as the surface resists wetting.

▲ The ammonium carbonate should be added to the acetic acid in small quantities to avoid very vigorous effervescence.

▲ Oxalic acid is harmful if taken internally. Contact with the eyes and skin should be avoided.

Strips of soft cotton cloth, or other absorbent material, are soaked with the solution. These are applied to the surface of the object and stippled into place using the end of a stiff brush which has been moistened with the solution. They are left on the surface until very nearly dry, which may take up to twenty-four hours, and then removed. If patina development is to be encouraged, the surface is then moistened by wiping the solution on using a soft cloth. If the ground colour with a minimum of patina is required, then the surface is allowed to dry without applying any more solution. When the surface is completely dry, it may be wax finished.

2.208   Blue-green patina on variegated dark brown Semi-matt

Ammonium sulphate	90 gm
Copper nitrate	90 gm
Ammonia (.880 solution)	1 cm³
Water	1 litre

Cloth technique (Twenty hours)

The disposition of the blue-green patina on polished surfaces tends to follow the texture of the applied cloth.

Tests carried out involving direct application of this solution were not successful. The surface stubbornly resists 'wetting', even with additions of wetting agents to the solution.

Soft cotton cloth which has been soaked with the solution is applied to the surface of the object, and stippled into place with a stiff bristle-brush. The object is then left for a period of about twenty hours. The cloth should be removed when it is very nearly dry, and the object then left to dry in air without washing. The blue-green patina tends to develop during the drying period. (If a patina-free ground colour is required, the cloth should be removed at an earlier stage when it is still wet, and the object thoroughly washed. The cloth application may be repeated, for a more developed ground colour.) When treatment is complete and the surface thoroughly dried out, it may be wax finished.

3 Copper (rolled sheet) and copper-plate

The copper sheet to which the recipes and results refer is the generally available 'commercial' grade, C 101 (B.S. 2870). The main colour heading for each recipe refers to the results obtained on polished surfaces, unless otherwise stated, and includes a reference to both the colour and surface quality. The results obtained on satin-finished surfaces, where these are significantly different, are given in the marginal notes.

Tests on copper-plated surfaces were carried out on copper-plated yellow brass. The results obtained are given in the marginal notes accompanying the copper recipes. Any bi-metallic colouring effects produced are also noted.

	Recipe numbers
Red Purple/purple-brown/orange/orange-brown	3.1 – 3.35
Brown Light bronzing/brown/dark brown	3.36 – 3.89
Black Black/dark slate/grey	3.90 – 3.126
Green patina Blue-green/green/blue	3.127 – 3.159
Lustre colours	3.160 – 3.174
Special techniques	3.175 – 3.212
Partial-plate colours Copper-plate on yellow brass	3.213 – 3.228

3.1*   Red Semi-matt

<table>
<tr><td></td><td>A</td><td>Copper sulphate</td><td>25 gm</td></tr>
<tr><td></td><td></td><td>Water</td><td>1 litre</td></tr>
<tr><td></td><td>B</td><td>Ammonium chloride</td><td>0.5 gm</td></tr>
</table>

Pl. VI, Pl. XV

Boiling immersion (A) (Fifteen minutes)
Boiling immersion (A+B) (Ten minutes)

Plate VI shows the result on copper-plated surfaces; plate XV shows the result on copper.

The article is immersed in the boiling copper sulphate solution for about fifteen minutes or until the colour is well developed. It is then removed to a bath of hot water, while the ammonium chloride is added to the colouring solution, and then re-immersed. The colour is brightened and tends to become more red. After about ten minutes the article is removed and washed in hot water. If the immersion is prolonged a bloom of greyish-green patina will tend to form, which may need to be removed with a soft bristle brush. The article is dried in sawdust and finally wax finished.

3.2*   Red/purplish-brown Semi-gloss

Copper sulphate	120 gm
Ammonia (.880 solution)	30 cm^3
Water	1 litre

Boiling immersion (Twenty minutes)

The effect produced is of a thin dark layer lying on top of an orange-brown ground. In the case of copper both colours are even. Tests carried out with copper-plated surfaces tended to produce irregularities in the dark layer allowing some small areas of orange-brown to show through.

For bi-metallic effects on copper-plated objects using this solution see recipe 3.218.

▲ A dense precipitate is formed which tends to sediment in the immersion vessel, causing 'bumping'. Glass vessels should not be used.

▲ Preparation of the solution, and colouring, should be carried out in a well ventilated area to prevent irritation of the eyes and respiratory system by the ammonia vapour. The skin, and particularly the eyes, should be protected from accidental splashes of this corrosive alkali.

Immersion in the boiling solution produces lustre colours after one minute, followed by the gradual change to a more opaque orange-pink colour which is well developed after seven or eight minutes. Continued immersion produces a darkening of this to an orange-brown and the gradual development of a purplish brown colour. After twenty minutes the object is removed and washed in hot water. After drying in sawdust it is wax finished.

3.3*   Red (slightly grained surface) Semi-matt

Copper sulphate	6.25 gm
Copper acetate	1.25 gm
Sodium chloride	2 gm
Potassium nitrate	1.25 gm
Water	1 litre

Boiling immersion (About one hour)

Pl. VI

If the colour develops unevenly then the object should be lightly bristle-brushed with pumice and then re-immersed.

The surface of sheet copper is slightly etched by the solution, producing a fine grained effect.

A reddish-brown colour is produced on copper-plate.

A bi-metallic colouring effect can be obtained on copper-plated surfaces, see recipe 3.214.

▲ Potassium nitrate is a powerful oxidising agent and should not be allowed to come into contact with combustible materials.

The article is immersed in the boiling solution and after a few minutes the surface darkens and becomes more opaque as the colour develops. The surface develops gradually as immersion continues, and after about one hour the article is removed and washed in hot water. It is dried in sawdust and may be wax finished when completely dry.

3.4*   Brownish-red/red Semi-gloss

<table>
<tr><td></td><td>A</td><td>Copper sulphate</td><td>125 gm</td></tr>
<tr><td></td><td></td><td>Ferrous sulphate</td><td>100 gm</td></tr>
<tr><td></td><td></td><td>Water</td><td>1 litre</td></tr>
<tr><td></td><td>B</td><td>Ammonium chloride</td><td>1 gm</td></tr>
</table>

Boiling immersion (A) (Fifteen minutes)
Boiling immersion (A+B) (Five minutes)

The surface tends to be variegated, both on copper and copper-plated surfaces.

A bi-metallic colouring effect can be produced on copper-plated yellow brass, see recipe 3.213.

The object must be thoroughly washed after immersion, to prevent the formation of a greenish bloom or patches of green. If there is a build-up of bloom during immersion, then this should be cleared by agitating the object, or by removing and rinsing in hot water if necessary.

If shorter immersion times are used before the addition of the ammonium chloride, then the results tend to be lighter and more glossy, but are more prone to unevenness.

The object is immersed in the boiling solution of copper sulphate and ferrous sulphate, and immersion continued for about fifteen minutes, until a brown colour has developed. The object is removed and rinsed in a tank of hot water, while the ammonium chloride is added to the colouring solution. The object is re-immersed in the colouring solution for about five minutes, causing a brightening and development of the surface to a red colour. It is then removed and washed very thoroughly in hot water. After drying in sawdust, it may be wax finished.

3.5*   Red (variegated) Gloss   *Pl. V*

Object heated and plunged into boiling water

The object is heated evenly to a full red/orange colour using a blow torch or preferably a kiln or muffle furnace. It is then immediately plunged into a bath of turbulently boiling water. When it has cooled to the temperature of the water, it is removed from the bath and washed in cold water. It may then be dried in sawdust and wax finished when dry.

The temperature of the object as it enters the water appears to be critical. The equipment should be arranged so that as little heat as possible is lost during the transfer from heat source to immersion in the boiling water.

Although good results are readily achieved with small scale items and thin materials, as the scale and wall thickness increases so it becomes more difficult to obtain an even colour.

Tests have shown that when sheet material is heated with a torch, the best colours are obtained on the reverse side to that being heated.

This technique is not suitable for use with copper-plated surfaces.

3.6   Reddish-purple Semi-matt

Copper nitrate	80 gm
Water	1 litre
Ammonia (.880 solution)	4–5 cm³

Boiling immersion (Thirty minutes)

The article is immersed in the boiling solution. After five minutes a pink tinge is apparent, followed by a gradual and even build-up of colour. After thirty minutes the article is removed and washed thoroughly in hot water before drying in sawdust. When dry the surface has a greyish bloom which becomes nearly invisible after wax finishing.

There is a tendency for darker patches to occur on both copper and copper-plated surfaces with this solution.

Increasing the ammonia content, up to about 10 cm³, produces a slightly lighter colour and more mottled surfaces and darker marks. The mottling produced on copper-plated surfaces at this concentration is even and effective.

Lower overall concentrations, or decreases in the quantity of added ammonia produce colours closer to mid-brown.

3.7   Reddish-brown/orange-brown Semi-gloss/semi-matt

Sodium chlorate	100 gm
Copper nitrate	25 gm
Water	1 litre

Boiling immersion (Forty minutes to one hour)

The object is immersed in the boiling solution, and the surface is quickly coloured with a dark lustre, which develops to a thin yellowish-orange colour. Continued immersion produces a gradual change to a more opaque orange-red or orange-brown colour. When the colour has developed evenly and to the required depth, after about forty minutes to one hour, the object is removed and washed in hot water. It is dried in sawdust, and may be wax finished when dry.

Prolonging the immersion beyond about one hour darkens the colour, but tends to produce a duller surface which is prone to spotting and unevenness.

▲ Sodium chlorate is a powerful oxidant and must not be allowed to come into contact with combustible material or acids. It should also be prevented from coming into contact with the eyes, skin or clothing.

3.8   Reddish-brown Gloss

Copper carbonate	160 gm
Sodium hydroxide	80 gm
Water	1 litre

Boiling immersion (Thirty minutes)

The sodium hydroxide is added to the water in small quantities, while stirring. Heat is evolved as it dissolves, warming the solution. When it has completely dissolved, the copper carbonate is added and the solution heated to boiling. The object is immersed in the boiling solution, and gradually acquires a pale grey colour which darkens to a greyish-mauve as immersion proceeds. After about thirty minutes the object is removed and dried in sawdust. When completely dry, it may be wax finished.

Similar results can be obtained using less concentrated solutions for correspondingly longer times, but these tend to be less even.

For results obtained with a long term immersion in this solution see recipe 3.9.

▲ Sodium hydroxide is a powerful caustic alkali, and the solid or solutions must be prevented from coming into contact with the eyes and skin. The sodium hydroxide, in small quantities, should be added to the water and *not* vice versa. Colouring should be carried out in a well ventilated area, as inhalation of the caustic vapours should be avoided.

3.9   Purplish-brown Semi-gloss

Copper carbonate	160 gm
Sodium hydroxide	80 gm
Water	1 litre

Hot/cold immersion (Several hours)

The sodium hydroxide is added to the water in small quantities, while stirring. Heat is evolved as it dissolves, warming the solution. When it has completely dissolved, the copper carbonate is added and the solution heated to boiling. The heat source is removed, and the article is immersed in the solution, and left for several hours without any further heating. When the colour has developed, the article is removed, washed in cold water and dried in sawdust. When completely dry, it may be wax finished.

A slightly brighter colour is produced than is obtained with a boiling immersion. The solution tends to separate into layers, the lower layer tending to produce a lighter colour on the metal surface. If this 'banding' is to be avoided, the object should be suspended well clear of the bottom of the vessel.

A bi-metallic colouring effect can be obtained with copper-plated yellow brass using this procedure, see recipe 3.217.

▲ Sodium hydroxide is a powerful caustic alkali, and the solid or solutions must be prevented from coming into contact with the eyes and skin. The sodium hydroxide, in small quantities, should be added to the water and *not* vice versa. Colouring should be carried out in a well ventilated area, to avoid inhalation of the caustic vapours.

3.10   Purple-brown   Gloss

Copper acetate	70 gm
Copper sulphate	40 gm
Potassium aluminium sulphate	6 gm
Acetic acid (10% solution)	10 cm³
Water	1 litre

Boiling immersion   (Fifteen minutes)

Tests tended to result in a slight unevenness of the surface colouring, taking the form of feint 'marbling'.

The object is immersed in the boiling solution. The colour development is gradual and even throughout the period of immersion, producing purple colours after about five minutes. The object is removed after fifteen minutes, washed thoroughly in hot water and dried in sawdust. When completely dry it is wax finished.

3.11   Dark reddish-brown   Semi-matt

Copper nitrate	200 gm
Sodium chloride	160 gm
Potassium hydrogen tartrate	120 gm
Ammonium chloride	40 gm
Water	1 litre

Cold immersion. Scratch-brushing   (Fifteen minutes)

The newly coloured surface is fragile when wet and handling should be avoided. It was found that the surface is particularly sensitive to household rubber gloves, which removed parts of the colour layer completely and caused staining.

The surface darkens unevenly and some near black areas can occur.

The article is immersed in the cold solution. The surface is immediately coloured and finely etched. After fifteen minutes the article is a bright pink colour. It is removed and scratch-brushed with the cold solution, and is then left to dry in air, without washing. Exposure to daylight will cause the surface to darken gradually to a dark reddish-brown. When colour change has ceased, the article may be wax finished.

3.12*   Purple   Semi-matt

Copper sulphate	25 gm
Ammonia (.880 solution)	3–5 cm³
Water	1 litre

Boiling immersion   (Fifteen to thirty minutes)

Increasing the ammonia content beyond the stated amount tends to inhibit colour development. At a rate of 10 cm³/litre ammonia, with the given concentration of copper sulphate, no colour is produced.

The article is immersed in the boiling solution, which produces a series of pink/green lustre colours after about one minute. Continued immersion causes a darkening of the surface to produce brown colours, which may be uneven initially, but which become more even in the later stages of immersion. When the purple colour has developed, which may take from fifteen to thirty minutes, the article is removed and washed in hot water. After drying in sawdust, it is wax finished.

3.13   Dull purple/purple-brown   Semi-matt/semi-gloss

Copper nitrate	200 gm
Water	1 litre

Boiling immersion   (Twenty minutes)

Tests carried out using lower concentrations of copper nitrate produced similar but lighter results, more grey-brown in colour.

Prolonged immersion in cold concentrated solutions produced very poor results, marked by surface streaking and patches of bare metal.

Tests carried out using hot solutions (eg. 200 gm/litre at 70°C) produced similar results to those obtained by boiling, but lighter in tone and less even.

Copper-plated surfaces yield very similar colours in all cases, but are generally more variegated and show a tendency to acquire additional darker surface marks.

The article is immersed in the boiling solution. A greenish tinge slowly appears during the first few minutes, which gradually darkens to a brown colour as immersion continues. A drab grey layer subsequently forms. After twenty minutes the article is removed and washed thoroughly in hot water to clear the grey layer and reveal the dull purple colour beneath. After thorough washing and drying in sawdust the surface has a characteristic grey bloom, which becomes almost invisible when the article is wax finished.

3.14   Reddish-brown   Gloss

Sodium hydroxide	100 gm
Sodium tartrate	60 gm
Copper sulphate	60 gm
Water	1 litre

Hot immersion   (Forty minutes)

A more pink colour with a transparent quality is produced on copper-plated surfaces. A dark brown or black colour tends to occur in a thin line along the edges of the surface.

▲ Sodium hydroxide is a powerful caustic alkali. The solid or solutions can cause severe burns. It must be prevented from coming into contact with the eyes or skin. When preparing the solution, small quantities of the sodium hydroxide should be added to large quantities of water and *not* vice versa. The solution will become warm as the sodium hydroxide is dissolved. When it is completely dissolved, the other ingredients are added.

The object is immersed in the hot solution (60°C) producing dull lustre colours after about one minute. The surface gradually becomes more opaque and the colour slowly turns brown as immersion continues. After about forty minutes, when the colour has developed, the object is removed and washed in hot water and either dried in sawdust or allowed to dry in air. It may be wax finished when dry.

3.15   Dark purplish-brown variegated with black Semi-matt

Copper sulphate	50 gm
Potassium chlorate	20 gm
Potassium permanganate	5 gm
Water	1 litre

Boiling immersion (Fifteen minutes)

The object is immersed in the boiling solution, and rapidly becomes coated with a chocolate-brown layer. Immersion is continued for fifteen minutes, after which the object is removed and gently rinsed in a bath of warm water. It is then allowed to dry in air, leaving a dark powdery layer on the surface. When dry this is gently brushed with a soft brush to remove any dry residue. This treatment will remove some of the top layer, partly revealing the underlying colour, and should be continued until the desired surface effect is obtained. A stiffer bristle brush may be required. After any residual dust has been removed with a soft brush, the object is wax finished.

Careful brushing will reveal an adherent chocolate brown beneath the loose dark layer. If the surface is overworked then this will tend to be removed leaving a reddish-brown on orange-brown variegated surface. This tends to be uneven, but the finish is more glossy.

A bi-metallic effect can be obtained on copper plated surfaces using this solution. See recipe 3.220.

▲ Potassium chlorate is a powerful oxidising agent, and must not be allowed to come into contact with combustible material. The solid or solutions should be prevented from coming into contact with the eyes, skin or clothing.

▲ The potassium chlorate and the potassium permanganate must not be mixed together in the solid state. They should each be separately dissolved in a portion of the water, and then mixed together.

▲ A nose and mouth face mask fitted with a fine dust filter should be worn when the dust is brushed from the surface.

3.16   Dark purplish-brown Semi-gloss

Copper sulphate	25 gm
Water	1 litre

Boiling immersion (Fifteen to thirty minutes)

The object is immersed in the boiling solution. After two or three minutes a brown colour develops which gradually darkens with continued immersion. When the surface has attained the dark purplish-brown colour, which may take from fifteen to thirty minutes, the object is removed and washed thoroughly in hot water. After drying in sawdust it is wax finished.

Tests have shown that the time taken for the full colour to develop is variable.

Tests with more concentrated solutions (up to 150 gm/litre of copper sulphate) produced similar results.

Scratch-brushing with the hot solution may be used to even the surface or as an after treatment. It should be carried out gently, producing a finely scratched surface with a warmer tone.

On copper-plated surfaces the colour is generally redder, but there is a tendency to patchiness and unevenness of tone.

3.17   Purplish-brown and orange-brown mottle Semi-matt

Copper nitrate	40 gm
Ammonia (.880 solution)	to form a paste

Applied paste (Several hours)

The copper nitrate is ground using a pestle and mortar, and a very small amount of ammonia added by the drop to form a crystalline 'paste'. This is applied to the object using a soft brush, and the surface then allowed to dry for several hours. When dry, the residue is washed away with cold water. The process may be repeated if necessary. After a final washing and air-drying, the object may be wax finished.

A brownish-purple surface mottled or spotted with a lighter more orange-brown is produced. Tests on copper-plated surfaces resulted in a less even mottling, the variegation tending to occur in patches.

A true paste will not form, but the moist crystalline mass that is obtained can be applied with a brush. This tends to become watery and may not dry out completely.

3.18   Reddish-brown (greenish-grey bloom) Semi-matt

Copper sulphate	120 gm
Potassium chlorate	60 gm
Water	1 litre

Hot immersion (Ten or fifteen minutes)

The object is immersed in the hot solution (70–80°C). A rapid series of lustre colours are produced which become more opaque and darken to an orange buff within two or three minutes. Continued immersion produces a gradual darkening to a dark brown colour. After about ten minutes a greenish layer begins to be deposited on the object, as a dense precipitate is formed in the solution. If this is not required then the surface should be gently bristle-brushed in warm water, and the object re-immersed. After about fifteen minutes the object is removed and washed. It should be bristle-brushed under warm water if the greenish layer has re-formed. After thorough washing it is dried in sawdust and wax finished when dry. A greyish or greenish-grey bloom tends to remain.

▲ A dense precipitate is formed as immersion progresses, causing 'bumping'. Glass vessels should not be used.

▲ Potassium chlorate is a powerful oxidising agent and must not be allowed to come into contact with combustible-material. Contact with the eyes, skin or clothing must be avoided.

3.19   Orange-brown Matt/semi-matt

Potassium chlorate	50 gm
Copper sulphate	25 gm
Water	1 litre

Hot immersion (Fifteen or twenty minutes)

The object is immersed in the hot solution (80°C) and is initially coloured with an uneven golden lustre finish which changes to a more opaque yellow-green. Continued immersion darkens the colour to orange and then finally to a more brown colour. After fifteen or twenty minutes, when the full colour has developed, the object is removed and washed in hot water. When it has been thoroughly dried in sawdust, it is wax finished. A green powdery film may develop when the sample is dried, which can be brushed away prior to waxing. The wax finish tends to make any residual green bloom invisible.

Tests showed a tendency for some colour variation at the edges of sheet material, and around surface features, eg. punch marks.

▲ Potassium chlorate is a powerful oxidising agent and should not be allowed to come into contact with combustible material. Contact with the eyes, skin and clothing must be avoided.

▲ A nose and mouth face mask fitted with a fine dust filter should be worn when brushing away any dry powder from the surface.

3.20★   Brownish-orange Semi-gloss/gloss   *Pl. V*

Copper sulphate	120 gm
Potassium chlorate	60 gm
Water	1 litre

Hot immersion (Five or six minutes)

The object is immersed in the hot solution (70–80°C), rapidly developing green lustre colours, which quickly become more opaque after about a minute. Continued immersion darkens the colour to a brownish orange. After five or six minutes the object is removed and washed thoroughly in hot water. It may be wax finished after thorough drying in sawdust.

Prolonging the immersion produces further darkening to a deep brown colour. See recipe 3.18.

▲ Potassium chlorate is a powerful oxidising agent and must not be allowed to come into contact with combustible materials. Contact with the eyes, skin or clothing must be avoided.

3.21   Orange Gloss

Copper acetate	7 gm
Ammonium chloride	3.5 gm
Acetic acid (6% solution)	100 cm³
Water	1 litre

Boiling immersion (Ten minutes)

The article is immersed in the boiling solution, and after three minutes a slightly lustrous orange colour develops. With continued immersion this colour darkens, until after ten minutes the article is removed. It is washed thoroughly in hot water, dried in sawdust and wax finished when dry.

The orange colour produced has a translucent quality. Surfaces coloured by this method are susceptible to some unevenness, particularly if the immersion is prolonged.

3.22   Red-orange Semi-matt

Ammonium carbonate	50 gm
Copper acetate	25 gm
Ammonium chloride	5 gm
Oxalic acid (crystals)	1 gm
Acetic acid (6% solution)	1 litre

Boiling immersion (One hour)

The article is immersed in the boiling solution. After about five minutes a patchy orange-yellow colouration with traces of green appears. Continued immersion in the boiling solution darkens the colour, which appears to be best after about one hour. The article should then be removed, washed in warm water and dried in sawdust. When dry, it may be wax finished.

Tests tended to result in streaky surfaces, which are difficult to avoid. Shorter immersion times produce less streaking, but a more varied, uneven surface colour. Prolonging the immersion time tends to increase the tendency for streaks to occur.

▲ When preparing the solution, the ammonium carbonate should be added in very small quantities. It reacts effervescently with the acid, and this reaction is very vigorous if large quantities are added.

▲ Oxalic acid is poisonous and harmful if taken internally.

3.23   Orange-brown Semi-gloss

Copper sulphate	25 gm
Nickel ammonium sulphate	25 gm
Potassium chlorate	25 gm
Water	1 litre

Boiling immersion (Twenty minutes)

Immersing the object in the boiling solution produces an immediate lustrous mottling of the surface. Continued immersion induces a change to a more even orange colour which darkens gradually. After twenty minutes the object is removed and washed in hot water. After thorough drying in sawdust it is wax finished.

▲ Nickel salts are a common cause of sensitive skin reactions.

▲ Potassium chlorate is a powerful oxidising agent and should not be allowed to come into contact with combustible material. The solid, and solutions, should not be allowed to come into contact with the eyes, skin or clothing.

3.24   Brownish-orange Semi-matt/matt

Copper sulphate	125 gm
Sodium chlorate	50 gm
Water	1 litre

Boiling immersion (Five to ten minutes)

The article is immersed in the boiling solution, which produces pink/green lustre colours very rapidly. These darken and become more opaque with continued immersion. When the orange-brown colour has developed, after about five or ten minutes, the article is removed and washed thoroughly in hot water. After drying in sawdust, it is wax finished.

Prolonging the immersion will darken the colour, but will also cause the deposition of a grey-green and very uneven layer. This layer is difficult to make even, and hard to remove when immersion is complete.

▲ Sodium chlorate is a powerful oxidising agent and must not be allowed to come into contact with combustible materials or acids. Inhaling the dust must be avoided, and the eyes and skin should be protected from contact with either the solid or solutions.

3.25   Orange-brown Semi-gloss

Barium chlorate	52 gm
Copper sulphate	40 gm
Water	1 litre

Hot immersion (Fifteen minutes)

Immersion in the hot solution (70°C) produces an immediate golden lustre colour, which changes gradually to pink/green and then to a brownish lustre after about three or four minutes. Continued immersion produces a greater opacity of colour and a gradual development to a yellow-brown or orange-brown colour. When the colour has fully developed, after fifteen minutes, the object is removed and washed in hot water. It is wax finished, after thorough drying in sawdust.

The lustre colours produced in the early stages of immersion are generally even, but are less bright than the equivalent colours produced by other methods.

▲ Barium compounds are poisonous and harmful if taken internally. Barium chlorate is a strong oxidising agent and should not be allowed to come into contact with combustible material. It should be prevented from coming into contact with the eyes, skin or clothing.

3.26   Orange-brown Gloss

Copper sulphate	100 gm
Ammonium chloride	0.5 gm
Water	1 litre

Hot immersion (Five to ten minutes)

The article is immersed in the hot solution (70–80°C) producing a pale pink-orange colouration within two or three minutes. This darkens with continued immersion. When an orange-brown colour has developed, after about five or ten minutes, the article is removed and washed in warm water. It is dried in sawdust. Exposure to daylight will cause a slight initial darkening of the colour. When the colour is stable, the article is wax finished.

Extending the immersion time beyond ten minutes tends to produce a darker and more matt layer, which is difficult to remove completely by bristle-brushing. It tends to 'cling', producing dull gritty patches.

3.27   Orange-brown (slight petrol lustre) Gloss

Copper sulphate	25 gm
Lead acetate	25 gm
Water	1 litre

Hot immersion   (Ten to fifteen minutes)

The article is immersed in the hot solution (80°C) producing green lustre colours after about one minute, which change gradually to produce an opaque orange after about six or seven minutes. This colour darkens with continued immersion. After ten or fifteen minutes the article is removed and washed thoroughly in hot water. After drying in sawdust, it is wax finished.

Small additions of ammonium chloride (at a rate of 2.5 gm/litre), introduced into the solution at a late stage in the immersion, produce a brightening of the colours. In this case however the colours thus produced darkened patchily on exposure to daylight, and the variation cannot be recommended.

▲ Lead acetate is highly toxic, and every precaution should be taken to prevent inhalation of the dust during weighing, and preparation of the solution. Colouring should be carried out in a well ventilated area as the vapour evolved must not be inhaled.

3.28   Orange-brown Gloss

Copper acetate	25 gm
Copper sulphate	19 gm
Water	1 litre

Hot immersion (Fifteen minutes)

The article is immersed in the hot solution (80°C). The colour develops gradually to produce a light terracotta after about five minutes, which darkens evenly as immersion continues. After ten minutes the surface is orange-brown in colour. The article is removed after fifteen minutes immersion, washed thoroughly and dried in sawdust. When dry, the surface has a characteristic grey bloom which is invisible after wax finishing.

There is some tendency for the surface to be slightly uneven. Tests carried out with copper tended to produce a slight, barely visible petrol lustre overlaying the orange-brown colour.

3.29*   **Orange-brown Gloss**   *Pl. VI*

'Rokusho'	3.25 gm
Copper sulphate	3.25 gm
Potassium aluminium sulphate	2 gm
Water	1 litre

Boiling immersion   (About one hour)

A mid-brown colour is produced on copper-plated surfaces, as shown in plate VI.

For description of 'Rokusho' and its method of preparation, see appendix 1.

The article is immersed in the boiling solution and after a few minutes the surface darkens and becomes more opaque as the colour develops. The surface develops gradually as immersion continues, and after about one hour the article is removed and washed in hot water. It is dried in sawdust and may be wax finished when completely dry.

3.30   **Orange-brown Gloss**   *Pl. VI*

'Rokusho'	5 gm
Copper sulphate	5 gm
Water	1 litre

Boiling immersion   (About one hour)

An orange-brown colour is also produced on copper-plate.

A bi-metallic colouring effect can be obtained on copper-plated surfaces, see recipe 3.221.

For description of 'Rokusho' and its method of preparation, see appendix 1.

Increasing the copper sulphate content to 20–30 gm/litre produces darker and more tan brown colours.

The article is immersed in the boiling solution and after a few minutes the surface darkens and becomes more opaque as the colour develops. The surface develops gradually as immersion continues, and after about one hour the article is removed and washed in hot water. It is dried in sawdust and may be wax finished when completely dry.

3.31*   **Dark orange-brown Matt**   *Pl. V*

Potassium chlorate	50 gm
Copper sulphate	25 gm
Nickel sulphate	25 gm
Water	1 litre

Boiling immersion   (Thirty to forty minutes)

The bloom of green patina is virtually invisible after wax finishing. The longer immersion time produces a better overall surface.

For bi-metallic effects on copper-plated surfaces using this solution see recipe 3.219.

▲ Potassium chlorate is a powerful oxidising agent and must not be allowed to come into contact with combustible material. The solid or solutions must be prevented from coming into contact with the eyes, skin or clothing.

▲ Nickel compounds are a common cause of sensitive skin reactions. Precautions should be taken to prevent inhalation of the dust, during preparation of the solution.

After two or three minutes' immersion in the boiling solution an orange colour develops on the surface of the object, which gradually darkens with continued immersion. The full colour develops in about thirty minutes, together with a green tinge which begins to appear after about twenty minutes. If this green colour is to be kept to a minimum, the object should be removed after thirty minutes. Colour development and surface quality are generally better with a longer immersion, the green tending to enhance the surface. The object is removed after about forty minutes, washed in warm water, and either dried in sawdust or allowed to dry in air. When dry, it is wax finished.

3.32   **Reddish-brown (light green patina) Semi-matt**

Copper sulphate	125 gm
Sodium chlorate	50 gm
Ferrous sulphate	25 gm
Water	1 litre

Boiling immersion   (Thirty minutes)

If the ground colour only is required, then the object should be intermittently bristle-brushed in hot water, during immersion. A residual greenish-grey bloom tends to remain.

▲ Sodium chlorate is a powerful oxidising agent and must not be allowed to come into contact with combustible material or acids. Inhalation of the dust must be avoided. The eyes and skin should be protected from contact with either the solid or solutions.

The object is immersed in the boiling solution. Orange colours develop after a few minutes and gradually darken with continued immersion. After about fifteen minutes a greenish layer begins to be deposited on the now brown ground colour. Agitating the object within the solution tends to inhibit the deposition of the green layer. After thirty minutes the object is removed and washed thoroughly in hot water. After drying in sawdust, it is wax finished.

3.33   Dark reddish-brown   Semi-gloss/semi-matt

Copper sulphate	100 gm
Copper acetate	40 gm
Potassium nitrate	40 gm
Water	1 litre

Boiling immersion   (Ten minutes)

The article is immersed in the boiling solution, which produces drab orange colours after one minute. These gradually darken and become more opaque with continued immersion. The article is removed when the colour has developed, after about ten minutes, and washed in hot water. After thorough drying in sawdust, it is wax finished.

▲ Potassium nitrate is a powerful oxidising agent and must not be allowed to come into contact with combustible material.

3.34   Reddish-brown   Semi-gloss/semi-matt

Copper sulphate	125 gm
Sodium acetate	12.5 gm
Water	1 litre

Boiling immersion   (Ten to fifteen minutes)

After one minute of immersion in the boiling solution a slightly lustrous orange/pink colour develops on the surface of the object. This gradually darkens, tending to become brown as immersion continues. When the colour has developed, after ten or fifteen minutes, the object is removed and washed in hot water. It is wax finished after thorough drying in sawdust.

The surfaces produced in tests tended to be slightly variegated, and the colour thinner at the edges of sheet material.

3.35*   Reddish brown   Semi-matt   *Pl. V*

Copper sulphate	50 gm
Ferrous sulphate	5 gm
Zinc sulphate	5 gm
Potassium permanganate	2.5 gm
Water	1 litre

Boiling immersion   (Thirty minutes)

The object is immersed in the gently boiling solution. A dark brown layer forms on the surface, which should not be disturbed. Immersion is continued for fifteen minutes, after which time the object is removed and the dark layer washed away in hot water using the bristle-brush. The object is then re-immersed for a further fifteen minutes and then removed and bristle-brushed gently with hot water. After thorough rinsing it is dried in sawdust, and wax finished when completely dry.

In tests with copper-plated surfaces a darker purplish-brown was obtained.

Lighter colours with a more green bias can be produced using this solution. See recipe 3.39.

3.36   Light brown bronzing   Semi-matt

Copper acetate	80 gm
Water	1 litre

Hot immersion. Scratch-brushing   (Thirty to forty minutes)

The object is immersed in the hot solution (50–60°C) and the surface colour allowed to develop through a series of lustrous colours, to a brownish colour. After thirty minutes it is removed and scratch-brushed with the hot solution, which initially removes some of the colour formed during immersion. Scratch-brushing is continued for several minutes until the surface colour has developed. The object is then washed in warm water and allowed to dry in air. When completely dry it is wax finished.

Tests carried out using the technique of direct scratch-brushing, without prior immersion, produced poor results with little colour development.

3.37   Light bronzing   Semi-gloss

Copper sulphate	8 gm
Copper acetate	3.25 gm
Potassium aluminium sulphate	2 gm
Water	1 litre
Acetic acid (6% solution)	20 cm³

Boiling immersion   (About one hour)

The article is immersed in the boiling solution and after a few minutes the surface darkens and becomes more opaque as the colour develops. The surface develops gradually as immersion continues, and after about one hour the article is removed and washed in hot water. It is dried in sawdust and may be wax finished when completely dry.

The results produced on sheet copper tended to be uneven and cannot generally be recommended.

A light bronzing with a reddish tint was produced on copper-plate. The results produced on a plated satin finish tended to be better than a polished surface.

A bi-metallic colouring effect can be obtained on copper-plated surfaces, see recipe 3.222.

3.38*   Brown (*or* variegated black on brown) Semi-matt   *Pl. V*

Potassium permanganate	2.5 gm
Copper sulphate	50 gm
Ferric sulphate	5 gm
Water	1 litre

Boiling immersion (Fifteen minutes)

The solution should not be boiled for any length of time prior to the immersion of the article, or the dark layer tends not to be deposited.

The article is immersed in the boiling solution, and the surface becomes coated with a superficial dark brown film, which should be removed by bristle-brushing in a bath of hot water. The article is re-immersed in the boiling solution and again becomes coated with dark brown.

If a plain brown surface is required, the brown film should be removed every few minutes as immersion continues. After about fifteen minutes the article is removed and washed thoroughly in hot water. After drying in air it may be wax finished.

If a variegated surface is required, then the brown film should be left on the surface as immersion continues. After fifteen minutes, the article is removed, and allowed to dry in air for a few minutes. The surface is then gently brushed with a dry bristle-brush, which will partially remove the brown layer, leaving a variegated surface. The article should be wax finished.

3.39   Mid-brown Semi-gloss/semi-matt

Copper sulphate	50 gm
Ferrous sulphate	5 gm
Zinc sulphate	5 gm
Potassium permanganate	2.5 gm
Water	1 litre

Boiling immersion (Ten to fifteen minutes)

Darker colours with a reddish bias can be produced using this solution, if the technique is modified. See recipe 3.35.

The object is immersed in the boiling solution, and becomes coated with a dark brown layer. After one or two minutes the object is removed and the dark layer brushed away under hot water, using a bristle-brush. It is then re-immersed for two minutes, and the bristle-brushing again carried out when the dark brown layer has re-formed. After two or three such immersions and bristle-brushings, the brown layer will tend not to be adherent when it re-forms, and can be removed by agitating the object while in the solution. Immersion is continued to ten or fifteen minutes, after which the object is removed and given a final bristle-brushing under hot water. It is thoroughly rinsed in warm water, dried in sawdust and wax finished when completely dry.

3.40   Mid-brown with greenish tinge (slight graining) Semi-matt

A	Copper sulphate	25 gm
	Ammonia (.880 solution)	30 cm³
	Water	1 litre
B	Potassium Hydroxide	16 gm
	Water	1 litre

Hot immersion (A)	(Ten minutes)
Hot immersion (B)	(One minute)

The colour tends to be slightly uneven, and there is a tendency for small streaks to occur. A slight grained effect is produced.

▲ Preparation of solution A, and colouring, should be carried out in a well ventilated area, to prevent irritation of the eyes and respiratory system by the ammonia vapour liberated.

▲ Potassium hydroxide, in solid form and in solution, is corrosive to the skin and particularly the eyes. Adequate precautions should be taken.

The article is immersed in the hot solution A (60°C) for about ten minutes, producing a slight etching effect and a darkening of the surface to a greenish-brown. It is then transferred to the hot solution B (60°C) for about one minute. It is removed, washed in warm water, allowed to dry in air, and then it is wax finished.

3.41   Reddish-brown (variegated) Semi-gloss   *Pl. VI*

Ferric nitrate	10 gm
Water	1 litre

Torch technique

A darker variegation can be obtained by finally dabbing with a stronger solution. (Ferric nitrate 40 gm/litre.) The sample shown in plate VI has been treated in this way.

▲ The fumes evolved as the solution is applied to the heated metal should not be inhaled. Adequate ventilation must be provided and a nose and mouth face mask worn, fitted with the correct filter.

▲ Ferric nitrate is corrosive and should not be allowed to come into contact with the eyes or skin. A face mask or goggles should be worn to protect the eyes from hot splashes as the solution is applied.

The object is heated with a blow torch and the solution applied sparingly with a scratch-brush, using small circular motions and working across the surface. The metal gradually darkens to a reddish-brown and the surface can be finished by lightly dabbing with a cloth dampened with the solution. When treatment is complete and the object has been allowed to cool and dry thoroughly, it may be wax finished.

3.42   Thin orange-brown Semi-matt

Copper nitrate	20 gm
Zinc sulphate	30 gm
Mercurous chloride	30 gm
Water	1 litre

Applied liquid   (Twice a day for ten days)

The solution is initially applied by vigorous dabbing with a soft cloth, to overcome resistance to wetting. Once some surface effect has been produced, the solution may be applied by wiping. The surface is allowed to dry completely after each application, which should be carried out twice a day for at least ten days. If the solution is used very sparingly then a patina-free ground colour tends to be produced. A more liberal use of the solution during the later stages of treatment will tend to encourage the formation of a blue-green patina. When treatment is complete, the article should be allowed to dry in air for several days before wax finishing.

Streaky and uneven surfaces tended to be produced in tests carried out on copper and copper-plate.

In tests, the blue-green patina was found to be difficult to produce.

▲ Mercurous chloride may be harmful if taken internally. There are dangers of cumulative effects.

3.43*   Orange-brown Semi-matt/semi-gloss

Copper nitrate	85 gm
Zinc nitrate	85 gm
Ferric chloride	3 gm
Hydrogen peroxide (3% solution)	1 litre

Applied liquid   (Twice a day for five days)

The cold solution is dabbed on to the surface of the object with a soft cloth, and the object left to dry thoroughly in air. This procedure is repeated twice a day for a period of about five days. An initial lustrous colouration is gradually replaced by a more opaque colour. Traces of blue-green patina may also appear during the later stages of treatment, if the application is not sufficiently sparing, which may be brushed away when dry if not required. The object is wax finished when completely dry.

Patina development can be encouraged by a less sparing use of the solution in the later stages of treatment.

▲ Hydrogen peroxide must be kept away from contact with combustible material. In the concentrated form it can cause burns to the eyes and skin, and when diluted will at least cause severe irritation.

3.44   Light orange-brown/ochre Semi-matt

Copper nitrate	35 gm
Ammonium chloride	35 gm
Calcium chloride	35 gm
Water	1 litre

Applied liquid to heated metal   (Repeat several times)

The article is heated either on a hotplate or with a torch to a temperature such that the solution sizzles gently when it is applied. The solution is applied with a soft brush, and dries out to produce a superficial powdery green layer. The procedure is repeated several times, each fresh application tending to remove the green powdery layer, until an underlying colour has developed. Any residual green is then removed by washing and the article is allowed to dry in air. When dry it may be wax finished.

If the solution is persistently applied as the surface is heated with a torch, a more adherent pale blue-green patina tends to form. However, solutions containing little or no chlorides are preferable for this effect.

3.45   Grey and buff variegated surface Semi-matt   *Pl. XVI*

Copper sulphate	20 gm
Zinc chloride	20 gm
Water	to form a paste

Applied paste   (Twenty-four hours)

The ingredients are ground to a thin creamy paste with a little water, using a pestle and mortar. The paste is applied sparingly to the surface of the object with a soft brush. It is then allowed to dry completely, and the residual dry paste washed off with cold water. The object is then allowed to dry in air for about twelve hours, and the procedure repeated if required. When treatment is complete, the object should be allowed to dry for several days at least. Exposure to daylight causes a variegated darkening of the ground colours. When the surface is completely dry and colour change has ceased, the object may be wax finished.

Repeated applications on copper surfaces tend to produce a darker variegation of black and brown, but are prone to surface defects such as spotting.

Repeated application on a copper-plated surface tends to breach the plating and produces patchy pink surfaces. A single sparing application is recommended, which tends to produce a greenish-brown surface which is finely reticulated. The procedure produces rather variable results.

3.46   Light brown on copper ground ('grained' surface) Semi-gloss

Potassium chlorate	50 gm
Copper sulphate	25 gm
Ferrous sulphate	25 gm
Water	1 litre

Boiling immersion (Twenty minutes)

The surface is etched by the solution to produce a 'grained' effect, similar to a directionally satinised finish in appearance.

▲ Potassium chlorate is a powerful oxidant and must not be allowed to come into contact with combustible materials. It should also be prevented from coming into contact with the eyes, skin or clothing.

The object is immersed in the boiling solution. The colour develops gradually during immersion, becoming darker and tending to acquire a bloom of greenish-grey patina during the latter stages. If the bloom becomes excessive it should be removed by gently bristle-brushing the surface. When the required colour has developed, after about twenty minutes, the object is removed and washed in hot water. It is dried in sawdust, and may be wax finished when dry.

3.47*   Mid-brown Gloss

Ammonium carbonate	50 gm
Copper acetate	25 gm
Acetic acid (6% solution)	1 litre

Boiling immersion (Thirty minutes)

Tests involving the addition of 5 gm of ammonium chloride and 2 gm of oxalic acid per litre to the solution when the colours have developed, produced poor results. The colour initially becomes lighter but is then subject to streaking and etching effects, producing very patchy matt yellow surfaces. The variation cannot be recommended. Very similar results were obtained with small additions of zinc chloride or copper chloride.

▲ The ammonium carbonate should be added to the acetic acid in small quantities. The reaction is effervescent, and if large amounts are added, the effervescence is very vigorous.

The object is immersed in the boiling solution, and the surface gradually darkens to a reddish-brown colour. After thirty minutes, when the colour has fully developed, the object is removed and washed in warm water. After careful drying in sawdust, it may be wax-finished.

3.48   Reddish-brown (slight petrol lustre) Semi-gloss

Copper sulphate	60 gm
Copper acetate	10 gm
Potassium aluminium sulphate	25 gm
Water	1 litre

Boiling immersion (Forty minutes)

A slightly pale reddish-brown 'figured' with darker lines was produced in tests. This also occurred with high copper brasses and is probably characteristic of this solution when used with copper-rich sheet materials.

The object is immersed in the boiling solution, producing a gradual darkening of the surface to a drab ochre/brown colour after ten minutes. A slight lustre may also be apparent. Continued immersion produces further darkening, although colour development is slow. After forty minutes the object is removed, washed in hot water and dried in sawdust. When dry, it is wax finished.

3.49   Orange-brown Semi-gloss/gloss

Ammonium acetate	50 gm
Copper acetate	30 gm
Copper sulphate	4 gm
Ammonium chloride	0.5 gm
Water	1 litre

Boiling immersion (Seven to ten minutes)

After one or two minutes of immersion in the boiling solution, the object takes on an olive colouration, which gradually darkens to a more ochre colour as immersion continues. The full colour tends to develop in about seven minutes, although it may take a little longer. Prolonging the immersion tends to cause the colour to become more pallid, and does not improve surface consistency. The object is removed and washed in hot water. After thorough drying in sawdust, it is wax finished.

3.50*   Orange-brown Semi-matt

Copper nitrate	200 gm
Water	1 litre

Applied liquid (Twice a day for five days)

The solution is initially dabbed on vigorously with a soft cloth, to ensure that the surface is evenly 'wetted'. Subsequent applications should be wiped on, and used very sparingly, to inhibit the formation of green patina. After each application the article is left to dry in air. This procedure is repeated twice a day for about five days. The article is then left to dry in air, and wax finished when completely dry.

3.51   Mid-brown (slightly lustrous) Semi-gloss

A	Copper sulphate	50 gm
	Water	1 litre
B	Potassium sulphide	12.5 gm
	Water	1 litre

Hot immersion (A) (Thirty seconds)
Cold immersion (B) (One minute)

The article is immersed in the hot copper sulphate solution (80°C) for about thirty seconds to produce a green/pink lustre colour. It is then transferred to the cold potassium sulphide solution and immersed for about one minute, producing a clouding of the surface and a smokey greenish-pink colour. This tends to be uneven and can be improved by applying the solutions alternately using a bristle-brush. When treatment is complete, the article is washed in warm water and allowed to dry in air. When dry, it is wax finished.

If the results produced in solution A are streaky, then the streaks will tend to be visible in some lights, after treatment with solution B. Application with the bristle-brush darkens and improves the results. Producing an adequate finish evenly is difficult, and the recipe cannot be generally recommended.

For results obtained using a scratch-brushed technique see recipe 3.116.

▲ Colouring should be carried out in a well ventilated area as some hydrogen sulphide gas will be liberated.

3.52   Mid-brown Satin/semi-matt

Sodium thiosulphate	250 gm
Water	1 litre
Copper sulphate	80 gm
Water	1 litre

Cold immersion (Fifteen hours)

The chemicals should be separately dissolved, and the two solutions then mixed together. The article is suspended in the cold solution and left for fifteen hours. The surface will undergo a series of colour changes to produce a black surface after two hours. After fifteen hours the article is removed and washed, removing the black layer to reveal the brown colouration beneath. After thorough washing and drying in sawdust the article is wax finished.

The black colour is adherent if the object is removed after about two hours. See recipe 3.108.

Prolonged immersion appears to cause a gradual softening of this layer as it builds, so that after a sufficient time it is easily washed off. Results suggest that it may be possible to determine an intermediate timing, where the black layer may be partially removed to produce a variegated black and brown surface.

3.53   Mid-brown Gloss

Copper sulphate	6.25 gm
Copper acetate	1.25 gm
Potassium nitrate	1.25 gm
Water	1 litre

Boiling immersion (About one hour)

The article is immersed in the boiling solution and after a few minutes the surface darkens and becomes more opaque as the colour develops. The surface develops gradually as immersion continues, and after about one hour the article is removed and washed in hot water. It is dried in sawdust and may be wax finished when completely dry.

In tests on polished surfaces the colour tended to be slightly lighter at the edges. Very similar results were produced on copper-plated surfaces.

A bi-metallic colouring effect can be produced on copper-plated surfaces, see recipe 3.223.

▲ Potassium nitrate is a powerful oxidising agent and must not be allowed to come into contact with combustible materials.

3.54   Black mottling on variegated brown Matt

A	Copper sulphate	50 gm
	Water	1 litre
B	Sodium thiosulphate	50 gm
	Lead acetate	12.5 gm
	Water	1 litre

Hot immersion (A)   (One minute)
Hot immersion (B)   (Two minutes)

The object is immersed in the hot copper sulphate solution (80°C) for one minute, producing a darkening of the surface. It is then transferred to solution B (80°C) for two minutes' immersion. These alternating immersions are repeated a number of times until an even black layer is produced on the surface. The object is then washed in warm water and allowed to dry in air. When dry, it is either rubbed gently with a soft cloth or bristle-brushed, and wax finished.

The black layer tends to be non-adherent. Careful drying and brushing can produce a partial black layer that is variegated or mottled, revealing areas of variegated brown. This recipe cannot be recommended for producing an even black colour, as suggested by some sources.

▲ Lead acetate is a highly toxic substance, and every precaution should be taken to prevent inhalation of the dust during preparation. The solution is very harmful if taken internally. Colouring should be carried out in a well ventilated area as the vapours evolved must not be inhaled.

3.55   Grey-brown mottle on reddish copper ground   Semi-gloss/semi-matt

A Copper nitrate	20 gm
Ammonium chloride	20 gm
Calcium hypochlorite	20 gm
Water	1 litre

B   Water

Cold immersion (A)   (Ten minutes)
Cold immersion (B)   (Twenty hours)

The article is immersed in solution A for about ten minutes, which produces a brightening of the surface to a slightly golden colour. The article is rinsed and transferred to a container of cold water, in which it is left for a period of about twenty hours. Small spots or patches of a darker colour develop slowly during the period of cold water immersion. When the surface is sufficiently developed, the article is removed and either dried in sawdust or allowed to dry in air. When dry, it is wax finished. The darker areas tend to darken slightly on exposure to daylight.

The initial immersion in the colouring solution should not be prolonged beyond the reddish-gold stage, or the results tend to be dull after cold water immersion.

▲ Care should be taken to avoid inhaling the fine dust that may be raised from the calcium hypochlorite (bleaching powder). It should also be prevented from coming into contact with the eyes or skin.

3.56   Greenish-brown on pink   Semi-matt

A Ammonium chloride	10 gm
Copper sulphate	10 gm
Water	1 litre

B Hydrogen peroxide (100 vols.)	200 cm³
Sodium chloride	5 gm
Acetic acid (glacial)	5 cm³
Water	1 litre

Cold immersion (A)   (Five minutes)
Hot immersion (B)   (A few seconds)

The object is immersed in solution A for about five minutes, producing a slight etching effect, and a pale pink colour typical of chloride-containing solutions. The object is then briefly immersed in solution B (at 60–70°C). The reaction tends to be violent, causing the solution to 'boil', and immersion should not be prolonged. A greenish-brown layer is produced. If this is uneven, the surface should be gently scratch-brushed and the object briefly re-immersed. Prolonged or repeated immersion in solution B will tend to strip the greenish-brown colour. When treatment is complete, the object is washed in warm water and allowed to dry in air, and may be wax finished when dry.

The surface finish is difficult to control, and the procedure cannot be generally recommended.

▲ Hydrogen peroxide can cause severe burns, and must be prevented from coming into contact with the eyes and skin. When preparing the solution, it should be added to the water prior to the addition of the other ingredients.

▲ Glacial acetic acid is very corrosive and must be prevented from coming into contact with the eyes or skin.

3.57   Metallic pink 'grained' surface   Semi-matt

Ferric oxide	25 gm
Nitric acid (70% solution)	100 cm³
Iron filings	5 gm
Water	1 litre

Cold immersion (Fifteen minutes)

The ferric oxide is dissolved in the nitric acid (producing toxic fumes), and the mixture is added to the water. The iron filings are finally added. The article is immersed in the solution, which produces an etched surface, and removed after about fifteen minutes. It is washed thoroughly in cold water and allowed to dry in air, and may be wax finished when dry.

The metal is etched by the solution, producing a grained effect, similar to a directionally polished satin finish. Copper-plated surfaces are rapidly etched by the solution to reveal the underlying metal.

▲ Preparation and colouring should be carried out in a fume cupboard. Toxic and corrosive brown nitrous fumes are evolved which must not be inhaled.

▲ Concentrated nitric acid is extremely corrosive, causing burns, and must be prevented from coming into contact with the eyes and skin.

3.58   Metallic pink 'grained' surface   Semi-matt

Copper carbonate	100 gm
Ammonium chloride	100 gm
Ammonia (.880 solution)	20 cm³
Water	1 litre

Hot immersion (Fifteen minutes)

The article is immersed in the hot solution (50°C) producing immediate etching effects, typical with chloride-containing solutions. After about fifteen minutes, the article is removed and washed thoroughly in warm water. It is allowed to dry in air, and may be wax finished when dry.

The metal is etched by the solution, producing a 'grained' surface similar to a directionally polished satin finish.

▲ When preparing the solution, care should be taken to avoid inhaling the ammonia vapour. The eyes and skin should be protected from accidental splashes of the concentrated ammonia solution.

3.59   Thin orange-brown Satin

Copper acetate	120 gm
Ammonium chloride	60 gm
Water	1 litre

Boiling immersion. Scratch-brushing (Ten minutes)

The object is immersed in the boiling solution, which rapidly etches the surface and produces a pink/orange colour typical of chloride solutions. After ten minutes, it is removed and scratch-brushed with the hot solution, removing most of the pink/orange layer. The object is then washed, allowed to dry in air, and is wax finished when dry.

The thin pink colour left after scratch-brushing darkens in daylight to an orange or orange-brown. The results produced tend to be very patchy and although the surface effect is interesting, it is very difficult to control and cannot be recommended.

This method is not suitable for use with copper-plated surfaces.

3.60   Dull pink (slightly grained) Semi-matt

Copper nitrate	80 gm
Nitric acid (10% solution)	100 cm³
Water	1 litre

Hot immersion (Five minutes)

The article is immersed in the hot solution (60–70°C), which causes etching of the surface and a gradual darkening. After five minutes it is removed and washed in warm water, and allowed to dry in air. When dry, it may be wax finished.

(Some sources suggest using this as a ground for a greenish patina, produced by applying a solution of ammonium carbonate (100 gm/litre) by wiping with a soft cloth and allowing to dry. Tests for extended periods produced only the slightest traces of a green tinge.)

▲ Nitric acid is corrosive and may cause severe irritation or burns. It should be prevented from coming into contact with the eyes or skin.

3.61   Dull pink and yellowish-green (etched grain) Matt

Copper sulphate	25 gm
Ammonium chloride	25 gm
Lead acetate	25 gm
Water	1 litre

Hot immersion (Fifteen minutes)

The object is immersed in the hot solution (80°C) producing slight etching effects and pale orange/pink colours. After fifteen minutes, the object is removed, washed in hot water and allowed to dry in air. Exposure to daylight causes a darkening of the colour. When the colour has stabilised and the object is completely dry, it is wax finished.

On copper surfaces the grain of the rolled sheet is etched, to produce a fine grained surface tinged with the overlying colours. The results tend to be patchy and are difficult to control. The technique cannot be generally recommended for either copper sheet or copper-plated surfaces.

▲ Lead acetate is very toxic and every precaution should be taken to prevent inhalation of the dust during weighing and preparation of the solution. The fumes evolved from the hot solution must not be inhaled.

3.62   Brown Semi-matt

Copper sulphate	25 gm
Sodium chloride	100 gm
Potassium polysulphide	10 gm
Water	1 litre

Hot immersion (Fifteen minutes)

The object is immersed in the hot solution (60°C), producing a slight matting of the surface due to etching, and the gradual development of dull brown colours. After about fifteen minutes, when the brown has developed, the object is removed. Relieving may be carried out at this stage, using the hot solution with the scratch-brush. The object is then removed, washed in warm water and allowed to dry in air. When dry, it is wax finished.

The resulting brown surface tends to be uneven. Gentle relieving produces a more even surface with a warmer tone. Overworking will break through to the bare metal.

▲ The colouring should be carried out in a well ventilated area, as some hydrogen sulphide gas will be liberated.

3.63   Brown bronzing with darker 'figure' Semi-matt

Copper sulphate	30 gm
Potassium permanganate	6 gm
Water	1 litre

Boiling immersion and scratch-brushing (Several minutes)

On immersion in the boiling solution, the object becomes coated with a brownish film, which develops to a dark purplish brown after one minute. This brown film is loosely adherent and sloughs off gradually to reveal uneven underlying grey-brown colours. After three or four minutes the object is removed and gently scratch-brushed with the hot solution. Overworking will cut through to the underlying metal. If the surface is unsatisfactory then the process is repeated. When the desired colour has been obtained, the object is washed in hot water, dried either in sawdust or in air, and wax finished when dry.

The results obtained in tests on both copper and copper-plated surfaces were variable, in both the degree of 'figure' obtained and in the overall depth of colour.

The development of the loose brown layer appears to be essential to successful colouring. If the solution is boiled for too long before the object is immersed, then a more dense precipitate is formed in the solution, and coating does not take place.

3.64   Variegated brown Semi-matt

Copper sulphate	200 gm
Ferric chloride	25 gm
Water	1 litre

Hot immersion (Thirty minutes)

Polished surfaces are etched to a slightly 'grained' finish. The colours produced are very uneven, and the procedure cannot be generally recommended.

▲ Ferric chloride is corrosive and should be prevented from coming into contact with the eyes and skin.

The object is immersed in the hot solution (60°C), which etches the surface and produces a pale pink colour, typical of chloride-containing solutions. Continued immersion produces a gradual change to a more brown colour. After thirty minutes the object is removed and washed in hot water, which tends to leach a slight yellowish-green colour from the surface. After thorough washing, the object is allowed to dry in air and exposed to daylight, which tends to cause an uneven darkening of the colour. When the surface development is complete and the object thoroughly dried out, which may take several days, it can be wax finished.

3.65   Light and dark brown (variegated) Semi-matt

Copper acetate	30 gm
Ammonium chloride	15 gm
Ferric oxide	30 gm
Water	1 litre

Hot immersion. Hot scratch-brushing (Ten to fifteen minutes)

An alternative technique which is more suited to larger objects is to heat locally with a torch and apply the pasty liquid with the scratch-brush, without prior immersion.

▲ A nose and mouth face mask fitted with a fine dust filter should be worn when brushing off the dry residue.

The solution is brought to the boil and then removed from the source of heat. The object is then immersed, and becomes coated with a thick brown layer. After ten minutes immersion the object is removed and gently heated on a hotplate, the surface being worked with the scratch-brush. More solution is applied and worked in with the scratch-brush, occasionally allowing the surface to dry off without scratch-brushing. When the colour has developed it is allowed to dry on the hotplate for a few minutes, removed and the dry residue removed with a stiff bristle-brush. When the surface is free of residue it is wax finished.

3.66   Reddish-brown Semi-gloss

Copper sulphate	500 gm
Copper acetate	100 gm
Acetic acid (30% solution)	25 cm^3
Water	1 litre

Hot scratch-brushing (Several minutes)

The colour produced tends to have a slightly dull or grey quality. Overworking with the scratch-brush tends to cut through to the metal, particularly in the region of surface features.

▲ Strong acetic acid is corrosive and should not be allowed to come into contact with the eyes and skin.

The article may be immersed in the hot solution (80°C) for about two minutes, or the solution may be applied liberally with a soft brush, to allow the colour to develop. The surface is then gently scratch-brushed with the hot solution until an even finish is produced. Relieved effects may be obtained by continued working, without the hot solution. The article is thoroughly washed in hot water, and either dried in sawdust, or allowed to dry in air. When dry it may be wax finished.

3.67   Dark-reddish brown (variegated) Semi-gloss

Ferric chloride	10 gm
Water	1 litre

Torch technique

▲ The fumes evolved as the solution is applied to the heated metal should not be inhaled. Adequate ventilation must be provided and a nose and mouth face mask fitted with the correct filter should be worn.

▲ Ferric chloride is corrosive and should not be allowed to come into contact with the eyes and skin. A face mask or goggles should be worn to protect the eyes from hot splashes as the solution is applied.

The object is heated with a blow torch and the solution applied sparingly with a scratch-brush using small circular motions and working across the surface. The metal gradually darkens to a reddish-brown tone. The variegated surface produced can be made more even by lightly dabbing with a cloth dampened with the solution. When treatment is complete, the object is allowed to cool and dry thoroughly, and may then be wax finished.

3.68   Dull reddish-brown Satin

Copper acetate	50 gm
Copper nitrate	50 gm
Water	1 litre

Immersion boiling (Thirty minutes)

Immersing the object in the boiling solution produces lustrous golden colours after one minute, which gradually change to more opaque matt pink/orange tints after five minutes. Continued immersion produces a gradual darkening to a terracotta. After thirty minutes the object is removed, washed in hot water and dried in sawdust. When completely dry it is wax finished.

3.69   Brown/dark brown Gloss

Copper carbonate	125 gm
Ammonia (.880 solution)	250 cm³
Water	750 cm³

Warm immersion (One hour)

The ammonia is added to the water, followed by the carbonate, which will partially dissolve. Some excess carbonate should remain in suspension. The solution is heated to about 50°C, and the object is immersed. The colour develops slowly during the period of immersion. After about one hour, when the colour has fully developed, the object is removed and washed thoroughly in warm water. It should be carefully dried in sawdust, and wax finished when dry.

For bi-metallic effects on copper-plated surfaces see recipe 3.228.

▲ This procedure should only be carried out where there is adequate ventilation. Ammonia vapour will be freely liberated when the solution is prepared, and during colouring, which can severely irritate the eyes and respiratory system. The skin, and particularly the eyes, must be protected against accidental splashes of this corrosive alkali.

3.70*   Dark reddish-brown Semi-matt   *Pl. VI*

Copper sulphate	125 gm
Ferrous sulphate	100 gm
Acetic acid (glacial)	6.5 cm³
Water	1 litre

Boiling immersion (Thirty to forty minutes)

The object is immersed in the boiling solution. The surface darkens gradually during the course of immersion, initially tending to be drab and rather uneven, but later becoming more even and generally darker and richer in colour. When the colour has developed, after thirty to forty minutes, the object is removed and washed in hot water. It is dried in sawdust, and wax finished when dry.

A bi-metallic effect can be produced on copper-plated surfaces using this solution. See recipe 3.227.

Plate VI shows the result on copper-plated surfaces.

▲ Acetic acid in the highly concentrated 'glacial' form is corrosive, and must not be allowed to come into contact with the skin, and more particularly the eyes. It liberates a strong vapour which is irritating to the respiratory system and eyes.

3.71   Dark purplish-brown Gloss

Ammonium carbonate	60 gm
Copper sulphate	30 gm
Oxalic acid (crystals)	1 gm
Acetic acid (6% solution)	1 litre

Hot immersion (Thirty minutes)

The object is immersed in the hot solution (80°C) producing light brownish colours after two or three minutes, which may be uneven initially. With continued immersion the colours become even and darken gradually. When the colour has fully developed, after about thirty minutes, the object is removed and washed in hot water. After thorough drying in sawdust, it may be wax finished.

Copper-plated surfaces tended to acquire a cloudy discolouration during the latter stages of immersion, overlaying the dark brown colour with a greenish tinge. A shorter immersion time of about twenty minutes produced better results.

A bi-metallic effect is produced with copper-plated surfaces. See recipe 3.224.

▲ Oxalic acid is poisonous and may be harmful if taken internally.

▲ The ammonium carbonate should be added to the acetic acid in small quantities. The reaction is effervescent, and if large amounts are added the effervescence becomes very vigorous.

3.72   Purplish-brown Semi-gloss

Sodium thiosulphate	90 gm
Potassium hydrogen tartrate	30 gm
Zinc chloride	15 gm
Water	1 litre

Hot immersion (Thirty minutes)

The article is immersed in the hot solution (50–60°C) which darkens the surface and gradually produces a bloom, which should be removed by bristle-brushing. Continued immersion gives rise to a brown colour, which slowly darkens. After about thirty minutes, when the colour has developed, the article is removed and washed in warm water. It is either dried in sawdust or allowed to dry in air, and may be wax finished when dry.

The surface tends to be cloudy and slightly uneven or streaked.

3.73*   Purplish-brown Gloss

A	Ammonium carbonate	180 gm
	Copper sulphate	60 gm
	Acetic acid (6% solution)	1 litre

B	Oxalic acid	1 gm
	Ammonium chloride	0.25 gm
	Acetic acid (6% solution)	500 cm³

Hot immersion (Fifteen minutes)

Solution A is boiled and allowed to evaporate gradually until its volume is reduced by half. Solution B is then added and the mixture heated to bring the temperature to 80°C. The article is immersed in the hot solution, the colour gradually darkening until it becomes brown after ten minutes. After fifteen minutes, when the colour has fully developed, the article is removed and rinsed carefully in warm water. The surface should not be disturbed at this stage. It is allowed to dry in air, and wax finished when completely dry.

The slight bloom that may occur during the later stages of immersion is virtually invisible after wax finishing.

▲ The ammonium carbonate should be added to the acetic acid in small quantities. The reaction is effervescent, and if large amounts are added the effervescence is very vigorous.

▲ Preparation of the solution, and colouring, should be carried out in a well ventilated area. Acetic acid vapour will be liberated when the solutions are boiled.

▲ Oxalic acid is harmful if taken internally, and irritating to the eyes and skin.

3.74*   Red-brown Satin/semi-gloss

Antimony trisulphide	30 gm
Ferric oxide	10 gm
Ammonium sulphide (16% solution) to form a paste	

Applied paste (sulphide paste) (Several minutes)

The ingredients are mixed to a thick paste using a pestle and mortar, by the very gradual addition of a small quantity of ammonium sulphide solution. The paste is applied to the whole surface of the article to be coloured, using a soft bristle-brush. A stiff nylon brush is then used to work the surface gradually in small areas. The nylon brush should be intermittently dipped in a mixture of the dry ingredients of the paste (in the same proportion as above), and the surface worked in this way until the paste becomes dry and falls from the the surface. When the whole surface is dry and completely free of any residue or dust, it may be burnished with a brass scratch-brush and wax finished.

A variegated surface can be obtained by brushing the paste onto the whole surface and allowing it to dry for about three hours, until it is superficially dry and cracking, and then burnishing it away with a nylon brush or a brass scratch-brush. This may be repeated, as necessary, to darken the colour.

▲ Ammonium sulphide solution is very corrosive and precautions should be taken to protect the eyes and skin while the paste is being prepared. It also liberates harmful fumes of ammonia and of hydrogen sulphide, and the preparation should therefore be carried out in a fume cupboard or well ventilated area. The paste once prepared is relatively safe to use but should be kept away from the skin and eyes.

▲ A nose and mouth face mask fitted with a fine dust filter should be worn when the residues are brushed away. The dust produced in this case contains antimony compounds, which are toxic.

3.75*   Dark purplish-brown/black Satin/semi-matt

Antimony trisulphide	12.5 gm
Sodium hydroxide	40 gm
Water	1 litre

Hot scratch-brushing (Several minutes)

The hot solution (60°C) is repeatedly applied liberally to the surface of the object with a soft brush, allowing time for the colour to develop. The surface is then worked lightly with the scratch-brush, while applying the hot solution and hot water alternately. The treatment is concluded by lightly burnishing the surface with the scratch-brush and hot water. After rinsing it is dried in sawdust, and wax finished when dry.

A lighter purplish-brown or reddish-brown is produced on copper-plated surfaces. This can be obtained without scratch-brushing.

▲ Sodium hydroxide is a powerful caustic alkali which can cause severe burns to the eyes and skin, in the solid form or in solution. When preparing the solution, small quantities of the sodium hydroxide should be added to a large quantity of water and not vice-versa. When the sodium hydroxide has completely dissolved, the antimony trisulphide is added.

3.76*   Orange-brown Semi-matt

Potassium sulphide	20 gm
Ammonium chloride	30 gm
Water	1 litre

Cold scratch-brushing (Several minutes)

The object is scratch-brushed with the cold solution, and worked in under hot water. A grey colouration is immediately produced. The surface is evened with the scratch-brush and then burnished with hot water, which brings out a warmer tone to the grey colour. The object is then allowed to dry completely, and is lightly burnished with a dry scratch-brush, which changes the colour to a dark brown and finally to a rich warm mid-brown. The object may be wax finished.

A dark grey or black colour can be produced with this solution, see recipe 3.90.

▲ The procedure should be carried out in a well ventilated area, as hydrogen sulphide gas is liberated by the solution.

3.77   Variegated dark brown/black on dark brown ground Semi-matt

Potassium polysulphide	4 gm
Water	1 litre

Torch technique

The metal is heated with a blow torch and the solution applied sparingly with a scratch-brush using small circular motions and working across the surface until a matt black colour appears. The amount of heat and solution used are critical. If too much solution is applied then the surface tends to strip, and if too little heat is used then the solution does not act on the metal. When the matt black colour has been obtained, it should be lightly brushed with a dry scratch-brush. This brings up the surface and highlights any faults, that can then be worked on with further heating and applications of solution. When treatment is complete, the object is allowed to cool and dry thoroughly, and may then be wax finished.

▲ The fumes evolved as the solution is applied to the heated metal should not be inhaled. Adequate ventilation must be provided.

▲ A face mask or goggles should be worn to protect the eyes from hot splashes as the solution is applied.

3.78*   Dark purplish-brown Satin/semi-gloss   *Pl. VI*

Antimony trisulphide	20 gm
Ferric oxide	20 gm
Potassium polysulphide	2 gm
Ammonia (.880 solution)	2 cm³
Water	to form a paste (approx 4 cm³)

Applied paste (sulphide paste) (Several minutes)

The potassium polysulphide is ground using a pestle and mortar and the other solid ingredients added and mixed. These are made up to a thick paste by the gradual addition of the ammonia and a small quantity of water. The paste is applied to the whole surface of the object to be coloured, using a soft bristle-brush. A stiff nylon brush is then used to work the surface gradually in small areas. The nylon brush should be intermittently dipped in a mixture of the dry ingredients of the paste (in the same proportion as above), and the surface worked in this way until the paste becomes dry and falls from the surface. When the whole surface is dry and completely free of any residue or dust, it may be burnished with a brass scratch-brush and wax finished.

A variegated surface can be obtained by brushing the paste on to the whole surface and allowing it to dry for about three hours, until it is superficially dry and cracking, and then burnishing it away with a nylon brush or a brass scratch-brush. This may be repeated as necessary to darken the colour.

The copper-plated sample shown in plate VI was prepared by applying a punched decorative texture, and abrading the raised portions to partly reveal the underlying brass. Application of the paste in the manner described above produced a bi-coloured effect rather than a plain brown surface.

▲ A nose and mouth face mask should be worn when the residues are brushed away.

3.79   Brownish-orange Gloss

Copper acetate	6 gm
Copper sulphate	1.5 gm
Sodium chloride	1.5 gm
Water	1 litre

Boiling immersion (About one hour)

The article is immersed in the boiling solution and after a few minutes the surface darkens and becomes more opaque as the colour develops. The surface develops gradually as immersion continues, and after about one hour the article is removed and washed in hot water. It is dried in sawdust and may be wax finished when completely dry.

A reddish-brown colour is produced on copper-plated surfaces.

A bi-metallic colouring effect can be obtained on copper-plated surfaces, see recipe 3.215.

3.80   Variegated black and brown Gloss

Copper sulphate	100 gm
Ammonium chloride	0.5 gm
Water	1 litre

Applied liquid to heated metal (Several minutes)

The object is gently heated on a hotplate and the solution applied either by brushing or by wiping with a soft cloth. The object should be heated to a temperature such that when the solution is applied, it sizzles gently. Application is continued until orange-brown colours have developed. Exposure to daylight will cause the surface to darken unevenly to a mixed black and brown. When changes in colour have ceased, the object is wax finished.

It is not possible to control the disposition of the black or brown colours on the surface.

3.81   Dark purplish-brown Semi-gloss

Sodium thiosulphate	6.25 gm
Ferric nitrate	50 gm
Water	1 litre

Hot immersion   (One minute)

During preparation, the nitrate solution is darkened by the addition of the thiosulphate, but slowly clears to give a straw-coloured solution.

Copper and copper-plated surfaces are sensitive to temperatures above 60°C, which tend to strip the surface colour.

A succession of lustre colours is rapidly produced when the article is immersed in the hot solution (50–55°C), changing to a more opaque purplish colour after about forty-five seconds. The colour darkens quickly and the article is removed after about one minute. After thorough washing it is allowed to dry in air, and then wax finished.

3.82   Cloudy grey/black on brown Semi-matt

Copper (cupric) chloride	50 gm
Water	1 litre

Hot immersion   (Ten minutes)

The results produced in tests were poor. The colour tended to be uneven and patchy and prone to spotting. The procedure cannot be generally recommended.

The object is immersed in the hot solution (70–80°C) producing etching effects and a pale salmon-pink colour. With continued immersion a denser and slightly more brown colour develops. After ten minutes, the object is removed and washed in hot water, removing the loose superficial layer. It is then allowed to dry in air, and exposed to daylight which darkens the colour. When all colour change has ceased, which may take several days, the object may be wax finished.

3.83   Pale orange ochre Semi-gloss

Copper nitrate	80 gm
Water	1 litre
Ammonia (.880 solution)	3 cm³

Applied liquid   (Ten days)

Copper and copper-plated surfaces are very resistant to wetting by this solution. This can be overcome by persistant dabbing and a very sparing use of the solution. The resulting surface tends to be irregularly mottled.

Flecks of blue-green patina are likely to form if the solution is used less sparingly, particularly in the later stages of treatment.

The solution is applied to the metal surface by dabbing with a soft cloth. The solution should be applied sparingly and may require vigorous dabbing in the initial stages to ensure adequate wetting. After each application the surface is allowed to dry completely in air. This procedure is repeated twice a day for ten days. The article should be allowed to dry for several days before wax finishing.

3.84   Variegated orange-brown and dark brown Semi-gloss

Ferric oxide	30 gm
Lead dioxide	15 gm
Water	to form a paste

Applied paste and heat   (Repeat several times)

Test results tended to be poor, producing unevenly variegated surfaces.

▲ Lead dioxide is toxic and very harmful if taken internally. The dust must be not inhaled. A nose and mouth face mask with a fine dust filter must be worn when brushing away the dry residue. Colouring must be carried out in a well ventilated area. The fumes evolved must not be inhaled.

The ingredients are mixed to a paste with a little water. The paste is applied to the surface with a soft brush, while the object is gently heated on a hotplate or with a torch. When the paste is completely dry, it is brushed off with a stiff bristle-brush. The procedure is repeated a number of times, until the colour has developed. After the final brushing, the object may be wax finished.

3.85   Reddish-brown and metallic lustre (unevenly variegated) Semi-matt

Copper nitrate	40 gm
Water	to form a paste

Applied paste   (Four to six hours)

A mixed surface of reddish-brown and metallic patches is produced. In tests it was found that the surface quality could not be controlled, patches of bare metal and staining tending to appear. Although the colours produced are interesting, the procedure cannot generally be recommended.

A true paste will not form, but the damp crystalline mass that is obtained can be applied with a brush. This tends to become watery and may not dry out.

The copper nitrate is ground using a pestle and mortar, and a very small amount of water added by the drop to form a crystalline 'paste'. The paste is applied using a soft brush, and the surface then allowed to dry for several hours. When dry, the residue is washed away with cold water. The process is repeated if necessary. After washing and air-drying, the object may be wax finished.

3.86*   Dark reddish-brown Gloss

Potassium sulphide	1 gm
Ammonium chloride	4 gm
Water	1 litre

Cold scratch-brushing (Five minutes)

The object is scratch-brushed with the cold solution for a short time and then washed in hot water. This alternation is repeated until the colour develops. The surface is then gently burnished with the scratch-brush in hot water, and finally with cold water. After thorough rinsing and drying in air the object may be wax finished.

A more concentrated hot solution can be used to produce a black colour. See recipe 3.88.

▲ The procedure should be carried out in a well ventilated area, as some hydrogen sulphide gas is liberated from the solution.

3.87   Dark brown/black (variegated) Semi-matt

Copper sulphate	6.25 gm
Copper acetate	1.25 gm
Potassium nitrate	1.25 gm
Sodium chloride	2 gm
Sulphur (flowers of sulphur)	3.5 gm
Acetic acid (6% solution)	1 litre

Boiling immersion (About one hour)

The article is immersed in the boiling solution and after a few minutes the surface darkens and becomes more opaque as the colour develops. The surface develops gradually as immersion continues, and after about one hour the article is removed and washed in hot water. Initially the colour is a variegated pinkish-brown, but this darkens in daylight to a dark brown or black colour, which also tends to be slightly variegated. When the dark colour has developed, the article may be wax finished.

Similar results are obtained on copper-plated surfaces. These are coloured reddish-brown initially, but darken to near black on exposure to daylight.

A bi-metallic colouring effect can be produced on copper-plated surfaces, see recipe 3.216.

▲ Potassium nitrate is a powerful oxidising agent and must not be allowed to come into contact with combustible materials. It should not be mixed with the flowers of sulphur in the solid state.

3.88   Blue-black (*or* dark brown) Semi-matt

Potassium sulphide	10 gm
Ammonium chloride	10 gm
Water	1 litre

Hot scratch-brushing (Five minutes)

The object is scratch-brushed with the hot solution (50–60°C) for several minutes. The solution is then progressively diluted with hot water as scratch-brushing continues. The object will develop a brown colour at first, which will become darker as scratch-brushing is prolonged, finally producing a semi-matt blue-black surface with a slightly frosted appearance. The object should finally be gently burnished by scratch-brushing with hot water. After thorough washing and drying in air the object may be wax finished.

For results obtained using a more concentrated solution, see recipe 3.95.

▲ The procedure should be carried out in a well ventilated area, as some hydrogen sulphide gas is liberated from the solution.

3.89*   Black mottle or variegation on reddish-brown Semi-gloss/satin   *Pl. V*

Potassium sulphide	10 gm
Ammonia (.880 solution)	1–2 cm³

Applied paste and scratch-brushing

The potassium sulphide is ground to a thin paste with a little ammonia using a pestle and mortar. The thin paste is stippled on to the surface of the object using a soft cloth, until the whole surface is covered. An immediate deep black colour is produced as the paste is applied. The black layer is allowed to remain on the surface undisturbed for one or two minutes, and is then scratch-brushed in hot water. The black layer is only partially adherent, and when worked with the scratch-brush tends to break up to give a variegated black on brown surface. Continued working consolidates the remaining black areas, and enriches the exposed portions of the surface to a reddish brown. If the results are unsatisfactory, then the whole surface should be wiped with the paste, the object thoroughly washed and the procedure repeated. When satisfactory results have been obtained, the object is washed in cold water and allowed to dry in air. It may be wax finished when dry.

The paste is not suitable for use with copper-plated surfaces, which are destructively attacked.

If the surface is worked with the scratch-brush too soon after application of the paste, then the surface tends to become steel grey in colour. If this occurs, the surface should be wiped with the paste and thoroughly washed, and the procedure re-started.

▲ The ingredients and the paste are corrosive and will cause burns or severe irritation to the eyes and skin. Ammonia vapour and some hydrogen sulphide gas will be liberated from the paste, and these must not be inhaled. Colouring should be carried out in a well ventilated area.

3.90*   Dark grey/black Semi-matt

Potassium sulphide	20 gm
Ammonium chloride	30 gm
Water	1 litre

Cold scratch-brushing (Several minutes)

The solution is laid on to the surface of the object with a scratch-brush, which is used very lightly to even the finish. The surface is then burnished with the scratch-brush under cold water. The object is then thoroughly dried, and may be wax finished when dry.

An orange-brown colour can be produced with this solution, see recipe 3.76.

▲ The procedure should be carried out in a well ventilated area, as hydrogen sulphide gas is liberated by the solution.

3.91*   Blue-black Satin/semi-gloss

Potassium sulphide	2 gm
Ammonium sulphate	0.5 gm
Water	1 litre

Cold scratch-brushing (Several minutes)

The object is scratch-brushed with the cold solution, alternating with hot and then cold water. This process is continued until the colour has developed. Any relieving that is required should then be carried out. The object is washed thoroughly in cold water and allowed to dry in air. When completely dry it is wax finished.

The surface is improved by gentle burnishing with the scratch-brush, first with hot water and then with cold water, after colouring.

3.92   Dark purplish-brown/black Semi-matt

Sodium sulphide	50 gm
Antimony trisulphide	50 gm
Water	1 litre
Acetic acid (10% solution)	100 cm^3

Hot scratch-brushing (Several minutes)

The solid ingredients are added to the water and dissolved by gradually raising the temperature to boiling. The solution is allowed to cool to 60°C and then acidulated gradually with the acetic acid, producing a dense brown precipitate. The object is scratch-brushed with this turbid solution, alternating with hot and then cold water. When the colour has developed it is washed in cold water, allowed to dry in air, and wax finished when dry.

In tests, similar colours were obtained on copper-plated surfaces.

In tests carried out on copper, the dark purplish-brown colour tended to be overlaid with a slight petrol lustre, which is visible from oblique angles. If the surface is overworked the colour will tend to a 'steely' dark slate grey.

▲ Colouring must be carried out in a well ventilated area, as hydrogen sulphide gas is readily evolved from the acidified sulphide solution.

▲ Sodium sulphide is corrosive and should be prevented from coming into contact with the eyes and skin.

3.93*   Dark purplish-brown/black Gloss   *Pl. V, Pl. VI*

Sodium hydroxide	25 gm
Potassium persulphate	10 gm
Water	1 litre

Hot immersion (One or two minutes)

The potassium persulphate is dissolved in the water, and the sodium hydroxide then added in small quantities while stirring. The solution is then brought up to the correct temperature (70–80°C) and the article immersed. An immediate lustrous red colouration is produced, which darkens rapidly to a very dark brown or black after about one or two minutes. When the desired depth of colour is reached, the article is removed and washed in warm water. It is dried in sawdust and may be wax finished when dry.

Plate VI shows the result on copper-plated surfaces.

▲ Potassium persulphate is a powerful oxidant and should not be allowed to come into contact with combustible material. The sodium hydroxide and the persulphate must not be allowed to come into contact in the solid state.

▲ Sodium hydroxide is a powerful caustic alkali which can cause severe burns. Both the solid and solutions should be prevented from coming into contact with the eyes and skin.

3.94*   Very dark reddish-brown/black

Ammonium persulphate	10 gm
Sodium hydroxide	30 gm
Water	1 litre

Boiling immersion (Two or three minutes)

The ammonium persulphate is dissolved in half the water, and the sodium hydroxide separately dissolved in the remaining water. The two solutions are then mixed together and the mixture heated to boiling. The article to be coloured is immersed in the boiling solution, which produces golden and then brown lustre colours. Continued immersion darkens these colours to an opaque but glossy finish. When the colour has developed, after about two or three minutes, the article is removed, washed in hot water and dried in sawdust. When dry, it may be wax finished.

A bi-metallic colouring effect can be obtained on copper-plated yellow brass, see recipe 3.225.

▲ Ammonium persulphate (peroxodisulphate) is a powerful oxidant, and must be kept out of contact with all combustible materials. It should be prevented from coming into contact with the eyes, skin or clothing.

▲ Sodium hydroxide is a powerful caustic alkali and must be prevented from coming into contact with the eyes and skin.

3.95* Dark reddish-brown/black Satin/semi-matt

Potassium sulphide	125 gm
Water	1 litre
Ammonia (.880 solution)	100 cm³

Cold application (Several minutes)

The potassium sulphide is ground to a paste with some of the water and a little ammonia, using a pestle and mortar. This paste is added to the rest of the water, and the remaining ammonia, to form a strong solution. The solution is wiped on to the surface of the object with a soft cloth. A black layer forms which is only partly adherent, and this is removed by washing and light bristle-brushing in water, revealing a reddish-brown colour. The procedure is repeated if necessary until a satisfactory colour is obtained. The object is then washed in cold water, allowed to dry in air, and wax finished when dry.

This procedure is not recommended for use on copper-plated surfaces.

If the surface is worked with a scratch brush, a steel grey colour tends to be produced, which darkens to black when the procedure is repeated. It is difficult to obtain the reddish brown colour once the steel grey colour has occurred.

▲ Preparation and colouring should be carried out in a well ventilated area. Some hydrogen sulphide gas will be liberated by the potassium sulphide solution. Ammonia vapour will also be liberated from the solution. These should not be inhaled. The solution should be prevented from coming into contact with the eyes and skin as it may cause burns or severe irritation.

3.96* Black Semi-matt/matt

Potassium sulphide	10 gm
Water	1 litre
Ammonia (.880 solution)	1 cm³

Warm immersion and bristle-brushing (About ten minutes)

The object is immersed in the warm solution (40°C) and becomes coated with a black film within one or two minutes. The surface is evened by bristle-brushing when this has developed, and immersion continued. After a further two minutes the object is removed and the surface evened by gently brushing with a brass brush. The object is re-immersed and left for several minutes to allow the matt black layer to develop. It is then removed and allowed to dry in air, without washing. When dry, it is gently rubbed with a soft cloth to remove any loose dust, and wax finished.

Treatment with this solution is not suitable for copper-plated surfaces, which tend to be destructively attacked.

As an alternative to immersion, the solution can be applied liberally to the surface using a soft brush, and without 'working' the surface.

▲ Some hydrogen sulphide gas will be liberated from the solution. Colouring should be carried out in a well ventilated area.

3.97 Variegated black Semi-matt

Potassium polysulphide	20 gm
Water	1 litre

Torch technique

The object is heated with a blow torch and the solution applied sparingly with a brass scratch-brush using small circular motions and working across the surface. A matt black colour quickly develops. Once this ground colour has been established, the surface can be variegated by lightly dabbing with a cloth dampened with the solution. When treatment is complete, the object is allowed to cool and dry thoroughly, and may then be wax finished.

▲ The fumes evolved as the solution is applied to the heated metal should not be inhaled. Adequate ventilation should be provided.

▲ A face shield or goggles should be worn to protect the eyes from hot splashes as the solution is applied.

3.98 Dark slate with purplish tint Semi-matt

Ammonium sulphide (16% solution)	100 cm³
Water	1 litre

Hot bristle-brushing (Several minutes)

The article is directly bristle-brushed with the hot solution (60°C) for several minutes, alternating with bristle-brushing under hot water. The procedure is repeated three or four times until the colour has developed and the article is then washed in warm water. After thorough drying in air or sawdust, it is wax finished.

If an uneven colour is produced, the surface can be worked with a scratch-brush to obtain an even light colour, and the bristle-brush application then repeated.

▲ Ammonium sulphide solution liberates a vapour containing ammonia and hydrogen sulphide gas, which should not be inhaled. Preparation and colouring must be carried out in a well ventilated area. The solution must be prevented from coming into contact with the eyes or skin, as it causes burns or severe irritation.

3.99 Purplish-black Semi-matt

Ammonium sulphide (16% solution)	100 cm³
Water	1 litre

Cold immersion and scratch-brushing (Ten to thirty minutes)

The article is immersed in the cold solution for a few minutes to allow the colour to develop, and then removed and lightly scratch-brushed with the cold solution, alternating with hot and cold water. It is then re-immersed for about ten minutes and then removed and scratch-brushed as before. This procedure is repeated until the desired colour is achieved. The article is finally lightly scratch-brushed alternately with hot and cold water. It is allowed to dry in air and wax finished.

The colour tends to be slightly uneven, and may need to be worked with the scratch-brush. Prolonged working tends to produce a dull slate colour.

As an alternative to immersion in the cold solution, the solution can be applied continuously to the surface with a soft brush. This tended to produce lighter bronzed finishes in tests.

▲ Ammonium sulphide solution liberates a vapour containing ammonia and hydrogen sulphide gas, which should not be inhaled. Preparation and colouring must be carried out in a well ventilated area. The solution must be prevented from coming into contact with the eyes and skin, as it causes burns or severe irritation.

3.100   Black/dark reddish brown Gloss/semi-gloss

| Barium sulphide | 10 gm |
| Water | 1 litre |

Hot immersion (Forty minutes)

The object is immersed in the hot solution (50°C) and the surface is gradually tinted by a succession of lustre colours, which develop and recede. These slowly darken during the latter half of the period of immersion until the final black colour develops, after about forty minutes. The object is removed and washed carefully in warm water, and allowed to dry in air. When the surface is thoroughly dry, the object may be wax finished.

For results obtained using a cold solution see recipe 3.101.

The surface tends to be fragile when removed from the hot solution, and should be treated with care until dry.

▲ Barium sulphide is poisonous and very harmful if taken internally. Some hydrogen sulphide gas will be liberated as the solution is heated, and colouring should be carried out in a well ventilated area.

3.101   Black/dark slate Semi-gloss/gloss

| Barium sulphide | 10 gm |
| Water | 1 litre |

Cold immersion (Two hours)

After a few minutes' immersion in the cold solution, the surface of the object is coloured a lustrous brown. Continued immersion produces a series of gradually changing lustre colours, and more opaque colours during the latter stages. When the full colours have developed, after about two hours, the object is removed and washed in cold water and allowed to dry in air. When dry, the object is wax finished.

For results obtained with hot solutions see recipe 3.100.

▲ Barium sulphide is poisonous and may be very harmful if taken internally.

3.102*   Black Semi-gloss/semi-matt

Sodium thiosulphate	20 gm
Antimony potassium tartrate	10 gm
Water	1 litre

Hot immersion (Forty-five minutes)

The article is immersed in the hot solution (60°C) for ten minutes, producing a slight darkening of the surface and an orange tinge. The temperature of the solution is then gradually raised to 80°C and immersion continued. After forty-five minutes the article is removed, washed in warm water and allowed to dry in air. When dry, it is wax finished.

The black may have a slightly brown tinge.

Timings for the immersion may be variable. If the desired colour is produced in a shorter time, the object should be removed at that stage. The immersion should not be continued longer than necessary.

▲ Antimony potassium tartrate is poisonous and may be harmful if taken internally.

▲ The vapour evolved from the hot solution should not be inhaled.

3.103   Dark slate/blue-black Semi-matt/semi-gloss

Sodium sulphide	30 gm
Sulphur (flowers of sulphur)	4 gm
Water	1 litre

Hot immersion and scratch-brushing (Fifteen minutes)

The solids are added to about 200 cm³ of the water, and boiled until they dissolve and yield a clear deep orange solution. This is added to the remaining water and the temperature adjusted to 60–70°C. The object is immersed in the hot solution and the colour allowed to develop for about ten minutes. It is then removed and scratch-brushed with the hot solution, until the surface is even and the colour developed. The object should be washed thoroughly in hot water and dried in sawdust. When dry, it may be wax finished.

A slight 'orange-peel' surface tended to occur in tests. Overworking with the scratch-brush should be avoided.

▲ Sodium sulphide in solid form and in solution must be prevented from coming into contact with the eyes and skin, as it can cause severe burns. Some hydrogen sulphide gas may be evolved as the solution is heated. Colouring should be carried out in a well ventilated area.

3.104   Blue-black Gloss

Potassium sulphate	100 gm
Potassium sulphide	8 gm
Ammonium chloride	8 gm
Water	1 litre

Boiling immersion and scratch-brushing (Several minutes)

Immersion in the boiling solution produces an immediate darkening of the surface with lustre colours, followed by the gradual development of more opaque colour. This tends to occur unevenly. After a few minutes immersion, the object is removed and scratch-brushed with the hot solution. When the surface has evened and the colour developed, the object is washed in hot water, and allowed to dry in air. When dry, it is wax finished.

The blue-black surface tends to have a reddish tinge.

▲ Colouring should be carried out in a well ventilated area, as some hydrogen sulphide gas will be liberated from the hot solution.

3.105* Black Semi-matt   *Pl. V*

Potassium sulphide	5 gm
Barium sulphide	10 gm
Ammonia (.880 solution)	20 cm³
Water	1 litre

Hot immersion and scratch-brushing (Twenty-five minutes)

The article is immersed in the hot solution (80–90°C), and a grey/black layer forms on the surface after about three minutes. This should be brass-brushed to clear any non-adherent material. Immersion is continued and the surface brass-brushed every few minutes for the first ten minutes. After twenty-five minutes the article is removed, washed in hot water and allowed to dry in air. When dry, it may be wax finished.

If the loose layers that form on the surface during the first half of the period of immersion are not removed, then a variegated and scaly slate-grey surface tends to develop.

▲ Barium sulphide is poisonous and may be very harmful if taken internally.

▲ Ammonia vapour will be liberated during preparation and colouring, which must be carried out in a well ventilated area. Some hydrogen sulphide gas may also be liberated. The ammonia solution must be prevented from coming into contact with the eyes and skin, as it will cause burns or severe irritation.

3.106 Black Semi-matt

Sodium thiosulphate	60 gm
Copper sulphate	42 gm
Potassium hydrogen tartrate	22 gm
Water	1 litre

Hot immersion (Fifteen minutes)

After one minute of immersion in the solution (at 50–60°C), a yellow-orange colour forms on the surface of the article and gradually turns to bright green. The article is removed from the solution after fifteen minutes and allowed to dry in air, without washing in water. When the article is completely dry it is gently rubbed with a soft cloth, which removes the green layer, revealing the underlying black colour. The article is then wax finished.

The green layer is superficial, and easily removed, but may leave a slight green tinge to the underlying colour. In some tests this took the form of a slightly uneven clouding of the surface.

3.107 Dark brown/black (slightly variegated) Semi-matt

Sodium thiosulphate	65 gm
Water	1 litre
Antimony trichloride	until cloudy

Hot immersion (Thirty minutes)

The sodium thiosulphate is dissolved in the water and the antimony trichloride added by the drop, to form a white precipitate. The solution begins to become orange and cloudy at the top, and then gradually throughout. The solution is heated to 50°C, and when the temperature is reached, the object is immersed. The surface initially passes through the lustre cycle—golden after three minutes, purple after four minutes, pale metallic blue after ten minutes. After about fifteen minutes grey/white streaks appear on the surface, which darkens to produce an appearance like a black powdery layer. The article is removed after thirty minutes and washed gently in warm water, removing the powdery black deposit to reveal the dark brown beneath. After drying in sawdust or in air, the article is wax finished.

A slight clouding of the surface with a more matt grey finish tends to occur. A surface which is matted or satin-grained prior to colouring, tends to produce a more even and slightly lighter brown colour.

▲ Antimony trichloride is poisonous if taken by mouth. It is also irritating to the skin, eyes and respiratory system, and contact must be avoided.

▲ The vapour evolved from the hot solution must not be inhaled.

3.108 Black/dark slate Gloss/semi-gloss

Sodium thiosulphate	125 gm
Water	500 cm³
Copper sulphate	40 gm
Water	500 cm³

Cold immersion (Two hours)

The chemicals should be separately dissolved, and the two solutions then mixed together. The article is suspended in the cold solution, and after ten minutes' immersion shows signs of darkening. After twenty minutes a reddish-purple lustre appears and gradually changes to a blue-green lustre after thirty minutes. The sequence of lustre colours is repeated, becoming more opaque and producing a maroon colour after about one hour. With continued immersion the colour slowly darkens to black. After about two hours the article is removed, washed thoroughly and allowed to dry in air. When dry it may be wax finished.

The lustre colours produced during the immersion tend to be less clear than those obtained with some other thiosulphate solutions.

The dark slate/black colour produced tends to have a slightly cloudy appearance, although it is generally even.

3.109 Dark slate Semi-gloss

Ammonium sulphate	30 gm
Sodium hydroxide	40 gm
Water	1 litre

Boiling immersion (Thirty minutes)

The object is immersed in the boiling solution and the surface gradually darkens to produce a black colour, and acquires a slightly crystalline appearance. After thirty minutes the object is removed, washed in warm water and allowed to dry in air. Relieving is then carried out, and the object wax finished.

The dark slate layer tends to be thin and only tenuously adherent before finishing. The layer can be worked with a bristle-brush to reveal areas of bronzed copper surface beneath. Test results suggest that relieving effects can be obtained with this procedure.

▲ Sodium hydroxide is a powerful caustic alkali, the solid and strong solutions causing severe burns to the eyes and skin. When preparing the solution, small quantities of the sodium hydroxide should be added to large quantities of water and *not* vice-versa. The reaction is exothermic and will warm the solution. When the sodium hydroxide is completely dissolved, the ammonium sulphate is added and the solution may be heated.

3.110 Dark slate Semi-gloss

A	Copper nitrate	100 gm
	Water	1 litre
B	Potassium sulphide	5 gm
	Water	1 litre
	Hydrochloric acid (35%)	5 cm³

Boiling immersion (A) (Fifteen seconds)
Cold immersion (B) (Thirty minutes)

The object is immersed in the boiling solution A for fifteen seconds and then transferred to solution B. The surface darkens initially to produce a series of uneven petrol blue/purple interference colours, and then gradually changes to a slate grey. After thirty minutes the object is removed and rinsed, relieved by gentle scratch-brushing or pumice if desired, and finally washed and dried in sawdust. When completely dry it is wax finished.

This recipe is suitable for producing antiquing and relieved effects only, on polished surfaces, and cannot be generally recommended.

▲ Solution B should only be prepared and used in a fume cupboard or a very well ventilated area. When the acid is added to the potassium sulphide solution, quantities of hydrogen sulphide gas are evolved. This is toxic and can be dangerous even in moderate concentrations.

▲ Hydrochloric acid is corrosive, and should be prevented from coming into contact with the eyes and skin.

3.111 Black Semi-gloss

Sodium thiosulphate	50 gm
Ferric nitrate	12.5 gm
Water	1 litre

Hot immersion (Twenty minutes)

The article is immersed in the hot solution (60–70°C) and after about one minute the surface is coloured with a purple-blue lustre, which gradually recedes. After five minutes a brown colour slowly appears which changes to a slate grey with continued immersion. After about twenty minutes the article is removed, washed in hot water and allowed to dry in air. The surface may be fragile at this stage and should be handled as little as possible. When completely dry the article is wax finished.

If the temperature is too high the colour layer will be patchy and more fragile when removed from solution, revealing a tan colour beneath the black.

3.112 Dark brown/black Semi-matt/semi-gloss

Nickel sulphate	35 gm
Copper sulphate	25 gm
Potassium permanganate	5 gm
Water	1 litre

Hot immersion (Five to thirty minutes)

The object is immersed in the hot solution (80–90°C) and the surface immediately becomes covered with a dark brown layer. After four or five minutes it is removed and gently bristle-brushed with hot water. The dark layer is removed, revealing a brown 'bronzing' beneath. If darker colours are required then the process should be repeated. If, on the other hand, a full very dark brown/black colour with a more matt finish is required, the object should be re-immersed, and the dark brown layer will then re-form and should be left on the surface as immersion is continued. After thirty minutes the object is removed, and gently bristle-brushed under hot water. After thorough rinsing, and drying in sawdust, it is wax finished.

Repeated bristle-brushing and immersion produces a more glossy surface and a lighter brown than is produced by uninterrupted immersion.

A bi-metallic effect can be obtained on copper-plated surfaces using this solution. See recipe 3.226.

▲ Nickel compounds are a common cause of sensitive skin reactions. Precautions should also be taken to avoid inhalation of the dust when preparing the solution.

3.113* Black Matt

Potassium permanganate	5 gm
Copper sulphate	50 gm
Ferric sulphate	5 gm
Water	1 litre

Boiling immersion (Twenty minutes)

The article is immersed in the boiling solution. A black layer forms on the surface, which should not be disturbed. Immersion is continued for about twenty minutes, and the article is then removed and allowed to dry in air, without washing. The surface should not be handled at this stage. After several hours' drying, the article may be wax finished.

Identical results were obtained for both copper and copper-plated surfaces. If the black layer is brushed when wet, it tends to be partially removed, revealing an underlying pinkish-brown colour.

3.114 Dark greenish-grey/black (variegated) on brown Semi-matt/matt

Nickel ammonium sulphate	50 gm
Sodium thiosulphate	50 gm
Water	1 litre

Hot immersion (Fifteen minutes)

A purplish grey colour quickly forms on the surface of the object on immersion, becoming patchy and metallic grey after one minute. After two minutes the object should be removed from the solution and lightly scratch-brushed with water, to remove the non-adherent deposits that build up on the surface. The object is then re-immersed and the scratch-brushing repeated twice at intervals of five minutes. After fifteen minutes the object is again lightly scratch-brushed, briefly re-immersed and then washed in hot water. After thorough drying in sawdust, it may be wax finished.

Tests produced variable results, the patches of variegation tending to be green-grey, but sometimes nearly black. On copper-plated surfaces there was a tendency for some bare metal to appear at the edges of sheet material.

In preparing the solution, the two chemicals should be separately dissolved in two portions of the water. The thiosulphate solution is warmed, and the nickel ammonium sulphate solution added gradually. The light green solution is used at a temperature of 60–80°C.

▲ Nickel salts are a common cause of sensitive skin reactions.

▲ The vapour evolved from the hot solution should not be inhaled.

3.115* Black Semi-matt

Potassium polysulphide	10 gm
Sodium chloride	10 gm
Water	1 litre

Hot immersion (A few minutes)

The object is immersed in the hot solution (60°C), which immediately blackens the surface. Light bristle-brushing can be used to help in producing an even surface finish. After a few minutes, the object is removed and rinsed in warm water. It is allowed to dry in air, and wax finished when dry.

The black colour produced tends to be uneven. If the surface is scratch-brushed then the black is partially removed, leaving a brown surface, textured with black.

▲ Colouring should be carried out in a well ventilated area, as hydrogen sulphide gas is liberated from the hot solution.

3.116 Black/brown (variegated) Semi-matt

A	Copper sulphate	50 gm
	Water	1 litre
B	Potassium sulphide	12.5 gm
	Water	1 litre

Hot immersion. Scratch-brushing
Cold immersion. Scratch-brushing

The article is immersed in the hot copper sulphate solution for about thirty seconds, transferred to the cold potassium sulphide solution, and immersed for about one minute, then removed and scratch-brushed with each solution in turn. The surface is finished by light scratch-brushing in the potassium sulphide solution. The article is washed thoroughly in warm water and allowed to dry in air before wax finishing.

The results produced in tests were generally unsatisfactory. Although the colour is dark, it is virtually impossible to produce either an even or an evenly variegated surface. The scratch-brush tends to break through and ruin the surface.

▲ Colouring should be carried out in a well ventilated area, as some hydrogen sulphide gas will be liberated.

3.117 Black (variegated) Semi-matt

Copper nitrate	200 gm
Zinc chloride	200 gm
Water	1 litre

Cold application (Twice a day for five days)

The solution is wiped on to the article with a soft cloth and allowed to dry in air. When it is completely dry, the surface is gently rubbed smooth with a soft dry cloth to remove any loose dust, and a further application of solution wiped on. This procedure is repeated twice a day for about five days. The solution takes effect immediately, giving rise to an etching effect and pale cream and pink colours typical with chloride solutions. These colours darken on exposure to daylight. When treatment is complete, the article is left to dry thoroughly in air, and when completely dry, it is wax finished.

The brown surface gradually turns black on exposure to daylight, producing an overall black colour with some lighter brown markings giving a 'cracked lacquer' effect, in tests.

3.118   Black on brownish ground Semi-matt

A	Barium sulphide	2 gm
	Potassium sulphide	2 gm
	Ammonium sulphide (16% solution)	2 cm³
	Water	1 litre

B	Copper nitrate	20 gm
	Ammonium chloride	20 gm
	Calcium hypochlorite	20 gm
	Water	1 litre

Cold immersion (A)   (Thirty seconds)
Cold immersion (B)   (A few minutes)

The article is immersed in the mixed sulphide solution, which immediately blackens the surface, and removed after about thirty seconds. If the surface is uneven then it can be lightly brushed with a scratch-brush and re-immersed for a similar length of time. The article is washed and transferred to solution B, which changes the colour to a dark slate or grey-brown. After a few minutes the article is removed and allowed to dry in air. When dry it may be wax finished.

The black colour tends to be partially adherent, producing a variegated finish of black on a brown-toned metal surface.

The secondary treatment in solution B tended to produce results that were only tenuously adherent, and cannot be generally recommended.

Similar poor results were also obtained with an after-treatment with a solution where the calcium hypochlorite is replaced by an equal quantity of calcium chloride.

▲ Ammonia vapour and hydrogen sulphide gas are liberated from the mixed sulphide solution. Colouring should be carried out in a fume cupboard or very well ventilated area. It should be prevented from coming into contact with the eyes and skin.

▲ Barium compounds are toxic, and harmful if taken internally.

▲ The fine dust from the calcium hypochlorite (bleaching powder) must not be inhaled.

3.119   Uneven dark lustre Semi-gloss/semi-matt

Butyric acid	20 cm³
Sodium chloride	17 gm
Sodium hydroxide	7 gm
Sodium sulphide	5 gm
Water	1 litre

Boiling immersion (One hour)

The solid ingredients are added to the water and allowed to dissolve, and the butyric acid then added, and the solution heated to boiling. The object is immersed in the boiling solution, which rapidly produces a series of patchy lustre colours. Continued immersion gradually produces a more opaque dark purplish-green colour, which tends to be uneven. After about one hour the object is removed and washed in hot water. It is allowed to dry in air, and may be wax finished when dry.

The results produced tend to be uneven or patchy, and the procedure cannot be generally recommended.

▲ Butyric acid should be prevented from coming into contact with the eyes and skin, as it can cause severe irritation. It also has a foul odour which tends to cling to the hair and clothing. Preparation and colouring should be carried out in a well ventilated area.

▲ Both sodium hydroxide and sodium sulphide are corrosive, and must be prevented from coming into contact with the eyes and skin.

3.120   Dark brown/black Gloss

Copper nitrate	200 gm
Sodium chloride	200 gm
Water	1 litre

Boiling immersion (Thirty minutes)

An immediate etching of the surface occurs when the article is immersed in the boiling solution, followed by the rapid development of a salmon pink/orange colour. After thirty minutes the article is removed, washed in hot water, and allowed to dry in air. The article is then exposed to daylight, turning regularly to ensure that all parts of the surface are exposed. The surface will darken rapidly at first to a brown colour, and then more slowly to black. The article may subsequently be wax finished.

This technique is not recommended for use with copper-plated surfaces, which will not withstand prolonged immersion in this solution.

The surface may darken irregularly producing an appearance similar to a heavily smoked surface.

Tests were carried out using lower concentrations of these ingredients, and also with additions of ammonium chloride and potassium hydrogen tartrate as suggested by some sources. The results were very similar in all cases.

3.121   Black and orange-brown (variegated) Semi-matt

Copper acetate	40 gm
Copper sulphate	200 gm
Potassium nitrate	40 gm
Sodium chloride	65 gm
Acetic acid (6% solution)	1 litre

Cold immersion. Scratch-brushing (Ten minutes)

The object is suspended in the cold solution, producing an immediate etching effect. After five minutes a pale cream or pink colour develops unevenly. After ten minutes the object is removed and scratch-brushed with the cold solution to even the surface. It should then be allowed to dry in air without washing. Any wet areas on the surface should be very gently brushed out with a soft brush, to prevent streaking or pooling. When dry, the object is exposed evenly to daylight, causing a darkening of the colours. When the colour has stabilised the object is wax finished.

The surface produced after exposure to daylight is very patchy and uneven. It does not appear to be possible to control the degree of variegation or the surface quality, and the recipe cannot generally be recommended.

▲ Potassium nitrate is a powerful oxidising agent and should not be allowed to come into contact with combustible material.

3.122   Brown and black (variegated) Semi-matt

Copper acetate	40 gm
Copper sulphate	200 gm
Potassium nitrate	40 gm
Sodium chloride	65 gm
Flowers of sulphur	110 gm
Acetic acid (6% solution)	1 litre

Hot immersion. Scratch-brushing (Five minutes)

The article is immediately etched on immersion in the hot solution (70–80°C). A pink/orange colour develops within two or three minutes. After five minutes the article is removed, and the uneven surface scratch-brushed with the hot solution to improve its consistency. It is then washed in hot water and left to dry in air. When dry the surface is exposed evenly to daylight, causing the colour to darken. When all colour change has ceased, it is wax finished.

The results produced after exposure to daylight are uneven. A slight graining effect occurred in tests, due to the etching action of the solution. The overall colour and surface quality are very difficult to control and the recipe cannot be generally recommended.

▲ Potassium nitrate is a powerful oxidising agent and should not be allowed to come into contact with combustible material. It must not be brought into contact with the sulphur in a dry state, but should be completely dissolved before the flowers of sulphur are added.

3.123   Black variegation on brown tinged metal Semi-matt

Copper nitrate	20 gm
Zinc sulphate	30 gm
Mercuric chloride	30 gm
Water	1 litre

Cold application (Several minutes)

The solution is wiped on to the surface of the object with a soft cloth, producing a slightly adherent grey-black layer. Continued rubbing with the cloth clears this layer and produces a surface that is evenly coated with mercury. This tends to be non-adherent when dry. The solution is then applied to the surface by dabbing with a soft cloth, and allowed to dry in air. The procedure may be repeated if necessary, omitting the initial wiping. When the colour has developed, the object is washed in cold water and allowed to dry in air. When dry it may be wax finished.

Darker and more coherent variegation was obtained on satin finished surfaces. On highly polished faces the black variegation tended to be thin and patchy, giving a 'smoked' appearance. Some traces of thin blue-green patina were obtained on both polished and satin surfaces.

▲ Mercuric chloride is very toxic and must be prevented from coming into contact with the skin and eyes. The dust must not be inhaled. If taken internally, very serious effects are produced.

▲ In view of the high risks involved, and limited colouring potential, this solution cannot be generally recommended.

3.124   Slate grey Semi-matt

A	Potassium nitrate	100 gm
	Water	1 litre
B	Copper sulphate	25 gm
	Ferrous sulphate	20 gm
	Acetic acid (glacial)	1 cm³

Boiling immersion (Thirty minutes)

The object is immersed in solution A for a period of about fifteen minutes, which produces a slight brightening of the surface. The object is removed to a bath of hot water, while the copper sulphate, ferrous sulphate and acetic acid are separately added to solution A. The object is re-immersed in the boiling solution, gradually producing a greyish-brown colour, for about fifteen minutes. It is washed in hot water, allowed to dry in air and then wax finished.

The results obtained in tests were poor. The surface is prone to patches and streaks, which reveal the bare metal.

▲ Potassium nitrate is a powerful oxidant and must be prevented from coming into contact with combustible materials.

▲ Glacial acetic acid is very corrosive and must be prevented from coming into contact with the eyes and skin.

3.125   Variegated metallic light grey on light brown Semi-matt

Copper nitrate	300 gm
Water	100 cm³
Silver nitrate	2 gm
Water	10 cm³

Applied liquid (Several minutes)

The copper nitrate and the silver nitrate are separately dissolved in the two portions of water, and then mixed together. The solution is warmed to about 40°C. The article to be coloured is immersed for a few minutes in a cold solution of sulphuric acid (50% solution), and then washed. The warm colouring solution is applied to the article by wiping with a soft cloth, producing a superficial grey layer which is washed away with warm water, revealing the underlying surface effect. When the article is completely dry, it may be wax finished.

A variegated partial chemical plating on a light brown coloured ground is produced.

▲ Silver nitrate should be prevented from coming into contact with the skin, and particularly the eyes which may be severely irritated.

▲ Sulphuric acid is very corrosive, causing severe burns, and must be prevented from coming into contact with the eyes and skin.

3.126   Metallic grey (slightly frosted surface) Semi-gloss/gloss

Sodium thiosulphate	300 gm
Antimony potassium tartrate	15 gm
Ammonium chloride	20 gm
Water	1 litre

Hot immersion (Twenty minutes)

Immersion in the hot solution (60°C) produces a series of lustre colours on the surface of the object. These are pale and pass from purple to brown and blue, becoming a very pale blue grey after about two minutes. This is followed by a gradual darkening of the surface, producing dark greys and browns. After about twenty minutes the object is removed, washed in warm water and dried in sawdust. It may be wax finished when dry.

A dark metallic grey colour is produced, with a high gloss. Tests tended to produce results in which this was overlaid by a duller light grey frosted effect. Some small areas of localised brown lustre may be visible in some lights.

▲ Antimony potassium tartrate is poisonous and may be harmful if taken internally.

▲ The vapour evolved from the hot solution must not be inhaled.

3.127* Blue-green patina on orange-brown ground Semi-matt   *Pl. V, Pl. VI*

Copper nitrate	110 gm
Water	110 cm³
Ammonia (.880 solution)	440 cm³
Acetic acid (6% solution)	440 cm³
Ammonium chloride	110 gm

Applied liquid (Five days)

A darker and more purple-brown ground is obtained on copper-plated surfaces. See plate VI.

▲ Colouring must be carried out in a well ventilated area. Ammonia vapour will be liberated from the strong solution, which will cause irritation of the eyes and respiratory system. The solution must be prevented from coming into contact with the eyes or skin, as it can cause burns or severe irritation.

The solution is applied to the surface by dabbing with a soft cloth, and the article then left to dry in air. The metal quickly darkens and a whiteish blue-green powdery patina forms gradually on the surface. When dry, this is rubbed with a soft dry cloth to remove any loose material and to smooth the surface, prior to the next application of solution. This procedure is repeated over a period of five days, during which time the patina darkens and becomes integral with the surface. When treatment is complete and the surface thoroughly dry, the article may be wax finished.

**3.128* Blue-green patina on variegated brown/black ground
Semi-matt**   *Pl. V*

Copper sulphate	20 gm
Copper acetate	20 gm
Ammonium chloride	10 gm
Acetic acid (6% solution)	to form a paste

Applied paste (Several days)

The ingredients are ground to a creamy paste with a little acetic acid, using a pestle and mortar. The paste is applied to the object with a soft brush, giving quite a thick coating which is then allowed to dry for a day. The dry residue is then washed away under cold water, using a soft brush. A thin layer of paste is then wiped onto the object with a cloth, and the object left to dry for a day. The residue is again washed off. This procedure of applying thin layers and drying is repeated in the same way until a good variegated patina is produced. The object should be allowed to dry thoroughly when treatment is complete, and may then be wax finished.

3.129* Blue-green patina on red ground Semi-matt   *Pl. V*

Copper nitrate	200 gm
Sodium chloride	200 gm
Water	1 litre

Applied liquid (Twice a day for five days)

A dabbing technique was found to be better than wiping which tends to inhibit the formation of the patina.

If there is an initial resistance to 'wetting' the liquid can be applied by scratch-brushing in the early stages.

In damp weather conditions the surface may take several weeks to dry out completely. Although apparently dry, the surface may 'sweat' intermittently. The drying process should not be rushed and wax finishing only carried out when it is certain that chemical action has ceased.

The solution is applied to the object by dabbing with a soft cloth and left to dry in air. This process is repeated twice a day for five days. The object is then allowed to dry in air without treatment for a further five days. The ground colour develops gradually throughout this period and after two or three days some blue-green patina forms on the surface. This continues to build up as long as the surface is moist. Finishing should only be carried out when it is certain that the surface is dry and patina development complete.

3.130 Blue green patina on red ground Matt

Copper nitrate	200 gm
Sodium chloride	200 gm
Water	1 litre

Applied liquid to heated metal (Several minutes)

The results produced are very similar to those obtained by a long term application of the solution to the unheated metal. See recipe 3.129.

Practice is required to obtain the right temperature to produce an even surface.

The procedure is not suitable for copper-plated surfaces, which tend to be destructively attacked.

The article is gently heated on a hotplate or with a blow torch, and the solution applied with a soft brush. The temperature of the surface should be such that it sizzles gently as the solution is applied. This procedure is repeated a number of times, each application being allowed to dry completely before the next is applied. Initially a superficial powdery layer is formed that is not adherent, accompanied by the gradual development of an underlying reddish ground. As the treatment is repeated so the patina tends to become more developed. When the surface is sufficiently developed, the article is allowed to cool and may be wax finished.

3.131*   Blue-green patina on mid-brown ground   Semi-matt/matt

Ammonium chloride	35 gm
Copper acetate	20 gm
Water	1 litre

Cold application   (Several days)

The ingredients are ground with a little of the water using a pestle and mortar. They are then added to the remaining water. The solution is applied by dabbing and wiping, using a soft cloth. The solution should be applied sparingly, to leave an evenly moist surface. The object is then allowed to dry in air. This procedure is repeated once a day for several days, producing a gradual development of the ground colour and blue-green patina. When treatment is complete, it should be left to dry for several days, during which there is further patina development. When it is completely dry, and there is no further surface change, the object is wax finished.

It is essential to ensure that all patina development is complete, and the surface is completely dry, before wax finishing is carried out. The final drying period may have to be extended to a matter of weeks in damp or humid conditions.

3.132*   Blue-green patina on brown/black ground   Semi-matt   *Pl. VI*

Copper nitrate	100 gm
Nitric acid (70% solution)	40 cm³
Water	1 litre

Torch technique

The object is heated with a blow torch and the solution applied sparingly with a bristle brush or paint brush. The liquid quickly turns dark brown, and becomes yellowish where it is not being directly heated. Continued heating makes the surface blacken and areas of blue-green patina begin to form. The blue-green patina tends to be superficial at this stage and the surface should be further heated and stippled with solution until a good dark brown or black ground has been established. The object is then gently heated and dabbed with an almost dry brush or cloth, barely damp with the solution, until an evenly distributed blue-green patina is obtained. When treatment is complete, the object is allowed to cool and dry thoroughly, and may then be wax finished.

▲ Nitric acid is very corrosive and must not be allowed to come into contact with the eyes, skin or clothing. A face shield or goggles should be worn to protect the eyes from hot splashes of solution as it is applied to the heated metal. .

▲ The fumes evolved as the solution is applied to the heated metal should not be inhaled. Adequate ventilation must be provided, and a nose and mouth face mask fitted with the correct filter should be worn.

3.133*   Blue-green patina (blue-green patina on black ground)   Semi-matt

Copper nitrate	200 gm
Water	1 litre

Torch technique

The metal surface is heated with a blow torch, and the solution applied with a soft brush until it is covered with an even blue-green patina. If the surface is then heated, without further application of the solution, it will become black. The solution may then be applied to this black ground, heating the surface as necessary, to form a blue-green patina on the ground colour. The surface quality obtained can be varied by the precise method of application—a stippled surface may be produced by stippling with a relatively dry brush, or a more mottled or marbled effect obtained using the same technique with a brush that is more 'loaded' with the solution. When the required finish has been achieved, the object is allowed to dry out, and may be wax finished when dry.

▲ Colouring should be carried out in a well ventilated area, so that inhalation of the vapours is avoided.

▲ A face shield or goggles should be worn to protect the eyes from hot splashes of solution.

3.134   Greenish patina on orange-brown ground   Semi-matt   *Pl. VI*

Copper nitrate	30 gm
Zinc chloride	30 gm
Water	to form a paste

Applied paste   (Two or three hours)

The ingredients are mixed by grinding together using a pestle and mortar. Water should be added by the drop, to form a creamy paste. Excess water should be avoided, as the paste tends to thin rather suddenly. The paste is applied to the object with a soft brush, and allowed to dry for two or three hours. The residual paste is gently washed away with cold water, and the object then allowed to dry in air for several days. Exposure to daylight will cause a variegated darkening of the ground colours. When the surface is completely dry, and colour change has ceased, it may be wax finished.

The finish tends to be very variegated. Extended times or repeated applications should not be used in the case of copper-plated surfaces, as this tends to cause some exposure of the underlying metal.

Repeated applications do not improve the results, and tend to require far longer final drying periods.

Plate VI shows the result on copper-plated surfaces.

3.135★ Olive green on black ground Matt

A	Copper nitrate	200 gm
	Water	1 litre

B	Potassium polysulphide	50 gm
	Water	1 litre

Torch technique (A)
Torch technique (B)

The article is heated with a blow torch and the copper nitrate solution applied with a soft brush until an even thin layer of blue-green patina covers the surface. This is again heated, without further application of the solution, until it turns black. Any loose residue should be brushed off with a soft dry brush. The copper nitrate is then again applied, heating as necessary, until the blue-green colour re-appears and is evenly distributed. While the surface is still hot, the polysulphide solution is applied until the required depth of colour is achieved. When treatment is complete, the article is allowed to dry for some time, and may then be wax finished.

The concentrations of both solutions can be varied within quite wide limits.

▲ Colouring should be carried out in a well ventilated area, so that inhalation of the vapours and of the hydrogen sulphide gas that are evolved is avoided.

▲ A face shield or goggles should be worn to protect the eyes from hot splashes of solution.

3.136★ Blue-green patina on red/red-brown variegated ground Semi-matt

Ammonium chloride	40 gm
Ammonium carbonate	120 gm
Sodium chloride	40 gm
Water	1 litre

Cold immersion and applied liquid (One hour)

The object to be coloured is immersed in the cold solution for about one hour, producing a slight etching effect and a gradual development of the ground colour. It is then removed and allowed to dry thoroughly in air. Any excessive surface moisture should be gently brushed out with a soft brush, to prevent 'runs' or 'pooling' of the solution and to leave an evenly moistened surface. The patina develops slowly as the surface dries out. If further patina development is desired, then the solution should be applied sparingly with a soft cloth and the object again left to dry thoroughly. When patina development is complete and the surface dry, the object may be wax finished.

A variegated red and reddish-brown ground is produced. Patina development tends to be slight, taking the form of thin 'watery' areas overlaying the ground.

It is essential to ensure that patina development is complete, and the surface dry, before wax finishing. In damp or humid conditions the drying time may need to be extended.

3.137 Blue-green patina (orange-brown ground) Matt   *Pl. V*

Ammonium carbonate	24 gm
Potassium hydrogen tartrate	6 gm
Sodium chloride	6 gm
Copper sulphate	6 gm
Acetic acid (6% solution)	to form a paste

Applied paste (Several days)

The solid ingredients are made to a paste with a little of the acetic acid, using a pestle and mortar. The paste is applied with a soft brush. The object is then left to dry for about two days, and the application is then repeated. It is then left to dry thoroughly in air, which generally takes several days. When completely dry, the surface is brushed gently with a bristle-brush to remove any loose material, rubbed with a soft cloth to smooth the surface, and wax finished.

If the paste is made up with stronger solutions of acetic acid (15–30%) then there is a greater tendency for the blue green patina to be non-adherent, partially revealing orange-brown ground colours.

▲ A nose and mouth face mask fitted with a fine dust filter should be worn when brushing away the dry residue.

3.138★ Blue-green patina on dark brown ground Semi-matt

Copper nitrate	80 gm
Water	1 litre
Ammonia (.880)	3 cm³

Applied liquid (Three days)

The solution is applied to the object by spraying with a moderately fine atomising spray to produce a misty coating. It is essential to apply only a fine misty coating and to avoid any pooling of the solution or runs on the surface. The object is then left to dry. This procedure is repeated twice a day for three days, after which the surface should be allowed to dry out thoroughly for several days before finishing with wax.

Prolonging the treatment beyond three days does not appear to produce any further development. In tests extending to twenty days no notable changes could be detected after three days.

▲ Inhalation of the fine spray is harmful. It is essential to wear a suitable nose and mouth face mask provided with the correct filter, and to ensure adequate ventilation.

3.139* Blue-green patina on red-brown/maroon ground Semi-matt

Copper nitrate	200 gm
Water	1 litre

Applied liquid (Five days)

The article is dipped in the cold solution for a few seconds, drained and allowed to dry in air. This procedure is repeated twice a day for about five days, during which time the surface darkens gradually to a red-brown. After about two days, patches of powdery blue-green patina also begin to appear on the surface. When treatment is complete, the article should be left to dry in air for a period of at least three days to allow the patina to develop fully and dry out. When dry, the article is wax finished.

Some sources suggest a scratch-brushed application. Tests carried out using this technique produced similar but lighter results over the same time.

Tests carried out in which the metal was briefly 'pickled' in a 10% solution of nitric acid prior to the initial dip, as suggested by some sources, gave very similar results with somewhat duller surfaces.

The patina tends to develop in streaks and patches corresponding to the draining of the solution from the surface. This can be minimised by 'brushing out' with a soft brush after dipping, to prevent runs and pooling.

If the ground colour alone is required, then after two days of dipping and drying, any patina should be brushed off scrupulously and application continued by wiping-on very sparingly. A patina-free ground is difficult to achieve.

3.140* Variegated (Pale green and olive patina on brown/black ground) Semi-matt/semi-gloss

Pl. V

Copper nitrate	20 gm
Sodium chloride	16 gm
Potassium hydrogen tartrate	12 gm
Ammonium chloride	4 gm
Water	to form a paste

Applied paste (Four hours)

The ingredients are ground to a paste with a pestle and mortar by the very gradual addition of a small quantity of cold water. The creamy paste is applied with a soft brush, and is left to dry. After four hours the dry residue is removed with a bristle-brush. The object is washed in cold water and allowed to dry in air. Exposure to daylight will darken the orange-brown ground to produce mixed green patination on a dark brown and black variegated ground. When the colour has stabilised, the object may be wax finished.

To produce this effect on copper-plated surfaces, the residue should be brushed away after two hours. If left for four hours the plate may be partially removed revealing patches of pink.

▲ A nose and mouth face mask fitted with a fine dust filter should be worn when brushing away the dry residue.

3.141 Variegated pale green/orange/dark brown on reddish ground Semi-matt

Copper nitrate	15 gm
Zinc nitrate	15 gm
Ferric chloride	0.5 gm
Hydrogen peroxide (100 vols)	to form a paste

Applied paste (Four hours)

The ingredients are ground using a mortar and pestle, and the hydrogen peroxide added in small quantities to form a thin creamy paste. The paste is applied to the object with a soft brush, and left to dry in air. After four hours, the residue is washed away with cold water and the object is allowed to dry in air for several hours. When completely dry, it may be wax finished.

A multicoloured stippled finish on a reddish ground is produced on both copper and copper-plated surfaces.

The surface of the object should be handled as little as possible until dry, and should not be brushed during washing.

▲ Hydrogen peroxide should not be allowed to come into contact with combustible material. It should be prevented from coming into contact with the eyes and skin, as it may cause severe irritation. It is very harmful if taken internally. The vapour evolved as it is added to the solid ingredients of the paste must not be inhaled.

3.142 Pale blue and dark green 'watery' surface Semi-matt

Pl. V

Ammonium carbonate	180 gm
Copper sulphate	60 gm
Copper acetate	20 gm
Oxalic acid	1.5 gm
Ammonium chloride	0.5 gm
Acetic acid (10% solution)	1 litre

Boiling immersion (Twenty-five minutes)

The object is immersed in the boiling solution, which produces some etching effects and light colouring after a few minutes. As the immersion continues streaky greenish films gradually develop, beginning to become dark after about ten minutes. The object is removed after twenty-five minutes and the hot solution applied with a soft brush, to leave the surface evenly moist. The object should then be allowed to dry thoroughly in air for several hours, without washing. When dry, it is wax finished.

A dark green colour develops on an underlying brown ground. As the surface dries a pale blue colour gradually develops in patches.

The procedure is not recommended for use with copper-plated surfaces. In tests, these were vigorously attacked and removed by the boiling solution.

▲ The ammonium carbonate should be added to the acetic acid in small quantities. The reaction is effervescent, and this may be very vigorous if large amounts are added.

▲ Oxalic acid is harmful if taken internally, and irritating to the eyes and skin.

3.143* Yellow-green/blue-green patina on red-orange ground Semi-matt

Copper nitrate	20 gm
Ammonium chloride	20 gm
Calcium hypochlorite	20 gm
Water	1 litre

Cold immersion (Twenty hours)

The article is suspended in the cold solution. After an initial 'bright dip' effect which enhances the metal surface to a reddish-gold colour, there is a gradual build up of colour and an etching effect that is typical with chloride solutions. The article is left in the solution for twenty hours and then removed, rinsed in cold water and allowed to dry in air. Drying in air produces a green powdery layer which is only partially adherent on polished surfaces. When the green layer has developed and the surface is completely dry, the object is wax finished.

After the long period of immersion the ground colour is a light brown, and darkens on exposure to daylight as the patina develops. Once the ground is fully dark, more patina can be induced by applying some of the solution sparingly with a soft cloth and allowing the surface to dry. This may be repeated if necessary.

It is essential to ensure that the surface is dry and patina development complete before wax finishing. In damp or humid conditions the drying time may need to be extended.

▲ Calcium hypochlorite (bleaching powder) can cause irritation of the eyes, skin and respiratory system. It is essential to avoid inhaling the fine dust. Both the powder and solutions should be prevented from coming into contact with the skin.

3.144 Green/bluish green (variegated) Semi-matt

Ammonia (.880 solution)	
Sodium chloride	20 gm/litre

Vapour technique (Several days)

The object is placed or suspended in a container, into the bottom of which the sodium chloride and ammonia solution are introduced. The object must be placed well clear of the surface of the liquid, and in a position that will favour an even distribution of vapour around it. The object should be well clear of the walls and lid of the container. The container is sealed and the vapour allowed to act on the object for about two days. It is then removed, washed in cold water and allowed to dry. When dry, the surface is rubbed with a dry cloth to remove any loose material. The object is then replaced in the sealed container for a further two days. It is then removed, washed in cold water and allowed to dry. When dry, it is rubbed with a soft dry cloth and then wax finished.

The exact method used will depend on the size of the object being coloured. It is essential to plan the procedure carefully. Ammonia vapour will concentrate in the container and should not simply be released. See 'Metal colouring techniques,' 9.

After-treatments involving the use of vapour from concentrated acetic acid, suggested by some sources, tended to make the colour less adherent.

▲ Ammonia solution is highly corrosive and must be prevented from coming into contact with the eyes or skin. The vapour is extremely irritating to the eyes and respiratory system, and will irritate the skin. Adequate ventilation must be provided.

3.145 Grey-green and blue-green patina on dark brown ground Semi-matt

Copper nitrate	30 gm
Sodium chloride	30 gm
Ammonium chloride	15 gm
Potassium aluminium sulphate	7 gm
Acetic acid (6% solution)	1 litre

Boiling immersion (Ten minutes)

The article is immersed in the boiling solution. After two or three minutes a fine etching effect and a pink colour develops on the surface. Immersion is continued to ten minutes, when the article is removed. It is then dipped several times in the solution, and removed. Excess solution is prevented from pooling or running on the surface by gently and rapidly brushing with a soft bristle-brush, and the article is left to dry thoroughly in air. Initially a thin veil of greyish-blue patina forms on the pink ground, but this later appears grey-green in colour as the ground darkens to a grey-brown on exposure to daylight. Some blue-green patina may also form gradually. When the surface is completely dry, it may be wax finished.

The results are difficult to control. Minimising the amount of solution left on the surface after dipping encourages the grey-green patina and produces a good surface. Any excess solution left on the surface tends to encourage the additional formation of the blue-green patina and produces an irregular patchy surface.

It is essential to ensure that the surface is completely dry before waxing. Drying may need to be prolonged in damp weather.

The initial immersion time should be reduced to five minutes in the case of copper-plated surfaces.

3.146 Slight blue-green patina on variegated orange-brown ground Semi-matt

Copper sulphate	10 gm
Lead acetate	10 gm
Ammonium chloride	5 gm
Water	to form a paste

Applied paste (Several days)

The ingredients are ground to a creamy paste with a little water, using a pestle and mortar. The paste is applied to the object with a soft brush, and is left to dry out completely. When completely dry, after several hours, the dry residue is brushed away with a bristle brush. This process is repeated three or four times. When treatment has finished, the object should be allowed to stand in a dry place for several days to ensure thorough drying. After a final brushing with the bristle-brush, it is wax finished.

Patina development is very slight, tending to appear as an additional colour in the variegated orange-brown surface.

▲ Lead acetate is highly toxic, and precaution should be taken to prevent the inhalation of the dust during weighing, and preparation of the paste. A mask fitted with a fine dust filter must be worn when brushing away the dry residue. A clean working method is essential throughout.

3.147   Blue-green patina on reddish-brown ground Semi-matt

Ammonium carbonate	150 gm
Copper acetate	60 gm
Sodium chloride	50 gm
Potassium hydrogen tartrate	50 gm
Acetic acid (10% solution)	1 litre

Applied liquid   (Twice a day for several days)

The solution is dabbed onto the object with a soft cloth, and then allowed to dry thoroughly in air. This process is repeated twice a day for several days until the desired colour is achieved. Any powder or other loose material that forms during drying periods should be gently brushed away with a soft dry cloth, prior to the next application of solution. When treatment is complete the surface should be allowed to dry in air for several days. When completely dry, it may be wax finished.

Excessive use of the solution in the latter stages tends to break down the developed surface, revealing areas of pink beneath the reddish brown, and producing a surface variegated with patches of patina, when the dry material is brushed away with a cloth.

For results obtained using a more concentrated form of this solution, see recipe 3.158.

It is essential to ensure that patina development is complete, and the surface completely dry, before wax finishing. In damp or humid conditions, the drying time may need to be extended.

▲ The ammonium carbonate should be added to the acetic acid in small quantities. The reaction is effervescent and if large quantities are added, then the effervescence is very vigorous.

3.148   Green patina on an orange-brown ground. Semi-matt   *Pl. V*

Ammonium carbonate	150 gm
Sodium chloride	50 gm
Copper acetate	60 gm
Potassium hydrogen tartrate	50 gm
Water	1 litre

Applied liquid   (Several days)

The solution is applied generously to the surface of the article using a soft cloth. It is then allowed to dry in air. This procedure is repeated twice a day for several days, during which time the colour gradually develops. It is then left to dry in air, without further treatment, to allow the patina to develop. When patina development is complete and the surface is completely dry, the article is wax finished.

A variegated patina is produced, which tends to be green if the residual moisture is minimal before allowing the surface to dry. A wetter surface tends to produce more blue-green results. The orange or orange-brown ground is also slightly variegated.

3.149   Green and blue-green patina on variegated orange-brown ground
Semi-matt

Ammonium sulphate	105 gm
Copper sulphate	3.5 gm
Ammonia (.880 solution)	2 cm³
Water	1 litre

Applied liquid   (Several days)

The solution is applied to the surface of the object with a soft cloth. It is then allowed to dry in air. This procedure is repeated twice a day for several days until the ground colour has developed and traces of patina begin to appear. The solution should then be applied more sparingly by dabbing with a soft cloth. The object is allowed to dry slowly in a damp atmosphere to encourage the development of patina. It is finally allowed to dry in a warm dry atmosphere, and wax finished when completely dry.

The ground is very variegated, and the patina tends to develop in patches of both a blue-green and an olive-green colour.

This recipe, which has been used commercially to induce patination on copper roofing, requires weathering in the open air to achieve the best results. In tests carried out in workshop conditions, only slight patination and patchy ground colours could be obtained.

3.150   Green patina on brown ground Semi-matt

Copper nitrate	100 gm
Hydrochloric acid (35% solution)	10 cm³
Water	1 litre

Torch technique

The object is heated with a blow torch and the solution sparingly applied with a bristle-brush or paint brush. A heavy deposit of yellowish-green patina quickly forms, but this is rather non-adherent at this stage. Continued heating and the application of small amounts of solution by brushing gradually darkens the surface. When the brown ground colour has been established, the green patina is built up either by stippling with a nearly dry brush, or by using a cloth which is barely damp with the solution, heat being frequently applied to the area being worked. When a good patina has been obtained, the surface can be burnished with a dry cloth. Wax finishing is carried out when the object is cool and thoroughly dry.

In some cases the patina may have a tendency to 'sweat' shortly after treatment. If this occurs then the object should be left in a warm dry place for some time, and the wax only applied when the surface is thoroughly dry.

▲ Hydrochloric acid is corrosive and should not be allowed to come into contact with the eyes and skin. A face shield or goggles should be worn to protect the eyes from hot splashes as the solution is applied.

▲ The fumes evolved as the solution is applied to the heated metal should not be inhaled. Adequate ventilation must be provided and a nose and mouth face mask fitted with the correct filter should be worn.

3.151   Thin blue-green patina on pale brown ground Matt

Ammonium carbonate	150 gm
Ammonium chloride	100 gm
Water	1 litre

Applied liquid   (Several days)

The solution is applied to the object, by wiping it on to the surface with a soft cloth. The solution should be used sparingly, leaving the surface evenly moist. The object is then allowed to dry in air. This procedure is repeated once a day for several days, until the grey-green patina is sufficiently developed. When treatment is complete, the object should be allowed to dry out thoroughly for several days, and is wax finished when completely dry.

The finish tends to develop slowly, and produces areas of a thin blue-green patina on a pale brown ground

It is essential to ensure that the surface is completely dry, and patina development complete, before wax finishing. The drying may need to be extended in damp or humid atmospheric conditions.

3.152   Dark green on mid-brown ground Matt

Ammonium sulphate	105 gm
Copper sulphate	3.5 gm
Ammonia (.880 solution)	2 cm³
Water	1 litre

Torch technique

The metal surface is heated with a blow torch, and the solution applied with a soft brush, until a dark green colour on a mid-brown ground is obtained. If the metal is too hot when the solution is applied then impermanent light grey or pinkish-red colours are produced. Continued application of the solution as the metal cools will remove these colours and encourage the brown ground and the dark green patina. If the green patina is heated once it has developed, by playing the torch across the surface, then it will tend to become black. When the desired surface finish has been achieved, the object is left to cool and dry for some time, and may then be wax finished.

A wide variety of surface effects are obtained, which depend on the temperature of the metal, and the manner in which the solution is applied (eg liberally or sparingly; by 'stippling' or by 'painting' etc).

▲ Colouring should be carried out in a well ventilated area, to avoid inhalation of the vapours.

▲ A face shield or goggles should be worn to protect the eyes from hot splashes of solution.

3.153   Pale blue (or green) patina on light brown/ochre ground Semi-matt

Ammonium carbonate	180 gm
Copper sulphate	60 gm
Copper acetate	20 gm
Oxalic acid	1.5 gm
Ammonium chloride	0.5 gm
Acetic acid (10% solution)	1 litre

Torch technique

The ingredients are dissolved and the solution boiled for about ten minutes. It is then left to stand for several days. The mixture should be shaken immediately prior to use. The surface of the object to be coloured is heated with a blow torch, and the solution applied with a soft brush. This process is continued until a pale blue patina is produced, on a light brown or ochre ground. If the torch is then gently played across the surface, the pale blue will change to green. Varying effects can be obtained by 'stippling' or 'dragging' the brush on the surface. When the required finish is obtained, the object is left to cool and dry for some time and may then be wax finished.

Continuous application of liberal quantities of the solution will tend to inhibit patina formation, but encourages the development of the ground colour.

▲ The ammonium carbonate should be added to the acetic acid in small quantities. The reaction is effervescent, and if large amounts are added, this may be very vigorous.

▲ Colouring should be carried out in a well ventilated area, so that inhalation of the vapours is avoided.

▲ A face shield or goggles should be worn to protect the eyes from hot splashes of solution.

3.154   Blue patina on mid-brown ground Matt

Ammonium carbonate	300 gm
Acetic acid (10% solution)	35 cm³
Water	1 litre

Warm scratch-brushing   (Two or three minutes)

The ammonium carbonate is added to the water, while stirring, and the acetic acid then added very gradually. The solution is then warmed to 30–40°C. The solution is applied to the object liberally with a soft brush, and then immediately scratch-brushed. The surface is worked for a short time, while applying more solution, and then allowed to dry in air. Any excess moisture on the object should be brushed out with a soft bristle-brush, to prevent 'pooling' or 'running' and to leave an evenly moist surface. When the surface is completely dry, which may take several days, it is wax finished.

A variegated reddish mid-brown ground is produced, which is overlaid with patches of patina. The best effects that could be obtained in tests were where the patina was least developed, by ensuring that the surface is very sparingly moist before leaving it to dry.

The results obtained in tests with this solution were poor. Although the colours obtained are interesting, the surfaces were very prone to streaking and patchiness, and little control over this could be gained. The technique cannot therefore be generally recommended.

3.155 Blue-green patina on brown ground Semi-matt/matt

Copper acetate	30 gm
Copper carbonate	15 gm
Ammonium chloride	30 gm
Hydrochloric acid (18% solution)	2 cm³
Water	to form a paste

Applied paste (Two or three days)

The ingredients are ground to a paste with a little water and the hydrochloric acid, using a pestle and mortar. The paste is applied with a soft brush and left for several hours to dry out. The dry residue is removed with a stiff bristle-brush. The procedure is repeated until a brown ground has developed with a thin patina. The paste is then thinned to a liquid and applied by the same method, but with a soft brush to remove residue. When the surface is developed and dried out, the object may be wax finished.

A thick incrustation tends to form if the paste is made too thick. This is difficult to remove by brushing, when dry, but tends to flake eventually.

▲ A nose and mouth face mask should be worn when brushing away the dry residue.

3.156 Cloudy greenish-grey on light brown (variegated) Semi-gloss

Sodium dichromate	150 gm
Nitric acid (10% solution)	20 cm³
Hydrochloric acid (15% solution)	6 cm³
Ethanol	1 cm³
Water	1 litre

Applied liquid (Fifteen minutes)

The sodium dichromate is dissolved in the water, and the other ingredients are then added separately while stirring the solution. The article is immersed in the cold solution for a period of three or four minutes. A superficial layer that tends to form should be removed by washing in cold water. The object is re-immersed for a further ten minutes, and then removed, washed in cold water, and allowed to dry in air. When dry, it may be wax finished.

A darker brown colour tends to be produced on copper-plated surfaces. Both copper and copper plate tend to be clouded with a variegated grey green colour. This is generally more developed on a textured surface, than on a high polish.

A bi-metallic effect can be produced if the copper plate is cut through to an underlying yellow brass, which remains uncoloured.

▲ Sodium dichromate is a powerful oxidant, and must not be allowed to come into contact with combustible material. The dust must not be inhaled. The solid and solutions should be prevented from contact with the eyes and skin, or clothing. Frequent exposure can cause skin ulceration, and more serious effects through absorption.

▲ Hydrochloric and nitric acids are corrosive and must not come into contact with eyes, skin or clothing.

3.157 Slight blue-green patina on mottled orange-brown Semi-matt

Ammonium chloride	10 gm
Ammonium acetate	10 gm
Water	1 litre

Applied liquid (Twice a day for ten days)

The solution is applied to the surface by wiping with a soft cloth or by brushing with a soft brush. It is then allowed to dry in air. This procedure is repeated twice a day for about ten days. The article is then allowed to dry in air, without treatment, for a further five days. When completely dry, and when patina development has ceased, it may be wax finished.

Colour development is generally slow, the ground developing unevenly at first, and the patina appearing later during the period of air-drying. Patina development is more pronounced in surface features such as punched marks or textured areas.

3.158 Slight blue-green patina on dull brown ground Matt

Ammonium carbonate	600 gm
Sodium chloride	200 gm
Copper acetate	200 gm
Potassium hydrogen tartrate	200 gm
Acetic acid (10% solution)	1 litre

Applied liquid (Twice a day for several days)

The solution is applied to the article to be coloured, by wiping sparingly with a soft cloth. It should then be allowed to dry thoroughly in air. This procedure should be repeated twice a day for several days, until the ground colour has developed and the patina has formed. The article should then be left to dry in air for several days, during which time the patina will continue to develop. When patina development has ceased and the surface is completely dry, the article may be wax finished.

An uneven dull brown ground tends to be produced, on which patina development is only slight. The procedure cannot be generally recommended for copper surfaces. The solution should not be used with copper-plate, which is destructively attacked.

For results obtained with a more dilute form of this solution, see recipe 3.147.

It is essential to ensure that the surface is dry and patina development complete before wax finishing. In damp or humid conditions the drying time may need to be extended.

▲ The ammonium carbonate should be added to the acetic acid in small quantities. The reaction is effervescent and if large amounts are added the effervescence is very vigorous.

3.159 Slight blue-green patina on a brown ground Matt

A	Ammonium sulphate	80 gm
	Water	1 litre
B	Copper sulphate	100 gm
	Sodium hydroxide	10 gm
	Water	1 litre

Applied liquid (A) (Five to ten days)
Applied liquid (B) (Two or three days)

Solution A is dabbed vigorously on to the object with a soft cloth, leaving it sparingly moist, and left to dry completely. The procedure is repeated twice a day for up to ten days. Solution B is then similarly applied. When treatment is complete, the object is allowed to dry out in air for several days, and then wax finished.

In tests, the solution tended to retract into pools and failed to 'wet' the surface. The addition of wetting agents produced no improvement, and the surface developed in patches rather than evenly.

▲ Sodium hydroxide is a caustic alkali, and should be prevented from coming into contact with the eyes or skin.

3.160* Orange lustre Gloss

Pl. V

Potassium permanganate	10 gm
Water	1 litre

Hot immersion (Three to five minutes)

The article is immersed in the hot solution (90°C). A golden lustre colour develops within one minute, gradually becoming more intense. When the lustre colour is fully developed, which may take from three to five minutes, the article is removed and washed in hot water which is gradually cooled during washing. The article is finally washed in cold water before being carefully dried either in sawdust, or by gently blotting the surface with absorbent tissue paper to remove excess moisture and allowing it to dry in air. When dry, it is wax finished.

The effect is only obtained on polished, or directionally grained satin surfaces. On rough, as-cast and matt surfaces a dull orange or orange-brown is produced.

Extending the immersion time produces a slight darkening initially to a more pink colour, which rapidly fades to a pale silvery grey. These latter colours are usually uneven. Colours produced at longer immersion times tend to be more fragile when wet, and should not be handled at that stage.

3.161* Pink/green lustre Gloss

Pl. VI

Copper nitrate	115 gm
Tartaric acid (crystals)	50 gm
Sodium hydroxide	65 gm
Ammonia (.880 solution)	15 cm^3
Water	1 litre

Boiling immersion (Fifteen minutes)

Immersion in the boiling solution produces a series of lustre colours, which begins with a blue that develops after about one minute to an even green lustre after about two minutes. If the green colour is required then the object should be immediately removed and washed in hot water, and allowed to dry in air. It is wax finished when dry. Continuing the immersion will cause a gradual pink tinge to appear which subsequently darkens, to produce the final colour after about fifteen minutes. The object is removed and finished as noted above.

Although a number of colour changes occur, the two colours noted are the only ones in this lustre series which are both rich and even.

▲ Sodium hydroxide is a powerful corrosive alkali, and the solid or solutions must be prevented from coming into contact with the eyes and skin. It should be added to the solution in small quantities.

▲ Preparation and colouring should be carried out in a well ventilated area. Ammonia vapour will be liberated, which if inhaled causes irritation of the respiratory system. A combination of caustic vapours will be liberated from the solution as it boils.

▲ A dense precipitate is formed which may cause severe bumping. Glass vessels should not be used.

3.162* Pinkish-red lustre Gloss

Heating in a kiln (A few minutes)

The kiln is pre-heated to 350°C and the object placed inside in a position that will favour even heating. After a few minutes the surface passes through a series of lustre colours, the most notable being a pinkish-red lustre. The object should be removed from the kiln as the pink darkens to red, and left to cool. The colour will generally be found to be lighter than it appears in the confines of the kiln. Finally the object may be wax finished.

The lustre is most vivid on highly polished surfaces.

The object must be perfectly clean and grease free to obtain an even unblemished surface.

The tests relating to this recipe were carried out using an electric kiln. The effects produced with a gas kiln are not known.

3.163* Lustre series Gloss

Sodium thiosulphate	280 gm
Lead acetate	25 gm
Water	1 litre
Potassium hydrogen tartrate	30 gm

Warm immersion (Various)

The sodium thiosulphate and the lead acetate are dissolved in the water, and the temperature is adjusted to 30–40°C. The potassium hydrogen tartrate is added immediately prior to use. The article to be coloured is immersed in the warm solution and a series of lustre colours are produced in the following sequence: golden-yellow/orange; purple; blue; pale blue/pale grey. The colour sequence takes about ten minutes to complete at 40°C. (Hotter solutions are faster, cooler solutions slower.) A second series of colours follows in the same sequence, but these tend to be less distinct. When the desired colour has been reached, the article is removed and immediately washed in cold water. It is allowed to dry in air, any excess moisture being removed by dabbing with an absorbent tissue. When dry it may be wax finished, or coated with a fine lacquer.

Citric acid may be used instead of potassium hydrogen tartrate, and in the same quantity. Tests carried out with both alternatives produced identical results.

▲ Lead acetate is very toxic and will give rise to serious conditions if taken internally. The dust, and the vapour evolved from the hot solution, must not be inhaled.

3.164   Dark red lustre Gloss

Sodium hydroxide	25 gm
Potassium persulphate	10 gm
Water	1 litre

Hot immersion (One minute)

The potassium persulphate is dissolved in the water, and the sodium hydroxide then added in small quantities while stirring. The solution is then brought up to the correct temperature (70–80°C) and the article immersed. A golden lustre colour is produced, which rapidly darkens to red. When the desired colour is obtained (within one minute), the article is immediately removed and immersed in a bath of warm water. It is then washed in cold water and dried in sawdust, and may be wax finished when dry.

Prolonging the immersion will rapidly change the colour to a more opaque, very dark purplish-brown/black, see recipe 3.93.

Very similar results are obtained with copper-plated surfaces.

▲ Potassium persulphate is a powerful oxidant and should not be allowed to come into contact with combustible material. The sodium hydroxide and the persulphate must not be allowed to come into contact in the solid state.

▲ Sodium hydroxide is a powerful caustic alkali which can cause burns. Both the solid and solutions should be prevented from coming into contact with the eyes and skin.

3.165*   Orange-brown with pink lustre Gloss

Potassium permanganate	10 gm
Sodium hydroxide	25 gm
Water	1 litre

Boiling immersion (Thirty minutes)

The object is immersed in the boiling solution, and after two or three minutes the surface shows signs of darkening. After ten minutes this develops to an uneven dark lustrous surface. Continued immersion causes some further darkening. After thirty minutes the object is removed, washed in warm water and allowed to dry in air. When dry, it may be wax finished.

Richer colours were obtained on copper rather than copper-plated surfaces. A similar but slightly lighter orange-brown with a pink tinge was produced on directionally grained satin surfaces.

▲ Sodium hydroxide is a powerful caustic alkali which can cause severe burns. Both the solid and solutions must be prevented from coming into contact with the eyes and skin. When preparing the solution, small quantities of sodium hydroxide should be added to large amounts of water and *not* vice-versa.

3.166   Lustre series Gloss

Sodium thiosulphate	125 gm
Lead acetate	35 gm
Water	1 litre

Hot immersion (Various)

The article is immersed in the hot solution (50–60°C), which produces a series of lustre colours in the following sequence: golden-yellow/orange; purple; blue; pale blue; pale grey. The timing tends to be variable, but it takes roughly five minutes for the purple stage to be reached, and about fifteen minutes for pale blue. (Tests failed to produce a further series of colours, the surface tending to remain pale grey after twenty minutes.) When the desired colour has been produced, the article is removed and immediately washed in cold water. It is allowed to dry in air, any excess moisture being removed with an absorbent tissue. When dry it may be wax finished or coated with a fine lacquer.

More dilute solutions produce the lustre colour sequence, but take a correspondingly longer time to produce.

▲ Lead acetate is very toxic and will give rise to serious conditions if taken internally. The dust, and the vapour evolved from the hot solution, must not be inhaled.

3.167   Lustre sequence Gloss

Antimony trisulphide	50 gm
Ammonia sulphide (16% solution)	100 cm³
Water	1 litre

Cold immersion (Two to ten minutes)

The article is immersed in the cold solution, producing a rapid sequence of lustre colours initially. After about two minutes the colour obtained is predominantly blue with some local red or golden surface variation. This changes gradually to a greenish lustre after four minutes, which slowly fades. The final colours will have developed after about six to eight minutes, and immersion beyond ten minutes produces no detectable change. When the desired colour has been reached, at any stage in the immersion, the article is removed, washed in cold water and allowed to dry in air. When dry, it is wax finished.

The colours produced at an early stage in the immersion are the most intense, but they are also subject to unpredictable variegated effects.

The best lustre colour obtained in tests occurred after about two minutes' immersion, consisting of an intense blue-purple, variegated with streaks of red lustre. Although the colour is good, the variegation cannot be controlled. During the later stages of immersion the colour becomes more opaque, and tends to a patchy slate grey after ten minutes.

▲ The preparation and colouring should be carried out in a well ventilated area. Ammonium sulphide solution liberates a mixed vapour of ammonia and hydrogen sulphide, which must not be inhaled, and which will irritate the eyes. The solution must be prevented from coming into contact with the eyes and skin, as it will cause burns or severe irritation.

3.168   Lustre series Gloss

Sodium thiosulphate	240 gm
Copper acetate	25 gm
Water	1 litre
Citric acid (crystals)	30 gm

Cold immersion (Various)

The sodium thiosulphate and the copper acetate are dissolved in the water, and the citric acid added immediately prior to use. The article is immersed in the cold solution and a series of lustre colours is produced in the following sequence: golden-yellow/orange; brown; purple; blue; pale grey. This sequence takes about thirty minutes to complete, the brown changing to purple after about fifteen minutes. It is followed by a second series of colours, in the same sequence, which takes a further thirty minutes to complete. The latter colours in the second series tend to be slightly more opaque. A third series of colours follows, if immersion is continued, but these tend to be indistinct and murky. (If the temperature of the solution is raised then the sequences are produced more rapidly. However, it was found in tests that the results tended to be less even. Temperatures in excess of about 35–40°C tend to cause the solution to decompose, with a concomitant loss of effect.) When the required colour has been produced, the article is removed and washed in cold water. It is allowed to dry in air, any excess moisture being removed by gentle blotting with an absorbent tissue. When dry it may be wax finished or coated with a fine lacquer.

3.169   Cloudy pale orange lustre Gloss

Potassium hydroxide	100 gm
Copper sulphate	30 gm
Sodium tartrate	30 gm
Water	1 litre

Warm immersion (Thirty minutes)

The potassium hydroxide is added gradually to the water and stirred. Heat is evolved as it dissolves, raising the temperature to about 35°C. The other ingredients are added and allowed to dissolve. There is no need to heat the solution. The object is immersed in the warm solution, and immersion continued until the lustre colour develops. The object is then removed and washed in warm water. It is allowed to dry in air, after any excess moisture has been removed by dabbing with absorbent tissue. When dry, it is wax finished.

A clear lustre colour could not be obtained on either copper or copper-plated surfaces in tests. Raising the temperature of the solution produced a slightly clearer surface and greater colour variegation on copper, but tends to induce darker marks. At higher temperatures the copper plated surfaces tested became patchy.

▲ Potassium hydroxide is a powerful corrosive alkali and contact of the solid or solutions with the eyes or skin must be prevented. When preparing the solution, small quantities of the potassium hydroxide should be added to large quantities of water, and *not* vice-versa.

3.170   Petrol green lustre Semi-matt

Sodium thiosulphate	65 gm
Copper sulphate	12 gm
Copper acetate	10 gm
Arsenic trioxide	5 gm
Sodium chloride	5 gm
Water	1 litre

Hot immersion (Thirty minutes)

Immersion in the hot solution (50–60°C) produces patchy lustre colours during the first fifteen minutes. Continued immersion produces a progressively more opaque and darker colour. After about thirty minutes, the article is removed and washed in warm water. It is allowed to dry in air, and may be wax finished when dry.

The surface tends to be very uneven. The procedure is difficult to control and cannot generally be recommended.

▲ Arsenic trioxide is very toxic and will give rise to very harmful effects if taken internally. The dust, and the vapour evolved from the hot solution, must not be inhaled.

3.171   Pink/green lustre on orange ground Gloss

Copper sulphate	25 gm
Sodium hydroxide	5 gm
Ferric oxide	25 gm
Water	1 litre

Hot immersion and applied liquid and bristle-brushing (A few minutes)

The copper sulphate is dissolved in the water and the sodium hydroxide added. The ferric oxide is added, and the mixture heated and boiled for about ten minutes. When it has cooled to about 70°C, the object is immersed in it for about ten seconds and becomes coated with a brown layer. The object is transferred to a hotplate and gently heated until the brown layer has dried. The dry residue is brushed off with a soft bristle-brush. The turbid solution is then applied to the metal with a brush, as the object continues to be heated on the hotplate. The residue is brushed off, when dry. The application is repeated two or three times, until a green/pink lustre is produced. After a final brushing with the bristle-brush, the object is wax finished.

The results tend to be variegated rather than even, and may appear patchy when viewed in some lights.

▲ Contact with the sodium hydroxide in the solid state must be avoided when preparing the solution. It is very corrosive to the skin and in particular to the eyes.

▲ A nose and mouth face mask fitted with a fine dust filter must be worn when brushing away any residue.

3.172   Reddish-brown lustre with blue tinge Gloss

Selenous acid	6 cm³
Water	1 litre
Sodium hydroxide solution (250 gm/litre)	50 drops

Warm immersion (Forty seconds)

The acid is dissolved in the water, and the sodium hydroxide added by the drop, while stirring. The solution is then heated to 25–30°C. The article to be coloured is immersed in the solution, producing a purplish-brown lustre after about fifteen seconds, which rapidly develops to a reddish-brown tinged with blue. The article is removed after about forty seconds, washed in cold water and allowed to dry in air. When dry it may be wax finished.

Prolonging the immersion causes a gradual fading of the colour. With long immersion times, ten to fifteen minutes, a black layer forms which is totally non-adherent and which flakes to reveal a tarnished metal surface.

▲ Selenous acid is poisonous and very harmful if swallowed or inhaled. Colouring must be carried out in a well ventilated area. Contact with the skin or eyes must be avoided to prevent severe irritation.

▲ Sodium hydroxide is a powerful caustic alkali. Contact with the eyes and skin must be prevented.

3.173   Dull pink lustre Gloss

Antimony trichloride	50 gm
Olive oil	to form a paste

Applied paste (Five minutes)

The antimony trichloride is ground to a creamy paste with a little olive oil using a pestle and mortar. The paste is brushed onto the surface of the object to be coloured. A purple lustre colour develops after about two minutes, changing to a pink lustre after about five minutes. A steel-blue appears if the paste is left for longer. When the desired colour has developed, the residual paste is washed from the surface with pure distilled turpentine. The object is finally rubbed with a soft cloth, and wax finished.

Tests tended to produce results which were slightly uneven. Some darker and redder areas may occur. Localised effects can be produced by the selective application of brush strokes.

If the paste is left to dry on the surface, it becomes difficult to remove without damaging the colour finish.

Tests carried out using aqueous solutions produced similar results, but these tended to be patchy.

▲ Antimony trichloride is highly toxic and every precaution must be taken to avoid contact with the eyes and skin. It is irritating to the respiratory system and eyes, and can cause skin irritation and dermatitis. It will cause very serious effects if taken internally.

3.174   Pink/green lustre on mottled brown ground Gloss

Copper sulphate	25 gm
Nickel ammonium sulphate	25 gm
Potassium chlorate	25 gm
Water	1 litre

Boiling immersion (One to two minutes)

The object is immersed in the boiling solution and rapidly removed when the mottled surface and lustre colour has developed. It should be washed immediately in hot water. After drying in sawdust, it is wax finished.

▲ Nickel salts are a common cause of sensitive skin reactions.

▲ Potassium chlorate is a powerful oxidising agent and should not be allowed to come into contact with combustible material. The solid, and solutions, should be prevented from coming into contact with the eyes, skin or clothing.

▲ The vapour evolved from the boiling solution must not be inhaled.

3.175*   Blue-green patina (stipple) on black ground Semi-matt

Copper nitrate	100 gm
Water	200 cm³
Ammonia (.880 solution)	400 cm³
Acetic acid (6% solution)	400 cm³
Ammonium chloride	100 gm

Sawdust technique (Twenty to thirty hours)

The copper nitrate is dissolved in the water and the ammonia gradually added. The acetic acid is then added, followed by the ammonium chloride. The article to be coloured is laid-up in sawdust which has been evenly moistened with this solution, and is then left for a period of twenty to thirty hours. When treatment is complete, the article is washed in cold water and allowed to dry in air. When completely dry, after several hours, it may be wax finished.

The surface is selectively etched, where the granular sawdust is in close contact with the metal, producing a stipple of etched spots which acquire a blue-green patina. The remainder of the surface is coloured dark brown or black. A wetter medium tends to result in more pronounced general etching, the loss of the black colour and inhibition of patina formation.

Copper-plated surfaces are broken through by this treatment to the underlying brass, and yield results that are very similar to those produced on the plain brass surfaces.

▲ Preparation and colouring should be carried out in a well ventilated area. Ammonia vapour and some acetic acid vapour will be liberated, which cause irritation of the eyes and respiratory system.

▲ The strong ammonia solution must be prevented from coming into contact with the eyes and skin, as it will cause burns or severe irritation.

3.176*   Greenish-blue patina (stipple) on black and brown ground Matt   *Pl. VI, Pl. XV*

Ammonium carbonate	120 gm
Ammonium chloride	40 gm
Sodium chloride	40 gm
Water	1 litre

Sawdust technique (Twenty to thirty hours)

The article to be coloured is laid-up in sawdust which has been evenly moistened with the solution, and is then left for a period of about twenty or thirty hours. The sawdust should be kept moist throughout the period of treatment. When treatment is complete, the article should be washed in cold water and allowed to dry in air. When it is completely dry, it may be wax finished.

The surface is selectively etched by the solution to produce a strongly textured ground which is coloured black. The patina develops in the form of stipple of incrustation. As treatment is prolonged, further etching tends to break through the black areas to a dull orange-brown ground.

Copper-plated surfaces are completely stripped by this treatment, and the results produced are those of the underlying metal. If the treatment is halted at an earlier stage a stipple of darkened copper-plate tends to remain. Plates VI and XV show this effect.

3.177*   Orange/reddish-brown mottle Semi-matt   *Pl. V, Pl. XVI*

Copper nitrate	200 gm
Water	1 litre

Sawdust technique (Two or three days)

The article to be coloured is laid up in sawdust which has been evenly moistened with the solution, and left for two or three days. Colour development may be monitored using a sample, but no deleterious effects were produced in tests, by unpacking and re-packing the object itself. When the colour has developed satisfactorily, the article is removed and washed in cold water. After several hours drying in air, it may be wax finished.

If the medium is allowed to dry out, then a green/blue-green patina develops.

3.178*   Blue-green patina (stipple) on dark brown ground Matt

Ammonium chloride	100 gm
Ammonium carbonate	150 gm
Water	1 litre

Sawdust technique (To twenty or thirty hours)

The article to be coloured is laid-up in sawdust which has been moistened evenly with the solution. It may then be left for periods of up to twenty or thirty hours. When treatment is complete, the object is washed in cold water and allowed to dry in air. When it is completely dry, it may be wax finished.

The surface is selectively etched by this treatment, producing a 'pitted' textured finish. The dark brown colour occurs where the etching is greatest, while the blue-green patina develops on the least etched areas. If a more moist medium is used, then a totally brown textured surface is produced.

Copper-plated surfaces tend to be 'stripped' by this treatment, producing effects that are attributable to the underlying metal. Halting the treatment at an earlier stage tends to leave a stipple of darkened copper plate.

3.179   Bronzed metal with etched stipple and blue-green patina Semi-matt

Copper nitrate	200 gm
Sodium chloride	100 gm
Water	1 litre

Sawdust technique (Twenty or thirty hours)

The article to be coloured is laid-up in sawdust which has been moistened evenly with the solution. It should then be left for a period of twenty or thirty hours. After twenty hours the object should be examined, and repacked for shorter periods of time, as necessary. The sawdust should be kept moist throughout the period of treatment. When treatment is complete, the article should be washed thoroughly in cold water and allowed to dry in air. When completely dry, after several hours, it may be wax finished.

The surface is selectively etched by the solution, where the granular sawdust is in close contact with the metal, causing an even stipple of etched areas which are coloured brown. The blue-green patina develops locally in the region of the etched areas. The remainder of the surface is not etched, but bronzed, gradually becoming dull as treatment is prolonged.

Tests carried out on copper-plated brass surfaces, produced similar results. In these cases however, the etched areas go through to the underlying brass, which acquires a yellow crystalline appearance.

3.180★   Brown with buff and red (stipple) Semi-matt

Copper nitrate	280 gm
Water	850 cm³
Silver nitrate	15 gm
Water	150 cm³

Sawdust technique (Twenty hours)

The copper nitrate and the silver nitrate are dissolved in separate portions of water, and the two solutions then mixed. The article to be coloured is laid-up in sawdust which has been evenly moistened with this mixed solution. It is left for a period of about twenty hours, ensuring that the sawdust remains moist throughout this time. When treatment is complete, the article is removed and washed in cold water. After allowing it to dry thoroughly in air for some time, it may be wax finished.

The surface is coloured brown by the treatment, and selectively etched where the granular sawdust is in close contact with the metal, producing a stipple of buff coloured etched spots. The red colour occurs as a fringe around the etched areas.

A similar effect is produced on copper-plated surfaces, but the surface is coloured yellowish-brown and the etched spots dark grey. The red fringes are far less bright. Longer treatments and moister media tend to expose the underlying metal.

▲ Silver nitrate must be prevented from coming into contact with the skin, and more particularly the eyes, as it may cause burns or severe irritation.

3.181★   Reddish brown (tinged with green patina) Semi-matt

Ammonium chloride	350 gm
Copper acetate	200 gm
Water	1 litre

Sawdust technique (Ten to twenty hours)

The object to be coloured is laid-up in sawdust which has been moistened evenly with the solution, and left for a period of from ten to twenty hours. After ten hours it should be examined and repacked if further treatment is required. It should subsequently be examined every few hours until the required surface finish is obtained. The sawdust should be kept moist during the full period of treatment. The object should then be removed, washed in cold water and allowed to dry in air. When completely dry, it may be wax finished.

The metal is etched and coloured brown by the solution, producing an even textured surface, which acquires a greenish tint from some slight patina development.

Copper-plated surfaces produced similar results, although the brown colour tended to be very dark. Longer treatment times tended to cause a breaching of the surface, to the underlying metal. This also occurred where a more moist medium was used.

3.182★   Bronzed metal stipple on etched orange-brown ground Semi-matt

Copper sulphate	100 gm
Lead acetate	100 gm
Ammonium chloride	10 gm
Water	1 litre

Sawdust technique (Ten to twenty hours)

The object to be coloured is laid-up in sawdust which has been evenly moistened with the solution. It is then left to develop for a period of up to about twenty hours. When treatment is complete, the object is removed, washed in cold water and allowed to dry in air. It may be wax finished when completely dry.

The surface is bronzed by the solution, and selectively etched to produce a network of orange-brown ground, giving a stippled effect. As treatment proceeds, the bronzed surface is progressively replaced by the etched ground, until a textured orange-brown surface is produced. A stipple of greenish-grey patina may also occur.

Copper-plated surfaces produce similar results during the early stages of treatment, but as etching continues, the underlying metal is exposed by 'pitting'. In the case of copper-plated yellow brass, this gave rise to an even stippled yellow-green patina in tests.

▲ Lead acetate is a toxic substance, and every precaution should be taken to prevent inhalation of the dust, when preparing the solution. It is very harmful if taken internally. A clean working method is essential throughout.

3.183★   Blue patina and black on orange-brown (stipple) Semi-matt   *Pl. VI*

Ammonium carbonate	150 gm
Copper acetate	60 gm
Sodium chloride	50 gm
Potassium hydrogen tartrate	50 gm
Water	1 litre

Sawdust technique (Ten to twenty hours)

The object to be coloured is laid-up in sawdust which has been evenly moistened with the solution. It is left for a period of from ten to twenty hours, colour development being monitored with a sample, so that the object remains undisturbed. When treatment is complete, the object is removed, washed in cold water and allowed to dry in air for several hours. When dry, it may be wax finished.

The surface is coloured brown by the solution and selectively etched where the granular sawdust is in close contact with the metal, producing a stipple of etched areas. These acquire a black and an orange-brown colour. The unetched surface becomes dark brown and tends to acquire a blue patina.

Copper-plated surfaces are subject to the same effects initially, but are rapidly cut through to the underlying metal. Plate VI shows the result on copper-plate after twenty hours.

3.184* Purplish-brown/black stipple on buff metal ground Semi-gloss/gloss

Butyric acid	20 cm³
Sodium chloride	17 gm
Sodium hydroxide	7 gm
Sodium sulphide	5 gm
Water	1 litre

Sawdust technique (Twenty hours)

The object to be coloured is laid-up in sawdust which has been evenly moistened with the solution, and left for a period of up to twenty hours. The timing of colour development tends to be unpredictable, and it is essential to include a sample of the same material as the object, when laying-up, so that progress can be monitored without disturbing the object. When treatment is complete, the object is removed and washed thoroughly with cold water. After a period of several hours air-drying, it may be wax finished.

▲ Butyric acid causes burns, and contact with the skin, eyes and clothing must be prevented. It also has a foul pungent odour, which tends to cling to the clothing and hair, and is difficult to eliminate.

▲ Sodium hydroxide is a powerful caustic alkali causing burns, and must be prevented from coming into contact with the eyes or skin.

▲ Sodium sulphide is corrosive and causes burns. It must be prevented from coming into contact with the eyes or skin. When preparing the solution it must not be mixed directly with the acid, or toxic hydrogen sulphide gas will be evolved. The acid should be added to the water and then neutralised with the sodium hydroxide, before the remaining ingredients are added.

3.185* Black and light brown stipple on dull pink ground Semi-matt

Ammonium chloride	16 gm
Sodium chloride	16 gm
Ammonia (.880 solution)	30 cm³
Water	1 litre

Sawdust technique (Twenty-four hours)

The object to be coloured is laid-up in sawdust which has been evenly moistened with the solution. It is then left for about twenty-four hours. The object should not be disturbed while colouring is in progress, as re-packing tends to cause a loss of surface quality and definition. Progress should be monitored by including a small sample of the object metal, when laying-up, which can be examined as necessary. When treatment is complete, the object is washed thoroughly in cold water and allowed to dry in air. It may be wax finished, when completely dry.

Some blue-green patina may occur in the form of a sparse stipple. Patina development is more pronounced in any surface features that are present, eg. punched marks.

The surface is locally etched by the solution where the granular sawdust is in close contact with the metal, causing fine pits in the surface. Areas surrounding these tend to be brown, interspersed with black areas where the etching action is least. If treatment is prolonged, the brown and black tends to gradually recede, exposing more etched pink ground. Shorter treatment times may be required for copper-plated surfaces.

▲ Ammonia solution causes burns or severe irritation, and must be prevented from coming into contact with the eyes or skin. The vapour should not be inhaled as it irritates the respiratory system. Colouring should be carried out in a well ventilated area.

3.186* Red/orange stipple on slightly lustrous metal ground Gloss   *Pl. V*

'Rokusho'	30 gm
Copper sulphate	5 gm
Water	1 litre

Sawdust technique (Several days)

The 'Rokusho' and the copper sulphate are added to the water, and the mixture boiled for a short time and then allowed to cool to room temperature. The object to be coloured is laid-up in sawdust which has been moistened with the solution. It is then left for several days. Colour development is very slow, and should be monitored using a sample so that the object remains undisturbed. When the colour has satisfactorily developed, the object is removed, washed in cold water and allowed to dry in air. It may be wax finished when dry.

The orange stipple produced tends to be translucent, becoming more opaque as treatment progresses. If the object is removed and repacked intermittently during treatment, the stipple becomes less definite and the surface takes on an overall cloudy orange appearance.

For description of 'Rokusho' and its method of preparation, see appendix 1.

3.187 Dark brown stipple on dull pink etch Matt

Ammonium carbonate	300 gm
Acetic acid (10% solution)	35 cm³
Water	1 litre

Sawdust technique (To twenty hours)

The ammonium carbonate is added to the water and the acetic acid then added gradually, while stirring. The object to be coloured is laid-up in sawdust which has been evenly moistened with this solution. It is then left for a period of up to twenty hours, ensuring that the sawdust remains moist throughout the period of treatment. After about ten hours, and at intervals of several hours subsequently, the object should be examined to check the progress. When the desired surface finish has been obtained, the object is removed, washed in cold water and allowed to dry in air. When thoroughly dry, it may be wax finished.

The surface is strongly etched, producing a textured surface which is coloured dark brown and black. The coloured surface is broken by a pitted etched pink, which rapidly becomes predominant as treatment proceeds.

Similar results are produced initially on copper-plated surfaces, but the plating is rapidly breached, revealing the underlying metal.

3.188 Dull pink/brown stipple on dull copper Semi-matt

Copper nitrate	20 gm
Ammonium chloride	20 gm
Calcium hypochlorite	20 gm
Water	1 litre

Sawdust technique (Twenty or thirty hours)

The object to be coloured is laid-up in sawdust which has been moistened evenly with the solution, and left for a period of twenty or thirty hours. The sawdust should be kept moist throughout this time. After treatment, the object is washed in cold water and allowed to dry in air. When thoroughly dry, it may be wax finished.

A lightly etched dull pink/brown stipple is produced on copper. The remainder of the surface is made slightly dull.

Similar results are produced on copper-plated surfaces. Prolonged treatment tends to break through to the underlying metal.

▲ The corrosive fine dust of the calcium hypochlorite (bleaching powder) must not be inhaled, as it can severely irritate the respiratory system. The solution must be prevented from coming into contact with the eyes, which would be severely irritated.

3.189 Brown stipple Semi-matt

Ammonium sulphate	85 gm
Copper nitrate	85 gm
Ammonia (.880 solution)	3 cm³
Water	1 litre

Sawdust technique (Thirty hours)

The article to be coloured is laid-up in sawdust which has been moistened evenly with the solution. It should then be left for a period of twenty or thirty hours. The sawdust should be kept moist throughout the period of treatment. After thirty hours the object should be examined, and repacked for shorter periods of time as necessary. When treatment is complete, the object should be washed thoroughly in cold water and allowed to dry in air. When dry it may be wax finished.

The surface is selectively etched where the granular sawdust is in close contact with the metal, producing a stipple of etched areas that become brown in colour. The remainder of the surface is little affected. Etching becomes more pronounced if a moister medium is used, and some slight green patina may develop in the region of the etched areas.

Similar results are produced on copper-plated surfaces, although prolonged treatment or wetter media tend to break through to the underlying metal.

3.190 Black stipple on pinkish-brown ground Semi-gloss

Potassium hydroxide	100 gm
Copper sulphate	30 gm
Sodium tartrate	30 gm
Water	1 litre

Sawdust technique (Ten hours)

The object is laid-up in sawdust which has been evenly moistened with the solution, and is then left undisturbed for a period of about ten hours. The sawdust should remain moist throughout this period. When treatment is complete, the object is removed and washed thoroughly in cold water. After it has been allowed to dry in air, it may be wax finished.

The black stipple tends to be lost if treatment is too long, or if the medium is too moist.

Very similar results are produced on copper-plated surfaces.

▲ Potassium hydroxide is a strong caustic alkali, and precautions should be taken to prevent it from coming into contact with the eyes and skin.

▲ When preparing the solution, the potassium hydroxide should be added to the water in small quantities, and *not* vice-versa. When it has dissolved completely, the other ingredients may be added.

3.191 Brown stipple on dull pink ground Semi-matt

Copper carbonate	300 gm
Ammonia (.880 solution)	200 cm³
Water	1 litre

Sawdust technique (To twenty or thirty hours)

The article to be coloured is laid-up in sawdust which has been evenly moistened with the above solution, and left for a period of up to twenty or thirty hours. The sawdust should remain moist throughout the period of treatment. When treatment is complete, the article is removed and washed in cold water. It is then allowed to dry in air for some time before wax finishing.

The surface is coloured brown by the treatment, and selectively etched back to a dull pink ground where the granular sawdust is in close contact with the metal, leaving a stipple of the brown surface on the dull ground. Some green patina may also occur.

A green stipple on a dull metal ground tends to be produced on copper-plated surfaces.

▲ The strong ammonia solution will liberate ammonia vapour, which irritates the eyes and respiratory system. Preparation and colouring must therefore be carried out in a well ventilated area.

▲ Ammonia solution causes burns and severe irritation, and must be prevented from coming into contact with the eyes and skin.

3.192 Blue-green patina (stipple) on reddish-brown ground Semi-matt

Ammonium chloride	15 gm
Potassium aluminium sulphate	7 gm
Copper nitrate	30 gm
Sodium chloride	30 gm
Acetic acid (6% solution)	1 litre

Sawdust technique (Twenty to thirty hours)

The object to be coloured is laid-up in sawdust which has been evenly moistened with the solution, and left for about twenty or thirty hours. The object is then removed and allowed to dry in air for several hours, without washing. When dry, the surface is brushed with a stiff bristle-brush to remove any particles of sawdust or loose material. The object is then left for a further period of several hours, after which it may be wax finished.

Shorter times tend to be required for copper-plated surfaces. A sample should be used to monitor colour development, and to check for any excessive corrosive action.

3.193   Slight stipple of blue-grey patina Semi-matt

| Di-ammonium hydrogen orthophosphate | 100 gm |
| Water | 1 litre |

Sawdust technique (Thirty hours)

The object to be coloured is laid-up in sawdust which has been evenly moistened with the solution. It is left for about thirty hours and then removed and washed in cold water. If further colour development is required the treatment should be repeated for additional shorter periods. When treatment is complete, the object is finally washed and allowed to dry in air, and wax finished when completely dry.

The surface is selectively etched, producing a stipple of pink etched areas, and leaving darkened unetched metal which develops a slight stipple of pale patina.

Tests carried out with this solution, using various other techniques, including cold and hot direct application, and hot and cold immersion, produced little or no effect.

3.194   Slight brown stipple on variegated brown Matt

Copper acetate	60 gm
Copper carbonate	30 gm
Ammonium chloride	60 gm
Water	1 litre
Hydrochloric acid (15% solution)	5 cm³

Sawdust technique (Ten to twenty hours)

The object to be coloured is laid-up in sawdust which has been evenly moistened with the solution. It is left for a period of about ten hours and then examined. If further treatment is required, it should be re-packed and examined at intervals of a few hours. When treatment is complete, the object is removed, washed thoroughly in cold water and allowed to dry in air. When completely dry, it may be wax finished.

The surface is bronzed initially, but is rapidly etched back to produce a very variegated matt brown surface. Some of the bronzed surface may remain in the form of a stipple. Some green patina tends to develop on the etched surface.

Similar effects are produced on copper-plated surfaces, but these are rapidly etched through, to reveal the underlying metal.

▲ Hydrochloric acid can cause burns or severe irritation, and must be prevented from coming into contact with the eyes and skin.

3.195   Slight blue-green patina/light bronze stipple on pink ground Semi-matt

Ammonium chloride	10 gm
Copper sulphate	4 gm
Potassium binoxalate	20 gm
Water	1 litre

Sawdust technique (Several days)

The object to be coloured is laid-up in sawdust which has been moistened with the solution, and left for several days. Colour development is very slow, and should be monitored using a sample so that the object remains undisturbed. When the colour has developed satisfactorily, the object is removed, washed in cold water and allowed to dry in air for several hours. It may be wax finished when dry.

The surface is locally etched where the granular sawdust is in close contact with the metal, producing an even stipple of 'bitten' areas which are coloured pink or brown. These are interspersed with areas of unetched metal surface which are coloured with a light bronze lustre. A light stipple of blue-green patina also tends to develop. An overall even stipple of these effects is produced. If treatment is too prolonged, the pink etched areas tend to predominate. Shorter treatment times are generally required for copper-plated surfaces.

▲ Potassium binoxalate is harmful if taken internally. It can also cause irritation, and should be prevented from coming into contact with the eyes or skin.

3.196   Metallic stipple on variegated brown ground Matt

Ammonium chloride	10 gm
Ammonia (.880 solution)	20 cm³
Acetic acid (30% solution)	80 cm³
Water	1 litre

Sawdust technique (One or two days)

The object to be coloured is laid-up in sawdust which has been evenly moistened with the solution, and left for one or two days. Colour development should be monitored using a sample, so that the object itself remains undisturbed. When the colour has developed satisfactorily, the object is removed, washed in cold water and allowed to dry in air for several hours. When dry, it may be wax finished.

The surface is selectively etched by the solution, producing a variegated light reddish-brown and dark brown ground, and leaving a scattering of unetched metal with a brown lustrous colour. The etching action is very marked, and surface development should be carefully monitored to avoid a dull 'scorched' appearance. Copper-plated surfaces tend to be etched through to the underlying metal, leaving some bright plate. Some slight blue-green patina may also occur, particularly in features such as punched marks.

▲ Preparation and colouring should be carried out in a well ventilated area. Ammonia vapour and some acetic acid vapour will be liberated, which can irritate the eyes and respiratory system. These solutions should be prevented from coming into contact with the eyes and skin, or severe irritation or burns may result.

3.197*   Dark brown mottle (some blue-green patina) Semi-matt   *Pl. VI*

Ammonium carbonate	120 gm
Ammonium chloride	40 gm
Sodium chloride	40 gm
Water	1 litre

Cotton-wool technique (Twenty to thirty hours)

Cotton-wool, moistened with the solution, is applied to the surface of the object to be coloured. It is important to ensure that the moist cotton-wool is in full contact with the surface of the object. It is then left for a period of about twenty to thirty hours, ensuring that the cotton-wool remains moist throughout the period of treatment. When treatment is complete, the object should be washed in cold water, and allowed to dry in air. When completely dry, it is wax finished.

A thin smooth dark brown layer forms on the surface, which is mottled or variegated with a lighter tone. Some blue-green patina tends to form, which is integral with the smooth brown layer, and has a cloudy appearance.

Copper-plated surfaces are completely stripped by this solution, the results produced in tests being attributable to the underlying metal. Plate VI shows the result on copper-plate after thirty hours.

3.198* Blue/greenish-blue variegated patina on brown ground Matt

Copper nitrate	100 gm
Water	200 cm³
Ammonia (.880 solution)	400 cm³
Acetic acid (6% solution)	400 cm³
Ammonium chloride	100 gm

Cotton-wool technique (Twenty to thirty hours)

The copper nitrate is dissolved in the water and the ammonia gradually added. The acetic acid is then added, followed by the ammonium chloride. Cotton-wool is wetted with the solution and applied to the surface of the object to be coloured. It is then left for a period of about twenty to thirty hours. The cotton-wool should be kept moist throughout the period of treatment. When treatment is complete, the object is washed in cold water and allowed to dry in air. When treatment is complete, the object is washed in cold water and allowed to dry in air. When dry, it is wax finished.

A very variegated surface is produced, with a finish that is smooth to the touch.

Copper-plated surfaces are stripped by this treatment, and the finishes produced are those of the underlying metal.

▲ Preparation and colouring should be carried out in a well ventilated area. Ammonia vapour and some acetic acid vapour will be liberated, which cause irritation of the eyes and respiratory system.

▲ The strong ammonia solution must be prevented from coming into contact with the eyes and skin, as it will cause burns or severe irritation.

3.199* Dark brown/black (variegated) tinged with bluish patina Matt/semi-matt

Ammonium chloride	100 gm
Ammonium carbonate	150 gm
Water	1 litre

Cotton-wool technique (To twenty or thirty hours)

Cotton-wool, moistened with the solution, is applied to the surface of the object to be coloured. It is then left for periods of up to twenty or thirty hours. The cotton-wool should be kept moist throughout the period of treatment. After about ten or fifteen hours the surface should be examined, by carefully exposing a small portion, to determine the extent of colour development. Subsequently a check should be made every few hours until the desired surface finish is obtained. When treatment is complete, the object is washed in cold water and allowed to dry in air. When completely dry, it may be wax finished.

The metal is etched by the solution, producing a dark brown surface with some black variegation and tinged with some bluish patina. If the medium is allowed to become dry, substantial incrustations of blue-green patina tend to occur.

Copper-plated surfaces tend to be stripped by this treatment, and the results produced are attributable to the underlying metal.

3.200 Mid-brown (some cloudy pale blue-green patina) Matt

Copper nitrate	200 gm
Sodium chloride	200 gm
Water	1 litre

Cotton-wool technique (Twenty hours)

Cotton-wool, moistened with the solution, is applied to the surface of the object to be coloured. It is then left for a period of about twenty hours. It is essential to ensure that the cotton-wool is in full contact with the surface and is kept moist throughout the period of treatment. Breaks in the colour and patches of patina may result if the cotton-wool is allowed to dry out or lift from the surface. When treatment is complete, the object should be washed thoroughly in cold water, and allowed to dry in air. When dry, it may be wax finished.

The surface is evenly etched and coloured mid-brown by the solution. Some cloudy areas of a pale blue-green patina tend to form, particularly if the surface is allowed to dry without washing.

Copper-plated surfaces are completely stripped by the solution, to reveal the etched surface of the underlying metal.

3.201* Red and light brown (mottled) Semi-matt

Copper nitrate	280 gm
Water	850 cm³
Silver nitrate	15 gm
Water	150 cm³

Cotton-wool technique (Twenty hours)

The copper nitrate and the silver nitrate are dissolved in separate portions of water, and the two solutions then mixed. Cotton-wool, moistened with this mixed solution, is applied to the surface of the object to be coloured. It is important to ensure that the cotton-wool is in full contact with the surface, and that it remains moist during the period of treatment. The object is removed after about twenty hours and washed in cold water. After thorough drying in air, it may be wax finished.

The red colour develops as a mottle on an etched light brown surface.

Similar results are produced on copper-plated surfaces. If the treatment is carried on for too long, the etching tends to penetrate to the underlying metal. Surface development should be monitored using a sample.

▲ Silver nitrate must be prevented from coming into contact with the skin, and more particularly the eyes, as it may cause burns or severe irritation.

3.202★ Light reddish-brown (variegated with green) Semi-matt

Ammonium chloride	350 gm
Copper acetate	200 gm
Water	1 litre

Cotton-wool technique (Ten to twenty hours)

Cotton-wool, moistened with the solution, is applied to the surface of the object. This is then left for a period of ten or twenty hours, ensuring that the cotton-wool remains moist. After the first few hours, the surface should be periodically examined by exposing a small portion, to check the progress. When the desired surface finish has been reached, the object is removed, washed in cold water and allowed to dry in air. When completely dry, it may be wax finished.

The surface is evenly etched to a fairly smooth light reddish-brown finish. This tends to be tinged with green producing a slightly variegated appearance. Some areas with a more developed bluish-green patina may occur.

Similar results are produced on copper-plated surfaces. The degree of variegation tends to be greater, and some ochre colours may occur where the underlying metal is partially exposed.

3.203★ Pale greenish-grey patina on grey-brown ground Semi-matt

Copper sulphate	100 gm
Lead acetate	100 gm
Ammonium chloride	10 gm
Water	1 litre

Cotton-wool technique (Ten to twenty hours)

Cotton-wool, moistened with the solution, is applied to the surface of the object to be coloured, and is left in place for a period of up to twenty hours. The cotton-wool should remain moist throughout the period of treatment. When treatment is complete, the object is removed and washed in cold water. It is then dried in air, and gradually darkens on exposure to daylight. When colour change has ceased, the object may be wax finished.

A slightly variegated grey-brown colour develops, and acquires a cloudy film of pale greenish patina. The ground colour darkens on exposure to daylight, becoming a dark grey with a brownish tinge. The cloudy patina is unaffected.

Similar results are produced on copper-plated surfaces, but as treatment is prolonged, the underlying metal is gradually exposed by the etching action of the solution.

▲ Lead acetate is a toxic substance, and every precaution should be taken to prevent inhalation of the dust when preparing the solution. It is very harmful if taken internally. A clean working method is essential throughout.

3.204★ Variegated orange-brown/purplish-brown Semi-matt   *Pl. VI*

Ammonium carbonate	150 gm
Copper acetate	60 gm
Sodium chloride	50 gm
Potassium hydrogen tartrate	50 gm
Water	1 litre

Cotton-wool technique (Ten to twenty hours)

Cotton-wool which has been moistened with the solution is applied to the surface of the article to be coloured. It is important to ensure that the cotton-wool is in full contact with the surface. It is left for a period of from ten to twenty hours, during which time the cotton-wool should remain moist. Colour development should be monitored with a sample, so that the article remains undisturbed. When treatment is complete, the article is removed and washed in cold water. After being allowed to dry in air for several hours, it may be wax finished.

The surface is etched back by the solution, and acquires a variegated brown colour.

Copper-plated surfaces are rapidly etched by the solution, initially producing similar effects to those on copper, but gradually acquiring colours relating to the underlying metal. Plate VI shows the result.

3.205 Dull brown with cloudy pale blue patina Matt

Ammonium carbonate	300 gm
Acetic acid (10% solution)	35 cm³
Water	1 litre

Cotton-wool technique (Twenty hours)

The ammonium carbonate is added to the water and the acetic acid then added gradually, while stirring. Cotton-wool, moistened with this solution, is applied to the surface of the object, and left for a period of about twenty hours. It is important to ensure that the cotton-wool is in full contact with the surface, and that it remains moist throughout. When treatment is complete, the object is washed in cold water and allowed to dry in air. When completely dry, it may be wax finished.

Copper-plated surfaces tend to be stripped by this treatment, producing effects that are attributable to the underlying metal.

3.206 Dull pink/brown (uneven) Matt

Copper nitrate	20 gm
Ammonium chloride	20 gm
Calcium hypochlorite	20 gm
Water	1 litre

Cotton-wool technique (Twenty hours)

Cotton-wool, moistened with the solution, is applied to the surface of the object to be coloured, and left for a period of about twenty hours. It is important to ensure that the cotton-wool is in close contact with the surface, and that it remains moist throughout the period of treatment. When treatment is complete, the object is removed and washed thoroughly in cold water. After a period of drying in air, it may be wax finished.

The surface is evenly etched by this treatment, but acquires an uneven dull pink/brown colour. Similar but slightly darker results are produced on copper-plated surfaces. Prolonged treatment tends to break through to the underlying metal. The technique cannot be generally recommended for copper or copper-plate.

▲ The corrosive fine dust of the calcium hypochlorite (bleaching powder) must not be inhaled, as it can severely irritate the respiratory system. The solution must be prevented from coming into contact with the eyes, which would be severely irritated.

3.207 Variegated lustre Gloss

Ammonium sulphate	85 gm
Copper nitrate	85 gm
Ammonia (.880 solution)	3 cm³
Water	1 litre

Cotton-wool technique (Thirty hours)

Cotton-wool, moistened with the solution, is applied to the surface of the object to be coloured. It is then left for a period of about thirty hours. It is essential to ensure that the moist cotton-wool is in full contact with the surface of the object, or breaks in the colour will occur. The cotton-wool should be kept moist throughout the period of treatment. When treatment is complete, the object should be washed thoroughly in cold water and allowed to dry in air. When dry, it is wax finished.

The technique is only suitable for use on polished surfaces. A golden lustre is produced which is variegated or mottled with pink and green lustre colours.

3.208 Variegated dark brown Gloss

Potassium hydroxide	100 gm
Copper sulphate	30 gm
Sodium tartrate	30 gm
Water	1 litre

Cotton-wool technique (Ten to twenty hours)

Cotton-wool, moistened with the solution, is applied to the surface of the object, which is then left for a period of from ten to twenty hours. It is important to ensure that the moist cotton-wool is in full contact with the surface, and that it remains moist throughout the period of treatment. When treatment is complete, the object is washed in cold water and allowed to dry in air. When dry, it may be wax finished.

A variegated dark brown colour is produced, which is gradually lost as treatment is prolonged, leaving a pink or light brown surface. Colour development should be monitored with a sample. An even dark brown surface is difficult to obtain, as local defects tend to occur, revealing the pink ground.

Similar results are obtained with copper-plated surfaces.

▲ Potassium hydroxide is a strong caustic alkali, and should be prevented from coming into contact with the eyes or skin.

▲ When preparing the solution, the potassium hydroxide should be added to the water in small quantities, and *not* vice-versa.

3.209 Mottled dull brown Semi-matt

Copper carbonate	300 gm
Ammonia (.880 solution)	200 cm³
Water	1 litre

Cotton-wool technique (Twenty hours)

Cotton-wool, moistened with the solution, is applied to the surface of the object to be coloured, which is then left for a period of twenty hours. It is important to ensure that the moist cotton-wool is in full contact with the surface, and that it remains moist throughout the period of treatment. When treatment is complete, the article is removed and washed in cold water. It is then allowed to dry in air for some time before wax finishing.

Tests on copper-plated surfaces produced a dull tarnish, and traces of black.

▲ The strong ammonia solution will liberate ammonia vapour, which irritates the eyes and respiratory system. Preparation and colouring must therefore be carried out in a well ventilated area.

▲ Ammonia solution causes burns and severe irritation, and must be prevented from coming into contact with the eyes and skin.

3.210   Greenish patina on dull brown Matt

Copper acetate	60 gm
Copper carbonate	30 gm
Ammonium chloride	60 gm
Water	1 litre
Hydrochloric acid (15% solution)	5 cm³

Cotton-wool technique (Ten to twenty hours)

Cotton-wool, moistened with the above solution, is applied to the surface of the article, and left for a period of about ten hours. The surface is then examined, by carefully lifting a small portion, to determine the stage it has reached. When the desired surface finish has been obtained, the article is washed thoroughly in cold water and allowed to dry in air. When completely dry, it may be wax finished.

A drab brown colour is initially produced on copper-plated surfaces, but the etching gradually breaks through to reveal the underlying metal. In the case of copper-plated yellow brass, this produced an interesting variegated surface of greenish-brown on frosted yellow.

▲ Hydrochloric acid can cause burns or severe irritation, and must be prevented from coming into contact with the eyes and skin.

3.211   Blue-green patina on pink-brown and grey ground Semi-matt   *Pl. XVI*

Ammonium carbonate	20 gm
Oxalic acid (crystals)	10 gm
Acetic acid (6% solution)	1 litre

Cloth technique (Twenty-four hours)

Strips of soft cotton cloth, or other absorbent material, are soaked with the solution. These are applied to the surface of the object and stippled into place using the end of a stiff brush which has been moistened with the solution. They are left on the surface until very nearly dry, which may take up to twenty-four hours, and then removed. If patina development is to be encouraged, the surface is then moistened by wiping the solution on using a soft cloth. If the ground colour and a minimum of patina are required, then the surface is allowed to dry without applying any more solution. When the surface is completely dry, it may be wax finished.

A mixed pinkish-brown and grey ground is produced which tends to take on the texture of the applied cloth. Cloudy areas of blue-green patina develop on this ground colour.

If the cloth is allowed to dry completely on the surface, it becomes very adherent and tends to damage the surface when removed.

A variety of tests were carried out using this solution. Immersion at temperatures from cold to boiling, produced no effect, even after prolonged immersion. Direct application of the cold or hot solution by dabbing and wiping also produced no effect, as the surface resists wetting.

▲ The ammonium carbonate should be added to the acetic acid in small quantities to avoid violent effervescence.

▲ Oxalic acid is harmful if taken internally. Contact with the eyes and skin should be avoided.

3.212   Blue-green patina on brown Semi-matt

Ammonium sulphate	90 gm
Copper nitrate	90 gm
Ammonia (.880 solution)	1 cm³
Water	1 litre

Cloth technique (Twenty hours)

Soft cotton cloth which has been soaked with the solution is applied to the surface of the object, and stippled into place with a stiff bristle-brush. The object is then left for a period of about twenty hours. The cloth should be removed when it is very nearly dry, and the object then left to dry in air without washing. The blue-green patina tends to develop during the drying period. (If a patina-free ground colour is required, the cloth should be removed at an earlier stage when it is still wet, and the object thoroughly washed. The cloth application may be repeated, for a more developed ground colour.) When treatment is complete and the surface thoroughly dried out, it may be wax finished.

The disposition of colour tends to follow the texture of the applied cloth.

Tests carried out involving direct application of this solution were not successful. The surface stubbornly resists 'wetting', even with additions of wetting agents to the solution.

3.213   Variegated red cut through to pink Semi-matt

A	Copper sulphate	125 gm
	Ferrous sulphate	100 gm
	Water	1 litre
B	Ammonium chloride	1 gm

Boiling immersion (A) (Fifteen minutes)
Boiling immersion (A+B) (Five minutes)

The object is immersed in the boiling solution of copper sulphate and ferrous sulphate, and immersion continued for about fifteen minutes, until a brown colour has developed. The object is removed and rinsed in a tank of hot water, while the ammonium chloride is added to the colouring solution. The object is re-immersed in the colouring solution for about five minutes, causing a brightening and development of the surface to a red colour. It is then removed and washed very thoroughly in hot water. After drying in sawdust, it may be wax finished.

The variegated red coloured copper-plate contrasts with the pink which is produced on the underlying yellow brass.

The object must be thoroughly washed after immersion, to prevent the formation of a greenish bloom or patches of green. If there is a build-up of bloom during immersion, then this should be cleared by agitating the object, or by removing and rinsing in hot water if necessary.

If shorter immersion times are used before the addition of the ammonium chloride, then the results tend to be lighter and more glossy, but are more prone to unevenness.

3.214   Reddish-brown cut through to 'frosted' pink   *Pl. VI*

Copper sulphate	6.25 gm
Copper acetate	1.25 gm
Sodium chloride	2 gm
Potassium nitrate	1.25 gm
Water	1 litre

Boiling immersion (About one hour)

The article is immersed in the boiling solution and after a few minutes the surface darkens and becomes more opaque as the colour develops. The surface develops gradually as immersion continues, and after about one hour the article is removed and washed in hot water. It is dried in sawdust and may be wax finished when completely dry.

The copper plate is coloured red to reddish-brown while the underlying yellow brass is etched to a 'frosted' pink.

Colour development should be carefully monitored visually as the edges of the plated areas tend to be eroded by the solution.

▲ Potassium nitrate is a powerful oxidising agent and must not be allowed to come into contact with combustible materials.

3.215   Reddish-brown cut through to frosted pink Semi-gloss/matt

Copper acetate	6 gm
Copper sulphate	1.5 gm
Sodium chloride	1.5 gm
Water	1 litre

Boiling immersion (About one hour)

The article is immersed in the boiling solution and after a few minutes the surface darkens and becomes more opaque as the colour develops. The surface develops gradually as immersion continues, and after about one hour the article is removed and washed in hot water. It is dried in sawdust and may be wax finished when completely dry.

The copper-plate is coloured reddish-brown while the underlying yellow brass is etched to a matt frosted pink.

Colour development should be carefully monitored visually as the edges of the plated areas tend to be eroded by the solution.

3.216   Dark brown/black cut through to 'frosted' pink Semi-matt/matt

Copper sulphate	6.25 gm
Copper acetate	1.25 gm
Potassium nitrate	1.25 gm
Sodium chloride	2 gm
Sulphur (flowers of sulphur)	3.5 gm
Acetic acid (6% solution)	1 litre

Boiling immersion (About one hour)

The article is immersed in the boiling solution and after a few minutes the surface darkens and becomes more opaque as the colour develops. The surface develops gradually as immersion continues, and after about one hour the article is removed and washed in hot water. The coloured areas of plate, which are purplish-brown initially, darken in daylight to near black. The underlying brass, which is etched to a 'frosted' pink, is not affected by daylight. When the plated areas have developed the dark colour, the article may be wax finished.

The copper-plate is initially coloured a purplish-brown, which darkens to near black in daylight. The underlying brass is etched to a frosted pink, which does not appear to darken.

▲ Potassium nitrate is a powerful oxidising agent and should not be allowed to come into contact with combustible materials.

3.217   Reddish-brown cut through to purple Semi-gloss   *Pl. VI*

Copper carbonate	160 gm
Sodium hydroxide	80 gm
Water	1 litre

Hot/cold immersion (Several hours)

The sodium hydroxide is added to the water in small quantities, while stirring. Heat is evolved as it dissolves, warming the solution. When it has completely dissolved, the copper carbonate is added and the solution heated to boiling. The heat source is removed, and the article suspended in the solution for several hours, without further heating. When the colour has developed, the article is removed, washed in cold water and dried in sawdust. When thoroughly dry, it may be wax finished.

The copper-plate is coloured reddish-brown, while the exposed yellow brass is coloured purple. (The purple colour is not obtained on yellow brass alone.)

The solution tends to separate into layers producing a lighter band of colour in the lower layer. If this 'banding' is to be avoided, the article should be suspended well clear of the bottom of the vessel.

▲ Sodium hydroxide is a powerful caustic alkali, and the solid or solutions must be prevented from coming into contact with the eyes and skin. The sodium hydroxide, in small quantities, should be added to the water and *not* vice-versa. Colouring should be carried out in a well ventilated area, as inhalation of the caustic vapours should be avoided.

3.218   Orange-brown cut through to dark purple Semi-matt

Copper sulphate	120 gm
Ammonia (.880 solution)	30 cm³
Water	1 litre

Boiling immersion (Twenty minutes)

After suitable surface preparation the plated article is immersed in the boiling solution. After the rapid development of lustre colours a more opaque orange develops on the plated portion and a darker more pink colour where the plating is cut through to the underlying metal. These colours darken with continued immersion. After about twenty minutes the object is removed, thoroughly washed in hot water and dried in sawdust. When dry it is wax finished.

The orange-brown of the plated surface is cut through to a very dark purple on the underlying brass. If the plate has been abraded down to the brass then a brighter and more red colour fringes the dark purple area.

▲ A dense precipitate is formed which tends to sediment in the immersion vessel, causing 'bumping'. Glass vessels should not be used.

▲ Preparation of the solution, and colouring, should be carried out in a well ventilated area to prevent irritation of the eyes and respiratory system by the ammonia vapour. The skin, and particularly the eyes, should be protected from accidental splashes of this corrosive alkali.

3.219   Orange-brown cut through to dark olive-green/black Matt   *Pl. VI*

Potassium chlorate	50 gm
Copper sulphate	25 gm
Nickel sulphate	25 gm
Water	1 litre

Boiling immersion (Thirty to forty minutes)

After two or three minutes' immersion in the boiling solution an orange colour develops on the surface of the object, which gradually darkens with continued immersion. The full colour develops in about thirty minutes, together with a green tinge which begins to appear after about twenty minutes. If this green colour is to be kept to a minimum, the object should be removed after thirty minutes. Colour development and surface quality are generally better with a longer immersion, the green tending to enhance the surface. The object is removed after about forty minutes, washed in warm water, and either dried in sawdust or allowed to dry in air. When dry, it is wax finished.

Plate VI shows the result obtained after thirty-five minutes' immersion.

▲ Potassium chlorate is a powerful oxidising agent and must not be allowed to come into contact with combustible material. The solid or solutions must be prevented from coming into contact with the eyes, skin or clothing.

▲ Nickel compounds are a common cause of sensitive skin reactions. Precautions should be taken to prevent inhalation of the dust, during preparation of the solution.

3.220   Dark purplish-brown cut through to light brown Semi-matt

Copper sulphate	50 gm
Potassium chlorate	20 gm
Potassium permanganate	5 gm
Water	1 litre

Boiling immersion (Fifteen minutes)

The object is immersed in the boiling solution, and rapidly becomes coated with a chocolate brown layer. Immersion is continued for fifteen minutes, after which the object is removed and gently rinsed in a bath of warm water. It is then allowed to dry in air, leaving a dark powdery layer on the surface. When dry this is gently brushed with a soft brush to remove any dry residue. This treatment will remove some of the top layer, partly revealing the underlying colour, and should be continued until the desired surface effect is obtained. A stiffer bristle-brush may be required. After any residual dust has been removed with a soft brush, the object is wax finished.

Careful light brushing will remove the dark layer to reveal the chocolate colour of the copper plate contrasting with the light brown of the underlying brass.

▲ Potassium chlorate is a powerful oxidising agent, and must not be allowed to come into contact with combustible material. The solid or solutions should be prevented from coming into contact with the eyes, skin or clothing.

▲ The potassium chlorate and the potassium permanganate must not be mixed together in the solid state. They should each be separately dissolved in a portion of the water, and then mixed together.

3.221   Orange-brown cut through to ochre Semi-gloss/gloss

'Rokusho'	5 gm
Copper sulphate	5 gm
Water	1 litre

Boiling immersion (About one hour)

The article is immersed in the boiling solution and after a few minutes the surface darkens and becomes more opaque as the colour develops. The surface develops gradually as immersion continues, and after about one hour the article is removed and washed in hot water. It is dried in sawdust and may be wax finished when completely dry.

For description of 'Rokusho' and its method of preparation, see appendix 1.

Increasing the copper sulphate content to 20–30 gm/litre produces darker and more tan brown colours.

3.222   Light bronzing with reddish tint cut through to purplish-brown

Copper sulphate	8 gm
Copper acetate	3.25 gm
Potassium aluminium sulphate	2 gm
Water	1 litre
Acetic acid (6% solution)	20 cm³

Boiling immersion (About one hour)

The article is immersed in the boiling solution and after a few minutes the surface darkens and becomes more opaque as the colour develops. The surface develops gradually as immersion continues, and after about one hour the article is removed and washed in hot water. It is dried in sawdust and may be wax finished when completely dry.

The copper-plate is lightly bronzed by the solution, while the underlying yellow brass takes on a purplish-brown colour.

3.223   Mid-brown cut through to purplish-brown Semi-gloss   *Pl. VI*

Copper sulphate	6.25 gm
Copper acetate	1.25 gm
Potassium nitrate	1.25 gm
Water	1 litre

Boiling immersion (About one hour)

The article is immersed in the boiling solution and after a few minutes the surface darkens and becomes more opaque as the colours develop. The surface develops gradually as immersion continues, and after about one hour the article is removed and washed in hot water. It is dried in sawdust and may be wax finished when completely dry.

The copper-plate is coloured mid-brown while the underlying yellow brass is coloured a darker purplish-brown.

▲ Potassium nitrate is a powerful oxidising agent and must not be allowed to come into contact with combustible materials.

3.224   Dark greenish-brown cut through to reddish-brown

Ammonium carbonate	60 gm
Copper sulphate	30 gm
Oxalic acid (crystals)	1 gm
Acetic acid (6% solution)	1 litre

Hot immersion (Twenty minutes)

The object is immersed in the hot solution (80°C) producing light brownish colours after two or three minutes, which may be uneven initially. With continued immersion the colours become even and darken gradually. When the colour has fully developed, after about twenty minutes, the object is removed and washed in hot water. After thorough drying in sawdust, it may be wax finished.

The contrast produced is slight, and becomes less apparent if immersion is prolonged. The dark brown colour of the copper-plated surface is cut through to a more reddish-brown on the underlying brass. No fringe colours are apparent.

▲ Oxalic acid is poisonous and may be harmful if taken internally.

▲ The ammonium carbonate should be added to the acetic acid in small quantities. The reaction is effervescent, and if large amounts are added the effervescence becomes very vigorous.

3.225   Very dark brown/black cut through to light bronze Semi-gloss/gloss

Ammonium persulphate	10 gm
Sodium hydroxide	30 gm
Water	1 litre

Boiling immersion (Two or three minutes)

The ammonium persulphate is dissolved in half the water, and the sodium hydroxide separately dissolved in the remaining water. The two solutions are then mixed together and the mixture heated to boiling. The article to be coloured is immersed in the boiling solution, which produces golden and then brown lustre colours. Continued immersion darkens these colours to an opaque but glossy finish. When the colour has developed, after about two or three minutes, the article is removed, washed in hot water and dried in sawdust. When dry, it may be wax finished.

The copper-plate is coloured a very dark brown or black, while the underlying yellow brass is slightly bronzed.

▲ Ammonium persulphate (peroxodisulphate) is a powerful oxidant, and must be kept out of contact with all combustible materials. It should be prevented from coming into contact with the eyes, skin or clothing.

▲ Sodium hydroxide is a powerful caustic alkali and must be prevented from coming into contact with the eyes and skin.

3.226   Dark brown cut through to black Matt/semi-matt

Nickel sulphate	35 gm
Copper sulphate	25 gm
Potassium permanganate	5 gm
Water	1 litre

Hot immersion (Five to thirty minutes)

The object is immersed in the hot solution (80–90°C) and the surface immediately becomes covered with a dark brown layer. After four or five minutes it is removed and gently bristle-brushed with hot water. The dark layer is removed revealing a brown 'bronzing' beneath. If darker colours are required then the process should be repeated. If, on the other hand a full very dark brown/black colour with a more matt finish is required, the object should be re-immersed, and the dark brown layer will then re-form and should be left on the surface as immersion is continued. After thirty minutes the object is removed, and gently bristle-brushed under hot water. After thorough rinsing, and drying in sawdust, it is wax finished.

The dark brown colour of the copper-plate is cut through to a black on the underlying brass, which is finely spotted with grey. A fringe of uncoloured copper occurs, around the black area.

▲ Nickel compounds are a common cause of sensitive skin reactions. Precautions should also be taken to avoid inhalation of the dust when preparing the solution.

3.227   Dark reddish-brown cut through to dark grey-brown Semi-matt

Copper sulphate	125 gm
Ferrous sulphate	100 gm
Acetic acid (glacial)	6.5 cm³
Water	1 litre

Boiling immersion (Thirty to forty minutes)

The object is immersed in the boiling solution. The surface darkens gradually during the course of immersion, initially tending to be drab and rather uneven, but later becoming more even and generally darker and richer in colour. When the colour has developed, after thirty to forty minutes, the object is removed and washed in hot water. It is dried in sawdust, and wax finished when dry.

The dark reddish-brown of the copper-plate is cut through to a dark grey-brown on the underlying brass, which is fringed with a light orange-brown.

▲ Acetic acid in the highly concentrated 'glacial' form is corrosive, and must not be allowed to come into contact with the skin, and more particularly the eyes. It liberates a strong vapour which is irritating to the respiratory system and eyes. Preparation of the solution should be carried out in a well ventilated area.

3.228   Brown cut through to black (Gloss)

Copper carbonate	125 gm
Ammonia (.880 solution)	250 cm³
Water	750 cm³

Warm immersion (One hour)

The ammonia is added to the water, followed by the carbonate, which will partially dissolve. Some excess carbonate should remain in suspension. The solution is heated to about 50°C, and the object is immersed. The colour develops slowly during the period of immersion. After about one hour, when the colour has fully developed, the object is removed and washed thoroughly in warm water. It should be carefully dried in sawdust, and wax finished when dry.

The brown coloured copper-plate is cut through to the underlying brass which is coloured black. The black area may be fringed with uncoloured copper-plate.

With copper-plated yellow brass, the plate can be left uncoloured and the yellow brass coloured black, if the solution is used cold for a brief immersion (a few minutes).

▲ This procedure should only be carried out where there is adequate ventilation. Ammonia vapour will be freely liberated when the solution is prepared, and during colouring, which can severely irritate the eyes and respiratory system. The skin, and particularly the eyes, must be protected against accidental splashes of this corrosive alkali.

4 Gilding metal (rolled sheet)

The red brass to which the recipes and results refer is CZ 101 (B.S. 2870).
This is commonly referred to as gilding metal and has the composition
copper 90%, zinc 10%.

   The main colour heading for each recipe refers to the results obtained
on polished surfaces, unless otherwise stated, and includes a reference to
both the colour and surface quality. The results obtained on satin-finished
surfaces, where these are significantly different, are given in the marginal
notes.

	Recipe numbers
Red Red/purplish-red/purplish-brown/orange-brown	4.1 – 4.22
Brown Light bronzing/brown/dark brown	4.23 – 4.92
Black Black/dark slate/grey	4.93 – 4.120
Green patina Blue-green/green	4.121 – 4.151
Lustre colours	4.152 – 4.169
Special techniques	4.170 – 4.207

4.1*   Purplish-red Semi-gloss   *Pl. VII*

Copper sulphate	120 gm
Ammonia (.880 solution)	30 cm³
Water	1 litre

Boiling immersion (Twenty minutes)

Immersion in the boiling solution produces lustre colours after one minute, followed by the gradual change to a more opaque orange-pink colour which is well developed after seven or eight minutes. Continued immersion produces a darkening of this to an orange-brown and the gradual development of a purplish-red colour. After twenty minutes the object is removed and washed in hot water. After drying in sawdust it is wax finished.

▲ A dense precipitate is formed which tends to sediment in the immersion vessel, causing 'bumping'. Glass vessels should not be used.

▲ Preparation of the solution, and colouring, should be carried out in a well ventilated area to prevent irritation of the eyes and respiratory system by the ammonia vapour. The skin, and particularly the eyes, should be protected from accidental splashes of this corrosive alkali.

4.2*   Red Semi-matt   *Pl. VIII*

A	Copper sulphate	25 gm
	Water	1 litre
B	Ammonium chloride	0.5 gm

Boiling immersion (A) (Fifteen minutes)
Boiling immersion (A+B) (Ten minutes)

The article is immersed in the boiling copper sulphate solution for about fifteen minutes or until the colour is well developed. It is then removed to a bath of hot water, while the ammonium chloride is added to the colouring solution, and then re-immersed. The colour is brightened and tends to become more red. After about ten minutes it is removed and washed in hot water. If the immersion is prolonged a bloom of greyish-green patina will tend to form, which may need to be removed with a soft bristle-brush. The article is dried in sawdust and finally wax finished.

4.3   Red Semi-gloss

Copper acetate	6 gm
Copper sulphate	1.5 gm
Sodium chloride	1.5 gm
Water	1 litre

Boiling immersion (About one hour)

The article is immersed in the boiling solution and after a few minutes the surface darkens and becomes more opaque as the colour develops. The surface develops gradually as immersion continues, and after about one hour the article is removed and washed in hot water. It is dried in sawdust and may be wax finished when completely dry.

Slightly lighter colours tended to be produced on satinised surfaces.

4.4   Red/brownish-red Semi-matt

A	Copper sulphate	125 gm
	Ferrous sulphate	100 gm
	Water	1 litre
B	Ammonium chloride	1 gm

Boiling immersion (A) (Fifteen minutes)
Boiling immersion (A+B) (Five minutes)

The object is immersed in the boiling solution of copper sulphate and ferrous sulphate, and immersion continued for about fifteen minutes, until a brown colour has developed. The object is removed and rinsed in a tank of hot water, while the ammonium chloride is added to the colouring solution. The object is re-immersed in the colouring solution for about five minutes, causing a brightening and development of the surface to a red colour. It is then removed and washed very thoroughly in hot water. After drying in sawdust, it may be wax finished.

The object must be thoroughly washed after immersion, to prevent the formation of a greenish bloom or patches of green. If there is a build-up of bloom during immersion, then this should be cleared by agitating the object, or by removing and rinsing in hot water if necessary.

If shorter immersion times are used before the addition of the ammonium chloride, then the results tend to be lighter and more glossy, but are more prone to unevenness.

4.5★   **Reddish-brown/orange-brown   Semi-gloss/semi-matt**

Sodium chlorate	100 gm
Copper nitrate	25 gm
Water	1 litre

Boiling immersion   (Forty minutes to one hour)

The object is immersed in the boiling solution, and the surface is quickly coloured with a dark lustre, which develops to a thin yellowish-orange colour. Continued immersion produces a gradual change to a more opaque orange-red or orange-brown colour. When the colour has developed evenly and to the required depth, after about forty minutes to one hour, the object is removed and washed in hot water. It is dried in sawdust, and may be wax finished when dry.

Prolonging the immersion beyond about one hour darkens the colour, but also tends to produce a duller surface which is more prone to spotting or unevenness.

▲ Sodium chlorate is a powerful oxidant and must not be allowed to come into contact with combustible material or acids. It should also be prevented from coming into contact with the eyes, skin or clothing.

4.6   **Red/orange-brown   Semi-matt**

Ammonium carbonate	50 gm
Ammonium chloride	5 gm
Copper acetate	25 gm
Oxalic acid (crystals)	1 gm
Acetic acid (6% solution)	1 litre

Boiling immersion   (One hour)

The object is immersed in the boiling solution and an uneven orange-yellow colouration with traces of green is produced after a few minutes. The surface gradually darkens with continued immersion producing a reddish-brown colour. After about one hour, the object is removed and washed in hot water. After thorough drying in sawdust, it may be wax finished.

The surfaces produced by this solution are prone to streaking and patchiness.

▲ The ammonium carbonate should be added to the acetic acid in small quantities. The reaction is effervescent and if large amounts are added, the effervescence becomes very vigorous.

▲ Oxalic acid is harmful if taken internally. Contact with the eyes and skin should be avoided.

4.7★   **Purplish-brown   Gloss/semi-gloss**   *Pl. VIII*

Copper nitrate	200 gm
Water	1 litre

Cold immersion   (Thirty-six hours)

The article is suspended in the cold solution, well clear of the bottom of the immersion vessel. The purplish-brown colouration develops very gradually and evenly during the period of immersion. After thirty-six hours the article is removed, washed thoroughly in cold water and dried in sawdust.

The article must be absolutely grease-free and thoroughly clean. The cold solution does not 'bite' the surface readily, and streaks and patches will occur if there is any surface contamination.

Lighter-coloured bands may occur on an object that is suspended too close to the bottom of the immersion vessel, due to the settling of a slight turbidity in the solution.

For results obtained using this solution at higher temperatures see recipe 4.65.

4.8★   **Reddish-purple/brown   Semi-gloss**

Copper nitrate	80 gm
Water	1 litre
Ammonia (.880 solution)	4–5 cm³

Boiling immersion   (Thirty minutes)

The article is immersed in the boiling solution. After five minutes a pink tinge is apparent, followed by a gradual build-up of colour. After thirty minutes the article is removed and washed thoroughly in hot water before drying in sawdust. When dry the surface has a greyish bloom which becomes nearly invisible after wax finishing.

Increasing the ammonia content to 10 cm³ produces a more transparent colour with irregular dark markings, and a very marked bloom when viewed from oblique angles.

Lower overall concentrations, or decreases in the quantity of ammonia added, produce drab mid-grey-brown colourations.

4.9★   **Purple/purplish-brown   Semi-matt**   *Pl. VII*

Copper sulphate	25 gm
Ammonia (.880 solution)	3–5 cm³
Water	1 litre

Boiling immersion   (Fifteen to thirty minutes)

The article is immersed in the boiling solution, which produces a series of pink/green lustre colours after about one minute. Continued immersion causes a darkening of the surface to produce brown colours, which may be uneven initially, but which become more even in the later stages of immersion. When the purple colour has developed, which may take from fifteen to thirty minutes, the article is removed and washed in hot water. After drying in sawdust, it is wax finished.

Increasing the ammonia content beyond the stated amount tends to inhibit colour development. At a rate of 10 cm³/litre ammonia, with the given concentration of copper sulphate, no colour is produced.

4.10   Brownish-purple mottled with lighter brown Semi-matt

Copper nitrate	40 gm
Ammonia (.880 solution)	to form a paste

Applied paste (Several hours)

The copper nitrate is ground using a pestle and mortar, and a very small amount of ammonia added by the drop to form a crystalline 'paste'. This is applied to the object using a soft brush, and the surface then allowed to dry for several hours. When dry, the residue is washed away with cold water. The process may be repeated if necessary. After a final washing and air-drying, the object may be wax finished.

Although the lighter coloured mottle tended to be even in tests, some more irregular patches also occurred.

A true paste will not form, but the moist crystalline mass that is obtained can be applied with a brush. This tends to become watery and may not dry out completely.

4.11*   Orange/brownish-orange Semi-gloss/gloss   *Pl. VII*

Copper sulphate	120 gm
Potassium chlorate	60 gm
Water	1 litre

Hot immersion (Five or six minutes)

The object is immersed in the hot solution (70–80°C), rapidly developing green lustre colours, which quickly become more opaque after about a minute. Continued immersion darkens the colour to orange. After five or six minutes the object is removed and washed thoroughly in hot water. It may be wax finished after thorough drying in sawdust.

Prolonging the immersion produces further darkening to a deep brown colour. See recipe 4.70.

▲ Potassium chlorate is a powerful oxidising agent and must not be allowed to come into contact with combustible materials. Contact with the eyes, skin or clothing must be avoided.

4.12   Orange/orange-brown Gloss

Copper acetate	7 gm
Ammonium chloride	3.5 gm
Acetic acid (6% solution)	100 cm³
Water	1 litre

Boiling immersion (Ten minutes)

The article is immersed in the boiling solution, and after three minutes a slightly lustrous orange colour develops. With continued immersion this colour becomes more opaque. The article is removed after about ten minutes. It is washed thoroughly in hot water and dried in sawdust. When dry it is wax finished.

The orange colour produced has a translucent quality. Surfaces coloured using this method are susceptible to some unevenness, particularly if the immersion is prolonged.

Handling the surface immediately after immersion should be avoided as the surface is easily marked at that stage.

4.13   Light orange-brown Gloss

Potassium chlorate	50 gm
Copper sulphate	25 gm
Ferrous sulphate	25 gm
Water	1 litre

Boiling immersion (Twenty minutes)

The object is immersed in the boiling solution. The colour develops gradually during immersion, becoming darker and tending to acquire a bloom of greenish-grey patina during the latter stages. If the bloom becomes excessive it should be removed by gently bristle-brushing the surface. When the required colour has developed, after about twenty minutes, the object is removed and washed in hot water. It is dried in sawdust, and may be wax finished when dry.

▲ Potassium chlorate is a powerful oxidant and must not be allowed to come into contact with combustible materials. It should also be prevented from coming into contact with the eyes, skin or clothing.

4.14   Orange-brown Semi-matt/matt

Copper sulphate	125 gm
Sodium chlorate	50 gm
Water	1 litre

Boiling immersion (Five to ten minutes)

The article is immersed in the boiling solution, which produces pink/green lustre colours very rapidly. These darken and become more opaque with continued immersion. When the orange-brown colour has developed, after about five or ten minutes, the article is removed and washed thoroughly in hot water. After drying in sawdust, it is wax finished.

Prolonging the immersion will darken the colour, but will also cause the deposition of a grey-green and very uneven layer. This layer is difficult to make even, and hard to remove when immersion is complete.

▲ Sodium chlorate is a powerful oxidising agent and must not be allowed to come into contact with combustible materials or acids. Inhaling the dust must be avoided, and the eyes and skin should be protected from contact with either the solid or solutions.

4.15*   Orange/orange-brown Semi-matt   *Pl. VIII*

Copper sulphate	25 gm
Nickel ammonium sulphate	25 gm
Potassium chlorate	25 gm
Water	1 litre

Boiling immersion (Twenty minutes)

▲ Nickel salts are a common cause of sensitive skin reactions.

▲ Potassium chlorate is a powerful oxidising agent and should not be allowed to come into contact with combustible material. The solid, and solutions, should not be allowed to come into contact with the eyes, skin or clothing.

Immersing the object in the boiling solution produces an immediate lustrous mottling of the surface. Continued immersion induces a change to a more even orange colour which darkens gradually. After twenty minutes the object is removed and washed in hot water. After thorough drying in sawdust it is wax finished.

4.16   Orange-brown Matt/semi-matt

Potassium chlorate	50 gm
Copper sulphate	25 gm
Water	1 litre

Hot immersion (Fifteen or twenty minutes)

Tests showed that there is a tendency to some colour variation at the edges of sheet material, and around surface features such as punched marks.

▲ Potassium chlorate is a powerful oxidising agent and should not be allowed to come into contact with combustible material. Contact with the eyes, skin and clothing must be avoided.

▲ A nose and mouth face mask fitted with a fine dust filter should be worn when brushing away any dry powder from the surface.

The object is immersed in the hot solution (80°C) and is initially coloured with an uneven golden lustre finish which changes to a more opaque yellow-green. Continued immersion darkens the colour to orange and then finally to a more brown colour. After fifteen or twenty minutes, when the full colour has developed, the object is removed and washed in hot water. When it has been thoroughly dried in sawdust, it is wax finished. A green powdery film may develop when the sample is dried, which can be brushed away prior to waxing. The wax finish tends to make any residual green bloom invisible.

4.17*   Orange-brown Matt

Potassium chlorate	50 gm
Copper sulphate	25 gm
Nickel sulphate	25 gm
Water	1 litre

Boiling immersion (Thirty to forty minutes)

The bloom of green patina is virtually invisible after wax finishing, integrating with the ground to produce an even matt finish. The longer immersion time was found to produce a better overall surface in tests.

▲ Potassium chlorate is a powerful oxidising agent and must not be allowed to come into contact with combustible material. The solid or solutions must be prevented from coming into contact with the eyes, skin or clothing.

▲ Nickel compounds are a common cause of sensitive skin reactions. Precautions should be taken to prevent inhalation of the dust, during preparation of the solution.

After two or three minutes immersion in the boiling solution an orange colour develops on the surface of the object, which gradually darkens with continued immersion. The full colour develops in about thirty minutes, together with a green tinge which begins to appear after about twenty minutes. If this green colour is to be kept to a minimum, the object should be removed after thirty minutes. Colour development and surface quality are generally better with a longer immersion, the green tending to enhance the surface. The object is removed after about forty minutes, washed in warm water, and either dried in sawdust or allowed to dry in air. When dry, it is wax finished.

4.18   Orange-brown Gloss

Copper sulphate	25 gm
Lead acetate	25 gm
Water	1 litre

Hot immersion (Ten to fifteen minutes)

Small additions of ammonium chloride (at a concentration of 2.5 gm/litre), introduced into the solution at a late stage in the immersion, produce a brightening of the colours. In this case however the colours thus produced darkened patchily on exposure to daylight, and the variation cannot be recommended.

▲ Lead acetate is highly toxic, and every precaution should be taken to prevent inhalation of the dust during weighing, and preparation of the solution. Colouring should be carried out in a well ventilated area.

The article is immersed in the hot solution (80°C) producing green lustre colours after about one minute, which change gradually to produce an opaque orange after about six or seven minutes. This colour darkens with continued immersion. After ten or fifteen minutes the article is removed and washed thoroughly in hot water. After drying in sawdust, it is wax finished.

4.19*   Orange-brown Gloss

'Rokusho'	3.25 gm
Copper sulphate	3.25 gm
Potassium aluminium sulphate	2 gm
Water	1 litre

Boiling immersion (About one hour)

Pl. VIII

For description of 'Rokusho' and its method of preparation, see appendix 1.

The article is immersed in the boiling solution and after a few minutes the surface darkens and becomes more opaque as the colour develops. The surface develops gradually as immersion continues, and after about one hour the article is removed and washed in hot water. It is dried in sawdust and may be wax finished when completely dry.

4.20   Orange-brown Gloss

'Rokusho'	5 gm
Copper sulphate	5 gm
Water	1 litre

Boiling immersion (About one hour)

For description of 'Rokusho' and its method of preparation, see appendix 1.

Increasing the copper sulphate content to 20–30 gm/litre produces darker and more tan brown colours.

The article is immersed in the boiling solution and after a few minutes the surface darkens and becomes more opaque as the colour develops. The surface develops gradually as immersion continues, and after about one hour the article is removed and washed in hot water. It is dried in sawdust and may be wax finished when completely dry.

4.21   Reddish-brown Semi-gloss

Copper acetate	50 gm
Copper nitrate	50 gm
Water	1 litre

Boiling immersion (Thirty minutes)

Immersing the object in the boiling solution produces lustrous golden colours after one minute, which gradually change to more opaque matt pink/orange tints after five minutes. Continued immersion produces a gradual darkening to terracotta. After thirty minutes the object is removed, washed in hot water and dried in sawdust. When completely dry it is wax finished.

4.22   Light patina on reddish-brown ground Semi-matt

Copper sulphate	125 gm
Sodium chlorate	50 gm
Ferrous sulphate	25 gm
Water	1 litre

Boiling immersion (Thirty minutes)

If the ground colour only is required, then the object should be intermittently bristle-brushed in hot water, during immersion. A residual greenish-grey bloom tends to remain.

▲ Sodium chlorate is a powerful oxidising agent and must not be allowed to come into contact with combustible material or acids. Inhalation of the dust must be avoided. The eyes and skin should be protected from contact with either the solid or solutions.

The object is immersed in the boiling solution. Orange colours develop after a few minutes and gradually darken with continued immersion. After about fifteen minutes a greenish layer begins to be deposited on the now brown ground colour. Agitating the object within the solution tends to inhibit the deposition of the green layer. After thirty minutes the object is removed and washed thoroughly in hot water. After drying in sawdust, it is wax finished.

4.23   Reddish-brown bronzing Semi-gloss

Potassium sulphide	125 gm
Water	1 litre
Ammonia (.880 solution)	100 cm³

Applied liquid and scratch-brushing (Several minutes)

▲ Preparation and colouring should be carried out in a well ventilated area. Some hydrogen sulphide gas will be liberated by the potassium sulphide solution. Ammonia vapour will also be liberated from the solution. These should not be inhaled. The solution should be prevented from coming into contact with the eyes and skin as it may cause burns or severe irritation.

The potassium sulphide is ground to a paste with some of the water and a little ammonia, using a pestle and mortar. This paste is added to the rest of the water, and the remaining ammonia, to form a strong solution. The solution is wiped on to the surface of the object with a soft cloth. The surface is rapidly blackened with a non-adherent film, which is rinsed off in cold water, revealing a light brown underlying colour. This procedure is repeated several times, and the surface finally finished by lightly scratch-brushing with a much diluted solution, and cold water. The object is rinsed in cold water, allowed to dry in air, and wax finished when dry.

4.24   Light reddish-brown bronzing Semi-matt

Sodium sulphide	30 gm
Sulphur (flowers of sulphur)	4 gm
Water	1 litre

Hot immersion and scratch-brushing (Fifteen minutes)

The solids are added to about 200 cm³ of the water, and boiled until they dissolve and yield a clear deep orange solution. This is added to the remaining water and the temperature adjusted to 60–70°C. The object is immersed in the hot solution and the colour allowed to develop for about ten minutes. It is then removed and scratch-brushed with the hot solution, until the surface is even and the colour developed. The object should be washed thoroughly in hot water and dried in sawdust. When dry, it may be wax finished.

The colour produced is light and has a grey sheen.

▲ Sodium sulphide in solid form and in solution must be prevented from coming into contact with the eyes and skin, as it can cause severe burns. Some hydrogen sulphide gas may be evolved as the solution is heated. Colouring should be carried out in a well ventilated area.

4.25   Light brown bronzing Gloss

Potassium sulphide	1 gm
Ammonium chloride	4 gm
Water	1 litre

Cold scratch-brushing (Five minutes)

The object is scratch-brushed with the cold solution for a short time and then washed in hot water. This alternation is repeated until the colour develops. The surface is then burnished gently with the scratch brush in hot water, and finally with cold water. After thorough rinsing and drying in air the object may be wax finished.

The colour tends to develop unevenly and may require prolonged careful working to obtain good results.

For results obtained with a more concentrated solution, see recipe 4.26.

▲ The procedure should be carried out in a well ventilated area, as some hydrogen sulphide gas is liberated from the solution.

4.26   Mid-brown Semi-gloss

Potassium sulphide	20 gm
Ammonium chloride	30 gm
Water	1 litre

Cold scratch-brushing (Several minutes)

The object is scratch-brushed with the cold solution, alternating with hot water. It should not be overworked at this stage, but a tenuous orange coloured layer should be allowed to build up. The surface is then scratch-brushed with hot water, which tends to remove the orange layer, leaving a pale surface. Burnishing with hot water will then bring up the mid-brown tone. The surface is finally burnished lightly under cold water, and allowed to dry in air. When dry it may be wax finished.

▲ The procedure should be carried out in a well ventilated area, as hydrogen sulphide gas is liberated by the solution.

4.27   Light bronzing Semi-gloss/gloss

Potassium sulphate	100 gm
Potassium sulphide	8 gm
Ammonium chloride	8 gm
Water	1 litre

Boiling immersion and scratch-brushing (Several minutes)

Immersion in the boiling solution produces an immediate darkening of the surface with lustre colours, followed by the gradual development of more opaque colour. This tends to occur unevenly. After a few minutes' immersion, the object is removed and scratch-brushed with the hot solution. When the surface has evened and the colour developed, the object is washed in hot water, and allowed to dry in air. When dry, it is wax finished.

Brasses with higher proportions of zinc do not respond well to this solution, producing only lightly tarnished surfaces.

▲ Colouring should be carried out in a well ventilated area, as some hydrogen sulphide gas will be liberated from the hot solution.

4.28★   Reddish-brown Semi-gloss

Ammonium sulphide (16% solution)	100 cm³
Water	1 litre

Cold immersion and scratch-brushing (Ten to thirty minutes)

The article is immersed in the cold solution for a few minutes to allow the colour to develop, and then removed and lightly scratch-brushed with the cold solution, alternating with hot and cold water. It is then re-immersed for about ten minutes and then removed and scratch-brushed as before. This procedure is repeated until the desired colour is achieved. The article is finally lightly scratch-brushed alternately with hot and cold water. It is allowed to dry in air and wax finished.

As an alternative to immersion in the cold solution, the solution can be applied continuously to the surface with a soft brush. This tended to produce lighter bronzed finishes in tests.

▲ Ammonium sulphide solution liberates a vapour containing ammonia and hydrogen sulphide gas, which should not be inhaled. Preparation and colouring must be carried out in a well ventilated area. The solution must be prevented from coming into contact with the eyes and skin, as it causes burns or severe irritation.

4.29   Reddish-brown (variegated) Semi-gloss

Ferric nitrate	10 gm
Water	1 litre

Torch technique

The object is heated with a blow torch and the solution applied sparingly with a scratch-brush, using small circular motions and working across the surface. The metal gradually darkens to a reddish-brown and the surface can be finished by lightly dabbing with a cloth dampened with the solution. When treatment is complete and the object has been allowed to cool and dry thoroughly, it may be wax finished.

A darker variegation can be obtained by finally dabbing with a stronger solution. (Ferric nitrate 40 gm/litre.)

▲ The fumes evolved as the solution is applied to the heated metal should not be inhaled. Adequate ventilation must be provided and a nose and mouth face mask worn, fitted with the correct filter.

▲ Ferric nitrate is corrosive and should not be allowed to come into contact with the eyes or skin. A face mask or goggles should be worn to protect the eyes from hot splashes as the solution is applied.

4.30   Pale bronzing/very light ochre (uneven) Semi-matt

Copper nitrate	80 gm
Nitric acid (10% solution)	100 cm³
Water	1 litre

Hot immersion (Five minutes)

The article is immersed in the hot solution (60–70°C), which causes etching of the surface and a gradual darkening. After five minutes it is removed and washed in warm water, and allowed to dry in air. When dry, it may be wax finished.

(Some sources suggest using this as a ground for a greenish patina, produced by applying a solution of ammonium carbonate (100 gm/litre) by wiping with a soft cloth and allowing to dry. Tests for extended periods produced only the slightest traces of a green tinge.)

The results obtained in tests were poor and the procedure cannot be generally recommended.

▲ Nitric acid is corrosive and may cause severe irritation or burns. It should be prevented from coming into contact with the eyes or skin.

4.31   Pale bronzing with pink tinge Semi-gloss

Potassium sulphide	2 gm
Ammonium sulphate	0.5 gm
Water	1 litre

Cold scratch-brushing (Several minutes)

The object is scratch-brushed with the cold solution, alternating with hot and then cold water. This process is continued until the colour has developed. Any relieving that is required should then be carried out. The object is washed thoroughly in cold water and allowed to dry in air. When completely dry it is wax finished.

The colour obtained is very pale, and difficult to produce evenly over a surface. The scratch-brush tends to 'bite' in places causing localised changes in colour which are difficult to work out.

4.32*   Mid-brown/reddish-brown Gloss

Copper acetate	70 gm
Copper sulphate	40 gm
Potassium aluminium sulphate	6 gm
Acetic acid (10% solution)	10 cm³
Water	1 litre

Boiling immersion (Fifteen minutes)

The object is immersed in the boiling solution. The colour development is gradual and even throughout the period of immersion, producing purple colours after about five minutes. The object is removed after fifteen minutes, washed thoroughly in hot water and dried in sawdust. When completely dry it is wax finished.

4.33*   Mid-brown Semi-matt   *Pl. VII*

Copper nitrate	200 gm
Water	1 litre

Applied liquid (Twice a day for five days)

The solution is initially dabbed on vigorously with a soft cloth, to ensure that the surface is evenly 'wetted'. Subsequent applications should be wiped on, and used very sparingly, to inhibit the formation of green patina. After each application the article is left to dry in air. This procedure is repeated twice a day for about five days. The article is then left to dry thoroughly in air, and wax finished when dry.

4.34*   Light orange-brown/ochre Semi-matt   *Pl. VIII*

Copper nitrate	35 gm
Ammonium chloride	35 gm
Calcium chloride	35 gm
Water	1 litre

Applied liquid to heated metal (Repeat several times)

If the solution is persistently applied as the surface is heated with a torch, a more adherent pale blue-green patina tends to form. Solutions containing little or no chlorides are preferable for this effect.

The article is either heated on a hotplate or with a torch to a temperature such that the solution sizzles gently when it is applied. The solution is applied with a soft brush, and dries out to produce a superficial powdery green layer. The procedure is repeated several times, each fresh application tending to remove the green powdery layer, until an underlying colour has developed. Any residual green is then removed by washing and the article is allowed to dry in air. When dry it may be wax finished.

4.35   Grey-green/buff Semi-matt

Copper nitrate	80 gm
Sodium chloride	40 gm
Ammonium chloride	10 gm
Potassium hydrogen tartrate	10 gm
Water	1 litre

Boiling immersion (Twenty minutes)

The results obtained with this recipe are more even than the similar results produced using a solution of copper nitrate and sodium chloride alone. See recipe 4.36.

An immediate fine etching of the surface occurs when the article is immersed in the boiling solution, followed by the gradual development of a salmon-pink colour. After twenty minutes the article is removed, washed in hot water, and allowed to dry in air. The article is then exposed to daylight, turning regularly to ensure that all parts of the surface are exposed. The surface darkens gradually to grey-green, and may be wax finished.

4.36   Dark greenish-grey/black Gloss

Copper nitrate	200 gm
Sodium chloride	200 gm
Water	1 litre

Boiling immersion (Thirty minutes)

The surface may darken irregularly, producing an appearance similar to a smoked surface.

An immediate etching of the surface occurs when the article is immersed in the boiling solution, followed by the rapid development of a salmon-pink colour. After thirty minutes the article is removed, washed in hot water, and allowed to dry in air. The article is then exposed to daylight, turning regularly to ensure that all parts of the surface are exposed. The surface will darken rapidly at first to an olive colour, and then more slowly to black. The article may subsequently be wax finished.

4.37*   Pale greenish ochre Semi-gloss/semi-matt

Copper nitrate	80 gm
Water	1 litre
Ammonia (.880 solution)	3 cm³

Applied liquid (Ten days)

Colour development is very slow and yields an even mottled surface.

Flecks of blue-green patina are inclined to form if the solution is used less sparingly, particularly in the later stages of treatment.

The solution is applied to the metal surface by dabbing with a soft cloth. The solution should be applied sparingly and may require vigorous dabbing in the initial stages to ensure adequate wetting. After each application the surface is allowed to dry completely in air. This procedure is repeated twice a day for ten days. The article should be allowed to dry for several days before wax finishing.

4.38   Light grey on ochre Matt

Bismuth nitrate	50 gm
Nitric acid (25% solution)	300 cm³
Water	700 cm³

Cold immersion (Twenty minutes)

The grey colour is more adherent on a matted or satin surface than on a highly polished finish. A near-white colour tends to be retained in surface features such as punched marks or textured areas.

The grey colour is very fragile when wet and the surface should not be handled at this stage.

▲ Nitric acid and sulphuric acid are corrosive and must be prevented from coming into contact with the eyes and skin.

The article to be coloured is briefly pickled in sulphuric acid (50% solution), and rinsed in cold water. It is then immersed in the acidified bismuth nitrate solution, which initially darkens the surface and subsequently produces a grey colouration. When the grey colour has fully developed, after about twenty minutes, the object is removed and rinsed in a bath of cold water. It is allowed to dry in air, and may be wax finished when dry.

4.39   Cloudy grey/black on brown Semi-matt

Copper (cupric) chloride	50 gm
Water	1 litre

Hot immersion (Ten minutes)

The object is immersed in the hot solution (70–80°C) producing etching effects and a pale salmon-pink colour. With continued immersion a denser and slightly more brown colour develops. After ten minutes, the object is removed and washed in hot water, removing the loose superficial layer. It is then allowed to dry in air, and exposed to daylight which darkens the colour. When all colour change has ceased, the object may be wax finished.

The results produced in tests were poor. The colour tended to be uneven and patchy and prone to spotting. The procedure cannot be generally recommended.

4.40   Greenish-brown Gloss

Copper sulphate	8 gm
Copper acetate	3.25 gm
Potassium aluminium sulphate	2 gm
Water	1 litre
Acetic acid (6% solution)	20 cm^3

Boiling immersion (About one hour)

The article is immersed in the boiling solution and after a few minutes the surface darkens and becomes more opaque as the colour develops. The surface develops gradually as immersion continues, and after about one hour the article is removed and washed in hot water. It is dried in sawdust and may be wax finished when completely dry.

4.41   Dull greenish-brown Semi-matt

A	Copper sulphate	25 gm
	Ammonia (.880 solution)	30 cm^3
	Water	1 litre
B	Potassium hydroxide	16 gm
	Water	1 litre

Hot immersion (A)   (Ten minutes)
Hot immersion (B)   (One minute)

The article is immersed in the hot copper sulphate solution (60°C) for about ten minutes, producing a slight etching effect and a darkening of the surface to a greenish-brown colour. The article is then transferred to the hot potassium hydroxide solution (60°C) for about one minute. It is removed, washed in warm water and allowed to dry in air. When dry it is wax finished.

The surfaces produced in tests tended to be streaky and the recipe cannot be generally recommended.

▲ Preparation of solution A, and colouring, should be carried out in a well ventilated area, to prevent irritation of the eyes and respiratory system by the ammonia vapour liberated.

▲ Potassium hydroxide, in solid form and in solution, is corrosive to the skin and particularly the eyes. Adequate precautions should be taken.

4.42   Light ochre (slightly frosted surface) Semi-matt

Copper carbonate	100 gm
Ammonium chloride	100 gm
Ammonia (.880 solution)	20 cm^3
Water	1 litre

Hot immersion (Fifteen minutes)

The article is immersed in the hot solution (50°C) producing immediate etching effects, typical with chloride-containing solutions. After about fifteen minutes, the article is removed and washed thoroughly in warm water. It is allowed to dry in air, and may be wax finished when dry.

▲ When preparing the solution, care should be taken to avoid inhaling the ammonia vapour. The eyes and skin should be protected from accidental splashes of the concentrated ammonia solution.

4.43   Light ochre (slightly frosted surface) Matt

Ferric oxide	25 gm
Nitric acid (70% solution)	100 cm^3
Iron filings	5 gm
Water	1 litre

Cold immersion (Fifteen minutes)

The ferric oxide is dissolved in the nitric acid (producing toxic fumes), and the mixture is added to the water. The iron filings are added last. The article is immersed in the solution, which produces an etched surface, and removed after about fifteen minutes. It is washed thoroughly in cold water and allowed to dry in air, and may be wax finished when dry.

▲ Preparation and colouring should be carried out in a fume cupboard. Toxic and corrosive brown nitrous fumes are evolved which must not be inhaled.

▲ Concentrated nitric acid is extremely corrosive, causing burns, and must be prevented from coming into contact with the eyes and skin.

4.44   Yellow ochre Satin/matt

Copper sulphate	25 gm
Ammonium chloride	25 gm
Lead acetate	25 gm
Water	1 litre

Hot immersion (Fifteen minutes)

The surface is finely etched by the solution to produce a smooth matt satin finish, which is sand-coloured. The etched surface is prone to staining and should be washed and dried carefully.

▲ Lead acetate is very toxic and every precaution should be taken to prevent inhalation of the dust during weighing and preparation of the solution. The vapour evolved from the hot solution must not be inhaled.

The object is immersed in the hot solution (80°C) producing slight etching effects and pale orange/pink colours. After fifteen minutes, the object is removed, washed in hot water and allowed to dry in air. Exposure to daylight causes a darkening of the colour. When the colour has stabilised and the object is completely dry, it is wax finished.

4.45   Greenish-grey on dull metal Semi-matt

Sodium dichromate	150 gm
Nitric acid (10% solution)	20 cm³
Hydrochloric acid (15% solution)	6 cm³
Ethanol	1 cm³
Water	1 litre

Cold immersion   (Fifteen minutes)

Results produced in tests tended to be poor. The greyish-green colour tends to be streaky or patchy, and the ground colour dull. A more developed variegated greenish surface was obtained on a directionally grained satin finish.

▲ Sodium dichromate is a powerful oxidant, and must not be allowed to come into contact with combustible material. The dust must not be inhaled. The solid and solutions should be prevented from coming into contact with the eyes and skin, or clothing. Frequent exposure can cause skin ulceration, and more serious effects through absorption.

▲ Both hydrochloric and nitric acids are corrosive and should be prevented from coming into contact with the eyes, skin or clothing.

The sodium dichromate is dissolved in the water, and the other ingredients are then added separately while stirring the solution. The article is immersed in the cold solution for a period of fifteen minutes. (If a superficial layer is formed it should be removed by washing in cold water.) The article is then removed and washed in cold water. It is allowed to dry in air and may be wax finished when dry.

4.46   Thin orange on pale grey Satin

Copper acetate	120 gm
Ammonium chloride	60 gm
Water	1 litre

Boiling immersion. Scratch-brushing (Ten minutes)

The surfaces produced in tests were interesting, but it was found to be difficult to control the variegation and results tended to be patchy. The technique cannot therefore be generally recommended.

The object is immersed in the boiling solution, which rapidly etches the surface and produces a salmon-pink colour typical of chloride solutions. After ten minutes it is removed and scratch-brushed with the hot solution, removing most of the salmon-pink layer. The object is then washed and allowed to dry in air. The colour darkens in daylight to an orange-yellow. When the colour has stabilised, the object may be wax finished.

4.47   Orange-brown Gloss

Copper sulphate	100 gm
Ammonium chloride	0.5 gm
Water	1 litre

Hot immersion (Five to ten minutes)

Extending the immersion time beyond ten minutes tends to produce a darker and more matt layer, which is difficult to remove completely by bristle-brushing. It tends to 'cling', producing dull gritty patches.

The article is immersed in the hot solution (70–80°C) producing a pale pink-orange colouration within two or three minutes. This darkens with continued immersion. When an orange-brown colour has developed, after about five or ten minutes, the article is removed and washed in warm water. It is dried in sawdust. Exposure to daylight will cause a slight initial darkening of the colour. When the colour is stable, the article is wax finished.

4.48   Orange-brown Gloss

Copper acetate	25 gm
Copper sulphate	19 gm
Water	1 litre

Hot immersion (Five or six minutes)

Prolonging the immersion beyond six minutes tends to produce unevenness and streaking on this metal.

The article is immersed in the hot solution (80°C). The colour develops gradually to produce a light terracotta after four minutes, which darkens evenly to produce an orange-brown after five or six minutes. The article is then removed, washed thoroughly, and dried in sawdust. When dry, the surface has a characteristic greyish bloom which becomes invisible after wax finishing.

4.49   Orange-brown Semi-gloss

Barium chlorate	52 gm
Copper sulphate	40 gm
Water	1 litre

Hot immersion (Fifteen minutes)

Immersion in the hot solution (70°C) produces an immediate golden lustre colour, which changes gradually to pink/green and then to a brownish lustre after about three or four minutes. Continued immersion produces a greater opacity of colour and a gradual development to a yellow-brown or orange-brown colour. When the colour has fully developed, after fifteen minutes, the object is removed and washed in hot water. It is wax finished, after thorough drying in sawdust.

The lustre colours produced in the early stages of immersion are generally even, but are less bright than the equivalent colours produced by other methods.

▲ Barium compounds are poisonous and harmful if taken internally. Barium chlorate is a strong oxidising agent and should not be allowed to come into contact with combustible material. It should be prevented from coming into contact with the eyes, skin or clothing.

4.50   Orange-brown/ochre Semi-gloss/gloss

Ammonium acetate	50 gm
Copper acetate	30 gm
Copper sulphate	4 gm
Ammonium chloride	0.5 gm
Water	1 litre

Boiling immersion (Seven to ten minutes)

After one or two minutes of immersion in the boiling solution, the object takes on an olive colouration, which gradually darkens to a more ochre colour as immersion continues. The full colour tends to develop in about seven minutes, although it may take a little longer. Prolonging the immersion tends to cause the colour to become more pallid, and does not improve surface consistency. The object is removed and washed in hot water. After thorough drying in sawdust, it is wax finished.

4.51   Orange-brown Semi-matt

Copper sulphate	6.25 gm
Copper acetate	1.25 gm
Sodium chloride	2 gm
Potassium nitrate	1.25 gm
Water	1 litre

Boiling immersion (About one hour)

The article is immersed in the boiling solution and after a few minutes the surface darkens and becomes more opaque as the colour develops. The surface develops gradually as immersion continues, and after about one hour the article is removed and washed in hot water. It is dried in sawdust and may be wax finished when completely dry.

If the colour develops unevenly then the surface should be bristle-brushed lightly with pumice and then re-immersed.

▲ Potassium nitrate is a powerful oxidising agent and should not be allowed to come into contact with combustible materials.

4.52★   Reddish-brown Semi-matt   *Pl. VII*

Copper sulphate	50 gm
Ferrous sulphate	5 gm
Zinc sulphate	5 gm
Potassium permanganate	2.5 gm
Water	1 litre

Boiling immersion (Thirty minutes)

When the object is immersed in the gently boiling solution, a dark brown layer forms on the surface, which should not be disturbed. Immersion is continued for fifteen minutes, after which time the object is removed and the dark layer washed away in hot water using the bristle brush. The object is then re-immersed for a further fifteen minutes and then removed and bristle-brushed gently with hot water. After thorough rinsing it is dried in sawdust, and wax finished when completely dry.

Lighter colours with a more green bias can be produced using this solution. See recipe 4.54.

4.53   Dark reddish-brown (variegated) Semi-gloss

| Ferric chloride | 10 gm |
| Water | 1 litre |

Torch technique

The object is heated with a blow torch and the solution applied sparingly with a scratch-brush using small circular motions and working across the surface. The metal gradually darkens to a reddish brown tone. The variegated surface produced can be made more even by lightly dabbing with a cloth dampened with the solution. When treatment is complete, the object is allowed to cool and dry thoroughly, and may then be wax finished.

▲ The fumes evolved as the solution is applied to the heated metal should not be inhaled. Adequate ventilation must be provided and a nose and mouth face mask fitted with the correct filter should be worn.

▲ Ferric chloride is corrosive and should not be allowed to come into contact with the eyes and skin. A face mask or goggles should be worn to protect the eyes from hot splashes as the solution is applied.

4.54   Dull olive-green Semi-gloss/semi-matt

Copper sulphate	50 gm
Ferrous sulphate	5 gm
Zinc sulphate	5 gm
Potassium permanganate	2.5 gm
Water	1 litre

Boiling immersion (Ten to fifteen minutes)

The object is immersed in the boiling solution, and becomes coated with a dark brown layer. After one or two minutes the object is removed and the dark layer brushed away under hot water, using a bristle-brush. It is then re-immersed for two minutes, and the bristle-brushing again carried out when the dark brown layer has re-formed. After two or three such immersions and bristle-brushings, the brown layer will tend not to be adherent when it re-forms, and can be removed by agitating the object while in the solution. Immersion is continued to ten or fifteen minutes, after which the object is removed and given a final bristle-brushing under hot water. It is thoroughly rinsed in warm water, dried in sawdust and wax finished when completely dry.

Darker colours with a reddish bias can be produced using this solution, if the technique is modified. See recipe 4.52.

4.55   Brown (or variegated black on brown) Semi-matt

Potassium permanganate	2.5 gm
Copper sulphate	50 gm
Ferric sulphate	5 gm
Water	1 litre

Boiling immersion (Fifteen minutes)

The article is immersed in the boiling solution, and the surface becomes coated with a superficial dark brown film, which should be removed by bristle-brushing in a bath of hot water. The article is re-immersed in the boiling solution and again becomes coated with dark brown.

If a plain brown surface is required, the brown film should be removed every few minutes as immersion continues. After about fifteen minutes the article is removed and washed thoroughly in hot water. After drying in air it may be wax finished.

If a variegated surface is required, then the brown film should be left on the surface as immersion continues. After fifteen minutes, the article is removed, and allowed to dry in air for a few minutes. The surface is then gently brushed with a dry bristle-brush, which will partially remove the brown layer, leaving a variegated surface. The article should be wax finished.

The solution should not be boiled for any length of time, prior to immersion of the object, or the black layer tends not to be deposited on the surface.

4.56   Brownish-ochre Semi-matt

Copper sulphate	25 gm
Sodium chloride	100 gm
Potassium polysulphide	10 gm
Water	1 litre

Hot immersion (Fifteen minutes)

The object is immersed in the hot solution (60°C), producing a slight matting of the surface due to etching, and the gradual development of dull brown colours. After about fifteen minutes, when the brown has developed, the object is removed. Relieving may be carried out at this stage, using the hot solution with the scratch-brush. The object is then removed, washed in warm water and allowed to dry in air. When dry, it is wax finished.

The resulting brownish ochre surface tends to be uneven. Gentle relieving with the scratch-brush produces a more even surface with a warmer tone. Care must be taken to avoid breaking through by overworking and revealing the bare metal.

▲ The colouring should be carried out in a well ventilated area, as some hydrogen sulphide gas will be liberated.

4.57   Mid-brown/reddish-brown Gloss

Copper carbonate	160 gm
Sodium hydroxide	80 gm
Water	1 litre

Boiling immersion (Thirty minutes)

The sodium hydroxide is added to the water in small quantities, while stirring. Heat is evolved as it dissolves, warming the solution. When it has completely dissolved, the copper carbonate is added and the solution heated to boiling. The object is immersed in the boiling solution, and gradually acquires a pale grey colour which darkens to a greyish-mauve as immersion proceeds. After about thirty minutes the object is removed and dried in sawdust. When completely dry, it may be wax finished.

An even mid-brown or reddish-brown colour is produced. Similar results are obtained with less concentrated solutions used for correspondingly longer periods of immersion, but these tend to be less even.

▲ Sodium hydroxide is a powerful caustic alkali, and the solid or solutions must be prevented from coming into contact with the eyes and skin. The sodium hydroxide, in small quantities, should be added to the water and not vice-versa. Colouring should be carried out in a well ventilated area, as inhalation of the caustic vapours should be avoided.

4.58   **Mid-brown   Gloss**

Copper sulphate	6.25 gm
Copper acetate	1.25 gm
Potassium nitrate	1.25 gm
Water	1 litre

Boiling immersion   (About one hour)

In tests on polished surfaces the colour tended to be slightly lighter at the edges.

▲ Potassium nitrate is a powerful oxidising agent and must not be allowed to come into contact with combustible materials.

The article is immersed in the boiling solution and after a few minutes the surface darkens and becomes more opaque as the colour develops. The surface develops gradually as immersion continues, and after about one hour the article is removed and washed in hot water. It is dried in sawdust and may be wax finished when completely dry.

4.59   **Orange-brown   Semi-gloss/semi-matt**

Copper sulphate	100 gm
Copper acetate	40 gm
Potassium nitrate	40 gm
Water	1 litre

Boiling immersion   (Ten minutes)

In tests carried out with this metal, the surface tended to be unevenly streaked with a very slightly lighter colour.

▲ Potassium nitrate is a powerful oxidising agent and must not be allowed to come into contact with combustible material.

The article is immersed in the boiling solution, which produces drab orange colours after one minute. These gradually darken and become more opaque with continued immersion. The article is removed when the colour has developed, after about ten minutes, and washed in hot water. After thorough drying in sawdust, it is wax finished.

4.60   **Mid-brown   Semi-matt**

Copper sulphate	125 gm
Sodium acetate	12.5 gm
Water	1 litre

Boiling immersion   (Ten to fifteen minutes)

The surfaces produced in tests tended to be slightly variegated, the colour tending to shy away from the edges of sheet material to leave a narrow line of bare metal.

After one minute of immersion in the boiling solution a slightly lustrous orange/pink colour develops on the surface of the object. This gradually darkens, tending to become brown as immersion continues. When the colour has developed, after ten or fifteen minutes, the object is removed and washed in hot water. It is wax finished after thorough drying in sawdust.

4.61   **Variegated dark brown on dark brown ground   Semi-matt**

Potassium polysulphide	4 gm
Water	1 litre

Torch technique

▲ The fumes evolved as the solution is applied to the heated metal should not be inhaled. Adequate ventilation must be provided.

▲ A face mask or goggles should be worn to protect the eyes from hot splashes as the solution is applied.

The metal is heated with a blow-torch and the solution applied sparingly with a scratch-brush using small circular motions and working across the surface until a matt black colour appears. The amount of heat and solution used are critical. If too much solution is applied then the surface tends to strip, and if too little heat is used then the solution does not act on the metal. When the matt black colour has been obtained, it should be lightly brushed with a dry scratch-brush. This brings up the surface and highlights any faults that can then be worked on with further heating and applications of solution. When treatment is complete, the object is allowed to cool and dry thoroughly, and may then be wax finished.

4.62   **Dark brown   Semi-matt**

Copper sulphate	125 gm
Ferrous sulphate	100 gm
Acetic acid (glacial)	6.5 cm³
Water	1 litre

Boiling immersion   (Thirty to forty minutes)

There is a tendency for matt areas to develop on the glossy underlying colour, particularly if the immersion is prolonged.

▲ Acetic acid in the highly concentrated 'glacial' form is corrosive, and must not be allowed to come into contact with the skin, and more particularly the eyes. It liberates a strong vapour which is irritating to the respiratory system and eyes. Preparation of the solution should be carried out in a well ventilated area.

The object is immersed in the boiling solution. The surface darkens gradually during the course of immersion, initially tending to be drab and rather uneven, but later becoming more even and generally darker and richer in colour. When the colour has developed, after thirty to forty minutes, the object is removed and washed in hot water. It is dried in sawdust, and wax finished when dry.

4.63*   Dark brown Semi-matt   *Pl. VIII*

Copper sulphate	30 gm
Potassium permanganate	6 gm
Water	1 litre

Boiling immersion and scratch-brushing (Several minutes)

The results obtained in tests were variable, both in the overall depth of colour obtained, and in the extent to which a dark 'figure' featured in the surface finish.

The development of the loose brown layer appears to be essential to successful colouring. If the solution is boiled for too long before the object is immersed, then a more dense precipitate is formed in the solution, and coating does not take place.

On immersion in the boiling solution, the object becomes coated with a brownish film, which develops to a dark purplish-brown after one minute. This brown colour film is loosely adherent and sloughs off gradually to reveal uneven underlying grey brown colours. After three or four minutes the object is removed and gently scratch-brushed with the hot solution. Overworking will cut through to the underlying metal. If the surface is unsatisfactory then the process is repeated. When the desired colour has been obtained, the object is washed in hot water, dried either in sawdust or in air, and wax finished when dry.

4.64*   Reddish-brown Semi-gloss   *Pl. VIII*

Antimony trisulphide	12.5 gm
Sodium hydroxide	40 gm
Water	1 litre

Hot scratch-brushing (Several minutes)

▲ Sodium hydroxide is a powerful caustic alkali which can cause severe burns to the eyes and skin, in the solid form or in solution. When preparing the solution, small quantities of the sodium hydroxide should be added to a large quantity of water and *not* vice-versa. When the sodium hydroxide has completely dissolved, the antimony trisulphide is added.

The hot solution (60°C) is repeatedly applied liberally to the surface of the object with a soft brush, allowing time for the colour to develop. The surface is then worked lightly with the scratch-brush, while applying the hot solution and hot water alternately. The treatment is concluded by lightly burnishing the surface with the scratch-brush and hot water. After rinsing it is dried in sawdust, and wax finished when dry.

4.65   Mid brown/purplish-brown Semi-matt/semi-gloss

Copper nitrate	200 gm
Water	1 litre

Boiling immersion (Twenty minutes)

Tests carried out with less concentrated solutions of copper nitrate produced lighter and more drab colours.

Similar mid grey-brown colourations were produced with concentrated hot solutions (70°C) but these were marred by dark marks that appeared on the polished surfaces.

Prolonged immersion in a concentrated cold solution produced distinctive results. See recipe 4.7.

The article is immersed in the boiling solution. A greenish tinge slowly appears during the first few minutes, which gradually darkens to a brown colour as immersion continues. A drab grey layer subsequently forms. After twenty minutes the article is removed and washed thoroughly in hot water to clear the grey layer and reveal the brown colour beneath. After thorough washing and drying in sawdust the surface has a characteristic grey bloom, which becomes almost invisible when the article is wax finished.

4.66   Dark purplish-brown Semi-gloss

Copper sulphate	25 gm
Water	1 litre

Boiling immersion (Fifteen to thirty minutes)

Tests have shown that the time taken for the full colour to develop is variable.

Tests with more concentrated solutions (up to 150 gm/litre of copper sulphate) produced very similar results.

Scratch-brushing with the solution may be used to even the surface or as an after-treatment to produce relieved effects. It should be carried out gently, producing a finely scratched surface with a warmer tone.

The object is immersed in the boiling solution. After two or three minutes a brown colour develops which gradually darkens with continued immersion. When the surface has attained the dark purplish-brown colour, which may take from fifteen to thirty minutes, the object is removed and washed thoroughly in hot water. After drying in sawdust, it is wax finished.

4.67   Light bronzing Semi-matt/Semi-gloss

Sodium sulphide	50 gm
Antimony trisulphide	50 gm
Water	1 litre
Acetic acid (10% solution)	100 cm³

Hot scratch-brushing (Several minutes)

It may require prolonged working to develop an even colour. In tests it was found that after some working the scratch-brush tended to 'cling' to the surface, causing unevenness. Intermittent treatment with cold water helps to overcome this.

▲ Sodium sulphide is corrosive and should be prevented from coming into contact with the eyes and skin.

▲ Addition of the acetic acid and colouring should be carried out in a well ventilated area, as toxic hydrogen sulphide gas will be evolved.

The solid ingredients are added to the water and dissolved by gradually raising the temperature to boiling. The solution is allowed to cool to 60°C and then acidulated gradually with the acetic acid, producing a dense brown precipitate. The object is scratch-brushed with this turbid solution, alternating with hot and then cold water. When the colour has developed it is washed in cold water, allowed to dry in air, and wax finished when dry.

4.68*   Dark brown Gloss

Ammonium carbonate	50 gm
Copper acetate	25 gm
Acetic acid (6% solution)	1 litre

Boiling immersion (Five minutes)

The object is immersed in the boiling solution, and the surface gradually darkens to a reddish-brown colour. After five minutes, when the colour has fully developed, the object is removed and washed in warm water. After careful drying in sawdust, it may be wax-finished.

After three or four minutes the dark brown colour has a slightly greenish bias. This becomes gradually less apparent. After six or seven minutes the bias is slightly red, but becoming less even. In tests, the optimum time was found to be about five minutes.

Tests involving the addition of 5 gm of ammonium chloride and 2 gm of oxalic acid per litre, to the solution when the colours have developed, produced poor results. The colour initially becomes lighter but is then subject to streaking and etching effects, producing very patchy matt yellow surfaces. The variation cannot be recommended. Very similar results were obtained with small additions of zinc chloride or copper chloride.

▲ The ammonium carbonate should be added to the acetic acid in small quantities. The reaction is effervescent, and if large amounts are added, the effervescence is very vigorous.

4.69   Reddish-brown (slightly variegated) Semi-gloss

Sodium thiosulphate	6.25 gm
Ferric nitrate	50 gm
Water	1 litre

Hot immersion (One minute)

A succession of lustre colours is rapidly produced when the article is immersed in the hot solution (50–60°C), changing to a more opaque purplish colour after about forty-five seconds. The colour darkens quickly and the article is removed after about one minute. After thorough washing it is allowed to dry in air, and then wax finished.

During preparation, the nitrate solution is darkened by the addition of the thiosulphate, but slowly clears to give a straw-coloured solution.

Higher temperatures should not be used. Tests carried out at 70–80°C caused stripping of the colour and strong etching effects.

4.70   Reddish-brown (greenish-grey bloom) Semi-matt

Copper sulphate	120 gm
Potassium chlorate	60 gm
Water	1 litre

Hot immersion (Ten or fifteen minutes)

The object is immersed in the hot solution (70–80°C). A rapid series of lustre colours is produced, which become more opaque and darken to an orange buff within two or three minutes. Continued immersion produces a gradual darkening to a dark brown colour. After about ten minutes a greenish layer begins to be deposited on the object, as a dense precipitate is formed in the solution. If this is not required then the surface should be gently bristle-brushed in warm water, and the object re-immersed. After about fifteen minutes the object is removed and washed. It should be bristle-brushed under warm water if the greenish layer has re-formed. After thorough washing it is dried in sawdust and wax finished when dry. A greyish or greenish-grey bloom tends to remain.

A dense precipitate is formed as immersion progresses, causing 'bumping'. Glass vessels should not be used.

▲ Potassium chlorate is a powerful oxidising agent and must not be allowed to come into contact with combustible material. Contact with the eyes, skin or clothing must be avoided.

4.71*   Reddish-brown Satin/semi-gloss   *Pl. VII*

Antimony trisulphide	30 gm
Ferric oxide	10 gm
Ammonium sulphide (16% solution) to form a paste	

Applied paste (sulphide paste) (Several minutes)

The ingredients are mixed to a thick paste using a pestle and mortar, by the very gradual addition of a small quantity of ammonium sulphide solution. The paste is applied to the whole surface of the article to be coloured, using a soft bristle-brush. A stiff nylon brush is then used to work the surface gradually in small areas. The nylon brush should be intermittently dipped in a mixture of the dry ingredients of the paste (in the same proportion as above), and the surface worked in this way until the paste becomes dry and falls from the surface. When the whole surface is dry and completely free of any residue or dust, it may be burnished with a brass scratch-brush and wax finished.

A variegated surface can be obtained by brushing the paste on to the whole surface and allowing it to dry for about three hours, until it is superficially dry and cracking, and then burnishing it away with a nylon brush or a brass scratch-brush. This may be repeated, as necessary, to darken the colour.

▲ Ammonium sulphide solution is very corrosive and precautions should be taken to protect the eyes and skin while the paste is being prepared. It also liberates harmful fumes of ammonia and of hydrogen sulphide, and the preparation should therefore be carried out in a fume cupboard or well ventilated area. The paste once prepared is relatively safe to use but should be kept away from the skin and eyes.

▲ A nose and mouth face mask fitted with a fine dust filter should be worn when the residues are brushed away. The dust produced in this case contains antimony compounds, which are toxic.

4.72*   Dark purplish-brown Satin/Semi-gloss

Antimony trisulphide	20 gm
Ferric oxide	20 gm
Potassium polysulphide	2 gm
Ammonia (.880 solution)	2 cm³
Water to form a paste	(approx 4 cm³)

Applied paste (sulphide paste) (Several minutes)

The potassium polysulphide is ground using a pestle and mortar and the other solid ingredients added and mixed. These are made up to a thick paste by the gradual addition of the ammonia and a small quantity of water. The paste is applied to the whole surface of the object to be coloured, using a soft bristle-brush. A stiff nylon brush is then used to work the surface gradually in small areas. The nylon brush should be intermittently dipped in a mixture of the dry ingredients of the paste (in the same proportion as above), and the surface worked in this way until the paste becomes dry and falls from the surface. When the whole surface is dry and completely free of any residue or dust, it may be burnished with a brass scratch-brush and wax finished.

A variegated surface can be obtained by brushing the paste onto the whole surface and allowing it to dry for about three hours, until it is superficially dry and cracking, and then burnishing it away with a nylon brush or a brass scratch-brush. This may be repeated if necessary to darken the colour.

▲ A nose and mouth face mask should be worn when the residues are brushed away, as they contain toxic antimony compounds.

4.73   Variegated brown Semi-gloss

Sodium thiosulphate	90 gm
Potassium hydrogen tartrate	30 gm
Zinc chloride	15 gm
Water	1 litre

Hot immersion (Thirty minutes)

The surface tends to be uneven and slightly cloudy.

The article is immersed in the hot solution (50–60°C) which darkens the surface and gradually produces a bloom, which should be removed by bristle-brushing. Continued immersion gives rise to a brown colour, which slowly darkens. After about thirty minutes, when the colour has developed, the article is removed and washed in warm water. It is either dried in sawdust or allowed to dry in air, and may be wax finished when dry.

4.74*   Reddish-brown Gloss

A	Ammonium carbonate	180 gm
	Copper sulphate	60 gm
	Acetic acid (6% solution)	1 litre
B	Oxalic acid	1 gm
	Ammonium chloride	0.25 gm
	Acetic acid (6% solution)	500 cm³

Hot immersion (Fifteen minutes)

The slight bloom that may occur during the later stages of immersion is virtually invisible after wax finishing.

▲ The ammonium carbonate should be added to the acetic acid in small quantities. The reaction is effervescent, and if large amounts are added the effervescence is very vigorous.

▲ Preparation of the solution, and colouring, should be carried out in a well ventilated area. Acetic acid vapour will be liberated when the solutions are boiled.

▲ Oxalic acid is harmful if taken internally, and irritating to the eyes and skin.

Solution A is boiled and allowed to gradually evaporate until its volume is reduced by half. Solution B is then added and the mixture heated to bring the temperature to 80°C. The article is immersed in the hot solution, the colour gradually darkening until it becomes brown after ten minutes. After fifteen minutes, when the colour has fully developed, the article is removed and rinsed carefully in warm water. The surface should not be disturbed at this stage. It is allowed to dry in air, and wax finished when completely dry.

4.75   Purplish-brown (slight petrol lustre) Semi-gloss

Copper sulphate	60 gm
Copper acetate	10 gm
Potassium aluminium sulphate	25 gm
Water	1 litre

Boiling immersion (Forty minutes)

A slightly purplish brown, figured with darker lines, was obtained in tests. This also occurred on copper, and is probably characteristic of this solution when used with copper-rich sheet materials.

The object is immersed in the boiling solution, producing a gradual darkening of the surface to a drab ochre/brown colour after ten minutes. A slight lustre may also be apparent. Continued immersion produces further darkening, although colour development is slow. After forty minutes the object is removed, washed in hot water and dried in sawdust, when dry, it is wax finished.

4.76*   Dark brown/black Semi-matt/semi-gloss   *Pl. VII*

Nickel sulphate	35 gm
Copper sulphate	25 gm
Potassium permanganate	5 gm
Water	1 litre

Hot immersion (Five to thirty minutes)

A lighter brown colour and a more glossy surface is produced by immersion with repeated bristle-brushing, than is produced by uninterrupted immersion.

▲ Nickel compounds are a common cause of sensitive skin reactions. Precautions should also be taken to avoid inhalation of the dust when preparing the solution.

The object is immersed in the hot solution (80–90°C) and the surface immediately becomes covered with a dark brown layer. After four or five minutes it is removed and gently bristle-brushed with hot water. The dark layer is removed revealing a brown 'bronzing' beneath. If darker colours are required then the process should be repeated. If, on the other hand a full very dark brown/black colour with a more matt finish is required, the object should be re-immersed, and the dark brown layer will then re-form and should be left on the surface as immersion is continued. After thirty minutes the object is removed, and gently bristle-brushed under hot water. After thorough rinsing, and drying in sawdust, it is wax finished.

4.77   Dark brown/black (variegated) Semi-matt

Copper sulphate	6.25 gm
Copper acetate	1.25 gm
Potassium nitrate	1.25 gm
Sodium chloride	2 gm
Sulphur (flowers of sulphur)	3.5 gm
Acetic acid (6% solution)	1 litre

Boiling immersion (About one hour)

▲ Potassium nitrate is a powerful oxidising agent and must not be allowed to come into contact with combustible materials. It should not be mixed with the flowers of sulphur in the solid state.

The article is immersed in the boiling solution. The surface darkens and gradually becomes more opaque as the colour develops. After about one hour the article is removed and washed in hot water. Initially the colour is a variegated pinkish-brown, but this darkens in daylight to a dark brown or black, which tends to be slightly variegated. When this has developed, the article may be wax finished.

4.78   Purplish-brown (slightly variegated) Gloss

Sodium hydroxide	100 gm
Sodium tartrate	60 gm
Copper sulphate	60 gm
Water	1 litre

Hot immersion (Forty minutes)

A thin line of dark brown or black tended to occur along the edges of the brown coloured surface.

▲ Sodium hydroxide is a powerful caustic alkali. The solid or solutions can cause severe burns. It must be prevented from coming into contact with the eyes or skin. When preparing the solution, small quantities of the sodium hydroxide should be added to large quantities of water and *not* vice versa. The solution will become warm as the sodium hydroxide is dissolved. When it is completely dissolved, the other ingredients are added.

The object is immersed in the hot solution (60°C) producing dull lustre colours after about one minute. The surface gradually becomes more opaque and the colour slowly turns brown as immersion continues. After about forty minutes, when the colour has developed, the object is removed and washed in hot water and either dried in sawdust or allowed to dry in air. It may be wax finished when dry.

4.79   Grey brown mottle on golden ochre Semi-gloss

A	Copper nitrate	20 gm
	Ammonium chloride	20 gm
	Calcium hypochlorite	20 gm
	Water	1 litre
B	Water	

Cold immersion (A)   (Ten minutes)
Cold immersion (B)   (Twenty hours)

The initial immersion in the colouring solution should not be prolonged beyond the bright golden stage, or the results will be dull after cold water immersion.

▲ Care should be taken to avoid inhaling the fine dust that may be raised from the calcium hypochlorite (bleaching powder). It should also be prevented from coming into contact with the eyes or skin.

The article is immersed in solution A for about ten minutes, which produces a brightening of the surface to a slightly golden colour. The article is rinsed and transferred to a container of cold water, in which it is left for a period of about twenty hours. Small spots or patches of a darker colour develop slowly during the period of cold water immersion. When the surface is sufficiently developed, the article is removed and either dried in sawdust or allowed to dry in air. When dry, it is wax finished. The darker areas tend to darken slightly on exposure to daylight.

4.80   Reddish-brown/dark brown mottle on dull yellow Semi-matt

Copper sulphate	50 gm
Potassium chlorate	20 gm
Potassium permanganate	5 gm
Water	1 litre

Boiling immersion (Fifteen minutes)

The object is immersed in the boiling solution, and rapidly becomes coated with a chocolate brown layer. Immersion is continued for fifteen minutes, after which the object is removed and gently rinsed in a bath of warm water. It is then allowed to dry in air, leaving a dark powdery layer on the surface. When dry this is gently brushed with a soft brush to remove any dry residue. This treatment will remove some of the top layer, partly revealing the underlying colour, and should be continued until the desired surface effect is obtained. A stiffer bristle-brush may be required. After any residual dust has been removed with a soft brush, the object is wax finished.

The results produced in tests tended to be variable. Although an interesting mottle was produced by brushing away the loosely adherent dark layers, this could not be obtained evenly. The technique is difficult to control, and drab grey-brown patches tend to occur.

▲ Potassium chlorate is a powerful oxidising agent, and must not be allowed to come into contact with combustible material. The solid or solutions should be prevented from coming into contact with the eyes, skin or clothing.

▲ The potassium chlorate and the potassium permanganate must not be mixed together in the solid state. They should each be separately dissolved in a portion of the water, and then mixed together.

4.81   Light brown (variegated) Semi-matt

A	Copper sulphate	50 gm
	Water	1 litre
B	Potassium sulphide	12.5 gm
	Water	1 litre

Hot immersion. Scratch-brushing (A)
Cold immersion. Scratch-brushing (B)

The article is immersed in the copper sulphate solution for about thirty seconds and transferred to the cold potassium sulphide solution, and immersed for about one minute. It is then removed and scratch-brushed with each solution in turn. The surface is finished with a light scratch-brushing in the potassium sulphide solution. The article is washed thoroughly in warm water and left to dry before wax finishing.

Test results were poor, yielding patchy surfaces and poor colour development. The solution does not 'work' well under the scratch-brush, which tends to break through and ruin areas of colour that have been built up. The recipe cannot be recommended.

▲ Colouring should be carried out in a well ventilated area, as some hydrogen sulphide gas will be liberated.

4.82   Olive-green/black (variegated) Semi-matt

Copper nitrate	200 gm
Zinc chloride	200 gm
Water	1 litre

Applied liquid (Twice a day for five days)

The solution is wiped on to the article with a soft cloth and allowed to dry in air. When it is completely dry, the surface is gently rubbed smooth with a soft dry cloth to remove any loose dust, and a further application of solution wiped on. This procedure is repeated twice a day for about five days. The solution takes effect immediately, giving rise to an etching effect and pale cream and pink colours typical with chloride solutions. These colours darken on exposure to daylight. When treatment is complete, the article is left to dry thoroughly in air, and when completely dry, it is wax finished.

The buff surface gradually darkens on exposure to daylight to produce an uneven variegation of pale olive and black. The black areas are marked by pale olive markings producing a 'cracked lacquer' effect, in tests.

4.83   Dark brown/ochre (variegated) Semi-matt

Copper nitrate	200 gm
Sodium chloride	160 gm
Potassium hydrogen tartrate	120 gm
Ammonium chloride	40 gm
Water	1 litre

Cold immersion. Scratch-brushing (Fifteen minutes)

Immerse the object in the cold solution. The surface is immediately coloured and finely etched. After fifteen minutes the article is a bright salmon pink colour, and is removed and scratch-brushed with the cold solution, and is then left to dry in air without washing. Exposure to daylight will cause the surface to darken gradually to an irregular variegated ochre and dark brown. When the colour has stabilised, the object may be wax finished.

The newly coloured surface is very fragile when wet and handling should be avoided. It was found that the surface is particularly sensitive to household rubber gloves, which removed parts of the colour layer completely and caused bad staining.

4.84   Drab olive on orange-brown Semi-matt

Copper nitrate	85 gm
Zinc nitrate	85 gm
Ferric chloride	3 gm
Hydrogen peroxide (3% solution)	1 litre

Applied liquid   (Twice a day for five days)

The cold solution is dabbed on to the surface of the object with a soft cloth, and the object left to dry thoroughly in air. This procedure is repeated twice a day for about five days. An initial lustrous colouration is gradually replaced by a more opaque colour. Traces of blue-green patina may also appear during the later stages of treatment, if the application is not sufficiently sparing, and these should be brushed away when dry if not required. The object is wax finished when completely dry.

Tests with this metal produced very uneven results, consisting of irregular patches of a drab olive on an orange-brown surface, and cannot generally be recommended.

Patina development can be encouraged by a less sparing use of the solution in the later stages of treatment.

▲ Hydrogen peroxide must be kept away from contact with combustible material. In the concentrated form it causes burns to the eyes and skin, and when diluted will at least cause severe irritation.

4.85   Variegated greenish-brown Semi-matt

Copper nitrate	20 gm
Zinc sulphate	30 gm
Mercurous chloride	30 gm
Water	1 litre

Applied liquid   (Twice a day for ten days)

The solution is initially applied by vigorous dabbing with a soft cloth, to overcome resistance to wetting. Once some surface effect has been produced, the solution may be applied by wiping. The surface is allowed to dry completely after each application, which should be carried out twice a day for at least ten days. If the solution is used very sparingly then a patina-free ground colour tends to be produced. A more liberal use of the solution during the later stages of treatment will tend to encourage the formation of a blue-green patina. When treatment is complete, the article should be allowed to dry in air for several days before wax finishing.

The surface tends to be uneven and slightly streaky.

In tests, the blue-green patina was found to be difficult to produce.

▲ Mercurous chloride may be harmful if taken internally. There are dangers of cumulative effects.

4.86   Light brown and ochre (variegated) Semi-matt

Copper acetate	30 gm
Ammonium chloride	15 gm
Ferric oxide	30 gm
Water	1 litre

Hot immersion/hot scratch-brushing   (Ten to fifteen minutes)

The solution is brought to the boil and then removed from the source of heat. The object is then immersed, and becomes coated with a thick brown layer. After ten minutes immersion the object is removed and gently heated on a hotplate, the surface being worked with the scratch-brush. More solution is applied and worked in with the scratch-brush, occasionally allowing the surface to dry off without scratch-brushing. When the colour has developed it is allowed to dry on the hotplate for a few minutes, and the dry residue removed with a stiff bristle-brush. When the surface is free of residue it is wax finished.

An alternative technique which is more suited to larger objects is to heat locally with a torch and apply the pasty liquid with the scratch-brush, without prior immersion.

▲ A nose and mouth face mask fitted with a fine dust filter should be worn when brushing off the dry residue.

4.87   Light brown variegated (satin-finished surfaces only) Semi-matt

Copper nitrate	20 gm
Zinc sulphate	30 gm
Mercuric chloride	30 gm
Water	1 litre

Applied liquid   (Several minutes)

The solution is wiped on to the surface of the object with a soft cloth, producing a slightly adherent grey-black layer. Continued rubbing with the cloth clears this layer and produces a surface that is evenly coated with mercury. This tends to be non-adherent when dry. The solution is then applied to the surface by dabbing with a soft cloth, and allowed to dry in air. The procedure may be repeated if necessary, omitting the initial wiping. When the colour has developed, the object is washed in cold water and allowed to dry in air. When dry it may be wax finished.

A light brown colour, variegated with some lighter areas and darker markings, was obtained on satin finished surfaces. On polished surfaces only thin brown patchy stains on a yellowed metal ground remained, after the non-adherent mercury layer was removed.

▲ Mercuric chloride is very toxic and must be prevented from coming into contact with the skin and eyes. The dust must not be inhaled. If taken internally, very serious effects are produced.

▲ In view of the high risks involved, and the very limited colouring potential, this solution cannot be generally recommended.

4.88   Slate grey Semi-matt

A	Potassium nitrate	100 gm
	Water	1 litre
B	Copper sulphate	25 gm
	Ferrous sulphate	20 gm
	Acetic acid (glacial)	1 cm³

Boiling immersion (Thirty minutes)

The object is immersed in the potassium nitrate solution for a period of about fifteen minutes, which produces a slight brightening of the surface. The object is removed to a bath of hot water, while the copper sulphate, ferrous sulphate and acetic acid are separately added to the colouring solution. The object is re-immersed in the boiling solution, which gradually produces a greyish-brown colour, and removed after about fifteen minutes. It is washed in hot water and allowed to dry in air. When dry it is wax finished.

The results obtained in tests were poor. The surface is prone to streaks and patches, which reveal the bare metal.

▲ Potassium nitrate is a powerful oxidant and must be prevented from coming into contact with combustible materials.

▲ Glacial acetic acid is very corrosive and must be prevented from coming into contact with the eyes and skin.

4.89   Thin reddish-brown Semi-matt

| Copper nitrate | 40 gm |
| Water | to form a paste |

Applied paste (Four to six hours)

The copper nitrate is ground using a pestle and mortar, and a very small amount of water added by the drop to form a crystalline 'paste'. The paste is applied using a soft brush, and the surface is then allowed to dry for several hours. When dry, the residue is washed away with cold water. The process is repeated if necessary. After washing and air-drying, the object may be wax finished.

Colouring tends to be slight and the procedure requires repetition. Some spotting and metallic patches may occur which are difficult to correct.

A true paste will not form, but the damp crystalline mass that is obtained can be applied with a brush. This tends to become watery and may not dry out.

4.90   Light brown bronzing (uneven) Semi-gloss

Ferric oxide	30 gm
Lead dioxide	15 gm
Water	to form a paste

Applied paste and heat (Repeat several times)

The ingredients are mixed to a paste with a little water. The paste is applied to the surface with a soft brush, while the object is gently heated on a hot plate or with a torch. When the paste is completely dry, it is brushed off with a stiff bristle-brush. The procedure is repeated a number of times, until the colour has developed. After the final brushing, the object may be wax finished.

The colour tends to develop slowly and unevenly. An orange-brown colour is produced in surface features such as punched marks, or surface texture.

▲ Lead dioxide is toxic and very harmful if taken internally. The dust must not be inhaled. A nose and mouth face mask with a fine dust filter must be worn when brushing away the dry residue.

4.91   Dull pink etch Semi-matt

A	Ammonium chloride	10 gm
	Copper sulphate	10 gm
	Water	1 litre
B	Hydrogen peroxide (100 vols.)	200 cm³
	Sodium chloride	5 gm
	Acetic acid (glacial)	5 cm³
	Water	1 litre

Cold immersion (A) (Five minutes)
Hot immersion (B) (A few seconds)

The object is immersed in solution A for about five minutes, producing a slight etching effect, and a pale pink colour typical of chloride-containing solutions. The object is then briefly immersed in solution B (at 60–70°C). The reaction tends to be violent, causing the solution to 'boil', and immersion should not be prolonged. A greenish-brown layer is produced. If this is uneven, the surface should be gently scratch-brushed and the object briefly re-immersed. Prolonged or repeated immersion in solution B will tend to strip the greenish-brown colour. When treatment is complete, the object is washed in warm water and allowed to dry in air, and may be wax finished when dry.

The surface finish is difficult to control, and the procedure cannot be generally recommended.

▲ Hydrogen peroxide can cause severe burns, and must be prevented from coming into contact with the eyes and skin. When preparing the solution, it should be added to the water prior to the addition of the other ingredients.

▲ Glacial acetic acid is very corrosive and must be prevented from coming into contact with the eyes or skin.

4.92   Light grey stipple on dull metal Semi-matt

Copper acetate	20 gm
Sodium tetraborate (borax)	5 gm
Potassium nitrate	5 gm
Mercuric chloride	2.5 gm
Olive oil	to form a paste

Applied paste (Two or three days)

The ingredients are ground to a thin paste with a little olive oil, using a pestle and mortar. The paste is applied to the object, either with a brush or a soft cloth, and is allowed to dry for several hours. When dry the object is brushed with a bristle-brush to remove the dry residue. This procedure is repeated as necessary. When treatment is complete, the surface is again brushed, and then cleaned with pure distilled turpentine. When this has dried, it is wax finished.

The colouring effect is very slight.

▲ Mercuric chloride is a highly toxic substance. It is essential to prevent any contact with the skin, and to prevent any chance of inhaling the dust. Gloves and masks must be worn. A nose and mouth face mask fitted with a fine dust filter must be worn when brushing away the dry residue.

▲ Potassium nitrate is a powerful oxidising agent and should not be allowed to come into contact with combustible material.

▲ In view of the limited colouring potential, and the high risks involved, this method cannot be recommended.

4.93   Dark brown/black (with slight petrol lustre) Gloss

Copper carbonate	200 gm
Ammonia (.880 solution)	250 cm³
Water	750 cm³

Cold immersion (Fifteen to thirty minutes)

The ammonia is added to the water, followed by the carbonate, which will partially dissolve. Some excess carbonate should remain in suspension. The article is suspended in the cold solution. The colour develops slowly, thoughout the period of immersion. When the desired depth of colour has been reached, the article is removed and washed in cold water. It should be dried carefully in fine sawdust, and wax finished when dry.

To obtain a black colour using this solution see recipe 4.94.

▲ This procedure should only be carried out where there is adequate ventilation. Ammonia vapour will be freely liberated when the solution is prepared, and in colouring, which can severely irritate the eyes and respiratory system. The skin, and particularly the eyes, must be protected against accidental splashes from this corrosive alkali.

4.94*   Blue-black/black Gloss   *Pl. VIII*

Copper carbonate	125 gm
Ammonia (.880 solution)	250 cm³
Water	750 cm³

Warm immersion (One hour)

The ammonia is added to the water, followed by the carbonate, which will partially dissolve. Some excess carbonate should remain in suspension. The solution is heated to about 50°C, and the object is immersed. The colour develops slowly during the period of immersion. After about one hour, when the colour has fully developed, the object is removed and washed thoroughly in warm water. It should be carefully dried in sawdust, and wax finished when dry.

▲ This procedure should only be carried out where there is adequate ventilation. Ammonia vapour will be freely liberated when the solution is prepared, and during colouring, which can severely irritate the eyes and respiratory system. The skin, and particularly the eyes, must be protected against accidental splashes of this corrosive alkali.

4.95*   Black (greenish tinge) Gloss

Ammonium carbonate	60 gm
Copper sulphate	30 gm
Oxalic acid (crystals)	1 gm
Acetic acid (6% solution)	1 litre

Hot immersion (Twenty-five minutes)

The object is immersed in the hot solution (80°C) producing light brownish colours after two or three minutes, which may be uneven initially. With continued immersion the colours become even and darken gradually. When the colour has fully developed, after about twenty-five minutes, the object is removed and washed in hot water. After thorough drying in sawdust, it may be wax finished.

▲ Oxalic acid is poisonous and may be harmful if taken internally.

▲ The ammonium carbonate should be added to the acetic acid in small quantities. The reaction is effervescent, and if large amounts are added the effervescence becomes very vigorous.

4.96*   Black Gloss   *Pl. XVI*

Sodium hydroxide	25 gm
Potassium persulphate	10 gm
Water	1 litre

Hot immersion (One minute)

The potassium persulphate is dissolved in the water, and the sodium hydroxide then added in small quantities while stirring. The solution is then brought up to the correct temperature (70–80°C) and the article immersed. The surface is immediately coloured a purplish-brown, which darkens to black within one minute. The article is then removed and washed in warm water. It is dried in sawdust and may be wax finished when dry.

▲ Potassium persulphate is a powerful oxidant and should not be allowed to come into contact with combustible material. The sodium hydroxide and the persulphate must not be allowed to come into contact in the solid state.

▲ Sodium hydroxide is a powerful caustic alkali which can cause severe burns. Both the solid and solutions should be prevented from coming into contact with the eyes and skin.

4.97* Black Gloss

Pl. VII

Ammonium persulphate	10 gm
Sodium hydroxide	30 gm
Water	1 litre

Boiling immersion (Two minutes)

▲ Ammonium persulphate (peroxodisulphate) is a powerful oxidant, and must be kept out of contact with all combustible materials. It should be prevented from coming into contact with the eyes, skin or clothing.

▲ Sodium hydroxide is a powerful caustic alkali and must be prevented from coming into contact with the eyes and skin.

The ammonium persulphate is dissolved in half the water, and the sodium hydroxide separately dissolved in the remaining water. The two solutions are then mixed together and the mixture heated to boiling. The article to be coloured is immersed in the boiling solution, which produces golden and then brown lustre colours. Continued immersion darkens these colours to an opaque but glossy finish. When the colour has developed, after about two minutes, the article is removed, washed in hot water and dried in sawdust. When dry, it may be wax finished.

4.98* Black (*or variegated blue on black*) Gloss/semi-gloss

Pl. VII

Ammonia (.880 solution)	
Sodium chloride	20 gm/litre

Vapour technique (About two days)

The exact method used will depend on the size of the object to be coloured. It is essential to plan the procedure carefully. Ammonia vapour will concentrate in the container and should not simply be released. See 'Metal colouring techniques', 9.

▲ Ammonia solution is highly corrosive and must be prevented from coming into contact with the eyes or skin. The vapour is extremely irritating to the eyes and respiratory system, and will irritate the skin. Adequate ventilation must be provided.

The object is placed or suspended in a container, into the bottom of which the sodium chloride and ammonia solution are introduced. The object must be placed well clear of the surface of the liquid, and in a position that will favour an even distribution of vapour around it. The object should be well clear of the walls and lid of the container. The container is sealed and the vapour allowed to act on the object for about one day. The object is then removed and washed in cold water. A variegated blue on a black ground will generally have developed at this stage. On polished surfaces, the blue can be removed by bristle-brushing to leave a gloss black, which can be dried and wax finished in the normal way. On satinised or textured surfaces the blue colour tends to be more tenacious. If the variegated blue on black is required then the object should be returned to the container after washing, for a further day. It is then removed, rinsed in cold water and allowed to dry. It may be wax finished when dry.

4.99 Black/dark purplish-grey Gloss/semi-gloss

Sodium thiosulphate	250 gm
Water	1 litre
Copper sulphate	80 gm
Water	1 litre

Cold immersion (Two hours)

The dark purplish-grey/black colour produced tends to have a slightly cloudy appearance, although it is generally even.

The lustre colours produced during the immersion tend to be less clear than those obtained using some other thiosulphate solutions.

The chemicals should be separately dissolved, and the two solutions then mixed together. The article is suspended in the cold solution, and after ten minutes' immersion shows signs of darkening. After twenty minutes a reddish-purple lustre appears and gradually changes to a blue-green lustre after thirty minutes. The sequence of lustre colours is repeated, becoming more opaque and producing a maroon colour after about one hour. With continued immersion the colour slowly darkens to black. After about two hours the article is removed, washed thoroughly and allowed to dry in air. When dry it may be wax finished.

4.100 Dark brown to black (slightly lustrous) Gloss

Sodium thiosulphate	20 gm
Antimony potassium tartrate	10 gm
Water	1 litre

Hot immersion (Forty-five minutes)

Tests tended to produce some unevenness, taking the form of duller areas with a slightly frosted surface, particularly at the edges.

Timings for the immersion may be variable. If the desired colour is produced in a shorter time, the object should be removed at that stage. The immersion should not be continued longer than necessary.

▲ Antimony potassium tartrate is poisonous and may be harmful if taken internally.

▲ The vapour evolved from the hot solution should not be inhaled.

The article is immersed in the hot solution (60°C) for ten minutes, producing a slight darkening of the surface and an orange tinge. The temperature of the solution is then gradually raised to 80°C and immersion continued. After forty-five minutes the article is removed, washed in warm water and allowed to dry in air. When dry, it is wax finished.

4.101   Dark purplish-grey/black Semi-gloss/gloss

Barium sulphide	10 gm
Water	1 litre

Hot immersion (Forty minutes)

The object is immersed in the hot solution (50°C) and the surface is gradually tinted by a succession of lustre colours, which develop and recede. These slowly darken during the latter half of the period of immersion until the final black colour develops, after about forty minutes. The object is removed and washed carefully in warm water, and allowed to dry in air. When the surface is thoroughly dry, the object may be wax finished.

The results obtained in tests tended to be slightly variegated, and the colour cloudy in appearance.

The surface tends to be fragile when removed from the hot solution, and should be treated with care until dry.

▲ Barium sulphide is poisonous and very harmful if taken internally. Some hydrogen sulphide gas will be liberated as the solution is heated, and colouring should be carried out in a well ventilated area.

4.102   Purplish/black Semi-matt

Ammonium sulphide (16% solution)	100 cm³
Water	1 litre

Hot bristle-brushing (Several minutes)

The article is directly bristle-brushed with the hot solution (60°C) for several minutes, alternating with bristle-brushing under hot water. The procedure is repeated three or four times until the colour has developed and the article is then washed in warm water. After thorough drying in air or sawdust, it is wax finished.

The bristle-brushing application can be improved, if uneven results are produced, by scratch-brushing with the solution until an even lighter coloured surface is obtained and then repeating the bristle-brush treatment.

Scratch-brushing with the hot solution produces lighter 'bronzing' colours, which are better if a cold solution is used, see recipe 4.28.

▲ Ammonium sulphide solution liberates a vapour containing ammonia and hydrogen sulphide gas, which should not be inhaled. Preparation and colouring must be carried out in a well ventilated area. The solution must be prevented from coming into contact with the eyes or skin, as it causes burns or severe irritation.

4.103   Dark brown and black (variegated) Semi-matt/matt

Sodium thiosulphate	60 gm
Copper sulphate	42 gm
Potassium hydrogen tartrate	22 gm
Water	1 litre

Hot immersion (Fifteen minutes)

After one minute of immersion in the solution (at 50–60°C), a yellow-orange colour forms on the surface of the article and gradually turns to bright green. The article is removed from the solution after fifteen minutes and allowed to dry in air, without washing in water. When the article is completely dry it is gently rubbed with a soft cloth, which removes the green layer, revealing the underlying variegated dark brown colour. The article is then wax finished.

The surfaces produced in tests were generally a very dark brown, with some areas that were virtually black.

The green layer is superficial, and easily removed, but may leave a slight green tinge to the underlying colour.

4.104   Dark slate Semi-gloss

Ammonium sulphate	30 gm
Sodium hydroxide	40 gm
Water	1 litre

Boiling immersion (Thirty minutes)

The object is immersed in the boiling solution and the surface gradually darkens to produce a black colour, and acquires a slightly crystalline appearance. After thirty minutes the object is removed, washed in warm water and allowed to dry in air. Relieving is then carried out, and the object wax finished.

The dark slate layer tends to be thin and only tenuously adherent before finishing. The layer can be worked with a bristle-brush to reveal areas of the metal surface beneath. Test results suggest that relieving effects can be obtained with this procedure.

▲ Sodium hydroxide is a powerful caustic alkali, the solid and strong solutions causing severe burns to the eyes and skin. When preparing the solution, small quantities of the sodium hydroxide should be added to large quantities of water and *not* vice-versa. The reaction is exothermic and will warm the solution. When the sodium hydroxide is completely dissolved, the ammonium sulphate is added and the solution may be heated.

4.105   Mottled metallic grey/black on brown Semi-matt/matt

Nickel ammonium sulphate	50 gm
Sodium thiosulphate	50 gm
Water	1 litre

Hot immersion (Fifteen minutes)

A purplish-grey colour quickly forms on the surface of the object on immersion, becoming patchy and metallic grey after one minute. After two minutes the object should be removed from the solution and lightly scratch-brushed with water, to remove the non-adherent deposits that build up on the surface. The object is then re-immersed and the scratch-brushing repeated twice at intervals of five minutes. After fifteen minutes the object is again lightly scratch-brushed, briefly re-immersed and then washed in hot water. After thorough drying in sawdust, it may be wax finished.

Tests produced a consistent mottled effect on polished surfaces. The colour of the mottling was variable from metallic grey to black, or mixtures of these colours. An even dull grey was produced on matted or grained surfaces.

In preparing the solution, the two chemicals should be separately dissolved in two portions of the water. The thiosulphate solution is warmed, and the nickel ammonium sulphate solution added gradually. The light green solution is used at a temperature of 60–80°C.

▲ Nickel salts are a common cause of sensitive skin reactions. The vapour evolved from the hot solution should not be inhaled.

4.106 Grey areas on black Semi-matt

Sodium thiosulphate	65 gm
Water	1 litre
Antimony trichloride	until cloudy

Hot immersion (Thirty minutes)

The sodium thiosulphate is dissolved in the water and the antimony trichloride added by the drop, to form a white precipitate. The solution begins to become orange and cloudy at the top, and then gradually throughout. The solution is heated to 50°C, and when the temperature is reached, the object is immersed. The surface initially passes through the lustre cycle—golden after three minutes, purple after four minutes, pale metallic blue after ten minutes. After about fifteen minutes grey/white streaks appear on the surface, which darkens to produce an appearance like a black powdery layer. The article is removed after thirty minutes and washed gently in warm water, removing the powdery black deposit, leaving an uneven grey and black surface. After drying in air or sawdust, the article is wax finished.

A tenuous and glossy black surface covered with matt grey cloudy patches is produced. These tend to reveal some underlying brown colours. The surface appears generally fragile and the recipe cannot therefore be recommended.

▲ Antimony trichloride is poisonous if taken by mouth. It is also irritating to the skin, eyes and respiratory system, and contact must be avoided.

▲ The vapour evolved from the hot solution should not be inhaled.

4.107 Dark grey with brown variegation Semi-gloss

Copper acetate	40 gm
Copper sulphate	200 gm
Potassium nitrate	40 gm
Sodium chloride	65 gm
Flowers of sulphur	110 gm
Acetic acid (6% solution)	1 litre

Hot immersion. Scratch-brushing (Five minutes)

The article is immediately etched on immersion in the hot solution (70–80°C). A pink/orange colour, typical with chloride-containing solutions, develops within two or three minutes. After five minutes the article is removed, and the uneven surface is scratch-brushed with the hot solution to improve the surface consistency. The article is then washed in hot water and allowed to dry in air. When dry the surface is exposed evenly to daylight, causing a darkening of the colour. When the colour has stabilised and no further signs of change can be detected, it is wax finished.

The surface produced after exposure to daylight tends to be uneven. The surface is predominantly dull black, with patches of brown. The effect is virtually impossible to control and the recipe cannot be generally recommended.

▲ Potassium nitrate is a powerful oxidising agent and should not be allowed to come into contact with combustible material. It must not be brought into contact with the sulphur in a dry state, but should be completely dissolved before the flowers of sulphur are added.

4.108 Dark slate Semi-gloss

A	Copper nitrate	100 gm
	Water	1 litre
B	Potassium sulphide	5 gm
	Water	1 litre
	Hydrochloric acid (35%)	5 cm^3

Boiling immersion (A) (Fifteen seconds)
Cold immersion (B) (Thirty minutes)

The object is immersed in the boiling solution A for fifteen seconds, then transferred to solution B. The surface darkens to produce a series of uneven petrol blue/purple interference colours, and gradually changes to a slate grey. After thirty minutes the object is removed and rinsed, relieved by gentle scratch-brushing or pumice if desired, and finally washed, dried in sawdust and wax finished.

This recipe is suitable for producing antiquing and relieved effects only, on polished surfaces, and cannot be generally recommended.

▲ Solution B should only be prepared and used in a fume cupboard or a very well ventilated area. When the acid is added to the potassium sulphide solution, quantities of hydrogen sulphide gas are evolved, which is toxic and can be dangerous even in moderate concentrations.

▲ Hydrochloric acid is corrosive, and should be prevented from coming into contact with the eyes and skin.

4.109 Black Semi-gloss

Sodium thiosulphate	50 gm
Ferric nitrate	12.5 gm
Water	1 litre

Hot immersion (Twenty minutes)

The article is immersed in the hot solution (60–70°C) and after about one minute the surface is coloured with a purple-blue lustre, which gradually recedes. After five minutes a brown colour slowly appears which changes to a slate grey with continued immersion. After about twenty minutes the article is removed, washed in hot water and allowed to dry in air. The surface may be fragile at this stage and should be handled as little as possible. When completely dry the article is wax finished.

If the temperature is too high the colour layer will be patchy and more fragile when removed from solution, revealing a tan colour beneath the black.

4.110   Uneven dark lustre/black   Semi-gloss/semi-matt

Butyric acid	20 cm³
Sodium chloride	17 gm
Sodium hydroxide	7 gm
Sodium sulphide	5 gm
Water	1 litre

Boiling immersion   (One hour)

The solid ingredients are added to the water and allowed to dissolve, and the butyric acid then added, and the solution heated to boiling. The object is immersed in the boiling solution, which rapidly produces a series of patchy lustre colours. Continued immersion gradually produces a more opaque dark purplish-green colour, which tends to be uneven. After about one hour the object is removed and washed in hot water. It is allowed to dry in air, and may be wax finished when dry.

The results produced tend to be uneven or patchy, and the procedure cannot be generally recommended.

▲ Butyric acid should be prevented from coming into contact with the eyes and skin, as it can cause severe irritation. It also has a foul odour which tends to cling to the hair and clothing. Preparation and colouring should be carried out in a well ventilated area.

▲ Both sodium hydroxide and sodium sulphide are corrosive, and must be prevented from coming into contact with the eyes and skin.

4.111   Dark slate on brown   Semi-matt

Sodium thiosulphate	65 gm
Copper sulphate	12 gm
Copper acetate	10 gm
Arsenic trioxide	5 gm
Sodium chloride	5 gm
Water	1 litre

Hot immersion   (Thirty minutes)

Immersion in the hot solution (50–60°C) produces patchy lustre colours during the first fifteen minutes. Continued immersion produces a progressively more opaque and darker colour. After about thirty minutes, the article is removed and washed in warm water. It is allowed to dry in air, and may be wax finished when dry.

The surface tends to be very uneven. The procedure is difficult to control and cannot generally be recommended.

▲ Arsenic trioxide is very toxic and will give rise to very harmful effects if taken internally. The dust, and the vapour evolved from the hot solution, must not be inhaled.

4.112   Variegated black and mid-brown   Semi-gloss

Copper acetate	80 gm
Water	1 litre

Hot immersion. Scratch-brushing   (Thirty to forty minutes)

The object is immersed in the hot solution (50–60°C) and the surface colour allowed to develop through a series of lustrous colours, to a brownish colour. After thirty minutes it is removed and scratch-brushed with the hot solution, which initially removes some of the colour formed during immersion. Scratch-brushing is continued for several minutes until the surface colour has developed. The object is then washed in warm water and allowed to dry in air. When completely dry it is wax finished.

Tests carried out using the technique of direct scratch-brushing, without prior immersion, produced poor results with little colour development.

The surface is prone to streaking during the period of immersion. The streaks are difficult to eliminate during scratch-brushing.

4.113   Black on red and ochre ground   Matt

Potassium permanganate	5 gm
Copper sulphate	50 gm
Ferric sulphate	5 gm
Water	1 litre

Boiling immersion   (Twenty minutes)

The article is immersed in the boiling solution. A black layer forms on the surface, which should not be disturbed. Immersion is continued for about twenty minutes, and the article is then allowed to dry in air, without washing. The surface should not be handled at this stage. After several hours drying, the article may be wax finished.

The matt black layers tends to be partially adherent, revealing an underlying reddish-brown and ochre mottled surface.

4.114   Variegated black and brown Gloss

Copper sulphate	100 gm
Ammonium chloride	0.5 gm
Water	1 litre

Applied liquid to heated metal (Several minutes)

It is not possible to control the disposition of the black or brown colours on the surface.

The object is gently heated on a hotplate and the solution applied either by brushing or by wiping with a soft cloth. The object should be heated to a temperature such that when the solution is applied, it sizzles gently. Application is continued until orange-brown colours have developed. Exposure to daylight will cause the surface to darken unevenly to a mixed black and brown. When changes in colour have ceased, the object is wax finished.

4.115   Black and ochre (variegated) Semi-matt

Copper acetate	40 gm
Copper sulphate	200 gm
Potassium nitrate	40 gm
Sodium chloride	65 gm
Acetic acid (6% solution)	1 litre

Cold immersion. Scratch-brushing (Ten minutes)

The surface produced after exposure to daylight is very uneven. It does not appear to be possible to control either the degree of variegation or the surface quality.

▲ Potassium nitrate is a powerful oxidising agent and should not be allowed to come into contact with combustible material.

The object is suspended in the cold solution, producing an immediate etching effect. After five minutes a pale cream or pink colour develops unevenly. After ten minutes the object is removed and scratch-brushed with the cold solution to even the surface. It should then be allowed to dry in air without washing. Any wet areas on the surface should be very gently brushed out with a soft brush, to prevent streaking or pooling. When dry, the object is exposed evenly to daylight, causing a darkening of the colours. When the colour has stabilised the object is wax finished.

4.116   Patchy slate on dull metal Semi-matt

A	Barium sulphide	2 gm
	Potassium sulphide	2 gm
	Ammonium sulphide (16% solution)	2 cm³
	Water	1 litre
B	Copper nitrate	20 gm
	Ammonium chloride	20 gm
	Calcium hypochlorite	20 gm
	Water	1 litre

Cold immersion (A) (Thirty seconds)
Cold immersion (B) (A few minutes)

The secondary treatment in solution B tended to produce results that were only tenuously adherent, and cannot be generally recommended.

Similar poor results were also obtained with an after-treatment of a solution where the calcium hypochlorite is replaced by an equal quantity of calcium chloride.

▲ Ammonia vapour and hydrogen sulphide gas are liberated from the mixed sulphide solution. Colouring should be carried out in a fume cupboard or very well ventilated area. It should be prevented from coming into contact with the eyes and skin.

▲ Barium compounds are toxic, and harmful if taken internally.

▲ The fine dust from the calcium hypochlorite (bleaching powder) must not be inhaled. Contact with the eyes and skin must be avoided.

The article is immersed in the mixed sulphide solution, which immediately blackens the surface, and removed after about thirty seconds. If the surface is uneven then it can be lightly brushed with a scratch-brush and re-immersed for a similar length of time. The article is washed and transferred to solution B, which changes the colour to a dark slate or grey-brown. After a few minutes the article is removed and allowed to dry in air. When dry it may be wax finished.

4.117   Variegated dark grey-brown and buff Semi-matt

Copper sulphate	200 gm
Ferric chloride	25 gm
Water	1 litre

Hot immersion (Thirty minutes)

A cloudy and uneven surface tends to be produced. The results are rather unpredictable and the procedure cannot be generally recommended.

▲ Ferric chloride is corrosive and should be prevented from coming into contact with the eyes and skin.

The object is immersed in the hot solution (60°C), which etches the surface and produces a pale pink colour, typical of chloride-containing solutions. Continued immersion produces a gradual change to a more brown colour. After thirty minutes the object is removed and washed in hot water, which tends to leach a slight yellowish-green colour from the surface. After thorough washing, the object is allowed to dry in air and exposed to daylight, which tends to cause an uneven darkening of the colour. When the surface development is complete and the object thoroughly dried out, which may take several days, it can be wax finished.

4.118   Variegated light metallic grey on buff Semi-matt

Copper nitrate	300 gm
Water	100 cm³
Silver nitrate	2 gm
Water	10 cm³

Applied liquid (Several minutes)

The copper nitrate and silver nitrate are separately dissolved in the two portions of water, and then mixed together. The solution is warmed to about 40°C. The article to be coloured is immersed for a few minutes in a cold solution of sulphuric acid (50% solution), and then washed. The warm colouring solution is applied to the article by wiping with a soft cloth, producing a superficial grey layer which is washed away with warm water, revealing the underlying surface effect. When the article is completely dry, it may be wax finished.

A variegated partial chemical plating on a buff coloured ground is produced.

▲ Silver nitrate should be prevented from coming into contact with the skin, and particularly the eyes which may be severely irritated.

▲ Sulphuric acid is very corrosive, causing severe burns, and must be prevented from coming into contact with the eyes and skin.

4.119   Black mottling on variegated light brown Matt

A	Copper sulphate	50 gm
	Water	1 litre
B	Sodium thiosulphate	50 gm
	Lead acetate	12.5 gm
	Water	1 litre

Hot immersion (A) (One minute)
Hot immersion (B) (Two minutes)

The object is immersed in the hot copper sulphate solution (80°C) for one minute, producing a darkening of the surface. It is then transferred to solution B (80°C) for two minutes' immersion. These alternating immersions are repeated a number of times until an even black layer is produced on the surface. The object is then washed in warm water and allowed to dry in air. When dry, it is either rubbed gently with a soft cloth or bristle-brushed, and wax finished.

The black layer tends to be non-adherent. Careful drying and brushing can produce a partial black layer that is patchy or mottled, revealing areas of variegated light brown. This recipe cannot be recommended for producing an even black colour, as suggested by some sources.

▲ Lead acetate is a highly toxic substance, and every precaution should be taken to prevent inhalation of the dust during preparation. The solution is very harmful if taken internally. Colouring should be carried out in a well ventilated area. The vapour evolved from the hot solution must not be inhaled.

4.120   Dark metallic grey (frosted grain) Gloss/semi-gloss

Sodium thiosulphate	300 gm
Antimony potassium tartrate	15 gm
Ammonium chloride	20 gm
Water	1 litre

Hot immersion (Twenty minutes)

Immersion in the hot solution (60°C) rapidly produces a series of lustre colours on the surface of the object. These are pale and pass from purple to brown and blue, becoming a very pale blue-grey after about two minutes. This is followed by a gradual darkening of the surface, producing dark greys and browns. After about twenty minutes the object is removed, washed in warm water and dried in sawdust. It may be wax finished when dry.

A metallic steel grey colour with a high gloss finish is produced. A more dull linear 'frosted grain' effect tends to be superimposed on this. Some small areas of a slightly darker tone may be visible in some lights.

▲ Antimony potassium tartrate is poisonous and may be harmful if taken internally. The vapour evolved from the hot solution not be inhaled.

4.121* **Blue-green patina on red-brown ground Semi-matt**   *Pl. VIII*

Copper nitrate	200 gm
Sodium chloride	200 gm
Water	1 litre

Applied liquid (Twice a day for five days)

The solution is dabbed on to the object with a soft cloth and left to dry in air. This process is repeated twice a day for five days. The object is then allowed to dry in air, without treatment, for a further five days. The ground colour develops gradually throughout this period and after two or three days some blue-green patina forms on the surface. This continues to build up as long as the surface is moist. Finishing should only be carried out when it is certain that the surface is dry and patina development complete.

A dabbing technique was found to be better than wiping which tends to inhibit the formation of the patina.

If there is an initial resistance to 'wetting' the liquid can be applied by scratch-brushing in the early stages.

In damp weather conditions the surface may take several weeks to dry completely. Although apparently dry, the surface may 'sweat' intermittently. The drying process should not be rushed and wax finishing only carried out when it is certain that chemical action has ceased.

4.122* **Blue-green patina on variegated brown ground Semi-matt**   *Pl. VIII*

Copper sulphate	20 gm
Copper acetate	20 gm
Ammonium chloride	10 gm
Acetic acid (6% solution)	to form a paste

Applied paste (Several days)

The ingredients are ground to a creamy paste with a little acetic acid, using a pestle and mortar. The paste is applied to the object with a soft brush, giving quite a thick coating which is then allowed to dry for a day. The dry residue is then washed away under cold water, using a soft brush. A thin layer of paste is then wiped onto the object with a cloth, and the object left to dry for a day. The residue is again washed off. This procedure of applying thin layers and drying is repeated in the same way until a good variegated patina is produced. The object should be allowed to dry thoroughly when treatment is complete, and may then be wax finished.

4.123* **Blue-green patina on brown ground Semi-matt**   *Pl. VII*

Copper nitrate	110 gm
Water	110 cm³
Ammonia (.880 solution)	440 cm³
Acetic acid (6% solution)	440 cm³
Ammonium chloride	110 gm

Applied liquid (Five days)

The solution is applied to the surface by dabbing with a soft cloth, and the article then left to dry in air. The metal quickly darkens and a whiteish blue-green powdery patina forms gradually on the surface. When dry, this is rubbed with a soft dry cloth to remove any loose material and to smooth the surface, prior to the next application of solution. This procedure is repeated over a period of five days, during which time the patina darkens and becomes integral with the surface. When treatment is complete and the surface thoroughly dry, the article may be wax finished.

The surface produced is dark and variegated.

▲ Colouring must be carried out in a well ventilated area. Ammonia vapour will be liberated from the strong solution, which will cause irritation of the eyes and respiratory system. The solution must be prevented from coming into contact with the eyes and skin, as it can cause burns or severe irritation.

4.124* **Blue-green patina on pale brown ground**   *Pl. VIII, Pl. VII*
       Semi-matt/matt

Ammonium chloride	35 gm
Copper acetate	20 gm
Water	1 litre

Applied liquid (Several days)

The ingredients are ground with a little water using a pestle and mortar and then added to the remaining water. The solution is applied to the object by dabbing and wiping, using a soft cloth. It should be applied sparingly, to leave an evenly moist surface. The object is then allowed to dry in air. This procedure is repeated once a day for several days, producing a gradual development of the ground colour and blue-green patina. When treatment is complete, the object is left to dry for several days, during which time there is further patina development. When the object is completely dry, and there is no further surface change, it is wax finished.

More liberal application of the solution tends to result in a bluer patina, which may take the form of a slight incrustation. See plate VII.

It is essential to ensure that all patina development is complete, and the surface is completely dry, before wax finishing is carried out. The final drying period may have to be extended to a matter of weeks in damp or humid conditions.

**4.125★ Blue-green and yellow-green patina on pale orange-brown   Pl. VIII
Semi-matt**

Copper nitrate	30 gm
Sodium chloride	30 gm
Ammonium chloride	15 gm
Potassium aluminium sulphate	7 gm
Acetic acid (6% solution)	1 litre

Boiling immersion (Ten minutes)

The article is immersed in the boiling solution. After two or three minutes a fine etching effect and a pink colour develops on the surface. Immersion is continued to ten minutes, and the article is then removed. It is then dipped several times in the solution and removed. Excess solution is prevented from pooling or running on the surface by gently and rapidly brushing with a soft bristle-brush, and the article is left to dry thoroughly in air. The underlying ground colour darkens from pink to grey-brown on exposure to daylight, as the mixed patina develops. When colour development has ceased, the article may be wax finished.

The results are difficult to control. The technique will always produce an irregular and variegated surface, the even distribution of which is encouraged by brushing in all directions after dipping.

It is essential to ensure that the surface is completely dry before waxing. Drying may need to be prolonged in damp weather.

4.126★ Blue-green patina on brown ground Semi-matt

Copper nitrate	80 gm
Water	1 litre
Ammonia (.880 solution)	3 cm³

Applied liquid (Three days)

The solution is applied to the object by spraying with a moderately fine atomising spray to produce a misty coating. It is essential to apply only a fine misty coating and to avoid any pooling of the solution or runs on the surface. The object is then left to dry. This procedure is repeated twice a day for three days, after which the surface should be allowed to dry out thoroughly for several days before finishing with wax.

Prolonging the treatment beyond three days does not appear to produce any further development. In tests extending to twenty days no notable changes could be detected after three days.

▲ Inhalation of the fine spray is harmful. It is essential to wear a suitable nose and mouth face mask provided with the correct filter, and to ensure adequate ventilation.

4.127★ Blue-green patina on red-brown/maroon ground Semi-matt

Copper nitrate	200 gm
Water	1 litre

Applied liquid (Five days)

The article is dipped in the cold solution for a few seconds, drained and allowed to dry in air. This procedure is repeated twice a day for about five days, during which time the surface darkens gradually to a red-brown. After about two days, patches of powdery blue-green patina also begin to appear on the surface. When treatment is complete, the article should be left to dry in air for a period of at least three days to allow the patina to develop fully and dry out. When dry, the article is wax finished.

Some sources suggest a scratch-brushed application of the solution. Tests carried out using this technique produced similar but slightly lighter results over the same period of time.

Tests carried out in which the metal was briefly 'pickled' in a 10% solution of nitric acid prior to the initial dip, as suggested by some sources, produced very similar results with somewhat duller surfaces.

The patina tends to develop in streaks and patches corresponding to the draining of the solution from the surface. This can be minimised by 'brushing out' with a soft brush after dipping, to prevent runs and pooling.

If the ground colour alone is required, then after two days of dipping and drying, any patina should be brushed off scrupulously and application continued by wiping-on very sparingly. A completely patina-free ground is difficult to achieve.

4.128★ Blue-green patina (orange-brown ground) Matt

Ammonium carbonate	24 gm
Potassium hydrogen tartrate	6 gm
Sodium chloride	6 gm
Copper sulphate	6 gm
Acetic acid (6% solution)	to form a paste

Applied paste (Several days)

The solid ingredients are made to a paste with a little of the acetic acid, using a pestle and mortar. The paste is applied with a soft brush. The object is then left to dry for about two days, and the application is then repeated. It is then left to dry thoroughly in air, which generally takes several days. When completely dry, the surface is brushed gently with a bristle-brush to remove any loose material, rubbed with a soft cloth to smooth the surface, and wax finished.

If the paste is made up with stronger solutions of acetic acid (15–30%) then there is a greater tendency for the blue-green patina to be non-adherent, partially revealing orange-brown ground colours.

▲ A nose and mouth face mask fitted with a fine dust filter should be worn when brushing away the dry residue.

4.129   Blue-green patina on brown/black ground Semi-matt

Copper nitrate	100 gm
Nitric acid (70% solution)	40 cm³
Water	1 litre

Torch technique

The object is heated with a blow-torch and the solution applied sparingly with a bristle-brush or paint brush. The liquid quickly turns dark brown, and becomes yellowish where it is not being directly heated. Continued heating makes the surface blacken and areas of blue-green patina begin to form. The blue-green patina tends to be superficial at this stage and the surface should be further heated and stippled with solution until a good dark brown or black ground has been established. The object is then gently heated and dabbed with an almost dry brush or cloth, barely damp with the solution, until an evenly distributed blue-green patina is obtained. When treatment is complete, the object is allowed to cool and dry thoroughly, and may then be wax finished.

▲ Nitric acid is very corrosive and must not be allowed to come into contact with the eyes, skin or clothing. A face shield or goggles should be worn to protect the eyes from hot splashes of solution as it is applied to the heated metal.

▲ The fumes evolved as the solution is applied to the heated metal should not be inhaled. Adequate ventilation must be provided, and a nose and mouth face mask fitted with the correct filter should be worn.

4.130*   Blue-green patina (blue-green patina on black ground)

Copper nitrate	200 gm
Water	1 litre

Torch technique

The metal surface is heated with a blow-torch, and the solution applied with a soft brush until an even blue-green patina develops. If the surface is then heated, without further application of the solution, it will become black. The solution may then be applied to this black ground, heating the surface as necessary, to form a blue-green patina on the ground colour. The surface quality obtained can be varied by the precise method of application—a stippled surface may be produced by stippling with a relatively dry brush, or a more mottled or marbled effect obtained using the same technique with a brush that is more 'loaded' with the solution. When the required finish is achieved, the object is allowed to dry out, and then wax finished.

▲ Colouring should be carried out in a well ventilated area, so that inhalation of the vapours is avoided.

▲ A face shield or goggles should be worn to protect the eyes from hot splashes of solution.

4.131*   Olive green on black ground Matt

A	Copper nitrate	200 gm
	Water	1 litre
B	Potassium polysulphide	50 gm
	Water	1 litre

Torch technique (A)
Torch technique (B)

The article is heated with a blow-torch and the copper nitrate solution applied with a soft brush until an even thin layer of blue-green patina covers the surface. This is again heated, without further application of the solution, until it turns black. Any loose residue should be brushed off with a soft dry brush. The copper nitrate is then again applied, heating as necessary, until the blue-green colour re-appears and is evenly distributed. While the surface is still hot, the polysulphide solution is applied until the required depth of colour is achieved. When treatment is complete, the article is allowed to dry for some time, and may then be wax finished.

The concentrations of both solutions can be varied within quite wide limits.

▲ Colouring should be carried out in a well ventilated area, so that inhalation of the vapours and of the hydrogen sulphide gas that are evolved is avoided.

▲ A face shield or goggles should be worn to protect the eyes from hot splashes of solution.

4.132   Thin blue-green patina on pale brown ground Matt

Ammonium carbonate	150 gm
Ammonium chloride	100 gm
Water	1 litre

Applied liquid (Several days)

The solution is applied to the object, by wiping it on to the surface with a soft cloth. The solution should be used sparingly, leaving the surface evenly moist. The object is then allowed to dry in air. This procedure is repeated once a day for several days, until the grey-green patina is sufficiently developed. When treatment is complete, the object should be allowed to dry out thoroughly for several days, and is wax finished when completely dry.

Colour development is slow, tending to produce a thin variegation of the patina on a pale brown ground.

It is essential to ensure that the surface is completely dry and patina development complete, before wax finishing. The drying period may need to be extended in damp or humid atmospheric conditions.

4.133   Pale blue (*or* green) patina on light brown/ochre ground   Semi-matt

Ammonium carbonate	180 gm
Copper sulphate	60 gm
Copper acetate	20 gm
Oxalic acid	1.5 gm
Ammonium chloride	0.5 gm
Acetic acid (10% solution)	1 litre

Torch technique

The ingredients are dissolved and the solution boiled for about ten minutes. It is then left to stand for several days. The mixture should be shaken immediately prior to use. The surface of the object to be coloured is heated with a blow torch, and the solution applied with a soft brush. This process is continued until a pale blue-patina is produced, on a light brown or ochre ground. If the torch is then gently played across the surface, the pale blue will change to green. Varying effects can be obtained by 'stippling' or 'dragging' the brush on the surface. When the required finish is obtained, the object is left to cool and dry for some time and may then be wax finished.

Continuous application of liberal quantities of the solution will tend to inhibit patina formation, but encourages the development of the ground colour.

▲ The ammonium carbonate should be added to the acetic acid in small quantities. The reaction is effervescent, and if large amounts are added, this may be very vigorous.

▲ Colouring should be carried out in a well ventilated area, so that inhalation of the vapours is avoided.

▲ A face shield or goggles should be worn to protect the eyes from hot splashes of solution.

4.134   Yellow-green/blue-green on orange-brown ground   Semi-matt

Copper nitrate	20 gm
Ammonium chloride	20 gm
Calcium hypochlorite	20 gm
Water	1 litre

Cold immersion   (Twenty hours)

The article is suspended in the cold solution. After an initial 'bright dip' effect which enhances the metal surface to a reddish-gold colour, there is a gradual build-up of colour and an etching effect that is typical with chloride solutions. The article is left in the solution for twenty hours and then removed, rinsed in cold water and allowed to dry in air. Drying in air produces a green powdery layer which is only partially adherent on polished surfaces. When the green layer has developed and the surface is completely dry, the object is wax finished.

The results produced in tests were very uneven, with streaking and patchiness of both the patina and the ground colour. Patina development can be increased if the surface is wiped with the cold solution and allowed to dry, after the main treatment when the ground colour has developed. This can be repeated as necessary.

It is essential to ensure that the surface is dry and patina development complete before wax finishing.

Tests on brasses with a higher proportion of zinc (yellow brass and muntz brass) produced strongly etched surfaces with no patina development.

▲ Calcium hypochlorite (bleaching powder) can cause irritation of the eyes, skin and respiratory system. It is essential to avoid inhaling the fine dust. Both the powder and solutions should be prevented from contact with the eyes or skin.

4.135★   Green and blue-green patina on orange-brown ground   Semi-matt

Ammonium carbonate	150 gm
Sodium chloride	50 gm
Copper acetate	60 gm
Potassium hydrogen tartrate	50 gm
Water	1 litre

Applied liquid   (Several days)

The solution is applied generously to the surface of the article using a soft cloth. It is then allowed to dry in air. This procedure is repeated twice a day for several days, during which time the colour gradually develops. It is then left to dry in air, without further treatment, to allow the patina to develop. When patina development is complete and the surface is completely dry, the article is wax finished.

The green patina tends to develop initially whether the surface is allowed to remain wet or only slightly moist. However, where a substantial green is produced, this tends to become blue-green and slightly non-adherent at the fringes. Better results were produced in tests, where excess moisture was brushed out with a soft brush.

4.136   Blue-green patina on orange/reddish-brown variegated ground   Semi-matt   *Pl. VII*

Ammonium chloride	40 gm
Ammonium carbonate	120 gm
Sodium chloride	40 gm
Water	1 litre

Cold immersion and applied liquid   (Several hours)

The object to be coloured is immersed in the cold solution for about one hour, producing a slight etching effect and a gradual development of the ground colour. It is then removed and allowed to dry thoroughly in air. Any excessive surface moisture should be gently brushed out with a soft brush, to prevent 'runs' or 'pooling' of the solution and to leave an evenly moistened surface. The patina develops slowly as the surface dries out. If further patina development is desired, then the solution should be applied sparingly with a soft cloth and the object again left to dry thoroughly. When patina development is complete and the surface absolutely dry, the object may be wax finished.

A variegated orange and reddish-brown ground colour is produced. Patina development tends to be slight, taking the form of thin areas overlaying the ground.

It is essential to ensure that patina development is complete, and the surface dry, before wax finishing. In damp or humid conditions the drying time may need to be extended.

4.137　Dark green and pale blue 'watery' surface　Semi-matt

Ammonium carbonate	180 gm
Copper sulphate	60 gm
Copper acetate	20 gm
Oxalic acid	1.5 gm
Ammonium chloride	0.5 gm
Acetic acid (10% solution)	1 litre

Boiling immersion　(Twenty-five minutes)

The object is immersed in the boiling solution, which produces some etching effects and light colouring after a few minutes. As the immersion continues streaky greenish films gradually develop, beginning to become dark after about ten minutes. The object is removed after twenty-five minutes and the hot solution applied with a soft brush, to leave the surface evenly moist. The object should then be allowed to dry thoroughly in air for several hours, without washing. When dry, it is wax finished.

The surface dries to a dark green initially, and pale blue patches gradually develop, to give a surface with a 'watery' or glazed appearance.

▲ The ammonium carbonate should be added to the acetic acid in small quantities. The reaction is effervescent, and this may be very vigorous if large amounts are added.

▲ Oxalic acid is harmful if taken internally and irritating to the eyes and skin.

4.138　Greenish patina on pink ground　Matt

Copper nitrate	30 gm
Zinc chloride	30 gm
Water	to form a paste

Applied paste　(Two or three hours)

The ingredients are mixed by grinding together using a pestle and mortar. Water should be added by the drop, to form a creamy paste. An excess of water should be avoided, as the paste tends to thin rather suddenly. The paste is applied to the object with a soft brush, and allowed to dry for two or three hours. The residual paste is gently washed away with cold water, and the object then allowed to dry in air for several days. Exposure to daylight will cause a variegated darkening of the ground colours. When the surface is completely dry, and colour change has ceased, the object may be wax finished.

The results tend to be very variegated and the patina may form in patches.

Repeated applications do not improve the results, and tend to require far longer final drying periods.

4.139　Green patina on brown ground　Semi-matt　　　　*Pl. VII*

Copper nitrate	100 gm
Hydrochloric acid (35% solution)	10 cm³
Water	1 litre

Torch technique

The object is heated with a blow-torch and the solution sparingly applied with a bristle-brush or paint brush. A heavy deposit of yellowish-green patina quickly forms, but this is rather non-adherent at this stage. Continued heating and the application of small amounts of solution by brushing gradually darkens the surface. When the brown ground colour has been established, the green patina is built up either by stippling with a nearly dry brush, or by using a cloth which is barely damp with the solution, heat being frequently applied to the area being worked. When a good patina has been obtained, the surface can be burnished with a dry cloth. Wax finishing is carried out when the object is cool and thoroughly dry.

In some cases the patina may have a tendency to 'sweat' shortly after treatment. If this occurs then the object should be left in a warm dry place for some time, and the wax only applied when the surface is thoroughly dry.

▲ Hydrochloric acid is corrosive and should not be allowed to come into contact with the eyes and skin. A face shield or goggles should be worn to protect the eyes from hot splashes as the solution is applied.

▲ The fumes evolved as the solution is applied to the heated metal should not be inhaled. Adequate ventilation must be provided and a nose and mouth face mask fitted with the correct filter should be worn.

4.140*　Orange and yellow-green (variegated)　Semi-matt　　　　*Pl. VIII*

Copper sulphate	10 gm
Lead acetate	10 gm
Ammonium chloride	5 gm
Water	to form a paste

Applied paste　(Several days)

The ingredients are ground to a creamy paste with a little water, using a pestle and mortar. The paste is applied to the object with a soft brush, and is left to dry out completely. When completely dry, after several hours, the dry residue is brushed away with a bristle-brush. This process is repeated three or four times. When treatment has finished, the object should be allowed to stand in a dry place for several days to ensure thorough drying. After a final brushing with the bristle-brush, it is wax finished.

Patina development is very slight, and tends to integrate with the ground when the surface is finished, to produce an orange and yellow-green variegation.

▲ Lead acetate is highly toxic, and every precaution should be taken to prevent the inhalation of the dust during weighing, and preparation of the paste. A mask fitted with a fine dust filter must be worn when brushing away the dry residue. A clean working method is essential throughout.

4.141   Variegated (green patina on buff and grey-brown) Semi-matt

Copper nitrate	20 gm
Sodium chloride	16 gm
Potassium hydrogen tartrate	12 gm
Ammonium chloride	4 gm
Water	to form a paste

Applied paste (Four hours)

The ingredients are ground to a paste with a pestle and mortar by the very gradual addition of a small quantity of cold water. The creamy paste is applied with a soft brush, and is left to dry. After four hours the dry residue is removed with a bristle-brush. The object is washed in cold water and allowed to dry in air. Exposure to daylight will darken the buff-orange ground to produce some green patina on an olive and grey-brown ground. When colour development has ceased, the object may be wax finished.

▲ A nose and mouth face mask fitted with a fine dust filter should be worn when brushing away the dry residue.

4.142   Variegated pale green/buff/pink Semi-matt

Copper sulphate	20 gm
Zinc chloride	20 gm
Water	to form a paste

Applied paste (Twenty-four hours)

The ingredients are ground to a thin creamy paste with a little water, using a pestle and mortar. The paste is applied sparingly to the surface of the object with a soft brush. It is then allowed to dry completely, and the residual dry paste washed off with cold water. The object is then allowed to dry in air for about twelve hours, and the procedure repeated if required. When treatment is complete, the object should be allowed to dry for several days at least. Exposure to daylight causes a variegated darkening of the ground colours. When the surface is completely dry and colour change has ceased, the object may be wax finished.

Repeated applications tend to produce darker colours, but the surface is very prone to extreme patchiness.

4.143   Blue-green patina on light brown/grey ground Matt

Ammonium carbonate	300 gm
Acetic acid (10% solution)	35 cm³
Water	1 litre

Warm scratch-brushing (Two or three minutes)

The ammonium carbonate is added to the water, while stirring, and the acetic acid then added very gradually. The solution is then warmed to 30–40°C. The solution is applied liberally with a soft brush, and then immediately scratch-brushed. The surface is worked for a short time, while applying more solution, and then allowed to dry in air. Any excess moisture on the object should be brushed out with a soft bristle-brush, to prevent 'pooling' or 'running' and to leave an evenly moist surface. When the surface is completely dry, which may take several days, it is wax finished.

A variegated or patchy light brown and grey ground is produced, which is overlaid with areas of blue-green patina. The best effects that could be obtained in tests were where the patina was least developed, by ensuring that the surface was very sparingly moist before leaving it to dry.

The results obtained in tests with this solution were poor. Although the colours obtained are interesting, the surfaces were very prone to streaking and patchiness, and little control over this could be gained. The technique cannot therefore be generally recommended.

4.144   Dark green on mid-brown ground Matt

Ammonium sulphate	105 gm
Copper sulphate	3.5 gm
Ammonia (.880 solution)	2 cm³
Water	1 litre

Torch technique

The metal surface is heated with a blow torch, and the solution applied with a soft brush, until a dark green colour on a mid-brown ground is obtained. If the metal is too hot when the solution is applied then impermanent light grey or pinkish-red colours are produced. Continued application of the solution as the metal cools will remove these colours and encourage the brown ground and the dark green patina. If the green patina is heated once it has developed, by playing the torch across the surface, then it will tend to become black. When the desired surface finish has been achieved, the object is left to cool and dry for some time, and may then be wax finished.

A wide variety of surface effects are obtained, which depend on the temperature of the metal, and the manner in which the solution is applied (eg liberally or sparingly; by 'stippling' or by 'painting' etc).

▲ Colouring should be carried out in a well ventilated area, to avoid inhalation of the vapours.

▲ A face shield or goggles should be worn to protect the eyes from hot splashes of solution.

4.145   Blue-green patina on patchy brown ground Semi-matt

Ammonium sulphate	105 gm
Copper sulphate	3.5 gm
Ammonia (.880 solution)	2 cm³
Water	1 litre

Applied liquid (Several days)

This recipe, which has been used commercially to induce patination on copper roofing, requires weathering in the open air to achieve the best results. In tests carried out in workshop conditions, only slight patination and patchy ground colours could be obtained.

The solution is applied to the surface of the object with a soft cloth. It is then allowed to dry in air. This procedure is repeated twice a day for several days until the ground colour has developed and traces of patina begin to appear. The solution should then be applied more sparingly by dabbing with a soft cloth. The object is allowed to dry slowly in a damp atmosphere to encourage the development of patina. It is finally allowed to dry in a warm dry atosphere, and wax finished when completely dry.

4.146   Mottled orange-brown and light ochre (light green patina) Semi-gloss

Copper nitrate	15 gm
Zinc nitrate	15 gm
Ferric chloride	0.5 gm
Hydrogen peroxide (100 vols.)	to form a paste

Applied paste (Four hours)

Some pale green and orange colours may be produced, but these tend to be non-adherent revealing a mottled ground.

The surface of the object should be handled as little as possible until dry, and should not be brushed during washing.

▲ Hydrogen peroxide should not be allowed to come into contact with combustible material. It should be prevented from coming into contact with the eyes and skin, as it may cause severe irritation. It is very harmful if taken internally. The vapour evolved as it is added to the solid ingredients of the paste must not be inhaled.

The ingredients are ground using a mortar and pestle, and the hydrogen peroxide added in small quantities to form a thin creamy paste. The paste is applied to the object with a soft brush, and left to dry in air. After four hours, the residue is washed away with cold water and the object is allowed to dry in air for several hours. When completely dry, it may be wax finished.

4.147   Slight blue-green patina on dull brown ground Matt

Ammonium carbonate	600 gm
Sodium chloride	200 gm
Copper acetate	200 gm
Potassium hydrogen tartrate	200 gm
Acetic acid (10% solution)	1 litre

Applied liquid (Twice a day for several days)

An uneven dull olive/brown ground tends to be produced, on which patina development is slight. The results are difficult to control and the technique cannot be generally recommended.

For results obtained with a more dilute form of this solution, see recipe 4.148.

It is essential to ensure that the surface is dry and patina development complete before wax finishing. In damp or humid conditions the drying time may need to be extended.

▲ The ammonium carbonate should be added to the acetic acid in small quantities. The reaction is effervescent and if large amounts are added the effervescence is very vigorous.

The solution is applied to the article to be coloured, by wiping sparingly with a soft cloth. It should then be allowed to dry thoroughly in air. This procedure should be repeated twice a day for several days, until the ground colour has developed and the patina has formed. The article should then be left to dry in air for several days, during which time the patina will continue to develop. When patina development has ceased and the surface is completely dry, the article may be wax finished.

4.148   Slight blue-green patina on variegated reddish-brown ground Semi-matt

Ammonium carbonate	150 gm
Copper acetate	60 gm
Sodium chloride	50 gm
Potassium hydrogen tartrate	50 gm
Acetic acid (10% solution)	1 litre

Applied liquid (Twice a day for several days)

Excessive use of the solution tends to result in a dull olive layer developing on the variegated reddish-brown ground that inhibits patina development. It is difficult to obtain a good overall surface finish on gilding metal using this solution.

For results obtained using a more concentrated form of this solution, see recipe 4.147

It is essential to ensure that patina development is complete, and the surface completely dry, before wax finishing. In damp or humid conditions, the drying time may need to be extended.

▲ The ammonium carbonate should be added to the acetic acid in small quantities. The reaction is effervescent and if large quantities are added, then the effervescence is very vigorous.

The solution is dabbed on to the object with a soft cloth, and then allowed to dry thoroughly in air. This process is repeated twice a day for several days until the desired colour is achieved. Any powder or other loose material that forms during drying periods should be gently brushed away with a soft dry cloth, prior to the next application of solution. When treatment is complete the surface should be allowed to dry in air for several days. When completely dry, it may be wax finished.

4.149   Blue-green patina on brown ground Semi-matt/matt

Copper acetate	30 gm
Copper carbonate	15 gm
Ammonium chloride	30 gm
Hydrochloric acid (18% solution)	2 cm³
Water	to form a paste

Applied paste (Two or three days)

The ingredients are ground to a paste with a little water and the hydrochloric acid, using a pestle and mortar. The paste is applied to the object with a soft brush and then left for several hours to dry out. The dry residue is removed with a stiff bristle-brush. The procedure is repeated until a brown ground has developed with a thin bloom of patina. The paste is then thinned out to a liquid and applied using the same procedure, but with a soft brush to remove residue. When the surface is sufficiently developed and well dried out, the object may be wax finished.

A thick incrustation tends to form if the paste is made too thick. This is difficult to remove by brushing, when dry, but tends to flake eventually.

▲ A nose and mouth face mask should be worn when brushing away the dry residue.

4.150   Slight blue-green patina on mottled grey/ochre Semi-matt

Ammonium chloride	10 gm
Ammonium acetate	10 gm
Water	1 litre

Applied liquid (Twice a day for ten days)

The solution is applied to the surface by wiping with a soft cloth or by brushing with a soft brush. It is then allowed to dry in air. This procedure is repeated twice a day for about ten days. The article is then allowed to dry in air, without treatment, for a further five days. When completely dry, and when patina development has ceased, it may be wax finished.

Colour development is generally slow, the ground developing unevenly at first, and the patina appearing later during the periods of air-drying. Patina development is more pronounced in surface features such as punched marks or textured areas.

4.151   Slight blue-green patina on brown ground Matt

A	Ammonium sulphate	80 gm
	Water	1 litre
B	Copper sulphate	100 gm
	Sodium hydroxide	10 gm
	Water	1 litre

Applied liquid (A) (Five to ten days)
Applied liquid (B) (Two or three days)

The ammonium sulphate solution is applied to the surface of the object by dabbing vigorously with a soft cloth, leaving it sparingly moist. It is then allowed to dry completely. This procedure is repeated twice a day for up to ten days. Solution B is then applied in the same way for two or three days. When treatment is complete, the object is allowed to dry in air for several days. When completely dry, it may be wax finished.

In tests, the solution tended to retract into pools and failed to 'wet' the surface. The addition of wetting agents produced no improvement, and the surface developed in patches rather than evenly.

▲ Sodium hydroxide is corrosive and should be prevented from coming into contact with the eyes and skin.

4.152* Pale green lustre Gloss

Pl. VII

Potassium hydroxide	100 gm
Copper sulphate	30 gm
Sodium tartrate	30 gm
Water	1 litre

Warm immersion (About ten minutes)

The potassium hydroxide is added gradually to the water and stirred. Heat is evolved as it dissolves, raising the temperature to about 35°C. The other ingredients are added and allowed to dissolve. There is no need to heat the solution. The object is immersed in the warm solution, and immersion continued until the lustre colour develops. The object is then removed and washed in warm water. It is allowed to dry in air, after any excess moisture has been removed by dabbing with absorbent tissue. When dry, it is wax finished.

An even pale green lustre is produced at the stated temperature. If the temperature is raised, or the immersion prolonged then a more variegated pink and green surface is produced. This is accompanied by a tendency for some opaque areas and darker marks to appear unevenly on the surface.

▲ Potassium hydroxide is a powerful corrosive alkali, and contact of the solid or solutions with the eyes or skin must be prevented. It should be added to the water in small quantities.

4.153* Blue lustre Gloss

Copper nitrate	115 gm
Tartaric acid (crystals)	50 gm
Sodium hydroxide	65 gm
Ammonia (.880 solution)	15 cm^3
Water	1 litre

Boiling immersion (Eight minutes)

Immersion in the boiling solution produces a series of lustre colours, which begin with a blue that develops after about one minute to an even green lustre after about two minutes. If the green colour is required then the object should be immediately removed and washed in hot water, and allowed to dry in air. It is wax finished when dry. Continuing the immersion will cause a gradual darkening of the surface to produce a blue colour after about eight minutes. The object is removed and finished as noted above.

To obtain a pink/green lustre with this solution see recipe 4.154.

Although a number of colour changes occur, the two colours noted are the only ones in this lustre series which are both rich and even.

▲ Sodium hydroxide is a powerful corrosive alkali, and the solid or solutions must be prevented from coming into contact with the eyes and skin. It should be added to the solution in small amounts.

▲ Preparation and colouring should be carried out in a well ventilated area. Ammonia vapour will be liberated, which if inhaled causes irritation of the respiratory system. A combination of caustic vapours will be liberated from the solution as it boils.

▲ A dense precipitate is formed which may cause severe bumping. Glass vessels should not be used.

4.154* Pink/green lustre Gloss

Pl. VIII

Copper nitrate	115 gm
Tartaric acid (crystals)	50 gm
Sodium hydroxide	65 gm
Ammonia (.880 solution)	15 cm^3
Water	1 litre

Boiling immersion (Fifteen minutes)

Immersion in the boiling solution produces a series of lustre colours, which begin with a blue that develops after about one minute to an even green lustre after about two minutes. If the green colour is required then the object should be immediately removed and washed in hot water, and allowed to dry in air. It is wax finished when dry. Continuing the immersion will cause a blue colour to appear which subsequently changes to produce the final colour after about fifteen minutes. The object is removed and finished as noted above.

To obtain a blue colour using this solution see recipe 4.153.

Although a number of colour changes occur, the colours noted are the only ones in this lustre series which are both rich and even.

▲ Sodium hydroxide is a powerful corrosive alkali, and the solid or solutions must be prevented from coming into contact with the eyes and skin. It should be added to the solution in small amounts.

▲ Preparation and colouring should be carried out in a well ventilated area. Ammonia vapour will be liberated, which if inhaled causes irritation of the respiratory system. A combination of caustic vapours will be liberated from the solution as it boils.

▲ A dense precipitate is formed which may cause severe bumping. Glass vessels should not be used.

4.155* Green and pink lustre Gloss

Copper sulphate	120 gm
Potassium chlorate	60 gm
Water	1 litre

Hot immersion (Thirty seconds)

The article is immersed in the hot solution (70–80°C) and quickly removed after thirty seconds immersion, when the green lustre has developed. It is immediately washed thoroughly in hot water. After drying in sawdust, it is wax finished. The colour appears to change on drying, and again on waxing, due to the change in the interference colour caused by the presence of the layer of water, and then the layer of wax on the surface. The final colour is a green lustre variegated with pink.

Beyond the green lustre phase, the surface rapidly becomes orange-buff coloured. See also recipe 4.11.

▲ Potassium chlorate is a powerful oxidising agent and must not be allowed to come into contact with combustible material. Contact with the eyes, skin or clothing must be avoided.

4.156* Golden brown lustre Gloss

| Potassium permanganate | 10 gm |
| Water | 1 litre |

Hot immersion (Three to five minutes)

The article is immersed in the hot solution (90°C). A golden lustre colour develops within one minute, gradually becoming more intense. When the lustre colour is fully developed, which may take from three to five minutes, the article is removed and washed in hot water which is gradually cooled during washing. The article is finally washed in cold water before being carefully dried either in sawdust, or by gently blotting the surface with absorbent tissue paper to remove excess moisture and allowing it to dry in air. When dry, it is wax finished.

The effect is only obtained on polished or directionally grained satin surfaces. On rough, as-cast and matt surfaces a dull orange or orange-brown is produced.

Extending the immersion time produces a slight darkening initially to a more pink colour, which rapidly fades to a pale silvery grey. These latter colours are usually uneven. Colours produced at longer immersion times tend to be more fragile when wet, and should not be handled at that stage.

4.157 Lustre series Gloss

Sodium thiosulphate	125 gm
Lead acetate	35 gm
Water	1 litre

Hot immersion (Various)

The article is immersed in the hot solution (50–60°C), which produces a series of lustre colours in the following sequence: golden-yellow/orange; purple; blue; pale blue; pale grey. The timing tends to be variable, but it takes roughly five minutes for the purple stage to be reached, and about fifteen minutes for pale blue. (Tests failed to produce a further series of colours, the surface tending to remain pale grey after twenty minutes.) When the desired colour has been produced, the article is removed and immediately washed in cold water. It is allowed to dry in air, any excess moisture being removed with an absorbent tissue. When dry it may be wax finished or coated with a fine lacquer.

More dilute solutions produce the lustre colour sequence, but take a correspondingly longer time to produce.

▲ Lead acetate is very toxic and will give rise to serious conditions if taken internally. The dust, and the vapour evolved from the hot solution, must not be inhaled.

4.158 Dull pink lustre Gloss

| Antimony trichloride | 50 gm |
| Olive oil | to form a paste |

Applied paste (Five minutes)

The antimony trichloride is ground to a creamy paste with a little olive oil using a pestle and mortar. The paste is brushed on to the surface of the object to be coloured. A purple lustre colour develops after about two minutes, changing to a pink lustre after about five minutes. A steel blue appears if the paste is left for longer. When the desired colour has developed, the residual paste is washed from the surface with pure distilled turpentine. The object is finally rubbed with a soft cloth, and wax finished.

Tests tended to result in surfaces which were slightly uneven.

Localised effects can be produced by the selective application of brush strokes.

If the paste is left to dry on the surface, it becomes difficult to remove without damaging the colour finish.

Tests carried out using aqueous solutions produced similar results, but these tended to be patchy.

▲ Antimony trichloride is highly toxic and every precaution must be taken to avoid contact with the eyes and skin. It is irritating to the respiratory system and eyes, and can cause skin irritation and dermatitis.

4.159 Pink/green lustre on orange ground Gloss

Copper sulphate	25 gm
Sodium hydroxide	5 gm
Ferric oxide	25 gm
Water	1 litre

Hot immersion and applied liquid and bristle-brushing (A few minutes)

The copper sulphate is dissolved in the water and the sodium hydroxide added. The ferric oxide is added, and the mixture heated and boiled for about ten minutes. When it has cooled to about 70°C, the object is immersed in it for about ten seconds and becomes coated with a brown layer. The object is transferred to a hotplate and gently heated until the brown layer has dried. The dry residue is brushed off with a soft bristle-brush. The turbid solution is then applied to the metal with a brush, as the object continues to be heated on the hotplate. The residue is brushed off, when dry. The application is repeated two or three times, until a green/pink lustre is produced. After a final brushing with the bristle-brush, the object is wax finished.

The results tend to be variegated rather than even, and may appear patchy when viewed in some lights.

▲ Contact with the sodium hydroxide in the solid state must be avoided when preparing the solution. It is very corrosive to the skin and in particular to the eyes.

4.160   Green lustre (pink tinge) Gloss

Potassium sulphide	10 gm
Water	1 litre
Ammonia (.880 solution)	1 cm³

Hot immersion (About ten minutes)

The object is immersed in the warm solution (40°C) and the surface gradually takes on a lustrous quality, which develops to a green lustre after about ten minutes. The object is then removed, washed in cold water, and allowed to dry in air. When dry it may be wax finished.

The green lustre tends to be unevenly tinged with pink.

As an alternative to immersion, the solution can be applied liberally using a soft brush, without 'working' the surface. The results tend to be more variegated.

▲ Some hydrogen sulphide gas will be liberated from the solution. Colouring should be carried out in a well ventilated area.

4.161* Mid-brown (pink/blue lustre) Gloss

Potassium permanganate	10 gm
Sodium hydroxide	25 gm
Water	1 litre

Boiling immersion (Thirty minutes)

The object is immersed in the boiling solution, and after two or three minutes the surface shows signs of darkening. After ten minutes this develops to an uneven dark lustrous surface. Continued immersion causes some further darkening. After thirty minutes the object is removed, washed in warm water and allowed to dry in air. When dry, it may be wax finished.

Prolonged immersion causes the surface to be slightly dull. The object should be removed when the colour has developed.

A pale golden brown colour is produced on directionally grained satin surfaces.

▲ Sodium hydroxide is a powerful caustic alkali which can cause severe burns. Both the solid and solutions must be prevented from coming into contact with the eyes and skin. When preparing the solution, small quantities of sodium hydroxide should be added to large amounts of water and *not* vice versa.

4.162   Dark petrol lustre on dull golden ground Semi-gloss

A	Copper sulphide	50 gm
	Water	1 litre
B	Potassium sulphide	12.5 gm
	Water	1 litre

Hot immersion (A)   (Thirty seconds)
Cold immersion (B)   (One minute)

The article is immersed in the hot copper sulphate solution (80°C) for about thirty seconds to produce a green/pink lustre colour. It is then transferred to the cold potassium sulphide solution and immersed for about one minute, producing a clouding of the surface and a smokey greenish-pink colour. This tends to be uneven and can be improved by applying the solutions alternately using a bristle-brush. When treatment is complete, the article is washed in warm water and allowed to dry in air. When dry, it is wax finished.

If the results produced in solution A are streaky, then these will tend to be visible in some lights, after treatment with solution B. Application with the bristle-brush darkens and improves the results. Producing an adequate finish evenly is difficult, and the recipe cannot be generally recommended.

For results obtained using a scratch-brushed technique see recipe 4.81.

▲ Colouring should be carried out in a well ventilated area as some hydrogen sulphide gas will be liberated.

4.163   Lustre sequence Gloss

Antimony trisulphide	50 gm
Ammonium sulphide (16% solution)	100 cm³
Water	1 litre

Cold immersion (Two to ten minutes)

The article is immersed in the cold solution, producing a rapid sequence of lustre colours initially. After about two minutes the colour obtained is predominantly blue with some local red or golden surface variation. This changes gradually to a greenish lustre after four minutes, which slowly fades. The final colours will have developed after about six to eight minutes, and immersion beyond ten minutes produces no detectable change. When the desired colour has been reached, at any stage in the immersion, the article is removed, washed in cold water and allowed to dry in air. When dry, it is wax finished.

The colours produced at an early stage in the immersion are the most intense, but they are also subject to unpredictable variegated effects.

The best colours obtained in tests occurred after about two minutes immersion, consisting of a dark purple-brown lustre with a blue sheen. Although the colour is good, it tends to be variegated with streaks of a lighter brown. This streaking could not be controlled. Prolonging the immersion tends to produce paler colours, giving a pale purple after ten minutes.

▲ The preparation and colouring should be carried out in a well ventilated area. Ammonium sulphide solution liberates a mixed vapour of ammonia and hydrogen sulphide, which must not be inhaled, and which will irritate the eyes. The solution must be prevented from coming into contact with the eyes and skin, as it will cause burns or severe irritation.

▲ Antimony trisulphide is harmful if taken internally.

4.164   Pink/green lustre on mottled brown ground Gloss

Copper sulphate	25 gm
Nickel ammonium sulphate	25 gm
Potassium chlorate	25 gm
Water	1 litre

Boiling immersion (One to two minutes)

The object is immersed in the boiling solution and rapidly removed when the mottled surface and lustre colour have developed. It should be washed immediately in hot water. After drying in sawdust, it is wax finished.

▲ Nickel salts are a common cause of sensitive skin reactions.

▲ Potassium chlorate is a powerful oxidising agent and should not be allowed to come into contact with combustible material. The solid, and solutions, should be prevented from coming into contact with the eyes, skin or clothing.

4.165   Reddish-brown lustre with blue tinge Gloss

Selenous acid	6 cm^3
Water	1 litre
Sodium hydroxide solution (250 gm/litre)	50 drops

Warm immersion (Forty seconds)

The acid is dissolved in the water, and the sodium hydroxide added by the drop, while stirring. The solution is then heated to 25–30°C. The article to be coloured is immersed in the solution, producing a purplish-brown lustre after about fifteen seconds, which rapidly develops to a reddish brown tinged with blue. The article is removed after about forty seconds, washed in cold water and allowed to dry in air. When dry it may be wax finished.

Prolonging the immersion causes a gradual fading of the colour. With long immersion times, ten to fifteen minutes, a dark slate layer forms. This is only partly adherent and tends to flake away leaving a metal surface which is finely and densely pinpointed with dark grey. This effect is difficult to control and tends to result in a variegated or patchy finish.

▲ Selenous acid is poisonous and very harmful if swallowed or inhaled. Colouring must be carried out in a well ventilated area. Contact with the skin or eyes must be avoided to prevent severe irritation.

▲ Sodium hydroxide is a powerful caustic alkali. Contact with the eyes and skin must be prevented.

4.166   Lustre series Gloss

Sodium thiosulphate	240 gm
Copper acetate	25 gm
Water	1 litre
Citric acid (crystals)	30 gm

Cold immersion (Various)

The sodium thiosulphate and the copper acetate are dissolved in the water, and the citric acid added immediately prior to use. The article is immersed in the cold solution and a series of lustre colours are produced in the following sequence: golden-yellow/orange; brown; purple; blue; pale grey. This sequence takes about thirty minutes to complete, the brown changing to purple after about fifteen minutes. It is followed by a second series of colours, in the same sequence, which take a further thirty minutes to complete. The latter colours in the second series tend to be slightly more opaque. A third series of colours follows, if immersion is continued, but these tend to be indistinct and murky. (If the temperature of the solution is raised then the sequences are produced more rapidly. However, it was found in tests that the results tended to be less even. Temperatures in excess of about 35–40°C tend to cause the solution to decompose, with a concomitant loss of effect.) When the required colour has been produced, the article is removed and washed in cold water. It is allowed to dry in air, any excess moisture being removed by gentle blotting with an absorbent tissue. When dry it may be wax finished or coated with a fine lacquer.

4.167 Dark straw/brown lustre Gloss

| Barium sulphide | 10 gm |
| Water | 1 litre |

Hot immersion (Two or three minutes)

The object is immersed in the hot solution (50°C), and the surface quickly takes on a golden colour, which darkens to a straw-coloured or brown lustrous finish. The object should be quickly removed and rinsed thoroughly in warm water. It is then allowed to dry in air, and wax finished when completely dry.

If immersion is continued for too long, the colour will quickly become very pale and may acquire a blue tinge.

Cold solutions will produce similar results, given very extended immersion times. In tests it was found that prolonged immersion in a cold solution tends to produce less even results.

▲ Barium sulphide is poisonous and may be very harmful if taken internally. Some hydrogen sulphide gas will be liberated as the solution is heated. Colouring should be carried out in a well ventilated area.

4.168 Pale brown (lustre) Gloss

Potassium sulphide	2 gm
Ammonium sulphate	0.5 gm
Water	1 litre

Cold immersion (One or two minutes)

The article is immersed in the cold solution. A golden lustre colour develops fairly quickly, and in roughly one to two minutes will begin to change to a mustard colour. As soon as this occurs the object should be transferred quickly to a container of cold water and thoroughly rinsed. It should then be washed in clean cold water and dried in sawdust. When completely dry it is wax finished.

The technique is difficult to control. The colour produced tends to be variegated rather than even.

The lustre colour is only produced on highly polished surfaces. Matt or other unpolished surfaces yield only patchy dull brownish colours.

4.169 Lustre series Gloss

Sodium thiosulphate	280 gm
Lead acetate	25 gm
Water	1 litre
Potassium hydrogen tartrate	30 gm

Warm immersion (Various)

The sodium thiosulphate and the lead acetate are dissolved in the water, and the temperature is adjusted to 30–40°C. The potassium hydrogen tartrate is added immediately prior to use. The article to be coloured is immersed in the warm solution and a series of lustre colours are produced in the following sequence: golden-yellow/orange; purple; blue; pale blue/pale grey. The colour sequence takes about ten minutes to complete at 40°C. (Hotter solutions are faster, cooler solutions slower.) A second series of colours follows in the same sequence, but these tend to be less distinct. When the desired colour has been reached, the article is removed and immediately washed in cold water. It is allowed to dry in air, any excess moisture being removed by dabbing with an absorbent tissue. When dry it may be wax finished, or coated with a fine lacquer.

Citric acid may be used instead of potassium hydrogen tartrate, and in the same quantity. Tests carried out with both these alternatives produced identical results.

▲ Lead acetate is very toxic and will give rise to serious conditions if taken internally. The dust, and the vapour evolved from the warm solution, must not be inhaled.

4.170★ Greenish-blue patina (stipple) on black ground Matt

Ammonium carbonate	120 gm
Ammonium chloride	40 gm
Sodium chloride	40 gm
Water	1 litre

Sawdust technique (Twenty to thirty hours)

The article to be coloured is laid-up in sawdust which has been evenly moistened with the solution, and is then left for a period of about twenty or thirty hours. The sawdust should be kept moist throughout the period of treatment. When treatment is complete, the article should be washed in cold water and allowed to dry in air. When it is completely dry, it may be wax finished.

The surface is strongly etched, producing a black textured ground on which the patina develops as a fine stipple of incrustation. As treatment is prolonged, so the patina tends to become more continuous, giving rise to a variegated appearance.

4.171★ Bluish-green stipple and dark brown Matt

Ammonium chloride	100 gm
Ammonium carbonate	150 gm
Water	1 litre

Sawdust technique (To twenty or thirty hours)

The article to be coloured is laid-up in sawdust which has been moistened evenly with the solution. It may then be left for periods of up to twenty or thirty hours. When treatment is complete, the object is washed in cold water and allowed to dry in air. When it is completely dry, it may be wax finished.

The surface is selectively etched by this treatment, producing a 'pitted' textured surface which is coloured brown. The patina develops locally in spots, in the region of the etched 'pits'. If the medium is more moist, patina development tends to be inhibited and a textured brown surface is produced.

4.172★ Black and buff/light brown stipple Semi-gloss   *Pl. VII*

Ammonium chloride	16 gm
Sodium chloride	16 gm
Ammonia (.880 solution)	30 cm³
Water	1 litre

Sawdust technique (Twenty-four hours)

The object to be coloured is laid-up in sawdust which has been evenly moistened with the solution. It is then left for about twenty-four hours. The object should not be disturbed while colouring is in progress, as re-packing tends to cause a loss of surface quality and definition. Progress should be monitored by including a small sample of the object metal, when laying-up, which can be examined as necessary. When treatment is complete, the object is washed thoroughly in cold water and allowed to dry in air. It may be wax finished, when completely dry.

Some blue-green patina may occur, in the form of a sparse stipple. Patina development is more pronounced in any surface features that are present, eg punched marks.

The surface is locally etched by the solution where the granular sawdust is in close contact with the metal, causing fine pits. These tend to become coloured a light buff or brown. Surrounding areas, that are not locally etched, tend to be black.

▲ Ammonia solution causes burns or severe irritation, and must be prevented from coming into contact with the eyes or skin. The vapour should not be inhaled as it irritates the respiratory system. Colouring should be carried out in a well ventilated area.

4.173★ Purple-brown/golden-brown/buff (variegated mottle) Semi-gloss/gloss   *Pl. VII*

Butyric acid	20 cm³
Sodium chloride	17 gm
Sodium hydroxide	7 gm
Sodium sulphide	5 gm
Water	1 litre

Sawdust technique (Twenty hours)

The object to be coloured is laid-up in sawdust which has been evenly moistened with the solution, and left for a period of up to twenty hours. The timing of colour development tends to be unpredictable, and it is essential to include a sample of the same material as the object, when laying-up, so that progress can be monitored without disturbing the object. When treatment is complete, the object is removed and washed thoroughly with cold water. After a period of several hours air-drying, it may be wax finished.

▲ Butyric acid causes burns, and contact with the skin, eyes and clothing must be prevented. It also has a foul pungent odour, which tends to cling to the clothing and hair and is difficult to eliminate.

▲ Sodium hydroxide is a powerful caustic alkali causing burns, and must be prevented from coming into contact with the eyes or skin.

▲ Sodium sulphide is corrosive and causes burns. It must be prevented from coming into contact with the eyes or skin. When preparing the solution it must not be mixed directly with the acid, or toxic hydrogen sulphide gas will be evolved. The acid should be added to the water and then neutralised with the sodium hydroxide, before the remaining ingredients are added.

4.174*   Orange/reddish-brown mottle Semi-matt   *Pl. VIII*

Copper nitrate	200 gm
Water	1 litre

Sawdust technique (Two or three days)

Some green/blue-green patina tends to develop if the moist medium is allowed to dry out.

The article to be coloured is laid-up in sawdust which has been evenly moistened with the solution, and left for two or three days. Colour development may be monitored using a sample, but no deleterious effects were produced, in tests, by unpacking and re-packing the object itself. When the colour has developed satisfactorily, the article is removed and washed in cold water. After several hours drying in air, it may be wax finished.

4.175*   Slight grey-green patina/light bronze (stipple) on brown ground   Semi-matt   *Pl. VII*

Ammonium chloride	10 gm
Copper sulphate	4 gm
Potassium binoxalate	20 gm
Water	1 litre

Sawdust technique (Several days)

The surface is locally etched where the granular sawdust is in close contact with the metal, producing an even stipple of 'bitten' areas which are coloured buff or light orange. These are interspersed with areas of unetched surface, coloured a slightly lustrous light bronze. A light stipple of grey-green patina also tends to develop.

▲ Potassium binoxalate is harmful if taken internally. It can also cause irritation, and should be prevented from coming into contact with the eyes or skin.

The object to be coloured is laid-up in sawdust which has been moistened with the solution, and left for several days. Colour development is very slow, and should be monitored using a sample so that the object remains undisturbed. When the colour has developed satisfactorily, the object is removed, washed in cold water and allowed to dry in air for several hours. It may be wax finished when dry.

4.176*   Brown with buff and red (stipple) Semi-matt

Copper nitrate	280 gm
Water	850 cm³
Silver nitrate	15 gm
Water	150 cm³

Sawdust technique (Twenty hours)

The surface is coloured brown by the treatment, and selectively etched where the granular sawdust is in close contact with the metal, producing a stipple of buff-coloured etched spots. These are fringed with red.

▲ Silver nitrate must be prevented from coming into contact with the skin, and more particularly the eyes, as it may cause burns or severe irritation.

The copper nitrate and the silver nitrate are dissolved in separate portions of water, and the two solutions then mixed. The article to be coloured is laid-up in sawdust which has been evenly moistened with this mixed solution. It is left for a period of about twenty hours, ensuring that the sawdust remains moist throughout this time. When treatment is complete, the article is removed and washed in cold water. After allowing it to dry thoroughly in air for some time, it may be wax finished.

4.177*   Light orange stipple on greenish ground Gloss   *Pl. VIII*

'Rokusho'	30 gm
Copper sulphate	5 gm
Water	1 litre

Sawdust technique (Several days)

The light orange stipple produced tends to be translucent, becoming more opaque as treatment progresses. If the object is removed and re-packed intermittently during treatment, the stipple becomes less definite and the surface takes on an overall cloudy light orange appearance.

For description of 'Rokusho' and its method of preparation, see appendix 1.

The 'Rokusho' and the copper sulphate are added to the water, and the mixture boiled for a short time and then allowed to cool to room temperature. The object to be coloured is laid-up in sawdust which has been moistened with the solution. It is then left for several days. Colour development is very slow, and should be monitored using a sample so that the object remains undisturbed. When the colour has satisfactorily developed, the object is removed, washed in cold water and allowed to dry in air. It may be wax finished when dry.

4.178*   Bronzed metal stipple on etched orange-brown ground   Semi-matt   *Pl. VIII*

Copper sulphate	100 gm
Lead acetate	100 gm
Ammonium chloride	10 gm
Water	1 litre

Sawdust technique (Ten to twenty hours)

The surface is bronzed by the solution, and selectively etched, producing a network of orange-brown ground and giving a stippled appearance. As treatment proceeds, the etched ground becomes predominant. A stipple of pale green patina also tends to occur.

▲ Lead acetate is a toxic substance, and every precaution should be taken to prevent inhalation of the dust, when preparing the solution. It is very harmful if taken internally. A clean working method is essential throughout.

The object to be coloured is laid-up in sawdust which has been evenly moistened with the solution. It is then left to develop for a period of up to about twenty hours. When treatment is complete, the object is removed, washed in cold water and allowed to dry in air. It may be wax finished when completely dry.

4.179   Dark brown on mid-brown (variegated) Semi-matt

Ammonium chloride	350 gm
Copper acetate	200 gm
Water	1 litre

Sawdust technique (Ten to twenty hours)

The object to be coloured is laid-up in sawdust which has been moistened evenly with the solution, and left for a period of from ten to twenty hours. After ten hours it should be examined and repacked if further treatment is required. It should subsequently be examined every few hours until the required surface finish is obtained. The sawdust should be kept moist during the full period of treatment. The object should then be removed, washed in cold water and allowed to dry in air. When completely dry, it may be wax finished.

The metal is selectively etched by this treatment, producing a network of 'pits' that are coloured light brown. The unetched areas of the surface take on a dark brown colour. Longer treatment times, or more moist media, tend to produce more extensive etching and a gradual loss of the dark brown. Some green patina development may also occur.

4.180   Black stipple on lightly bronzed ground Semi-matt

Copper carbonate	300 gm
Ammonia (.880 solution)	200 cm³
Water	1 litre

Sawdust technique (To twenty or thirty hours)

The article to be coloured is laid-up in sawdust which has been evenly moistened with the above solution, and left for a period of up to twenty or thirty hours. The sawdust should remain moist throughout the period of treatment. When treatment is complete, the article is removed and washed in cold water. It is then allowed to dry in air for some time before wax finishing.

Colour development should be monitored by using a sample.

▲ The strong ammonia solution will liberate ammonia vapour, which irritates the eyes and respiratory system. Preparation and colouring must therefore be carried out in a well ventilated area.

▲ Ammonia solution causes burns and severe irritation, and must be prevented from coming into contact with the eyes and skin.

4.181   Light brown/orange stipple Semi-matt

Ammonium sulphate	85 gm
Copper nitrate	85 gm
Ammonia (.880 solution)	3 cm³
Water	1 litre

Sawdust technique (Thirty hours)

The article to be coloured is laid-up in sawdust which has been moistened evenly with the solution. It should then be left for a period of twenty or thirty hours. The sawdust should be kept moist throughout the period of treatment. After thirty hours the object should be examined, and re-packed for shorter periods of time as necessary. When treatment is complete, the object should be washed thoroughly in cold water and allowed to dry in air. When dry it may be wax finished.

The surface is selectively etched where the granular sawdust is in close contact with the metal, producing a stipple of etched areas that take on a light brown or orange colour. The remainder of the surface is little affected. Etching becomes more pronounced if a moister medium is used, and some slight green patina may develop in the region of the etched areas.

4.182   Bronzed metal with etched stipple and blue-green patina Semi-matt

Copper nitrate	200 gm
Sodium chloride	100 gm
Water	1 litre

Sawdust technique (Twenty or thirty hours)

The article to be coloured is laid-up in sawdust which has been moistened evenly with the solution. It should then be left for a period of twenty or thirty hours. After twenty hours the object should be examined, and re-packed for shorter periods of time, as necessary. The sawdust should be kept moist throughout the period of treatment. When treatment is complete, the article should be washed thoroughly in cold water and allowed to dry in air. When completely dry, after several hours, it may be wax finished.

The surface is selectively etched by the solution, where the granular sawdust is in close contact with the metal, causing an even stipple of etched areas that are ochre coloured. The blue-green patina develops locally in the region of the etched areas. The remainder of the surface is bronzed, but becomes more dull and ochre-coloured as treatment is prolonged.

4.183   Black with blue patina on light brown ground (stipple)   *Pl. VII*
Semi-matt

Ammonium carbonate	150 gm
Copper acetate	60 gm
Sodium chloride	50 gm
Potassium hydrogen tartrate	50 gm
Water	1 litre

Sawdust technique (Ten to twenty hours)

The object to be coloured is laid-up in sawdust which has been evenly moistened with the solution. It is left for a period of from ten to twenty hours, colour development being monitored with a sample, so that the object remains undisturbed. When treatment is complete, the object is removed, washed in cold water and allowed to dry in air for several hours. When dry, it may be wax finished.

The surface is blackened by the solution and selectively etched where the granular sawdust is in close contact with the metal, producing a stipple of etched areas, which are coloured predominantly light brown. The black unetched surface tends to acquire some blue patina.

4.184   Slight stipple of blue-grey patina Semi-matt

Di-ammonium hydrogen orthophosphate	100 gm
Water	1 litre

Sawdust technique (Thirty hours)

The object to be coloured is laid-up in sawdust which has been evenly moistened with the solution. It is left for about thirty hours and then removed and washed in cold water. If further colour development is required the treatment should be repeated for additional shorter periods. When treatment is complete, the object is finally washed and allowed to dry in air, and wax finished when completely dry.

The surface is selectively etched by the solution, leaving areas of unetched metal which develop a stipple of pale patina.

Tests carried out with this solution, using various other techniques, including cold and hot direct application, and hot and cold immersion, produced little or no effect.

4.185   Brown stipple on variegated buff ground Semi-matt/matt

Copper acetate	60 gm
Copper carbonate	30 gm
Ammonium chloride	60 gm
Water	1 litre
Hydrochloric acid (15% solution)	5 cm³

Sawdust technique (Ten to twenty hours)

The object to be coloured is laid-up in sawdust which has been evenly moistened with the solution. It is left for a period of about ten hours and then examined. If further treatment is required, it should be re-packed and examined at intervals of a few hours. When treatment is complete, the object is removed, washed thoroughly in cold water and allowed to dry in air. When completely dry, it may be wax finished.

A stippled orange and dark brown glossy surface is produced initially, but this is rapidly etched back to produce a very variegated buff and yellow-brown matt surface. Some greenish patina may also occur on the etched areas.

▲ Hydrochloric acid can cause burns or severe irritation, and must be prevented from coming into contact with the eyes and skin.

4.186   Metallic stipple on buff/black ground Matt

Ammonium chloride	10 gm
Ammonia (.880 solution)	20 cm³
Acetic acid (30% solution)	80 cm³
Water	1 litre

Sawdust technique (One or two days)

The object to be coloured is laid-up in sawdust which has been evenly moistened with the solution, and left for one or two days. Colour development should be monitored using a sample, so that the object itself remains undisturbed. When the colour has developed satisfactorily, the object is removed, washed in cold water and allowed to dry in air for several hours. When dry, it may be wax finished.

The surface is selectively etched by the solution, producing a variegated buff ground, and leaving an even stipple of slightly lustrous unetched metal. As treatment progresses, the buff ground tends to acquire black 'scorched' patches, and the unetched metal surface tends to become similar in colour. At this stage the overall appearance is drab, and the metallic stipple can only be seen in oblique light.

Some slight blue-green patina development may occur in surface features such as punched marks.

▲ Preparation and colouring should be carried out in a well ventilated area. Ammonia vapour and some acetic acid vapour will be liberated, which can irritate the eyes and respiratory system. These solutions should be prevented from coming into contact with the eyes and skin, or severe irritation or burns may result.

4.187   Black and greenish-yellow stipple on mid-brown Gloss

Potassium hydroxide	100 gm
Copper sulphate	30 gm
Sodium tartrate	30 gm
Water	1 litre

Sawdust technique (Twenty hours)

The object is laid-up in sawdust which has been evenly moistened with the solution, and is then left undisturbed for a period of about twenty hours. The sawdust should remain moist throughout this period. When treatment is complete, the object is removed and washed thoroughly in cold water. After it has been allowed to dry in air, it may be wax finished.

The black and greenish-yellow stipple tends to be lustrous.

▲ Potassium hydroxide is a strong caustic alkali, and precautions should be taken to prevent it from coming into contact with the eyes and skin.

▲ When preparing the solution, the potassium hydroxide should be added to the water in small quantities, and *not* vice-versa. When it has dissolved completely, the other ingredients may be added.

4.188   Light buff stipple on dull metal ground Semi-matt

Copper nitrate	20 gm
Ammonium chloride	20 gm
Calcium hypochlorite	20 gm
Water	1 litre

Sawdust technique (Twenty or thirty hours)

The object to be coloured is laid-up in sawdust which has been moistened evenly with the solution, and left for a period of twenty or thirty hours. The sawdust should be kept moist throughout this time. After treatment, the object is washed in cold water and allowed to dry in air. When thoroughly dry, it may be wax finished.

▲ The corrosive fine dust of the calcium hypochlorite (bleaching powder) must not be inhaled, as it can severely irritate the respiratory system. The solution must be prevented from coming into contact with the eyes, which would be severely irritated.

4.189   Blue-green patina (stipple) on black ground Semi-matt

Copper nitrate	100 gm
Water	200 cm³
Ammonia (.880 solution)	400 cm³
Acetic acid (6% solution)	400 cm³
Ammonium chloride	100 gm

Sawdust technique (Twenty to thirty hours)

The copper nitrate is dissolved in the water and the ammonia gradually added. The acetic acid is then added, followed by the ammonium chloride. The article to be coloured is laid-up in sawdust which has been evenly moistened with this solution, and is then left for a period of twenty to thirty hours. When treatment is complete, the article is washed in cold water and allowed to dry in air. When completely dry, after several hours, it may be wax finished.

The surface is selectively etched where the granular sawdust is in close contact with the metal, producing a stipple of etched spots which acquire a blue-green patina. The remainder of the surface is coloured dark brown or black. A wetter medium tends to result in more general etching, the loss of the black colour and inhibition of patina formation.

Colour development should be monitored using a sample of the metal, which is laid-up in sawdust with the article being coloured.

▲ Preparation and colouring should be carried out in a well ventilated area. Ammonia vapour and some acetic acid vapour will be liberated, which cause irritation of the eyes and respiratory system.

▲ The strong ammonia solution must be prevented from coming into contact with the eyes and skin, as it will cause burns or severe irritation.

4.190   Black and blue patina (stipple) on buff ground Semi-matt

177	Ammonium carbonate	300 gm
	Acetic acid (10% solution)	35 cm³
	Water	1 litre

Sawdust technique (To twenty hours)

The ammonium carbonate is added to the water and the acetic acid then added gradually, while stirring. The object to be coloured is laid-up in sawdust which has been evenly moistened with this solution. It is then left for a period of up to twenty hours, ensuring that the sawdust remains moist throughout the period of treatment. After about ten hours, and at intervals of several hours subsequently, the object should be examined to check the progress. When the desired surface finish has been obtained, the object is removed, washed in cold water and allowed to dry in air. When thoroughly dry, it may be wax finished.

The metal is strongly etched to produce a textured surface that is coloured predominantly black, and acquires a blue patina. This is broken by areas of a dull pink etched surface, which becomes more extensive as treatment proceeds.

4.191   Blue-green patina (stipple) on brown/orange ground Semi-matt

Ammonium chloride	15 gm
Potassium aluminium sulphate	7 gm
Copper nitrate	30 gm
Sodium chloride	30 gm
Acetic acid (6% solution)	1 litre

Sawdust technique (Twenty to thirty hours)

The object to be coloured is laid-up in sawdust which has been evenly moistened with the solution, and left for about twenty or thirty hours. The object is then removed and allowed to dry in air for several hours, without washing. When dry, the surface is brushed with a stiff bristle-brush to remove any particles of sawdust or loose material. The object is then left for a further period of several hours, after which it may be wax finished.

4.192★   Pale blue/brown/black (variegated) Semi-matt

Ammonium carbonate	120 gm
Ammonium chloride	40 gm
Sodium chloride	40 gm
Water	1 litre

Cotton-wool technique (Twenty to thirty hours)

Cotton-wool, moistened with the solution, is applied to the surface of the object to be coloured. It is important to ensure that the moist cotton-wool is in full contact with the surface of the object. It is then left for a period of about twenty to thirty hours. It is essential that the cotton-wool remains moist throughout the period of treatment. When treatment is complete, the object should be washed in cold water, and allowed to dry in air. When completely dry, it is wax finished.

A very variegated surface is produced, with a finish that is smooth to the touch. It consists of smooth thin deposited layers of pale blue on a mixed black and brown ground.

4.193★   Blue-green patina on brown ground (variegated) Matt/semi-matt

Ammonium chloride	100 gm
Ammonium carbonate	150 gm
Water	1 litre

Cotton-wool technique (To twenty or thirty hours)

Cotton-wool, moistened with the solution, is applied to the surface of the object to be coloured. It is then left for periods of up to twenty or thirty hours. The cotton-wool should be kept moist throughout the period of treatment. After about ten or fifteen hours the surface should be examined, by carefully exposing a small portion, to determine the extent of colour development. Subsequently a check should be made every few hours until the desired surface finish is obtained. When treatment is complete, the object is washed in cold water and allowed to dry in air. When completely dry, it may be wax finished.

The metal is evenly etched to a relatively smooth brown finish. The patina develops as a smooth deposited layer, which tends to occur in patches, producing a very variegated surface. If the medium is too moist, etching is more pronounced and the patina generally fails to develop.

4.194* Purplish-red and light brown (mottled) Semi-matt

Copper nitrate	280 gm
Water	850 cm³
Silver nitrate	15 gm
Water	150 cm³

Cotton-wool technique (Twenty hours)

The red colour develops as a mottle on an etched light brown surface. Surface development should be monitored using a sample, to ensure that excessive etching does not take place.

▲ Silver nitrate must be prevented from coming into contact with the skin, and more particularly the eyes, as it may cause burns or severe irritation.

The copper nitrate and the silver nitrate are dissolved in separate portions of water, and the two solutions then mixed. Cotton-wool, moistened with this mixed solution, is applied to the surface of the object to be coloured. It is important to ensure that the cotton-wool is in full contact with the surface, and that it remains moist during the period of treatment. The object is removed after about twenty hours and washed in cold water. After thorough drying in air, it may be wax finished.

4.195* Mottled greenish-grey/buff Semi-matt   *Pl. VIII*

Copper sulphate	100 gm
Lead acetate	100 gm
Ammonium chloride	10 gm
Water	1 litre

Cotton-wool technique (Ten to twenty hours)

The surface darkens slightly on exposure to daylight.

▲ Lead acetate is a toxic substance, and every precaution should be taken to prevent inhalation of the dust when preparing the solution. It is very harmful if taken internally. A clean working method is essential throughout.

Cotton-wool, moistened with the solution, is applied to the surface of the object to be coloured, and is left in place for a period of up to twenty hours. The cotton-wool should remain moist throughout the period of treatment. When treatment is complete, the object is removed and washed in cold water. It is then dried in air, and gradually darkens on exposure to daylight. When colour change has ceased, the object may be wax finished.

4.196* Light brown (variegated with green) Semi-matt

Ammonium chloride	350 gm
Copper acetate	200 gm
Water	1 litre

Cotton-wool technique (Ten to twenty hours)

The surface is evenly etched to a fairly smooth light brown finish. This tends to be tinged with green, producing a slightly variegated appearance. Some areas with a more developed bluish-green patina may occur.

Cotton-wool, moistened with the solution, is applied to the surface of the object. This is then left for a period of ten or twenty hours, ensuring that the cotton-wool remains moist. After the first few hours, the surface should be periodically examined by exposing a small portion, to check the progress. When the desired surface finish has been reached, the object is removed, washed in cold water and allowed to dry in air. When completely dry, it may be wax finished.

4.197* Black Gloss

Copper carbonate	300 gm
Ammonia (.880 solution)	200 cm³
Water	1 litre

Cotton-wool technique (Twenty hours)

▲ The strong ammonia solution will liberate ammonia vapour, which irritates the eyes and respiratory system. Preparation and colouring must therefore be carried out in a well ventilated area.

▲ Ammonia solution causes burns and severe irritation, and must be prevented from coming into contact with the eyes and skin.

Cotton-wool, moistened with the solution, is applied to the surface of the object to be coloured, which is then left for a period of twenty hours. It is important to ensure that the moist cotton-wool is in full contact with the surface, and that it remains moist throughout the period of treatment. When treatment is complete, the article is removed and washed in cold water. It is then allowed to dry in air for some time before wax finishing.

4.198 Variegated lustre Gloss

Ammonium sulphate	85 gm
Copper nitrate	85 gm
Ammonia (.880 solution)	3 cm³
Water	1 litre

Cotton-wool technique (Thirty hours)

The technique is only suitable for use on polished surfaces. A golden lustre is produced which is variegated or mottled with pink and green lustre colours.

Cotton-wool, moistened with the solution, is applied to the surface of the object to be coloured. It is then left for a period of about thirty hours. It is essential to ensure that the moist cotton-wool is in full contact with the surface of the object, or breaks in the colour will occur. The cotton-wool should be kept moist throughout the period of treatment. When treatment is complete, the object should be washed thoroughly in cold water and allowed to dry in air. When dry, it is wax finished.

4.199   Dull ochre Matt

Copper nitrate	200 gm
Sodium chloride	200 gm
Water	1 litre

Cotton-wool technique (Twenty hours)

The surface is etched to a matt finish and coloured a dull ochre by the solution. Some blue-green patina may develop when the surface is dried.

Test results were dull and marked by darker stains. The technique cannot be generally recommended.

Cotton-wool, moistened with the solution, is applied to the surface of the object to be coloured. It is then left for a period of about twenty hours. It is essential to ensure that the cotton-wool is in full contact with the surface and is kept moist throughout the period of treatment. Breaks in the colour and patches of patina may result if the cotton-wool is allowed to dry out or lift from the surface. When treatment is complete, the object should be washed thoroughly in cold water, and allowed to dry in air. When dry, it may be wax finished.

4.200*   Greenish patina on mottled light-brown/buff Matt

Copper acetate	60 gm
Copper carbonate	30 gm
Ammonium chloride	60 gm
Water	1 litre
Hydrochloric acid (15% solution)	5 cm³

Cotton-wool technique (Ten to twenty hours)

In tests, the pale greenish patina occurred in the form of fine spots.

▲ Hydrochloric acid can cause burns or severe irritation, and must be prevented from coming into contact with the eyes and skin.

Cotton-wool, moistened with the above solution, is applied to the surface of the article, and left for a period of about ten hours. The surface is then examined, by carefully lifting a small portion, to determine the stage it has reached. When the desired surface finish has been obtained, the article is washed thoroughly in cold water and allowed to dry in air. When completely dry, it may be wax finished.

4.201   Metallic grey/blue lustre on brown ground Gloss

Potassium hydroxide	100 gm
Copper sulphate	30 gm
Sodium tartrate	30 gm
Water	1 litre

Cotton-wool technique (Twenty hours)

The brown ground is not apparent, except where breaches in the metallic lustre occur. An even lustrous surface is difficult to obtain. Colour development should be monitored with a sample.

▲ Potassium hydroxide is a strong caustic alkali, and should be prevented from coming into contact with the eyes or skin.

▲ When preparing the solution, the potassium hydroxide should be added to the water in small quantities, and *not* vice-versa.

Cotton-wool, moistened with the solution, is applied to the surface of the object, which is then left for a period of about twenty hours. It is important to ensure that the moist cotton-wool is in full contact with the surface, and that it remains moist throughout the period of treatment. When treatment is complete, the object is washed in cold water and allowed to dry in air. When dry, it may be wax finished.

4.202   Buff/yellowish-brown (uneven) Matt

Copper nitrate	20 gm
Ammonium chloride	20 gm
Calcium hypochlorite	20 gm
Water	1 litre

Cotton-wool technique (Twenty hours)

The surface produced tends to be dull and uneven. The technique cannot be generally recommended.

▲ The corrosive fine dust of the calcium hypochlorite (bleaching powder) must not be inhaled, as it can severely irritate the respiratory system. The solution must be prevented from coming into contact with the eyes, which would be severely irritated.

Cotton-wool, moistened with the solution, is applied to the surface of the object to be coloured, and left for a period of about twenty hours. It is important to ensure that the cotton-wool is in close contact with the surface, and that it remains moist throughout the period of treatment. When treatment is complete, the object is removed and washed thoroughly in cold water. After a period of drying in air, it may be wax finished.

4.203   Blue/greenish-blue variegated patina on dark ochre ground Semi-matt

Copper nitrate	100 gm
Water	200 cm³
Ammonia (.880 solution)	400 cm³
Acetic acid (6% solution)	400 cm³
Ammonium chloride	100 gm

Cotton-wool technique (Twenty to thirty hours)

A very variegated surface is produced, with a finish that is smooth to the touch.

▲ Preparation and colouring should be carried out in a well ventilated area. Ammonia vapour and some acetic acid vapour will be liberated, which cause irritation of the eyes and respiratory system.

▲ The strong ammonia solution must be prevented from coming into contact with the eyes and skin, as it will cause burns or severe irritation.

The copper nitrate is dissolved in the water and the ammonia gradually added. The acetic acid is then added, followed by the ammonium chloride. Cotton-wool is wetted with the solution and applied to the surface of the object to be coloured. It is then left for a period of about twenty to thirty hours. The cotton-wool should be kept moist throughout the period of treatment. When treatment is complete, the object is washed in cold water and allowed to dry in air. When dry, it is wax finished.

4.204* Ochre patches on mid-brown Semi-matt   *Pl. VII*

Ammonium carbonate	150 gm
Copper acetate	60 gm
Sodium chloride	50 gm
Potassium hydrogen tartrate	50 gm
Water	1 litre

Cotton-wool technique (Ten to twenty hours)

The surface is etched back by the solution and acquires a mid-brown colour. Some less-etched patches tend to occur, which become ochre in colour.

Cotton-wool which has been moistened with the solution is applied to the surface of the article to be coloured. It is important to ensure that the cotton-wool is in full contact with the surface. It is left for a period of from ten to twenty hours, during which time the cotton-wool should remain moist. Colour development should be monitored with a sample, so that the article remains undisturbed. When treatment is complete, the article is removed and washed in cold water. After being allowed to dry in air for several hours, it may be wax finished.

4.205 Variegated dark brown/black (cloudy and uneven) Matt

Ammonium carbonate	300 gm
Acetic acid (10% solution)	35 cm³
Water	1 litre

Cotton-wool technique (Twenty hours)

An uneven cloudy smooth black surface tends to be produced, which is broken to reveal some buff ground.

The ammonium carbonate is added to the water and the acetic acid then added gradually, while stirring. Cotton-wool, moistened with this solution, is applied to the surface of the object, and left for a period of about twenty hours. It is important to ensure that the cotton-wool is in full contact with the surface, and that it remains moist throughout. When treatment is complete, the object is washed in cold water and allowed to dry in air. When completely dry, it may be wax finished.

4.206 Blue-green patina on variegated brown ground Semi-matt   *Pl. VIII*

Ammonium sulphate	90 gm
Copper nitrate	90 gm
Ammonia (.880 solution)	1 cm³
Water	1 litre

Cloth technique (Twenty hours)

The disposition of colour tends to correspond to the texture of the applied cloth.

Tests carried out involving direct application of this solution were not successful. The surface stubbornly resists 'wetting', even with additions of wetting agents to the solution.

Soft cotton cloth which has been soaked with the solution is applied to the surface of the object, and stippled into place with a stiff bristle-brush. The object is then left for a period of about twenty hours. The cloth should be removed when it is very nearly dry, and the object then left to dry in air without washing. The blue-green patina tends to develop during the drying period. (If a patina-free ground colour is required, the cloth should be removed at an earlier stage when it is still wet, and the object thoroughly washed. The cloth application may be repeated, for a more developed ground colour.) When treatment is complete and the surface thoroughly dried out, it may be wax finished.

4.207 Blue-green patina on dark grey ground Semi-matt

Ammonium carbonate	20 gm
Oxalic acid (crystals)	10 gm
Acetic acid (6% solution)	1 litre

Cloth technique (Twenty-four hours)

A dark grey ground is produced, which tends to take on the texture of the applied cloth. Cloudy areas of blue-green patina develop on this ground.

If the cloth dries completely on the surface, it adheres, damaging the surface when removed.

A variety of tests were carried out using this solution. Immersion, at temperatures from cold to boiling, produced no effect, even after prolonged immersion. Direct application of the cold or hot solution by dabbing and wiping also produced no effect, as the surface resists wetting.

▲ The ammonium carbonate should be added to the acetic acid in small quantities to avoid violent effervescence.

▲ Oxalic acid is harmful if taken internally. Contact with the eyes and skin should be avoided.

Strips of soft cotton cloth, or other absorbent material, are soaked with the solution. These are applied to the surface of the object and stippled into place using the end of a stiff brush which has been moistened with the solution. They are left on the surface until very nearly dry, which may take up to twenty-four hours, and then removed. If patina development is to be encouraged, the surface is then moistened by wiping the solution on using a soft cloth. If the ground colour and a minimum of patina is required, then the surface is allowed to dry without applying any more solution. When the surface is completely dry, it may be wax finished.

5 Yellow brass (rolled sheet)

Two grades of yellow brass were tested: yellow brass with the composition copper 70%, zinc 30% (CZ 106. B.S. 2870) and muntz brass with the composition copper 60%, zinc 40% (CZ 108. B.S. 2870).

The main colour heading for each recipe refers to the results obtained on a polished yellow brass surface, unless otherwise indicated, and includes a reference to both the colour and surface quality. Results obtained on muntz brass and on satin finished surfaces, where these are significantly different, are given in the marginal notes.

	Recipe numbers
Red Reddish-purple/reddish-brown	5.1 – 5.16
Pink Pink-purple/pink-brown	5.17 – 5.33
Brown Light bronzing/brown/dark brown	5.34 – 5.96
Black Black/dark slate/grey	5.97 – 5.111
Green patina Blue-green/green/blue	5.112 – 5.141
Lustre colours	5.142 – 5.163
Special techniques	5.164 – 5.201

5.1*   Orange-red Semi-gloss   *Pl. IX*

Copper sulphate	120 gm
Ammonia (.880 solution)	30 cm³
Water	1 litre

Boiling immersion (Twenty minutes)

A dense precipitate is formed which tends to sediment in the immersion vessel, causing 'bumping'. Glass vessels should not be used.

▲ Preparation of the solution, and colouring, should be carried out in a well ventilated area to prevent irritation of the eyes and respiratory system by the ammonia vapour. The skin, and particularly the eyes, should be protected from accidental splashes of this corrosive alkali.

Immersion in the boiling solution produces lustre colours after one minute, followed by the gradual change to a more opaque orange-pink colour which is well developed after seven or eight minutes. Continued immersion produces a darkening of this to orange and the gradual development of an orange-red colour. After twenty minutes the object is removed and washed in hot water. After drying in sawdust it is wax finished.

5.2*   Reddish-purple/purple Semi-matt   *Pl. IX, Pl. X, Pl. XV*

Copper sulphate	25 gm
Ammonia (.880 solution)	3–5 cm³
Water	1 litre

Boiling immersion (Fifteen to thirty minutes)

Increasing the ammonia content beyond the stated amount tends to inhibit colour development. At a rate of 10 cm³/litre ammonia, with the given concentration of copper sulphate, no colour is produced.

Plate IX shows the effect on muntz brass (Cu 60:40 Zn); plate X shows the effect on yellow brass (Cu 70:30 Zn). Plate XV shows the effect on yellow brass after a longer immersion, to 50 minutes.

The article is immersed in the boiling solution, which produces a series of pink/green lustre colours after about one minute. Continued immersion causes a darkening of the surface to produce brown colours, which may be uneven initially, but which become more even in the later stages of immersion. When the purple colour has developed, which may take from fifteen to thirty minutes, the article is removed and washed in hot water. After drying in sawdust, it is wax finished.

5.3*   Reddish-purple Semi-gloss/semi-matt

| Copper nitrate | 200 gm |
| Water | 1 litre |

Cold immersion (Thirty-six hours)

The colour obtained appears to be particular to this grade of yellow brass (Cu 70:30 Zn). Tests carried out with brasses having both higher and lower proportions of zinc produced purplish-brown colours.

The article must be absolutely grease-free and thoroughly clean. The cold solution does not 'bite' readily, and streaks and patches will occur if there is any surface contamination.

Lighter-coloured bands may occur on an object that is suspended too close to the bottom of the immersion vessel, due to the settling of a slight turbidity in the solution.

For results obtained using this solution at higher temperatures see recipe 5.71.

The article is suspended in the cold solution, well clear of the bottom of the immersion vessel. The reddish-purple colouration develops very gradually and evenly during the period of immersion. After thirty-six hours the article is removed, washed thoroughly in cold water and dried in sawdust.

5.4*   Reddish-brown/purplish-brown Gloss   *Pl. X*

Copper acetate	70 gm
Copper sulphate	40 gm
Potassium aluminium sulphate	6 gm
Acetic acid (10% solution)	10 cm³
Water	1 litre

Boiling immersion (Thirty minutes)

Yellow brass (Cu 70:30 Zn) yields a reddish-brown, whereas brass with a higher proportion of zinc, muntz brass (Cu 60:40 Zn), produced a less bright purplish-brown colour.

The object is immersed in the boiling solution. The colour development is gradual and even throughout the period of immersion, producing purplish colours after about ten minutes. The object is removed after thirty minutes, washed thoroughly in hot water and dried in sawdust. When completely dry it is wax finished.

5.5   Dark purple/purplish-brown Semi-matt

| Copper acetate | 80 gm |
| Water | 1 litre |

Boiling immersion (Thirty to forty minutes)

In tests, a bluish superficial deposit tended to form, which was removed by lightly brushing the surface with a soft bristle-brush during immersion.

The article is immersed in the boiling solution, which initially produces lustre colours which gradually change to a more opaque brownish colour. This darkens gradually as immersion continues. After about thirty or forty minutes, when the colour has fully developed, the article is removed and washed in hot water. It is dried in sawdust and may be wax finished when dry.

5.6   Reddish-brown Semi-gloss

Copper acetate	50 gm
Copper nitrate	50 gm
Water	1 litre

Boiling immersion (Thirty minutes)

A richer colour is produced on yellow brass (Cu 70:30 Zn) than on muntz brass, which tends to develop a light mottled surface.

Immersing the object in the boiling solution produces lustrous golden colours after one minute, which gradually change to more opaque matt pink/orange tints after five minutes. Continued immersion produces a gradual darkening to a terracotta. After thirty minutes the object is removed, washed in hot water and dried in sawdust. When completely dry it is wax finished.

5.7*   Brownish-purple Gloss

Copper carbonate	160 gm
Sodium hydroxide	80 gm
Water	1 litre

Boiling immersion (Thirty minutes)

Similar results can be obtained using less concentrated solutions for correspondingly longer times, but the results tend to be less even. An even green lustre colour can be obtained with a brief immersion in a quarter-strength solution.

▲ Sodium hydroxide is a powerful caustic alkali, and the solid or solutions must be prevented from coming into contact with the eyes and skin. The sodium hydroxide, in small quantities, should be added to the water and *not* vice-versa. Colouring should be carried out in a well ventilated area, as inhalation of the caustic vapours should be avoided.

The sodium hydroxide is added to the water in small quantities, while stirring. Heat is evolved as it dissolves, warming the solution. When it has completely dissolved, the copper carbonate is added and the solution heated to boiling. The object is immersed in the boiling solution, and gradually acquires a pale grey colour which darkens to a greyish-mauve as immersion proceeds. After about thirty minutes the object is removed and dried in sawdust. When completely dry, it may be wax finished.

5.8   Reddish-brown Semi-gloss   *Pl. IX*

Copper sulphate	25 gm
Water	1 litre

Boiling immersion (Fifteen to twenty minutes)

The results obtained with muntz brass (Cu 60:40 Zn) were consistently darker and more purple than those obtained with yellow brass (Cu 70:30 Zn). Some dark olive patches tended to develop on muntz brass, particularly at the edges.

Tests with more concentrated solutions produced very similar results, and no improvement in surface quality.

The article is immersed in the boiling solution. After two or three minutes a brown colour develops which gradually darkens as immersion continues. After about fifteen or twenty minutes, when the full reddish-brown colour has developed, the article is removed and thoroughly washed in hot water. After drying in sawdust, it is wax finished.

5.9   Brownish-purple spotted with lighter brown Semi-matt

Copper nitrate	40 gm
Ammonia (.880 solution)	to form a paste

Applied paste (Several hours)

A true paste will not form, but the moist crystalline mass that is obtained can be applied with a brush. This tends to become watery and may not dry out completely.

The copper nitrate is ground using a pestle and mortar, and a very small amount of ammonia added by the drop to form a crystalline 'paste'. This is applied to the object using a soft brush, and the surface then allowed to dry for several hours. When dry, the residue is washed away with cold water. The process may be repeated if necessary. After a final washing and air-drying, the object may be wax finished.

5.10   Reddish-brown (slightly variegated) Semi-gloss

Copper nitrate	40 gm
Water	to form a paste

Applied paste (Four to six hours)

A generally even, watery reddish-brown colour is produced that is slightly variegated with lighter tones. In tests, repetition of the procedure was not required.

A true paste will not form, but the damp crystalline mass that is obtained can be applied with a brush. This tends to become watery and may not dry out.

The copper nitrate is ground using a pestle and mortar, and a very small amount of water added by the drop to form a crystalline 'paste'. The paste is applied using a soft brush, and the surface then allowed to dry for several hours. When dry, the residue is washed away with cold water. The process is repeated if necessary. After washing and air-drying, the object may be wax finished.

5.11   Purple-brown Gloss

Copper acetate	25 gm
Copper sulphate	19 gm
Water	1 litre

Hot immersion (Fifteen minutes)

The purple-brown is obtained on yellow brass (Cu 70:30 Zn). On brass with a higher proportion of zinc, eg muntz brass (Cu 60:40 Zn), a more dull grey-brown is produced.

The article is immersed in the hot solution (80˚C). The colour develops gradually to produce a light terracotta after about five minutes, which darkens evenly as immersion continues. After ten minutes the surface is orange-brown in colour. The article is removed after fifteen minutes immersion, washed thoroughly and dried in sawdust. When dry, the surface has a characteristic grey bloom which is invisible after wax finishing.

5.12   Red-brown Semi-gloss

Ammonium carbonate	50 gm
Copper acetate	25 gm
Ammonium chloride	5 gm
Oxalic acid (crystals)	1 gm
Acetic acid (6% solution)	1 litre

Boiling immersion (One hour)

Tests tended to result in streaky surfaces, which are difficult to avoid or correct. Shorter immersion times produce less streaking, and a more varied uneven surface colour. Prolonging the immersion time tends to increase the tendency for streaks to occur.

▲ When preparing the solution, the ammonium carbonate should be added in very small quantities. It reacts effervescently with the acid, and this reaction is very vigorous if large quantities are added.

The article is immersed in the boiling solution. After about five minutes a patchy orange-yellow colouration with traces of green appears. Continued immersion in the boiling solution darkens the colour, which appears to be best after about one hour. The article should then be removed, washed in warm water and dried in sawdust. When dry, it may be wax finished.

5.13   Reddish-pink Matt

A	Copper sulphate	25 gm
	Water	1 litre
B	Ammonium chloride	0.5 gm

Boiling immersion (A) (Fifteen minutes)
Boiling immersion (A+B) (Ten minutes)

The article is immersed in the boiling copper sulphate solution for about fifteen minutes or until the colour is well developed. It is then removed to a bath of hot water, while the ammonium chloride is added to the colouring solution, and then re-immersed. The colour is brightened and tends to become more red. After about ten minutes it is removed and washed in hot water. If the immersion is prolonged a bloom of greyish-green patina will tend to form, which may need to be removed with a soft bristle-brush. The article is dried in sawdust and finally wax finished.

5.14   Orange-brown Gloss

Copper sulphate	500 gm
Copper acetate	100 gm
Acetic acid (30% solution)	25 cm³
Water	1 litre

Hot scratch-brushing (Several minutes)

The colour produced has a slightly transparent quality, and tends to vary in tone across the surface.

▲ Strong acetic acid is corrosive and should not be allowed to come into contact with the eyes and skin.

The article is either immersed in the hot solution (80˚C) for about two minutes, or the solution may be applied liberally with a soft brush, to allow the colour to develop. The surface is then gently scratch-brushed with the hot solution until an even finish is produced. Relieved effects may be obtained by continued working, without the hot solution. The article is thoroughly washed in hot water, and either dried in sawdust, or allowed to dry in air. When dry it may be wax finished.

5.15   Pale purplish-brown Gloss

Copper acetate	7 gm
Ammonium chloride	3.5 gm
Acetic acid (6% solution)	100 cm³
Water	900 cm³

Boiling immersion (Twenty minutes)

The colour produced has a translucent and slightly watery quality. It is also very uneven in tone over the surface and tends to be overlaid with areas of lustre.

The variability and the lustrous overlay are more marked on muntz brass (Cu 60:40 Zn) although the colours produced are richer than on yellow brass (Cu 70:30 Zn).

The article is immersed in the boiling solution, and after about three minutes a slightly lustrous yellow colour develops. With continued immersion this colour becomes more opaque and darkens to a pale brown. After twenty minutes the article is removed, washed thoroughly in hot water and dried in sawdust. When dry it is wax finished.

5.16   Black on mottled orange-brown ground Matt

Potassium permanganate	5 gm
Copper sulphate	50 gm
Ferric sulphate	5 gm
Water	1 litre

Boiling immersion (Twenty minutes)

The black layer tends to be only partially adherent, revealing a mottled underlying colour. The black tends to occur as a variegation on this ground colour. If the surface is bristle-brushed when wet, after about five or ten minutes' immersion, the black can be completely removed, leaving the semi-gloss mottled ground.

The article is immersed in the boiling solution. A black layer forms on the surface, which should not be disturbed. Immersion is continued for about twenty minutes, and the article is then removed and allowed to dry in air, without washing. The surface should not be handled at this stage. After several hours drying, the article may be wax finished.

5.17*   Dark pink (slightly variegated) Matt

A	Copper sulphate	125 gm
	Ferrous sulphate	100 gm
	Water	1 litre
B	Ammonium chloride	1 gm

Boiling immersion (A) (Fifteen minutes)
Boiling immersion (A+B) (Five minutes)

The object must be thoroughly washed after immersion, to prevent the formation of a greenish bloom or patches of green. If there is a build-up of bloom during immersion, then this should be cleared by agitating the object, or by removing and rinsing in hot water if necessary.

If shorter immersion times are used before the addition of the ammonium chloride, then the results tend to be lighter and more glossy, but are more prone to unevenness.

The object is immersed in the boiling solution A for about fifteen minutes, until a brown colour has developed. The object is removed and rinsed in a tank of hot water, while the ammonium chloride is added to solution A. The object is re-immersed in the colouring solution for about five minutes, causing the surface to develop and brighten to a dark pink. It is then removed and washed very thoroughly in hot water. After drying in sawdust, it may be wax finished.

5.18   Pink with grey-white spotting Matt

Copper nitrate	35 gm
Ammonium chloride	35 gm
Calcium chloride	35 gm
Water	1 litre

Cold immersion (Fifteen minutes)

Some fine greyish-white spotting occurs on the pink etched surface as it dries out. This was particularly marked on the muntz brass (Cu 60:40 Zn) where a sparing amount of solution was allowed to dry on the surface. This also gave rise to watery-grey blotches, producing a greyish-white and watery-grey fringed mottling on the etched pink ground.

The object is immersed in the cold solution. A pink colour and a slight etched effect develop gradually on the surface within five minutes. After fifteen minutes the object is removed, washed in hot water, and allowed to dry in air. When completely dry the object is wax finished.

5.19   Cloudy grey on pink/brown Semi-matt

Copper (cupric) chloride	50 gm
Water	1 litre

Hot immersion (Ten minutes)

The results produced in tests were poor. The colour tended to be uneven and patchy and prone to spotting. The procedure cannot be generally recommended.

The object is immersed in the hot solution (70–80°C) producing etching effects and a pale salmon-pink colour. With continued immersion a denser and slightly more brown colour develops. After ten minutes, the object is removed and washed in hot water, removing the loose superficial layer. It is then allowed to dry in air, and exposed to daylight, which darkens the colour. When all colour change has ceased, which may take several days, the object may be wax finished.

5.20   Variegated light brown/grey on pink ground Matt

Copper sulphate	200 gm
Ferric chloride	25 gm
Water	1 litre

Hot immersion (Thirty minutes)

The surface is very variegated, producing some green and grey colouration, in addition to the noted colours, on yellow brass (Cu 70:30 Zn). Brass with a higher proportion of zinc eg muntz brass (Cu 60:40 Zn), acquires a mottled light grey and light brown on a pink ground, the surface quality tending to be better.

▲ Ferric chloride is corrosive and should be prevented from coming into contact with the eyes and skin.

The object is immersed in the hot solution (60°C), which etches the surface and produces a pale pink colour, typical of chloride containing solutions. Continued immersion produces a gradual change to a more brown colour. After thirty minutes the object is removed and washed in hot water, which tends to leach a slight yellowish-green colour from the surface. After thorough washing, the object is allowed to dry in air and exposed to daylight, which tends to cause an uneven darkening of the colour. When the surface development is complete and the object thoroughly dried out, which may take several days, it can be wax finished.

5.21   Variegated (pink and light brown) Semi-matt

Copper nitrate	20 gm
Sodium chloride	16 gm
Potassium hydrogen tartrate	12 gm
Ammonium chloride	4 gm
Water	to form a paste

Applied paste (Four hours)

▲ A nose and mouth face mask fitted with a fine dust filter should be worn when brushing away the dry residue.

The ingredients are ground to a paste with a pestle and mortar by the very gradual addition of a small quantity of cold water. The creamy paste is applied with a soft brush, and left to dry. After four hours the dry residue is removed with a bristle-brush. The object is washed in cold water and allowed to dry in air. Exposure to daylight will produce a slight darkening of the colours. When the colour has stabilised, the object may be wax finished.

5.22*   Mottled purple-pink Semi-matt   *Pl. X*

Copper nitrate	15 gm
Zinc nitrate	15 gm
Ferric chloride	0.5 gm
Hydrogen peroxide (100 vols.)	to form a paste

Applied paste (four hours)

The surface of the object should be handled as little as possible until dry, and should not be brushed during washing.

If the dry surface is brushed with a bristle-brush prior to wax finishing, some underlying ground colour may be exposed to produce a variegated pink and mottled ochre surface.

▲ Hydrogen peroxide should not be allowed to come into contact with combustible material. It should be prevented from coming into contact with the eyes and skin, as it may cause severe irritation. It is very harmful if taken internally. The vapour evolved as it is added to the solid ingredients of the paste must not be inhaled.

The ingredients are ground using a mortar and pestle, and the hydrogen peroxide added in small quantities to form a thin creamy paste. The paste is applied to the object with a soft brush, and left to dry in air. After four hours, the residue is washed away with cold water and the object is allowed to dry in air for several hours. When completely dry, it may be wax finished.

5.23   Red-brown/olive/blue-grey (variegated) Matt

Copper nitrate	200 gm
Sodium chloride	160 gm
Potassium hydrogen tartrate	120 gm
Ammonium chloride	40 gm
Water	1 litre

Cold immersion and scratch-brushing (Fifteen minutes)

The newly coloured surface is fragile while wet and handling should be avoided. It was found that the surface is particularly sensitive to household rubber gloves, which removed parts of the colour layer and caused staining.

This variegated red-brown surface could only be obtained on yellow brass (Cu 70:30 Zn). Muntz brass (Cu 60:40 Zn) darkened from an even light pink to a dark green patchy surface.

The results obtained in tests were very variable.

The object is immersed in the cold solution. The surface is immediately coloured and finely etched. After fifteen minutes the article is pink in colour, and is removed and scratch-brushed with the cold solution to even the surface and slightly lighten the colour. It is then left to dry in air without washing. Exposure to daylight will cause the surface to darken gradually to a red-brown variegated with olive and blue-grey. When the colour has stabilised, the object may be wax finished.

5.24   Pink plating and yellow-green (variegated) Semi-matt

Copper acetate	120 gm
Ammonium chloride	60 gm
Water	1 litre

Boiling immersion and scratch-brushing (Five minutes)

The variegated surfaces produced in tests were interesting, but the effects proved to be difficult to control and the results patchy. The method cannot therefore be generally recommended.

The pink layer is probably chemically plated copper, which when partially removed reveals a yellow-green colour to the underlying brass.

The object is immersed in the boiling solution, which etches the surface and produces a bright copper-pink colour typical with chloride solutions. After five minutes the object is removed and scratch-brushed with the hot solution, partially removing the uneven pink layer. The object is then washed, allowed to dry in air, and is wax finished when dry.

5.25   Variegated pink with grey spotting Matt

Copper sulphate	100 gm
Ammonium chloride	0.5 gm
Water	1 litre

Applied liquid to heated metal (Several minutes)

The object is gently heated on a hotplate and the solution applied either by brushing or with a soft cloth. The object should be heated to a temperature such that when the solution is applied, it sizzles gently. Application is continued until an overall pink colour is produced. Exposure to daylight will cause the surface to darken slightly to give a pink, variegated with darker areas. A scattering of fine grey spots will also appear. When colour changes have ceased, the object is wax finished.

5.26★   Pink and light brown (variegated) Semi-matt

Copper nitrate	200 gm
Zinc chloride	200 gm
Water	1 litre

Applied liquid (Twice a day for five days)

The solution is wiped on to the article with a soft cloth and allowed to dry in air. When it is completely dry, the surface is gently rubbed smooth with a soft dry cloth to remove any loose dust, and a further application of solution wiped on. This procedure is repeated twice a day for about five days. The solution takes effect immediately, giving rise to an etching effect and pale cream and pink colours typical with chloride solutions. These colours darken on exposure to daylight. When treatment is complete, the article is left to dry thoroughly in air, and when completely dry, it is wax finished.

5.27   Pink etch ('frosted') Matt

Copper acetate	6 gm
Copper sulphate	1.5 gm
Sodium chloride	1.5 gm
Water	1 litre

Boiling immersion (About one hour)

The article is immersed in the boiling solution and after a few minutes the surface is slightly etched and a frosted pink colour develops. The surface develops gradually as immersion continues, and after about one hour the article is removed and washed in hot water. It is dried in sawdust and may be wax finished when completely dry.

5.28   Pink etch ('frosted') Matt   *Pl. X*

Copper sulphate	6.25 gm
Copper acetate	1.25 gm
Sodium chloride	2 gm
Potassium nitrate	1.25 gm
Water	1 litre

Boiling immersion (About one hour)

If the surface develops unevenly then it should be bristle-brushed with pumice and the object then re-immersed.

▲ Potassium nitrate is a powerful oxidising agent and must not be allowed to come into contact with combustible materials.

The article is immersed in the boiling solution and after a few minutes the surface is slightly etched and a frosted pink colour develops. The surface develops gradually as immersion continues, and after about one hour the article is removed and washed in hot water. It is dried in sawdust and may be wax finished when completely dry.

5.29   Dull orange-ochre/pink Matt

Copper nitrate	200 gm
Sodium chloride	200 gm
Water	1 litre

Boiling immersion (Twenty minutes)

On yellow brass (Cu 70:30 Zn) a mottled matt dull rust colour is produced. On higher-zinc brasses, eg muntz brass (Cu 60:40 Zn), a variegated dull ochre and pink semi-matt surface is produced.

Tests carried out using less concentrated solutions and solutions containing small additions of ammonium chloride and potassium hydrogen tartrate produced similar but more patchy surfaces.

An immediate etching of the surface occurs when the article is immersed in the boiling solution, followed by the gradual development of an orange-ochre colour. After twenty minutes the article is removed, washed in hot water, and allowed to dry in air. Exposure to daylight will produce a slight darkening of the colour initially. Finally the article is wax finished.

5.30   Pink etch ('frosted') Matt

Copper sulphate	6.25 gm
Copper acetate	1.25 gm
Potassium nitrate	1.25 gm
Sodium chloride	2 gm
Sulphur (flowers of sulphur)	3.5 gm
Acetic acid (6% solution)	1 litre

Boiling immersion (About one hour)

▲ Potassium nitrate is a powerful oxidising agent and must not be allowed to come into contact with combustible materials. It must not be brought into contact with the sulphur in a dry state.

The article is immersed in the boiling solution and after a few minutes the surface is slightly etched and a frosted pink colour develops. The surface develops gradually as immersion continues, and after about one hour the article is removed and washed in hot water. It is dried in sawdust and may be wax finished when completely dry.

5.31   Olive-brown on pink ground Semi-matt

Copper acetate	40 gm
Copper sulphate	200 gm
Potassium nitrate	40 gm
Sodium chloride	65 gm
Flowers of sulphur	110 gm
Acetic acid (6% solution)	1 litre

Hot immersion and scratch-brushing (Five minutes)

The article is immediately etched on immersion in the hot solution (70–80°C). A pink/orange colour, typical with chloride-containing solutions, develops within two or three minutes. After five minutes the article is removed, and the uneven surface is scratch-brushed with the hot solution to improve the surface consistency. The article is then washed in hot water and allowed to dry in air. When dry the surface is exposed evenly to daylight, causing a darkening of the colour. When the colour has stabilised and no further signs of change can be detected, it is wax finished.

In tests carried out with yellow brass and muntz brass, it was found that the olive-brown colour developed very unevenly and tended to be adherent only in patches. It is difficult to control the technique sufficiently to produce a good surface, and the recipe cannot be generally recommended.

▲ Potassium nitrate is a powerful oxidising agent and should not be allowed to come into contact with combustible material. It must not be brought into contact with the sulphur in a dry state, but should be completely dissolved before the flowers of sulphur are added.

5.32   Greenish-brown on pink Semi-matt

A	Ammonium chloride	10 gm
	Copper sulphate	10 gm
	Water	1 litre
B	Hydrogen peroxide (100 vols.)	200 cm³
	Sodium chloride	5 gm
	Acetic acid (glacial)	5 cm³
	Water	1 litre

Cold immersion (A) (Five minutes)
Hot immersion (B) (A few seconds)

The object is immersed in solution A for about five minutes, producing a slight etching effect, and a pale pink colour typical of chloride-containing solutions. The object is then briefly immersed in solution B (at 60–70°C). The reaction tends to be violent, causing the solution to 'boil', and immersion should not be prolonged. A greenish-brown layer is produced. If this is uneven, the surface should be gently scratch-brushed and the object briefly re-immersed. Prolonged or repeated immersion in solution B will tend to strip the greenish-brown colour. When treatment is complete, the object is washed in warm water and allowed to dry in air, and may be wax finished when dry.

The surface finish is difficult to control, and the procedure cannot be generally recommended.

▲ Hydrogen peroxide can cause severe burns, and must be prevented from coming into contact with the eyes and skin. When preparing the solution, it should be added to the water prior to the addition of the other ingredients.

▲ Glacial acetic acid is very corrosive and must be prevented from coming into contact with the eyes or skin.

5.33   Pink and light brown (variegated) Semi-matt

Copper sulphate	25 gm
Ammonium chloride	25 gm
Lead acetate	25 gm
Water	1 litre

Hot immersion (Fifteen minutes)

The object is immersed in the hot solution (80°C) producing slight etching effects and pale orange/pink colours. After fifteen minutes, the object is removed, washed in hot water and allowed to dry in air. Exposure to daylight causes a darkening of the colour. When the colour has stabilised and the object is completely dry, it is wax finished.

The variegated surface could only be obtained on yellow brass (Cu 70:30 Zn). It is difficult to produce an adequate surface without localised stains. On muntz brass (Cu 60:40 Zn) a strong etching effect occurred producing a patchy spangled dull surface.

▲ Lead acetate is very toxic and every precaution should be taken to prevent inhalation of the dust during weighing and preparation of the solution. The vapour evolved from the hot solution must not be inhaled.

5.34*   Dull golden yellow Semi-matt

Sodium dichromate	150 gm
Nitric acid (10% solution)	20 cm³
Hydrochloric acid (15% solution)	6 cm³
Ethanol	1 cm³
Water	1 litre

Cold immersion (Two or three minutes)

The sodium dichromate is dissolved in the water, and the other ingredients are then added separately while stirring the solution. The article is immersed in the cold solution for a period of about three or four minutes, and is then removed and washed in cold water. It is allowed to dry in air, and may be wax finished when dry.

The golden colour tends to be fragile when wet, and should not be touched.

Prolonging the immersion tends to produce a fine satin finish to the barely-coloured metal.

▲ Sodium dichromate is a powerful oxidant, and must not be allowed to come into contact with combustible material. The dust must not be inhaled. The solid and solutions should be prevented from coming into contact with the eyes and skin, or clothing. Frequent exposure can cause skin ulceration, and more serious effects through absorption.

▲ Both hydrochloric and nitric acids are corrosive and should be prevented from coming into contact with the eyes, skin or clothing.

5.35   Yellow ochre (frosted surface) Matt

Ferric oxide	25 gm
Nitric acid (70% solution)	100 cm³
Iron filings	5 gm
Water	1 litre

Cold immersion (Fifteen minutes)

The ferric oxide is dissolved in the nitric acid (producing toxic fumes), and the mixture is added to the water. The iron filings are added last. The article is immersed in the solution, which produces an etched surface, and removed after about fifteen minutes. It is washed thoroughly in cold water and allowed to dry in air, and may be wax finished when dry.

A greenish-yellow colour, rather than yellow ochre, was obtained on muntz brass (Cu 60:40 Zn).

▲ Preparation and colouring should be carried out in a fume cupboard. Toxic and corrosive brown nitrous fumes are evolved which must not be inhaled.

▲ Concentrated nitric acid is extremely corrosive, causing burns, and must be prevented from coming into contact with the eyes and skin.

5.36*   Pale yellow (frosted surface) Semi-matt

Copper carbonate	100 gm
Ammonium chloride	100 gm
Ammonia (.880 solution)	20 cm³
Water	1 litre

Hot immersion (Fifteen minutes)

The article is immersed in the hot solution (50°C) producing immediate etching effects, typical with chloride-containing solutions. After about fifteen minutes, the article is removed and washed thoroughly in warm water. It is allowed to dry in air, and may be wax finished when dry.

▲ When preparing the solution, care should be taken to avoid inhaling the ammonia vapour. The eyes and skin should be protected from accidental splashes of the concentrated ammonia solution.

5.37   Grey-brown mottle or spotting on golden yellow Semi-gloss

A	Copper nitrate	20 gm
	Ammonium chloride	20 gm
	Calcium hypochlorite	20 gm
	Water	1 litre
B	Water	

Cold immersion (A) (Ten minutes)
Cold immersion (B) (Twenty hours)

The article is immersed in solution A for about ten minutes, which produces a brightening of the surface to a slightly golden colour. The article is rinsed and transferred to a container of cold water, in which it is left for a period of about twenty hours. Small spots or patches of a darker colour develop slowly during the period of cold water immersion. When the surface is sufficiently developed, the article is removed and either dried in sawdust or allowed to dry in air. When dry, it is wax finished. The darker areas tend to darken slightly on exposure to daylight.

The initial immersion in the colouring solution should not be prolonged beyond the light golden yellow stage, or the results will tend to be dull after cold water immersion.

▲ Care should be taken to avoid inhaling the fine dust that may be raised from the calcium hypochlorite (bleaching powder). It should also be prevented from coming into contact with the eyes or skin.

5.38   Dull greenish-yellow (slight spangled etch) Semi-matt

A	Copper sulphate	25 gm
	Ammonia (.880 solution)	30 cm³
	Water	1 litre
B	Potassium hydroxide	16 gm
	Water	1 litre

Hot immersion (A) (Ten minutes)
Hot immersion (B) (One minute)

The article is immersed in the hot copper sulphate solution (60°C) for about ten minutes, producing a slight etching effect and a darkening of the surface to a greenish-brown colour. The article is then transferred to the hot potassium hydroxide solution (60°C) for about one minute. It is removed, washed in warm water and allowed to dry in air. When dry it is wax finished.

A fine spangled surface is produced, which is marred by the streakiness of the colours. The technique cannot be generally recommended.

▲ Preparation of solution A, and colouring, should be carried out in a well ventilated area, to prevent irritation of the eyes and respiratory system by the ammonia vapour liberated.

▲ Potassium hydroxide, in solid form and in solution, is corrosive to the skin and particularly the eyes. Adequate precautions should be taken.

5.39   Grey spotting on dull metal surface Semi-matt

Copper acetate	20 gm
Sodium tetraborate (borax)	5 gm
Potassium nitrate	5 gm
Mercuric chloride	2.5 gm
Olive oil	to form a paste

Applied paste (Two or three days)

The ingredients are ground to a thin paste with a little olive oil, using a pestle and mortar. The paste is applied to the object, either with a brush or a soft cloth, and is allowed to dry for several hours. When dry, the object is brushed with a bristle-brush to remove the dry residue. This procedure is repeated as necessary. When treatment is complete, the surface is again brushed, and then cleaned with pure distilled turpentine. When this has dried, it is wax finished.

The colouring effect is very slight.

▲ Mercuric chloride is a highly toxic substance. It is essential to prevent any contact with the skin, and to prevent any chance of inhaling the dust. Gloves and masks must be worn. A nose and mouth face mask fitted with a fine dust filter must be worn when brushing away the dry residue.

▲ Potassium nitrate is a powerful oxidising agent and should not be allowed to come into contact with combustible material.

▲ In view of the limited colouring potential, and the high risks involved, this method cannot be recommended.

5.40*   Light bronzing Satin/semi-matt

Sodium sulphide	30 gm
Sulphur (flowers of sulphur)	4 gm
Water	1 litre

Hot immersion and scratch-brushing (Fifteen minutes)

The solids are added to about 200 cm³ of the water, and boiled until they dissolve and yield a clear deep orange solution. This is added to the remaining water and the temperature adjusted to 60–70°C. The object is immersed in the hot solution and the colour allowed to develop for about ten minutes. It is then removed and scratch-brushed with the hot solution, until the surface is even and the colour developed. The object should be washed thoroughly in hot water and dried in sawdust. When dry, it may be wax finished.

A pale warm golden brown bronzing with a near satin surface is produced. The colour is light, but a generally good surface quality was obtained in tests.

▲ Sodium sulphide in solid form and in solution must be prevented from coming into contact with the eyes and skin, as it can cause severe burns. Some hydrogen sulphide gas may be evolved as the solution is heated. Colouring should be carried out in a well ventilated area.

5.41   Pale golden bronzing Satin/semi-matt

Potassium sulphide	10 gm
Ammonium chloride	10 gm
Water	1 litre

Hot scratch-brushing (Five minutes)

The object is scratch-brushed with the hot solution (50–60°C) for several minutes. The solution is then progressively diluted with hot water as scratch-brushing continues. It should finally be gently burnished by scratch-brushing with hot water. After thorough rinsing and drying in air, the object may be wax finished.

The results produced are very pale. Tests carried out using different concentrations of the ingredients, and also using cold solutions produced paler results. Immersion without scratch-brushing produced patchy lustre colours only. Very similar results were obtained with a solution of potassium sulphide and sodium chloride.

▲ The procedure should be carried out in a well ventilated area, as some hydrogen sulphide gas is liberated from the solution.

5.42   Dull golden bronzing Semi-matt

Potassium sulphide	2 gm
Ammonium sulphate	0.5 gm
Water	1 litre

Cold scratch-brushing (Several minutes)

The object is scratch-brushed with the cold solution, alternating with hot and then cold water. This process is continued until the colour has developed. Any relieving that is required should then be carried out. The object is washed thoroughly in cold water and allowed to dry in air. When completely dry it is wax finished.

Imperfections in the bronzing are difficult to remedy, even with prolonged working, particularly on yellow brass (Cu 70:30 Zn).

5.43   Pale golden bronzing Semi-matt

Sodium sulphide	50 gm
Antimony trisulphide	50 gm
Water	1 litre
Acetic acid (10% solution)	100 cm³

Hot scratch-brushing (Several minutes)

The solid ingredients are added to the water and dissolved by gradually raising the temperature to boiling. The solution is allowed to cool to 60°C and then acidulated gradually with the acetic acid, producing a dense brown precipitate. The object is scratch-brushed with this turbid solution, alternating with hot and then cold water. When the colour has developed it is washed in cold water, allowed to dry in air, and wax finished when dry.

Immersion in the hot solution, without scratch-brushing, produces a bright golden brown on directionally polished satin surfaces, but only patchy finishes on polished surfaces.

▲ Sodium sulphide is corrosive and must be prevented from coming into contact with the eyes and skin.

▲ Addition of the acetic acid, and colouring, should be carried out in a well ventilated area, as toxic hydrogen sulphide gas will be evolved.

5.44   Pale greyish bronzing Semi-matt

Ammonium sulphide (16% solution)	100 cm³
Water	1 litre

Cold immersion and scratch-brushing (Ten to thirty minutes)

The article is immersed in the cold solution for a few minutes to allow the colour to develop, and then removed and lightly scratch-brushed with the cold solution, alternating with hot and cold water. It is then re-immersed for about ten minutes and then removed and scratch-brushed as before. This procedure is repeated until the desired colour is achieved. The article is finally lightly scratch-brushed alternately with hot and cold water. It is allowed to dry in air and wax finished.

It is difficult to obtain an overall even surface. The scratch-brush tends to 'drag' on the surface, causing patches which are difficult to remove.

As an alternative to immersion in the cold solution, the solution can be applied continuously to the surface with a soft brush. This tended to produce lighter bronzed finishes in tests.

▲ Ammonium sulphide solution liberates a vapour containing ammonia and hydrogen sulphide gas, which should not be inhaled. Preparation and colouring must be carried out in a well ventilated area. The solution must be prevented from coming into contact with the eyes and skin, as it causes burns or severe irritation.

5.45*   Light brown bronzing Semi-matt/semi-gloss   Pl. X

Antimony trisulphide	12.5 gm
Sodium hydroxide	40 gm
Water	1 litre

Hot scratch-brushing (Several minutes)

The hot solution (60°C) is repeatedly applied liberally to the surface of the object with a soft brush, allowing time for the colour to develop. The surface is then worked lightly with the scratch-brush, while applying the hot solution and hot water alternately. The treatment is concluded by lightly burnishing the surface with the scratch-brush and hot water. After rinsing it is dried in sawdust, and wax finished when dry.

▲ Sodium hydroxide is a powerful caustic alkali which can cause severe burns to the eyes and skin, in the solid form or in solution. When preparing the solution, small quantities of the sodium hydroxide should be added to a large quantity of water and *not* vice-versa. When the sodium hydroxide has completely dissolved, the antimony trisulphide is added.

5.46   Light reddish-brown Satin/semi-gloss

Antimony trisulphide	30 gm
Ferric oxide	10 gm
Ammonium sulphide (16% solution) to form a paste	

Applied paste (sulphide paste) (Several minutes)

The ingredients are mixed to a thick paste using a pestle and mortar, by the very gradual addition of a small quantity of ammonium sulphide solution. The paste is applied to the whole surface of the article to be coloured, using a soft bristle-brush. A stiff nylon brush is then used to work the surface gradually in small areas. The nylon brush should be intermittently dipped in a mixture of the dry ingredients of the paste (in the same proportion as above), and the surface worked in this way until the paste becomes dry and falls from the surface. When the whole surface is dry and completely free of any residue or dust, it may be burnished with a brass scratch-brush and wax finished.

A variegated surface can be obtained by brushing the paste onto the whole surface and allowing it to dry for about three hours, until it is superficially dry and cracking, and then burnishing it away with a nylon brush or a brass scratch-brush. This may be repeated, as necessary, to darken the colour.

▲ Ammonium sulphide solution is very corrosive and precautions should be taken to protect the eyes and skin while the paste is being prepared. It also liberates harmful fumes of ammonia and of hydrogen sulphide, and the preparation should therefore be carried out in a fume cupboard or well ventilated area. The paste once prepared is relatively safe to use but should be kept away from the skin and eyes.

▲ A nose and mouth face mask fitted with a fine dust filter should be worn when the residues are brushed away. The dust produced in this case contains antimony compounds, which are toxic.

5.47   Purplish-brown Satin/semi-gloss

Antimony trisulphide	20 gm
Ferric oxide	20 gm
Potassium polysulphide	2 gm
Ammonia (.880 solution)	2 cm³
Water	to form a paste (approx 4 cm³)

Applied paste (sulphide paste) (Several minutes)

The potassium polysulphide is ground using a pestle and mortar and the other solid ingredients added and mixed. These are made up to a thick paste by the gradual addition of the ammonia and a small quantity of water. The paste is applied to the whole surface of the object to be coloured, using a soft bristle-brush. A stiff nylon brush is then used to work the surface gradually in small areas. The nylon brush should be intermittently dipped in a mixture of the dry ingredients of the paste (in the same proportion as above), and the surface worked in this way until the paste becomes dry and falls from the surface. When the whole surface is dry and completely free of any residue or dust, it may be burnished with a brass scratch-brush and wax finished.

A variegated surface can be obtained by brushing the paste on to the whole surface and allowing it to dry for about three hours, until it is superficially dry and cracking, and then burnishing it away with a nylon brush or a brass scratch-brush. This may be repeated as necessary to darken the colour.

▲ A nose and mouth face mask should be worn when the residues are brushed away.

5.48   Light brown Semi-gloss

Sodium thiosulphate	90 gm
Potassium hydrogen tartrate	30 gm
Zinc chloride	15 gm
Water	1 litre

Hot immersion (Thirty minutes)

A transparent light brown colour is produced, which tends to be uneven.

The article is immersed in the hot solution (50–60°C) which darkens the surface and gradually produces a bloom, which should be removed by bristle-brushing. Continued immersion gives rise to a brown colour, which slowly darkens. After about thirty minutes, when the colour has developed, the article is removed and washed in warm water. It is either dried in sawdust or allowed to dry in air, and may be wax finished when dry.

5.49   Variegated pale bronzing Semi-gloss/semi-matt

Ferric oxide	30 gm
Lead dioxide	15 gm
Water	to form a paste

Applied paste and heat (Repeat several times)

The colour develops slowly, with a gradual increase in the light brown as the procedure is repeated. An orange-brown colour is produced in surface features such as punched marks or textured areas.

▲ Lead dioxide is toxic and very harmful if taken internally. The dust must not be inhaled. A nose and mouth face mask with a fine dust filter must be worn when brushing away the dry residue.

The ingredients are mixed to a paste with a little water. The paste is applied to the surface with a soft brush, while the object is gently heated on a hotplate or with a torch. When the paste is completely dry, it is brushed off with a stiff bristle-brush. The procedure is repeated a number of times, until the colour has developed. After the final brushing, the object may be wax finished.

5.50   Orange-brown Gloss

Copper sulphate	100 gm
Ammonium chloride	0.5 gm
Water	1 litre

Hot immersion (Five to ten minutes)

Extending the immersion time beyond ten minutes tends to produce a darker and more matt layer, which is difficult to remove completely. It tends to 'cling', producing dull gritty patches.

The article is immersed in the hot solution (70–80°C) producing a pale pink-orange colouration within two or three minutes. This darkens with continued immersion. When an orange-brown colour has developed, after about five or ten minutes, the article is removed and washed in warm water. It is dried in sawdust. Exposure to daylight will cause a slight initial darkening of the colour. When the colour is stable, the article is wax finished.

5.51   Reddish-brown (variegated) Semi-gloss   *Pl. IX*

Ferric nitrate	10 gm
Water	1 litre

Torch technique

A darker variegation can be obtained by finally dabbing with a stronger solution. (Ferric nitrate 40 gm/litre.) Plate IX shows this effect.

▲ The fumes evolved as the solution is applied to the heated metal should not be inhaled. Adequate ventilation must be provided and a nose and mouth face mask worn, fitted with the correct filter.

▲ Ferric nitrate is corrosive and should not be allowed to come into contact with the eyes or skin. A face mask or goggles should be worn to protect the eyes from hot splashes as the solution is applied.

The object is heated with a blow torch and the solution applied sparingly with a scratch-brush, using small circular motions and working across the surface. The metal gradually darkens to a reddish-brown and the surface can be finished by lightly dabbing with a cloth dampened with the solution. When treatment is complete and the object has been allowed to cool and dry thoroughly, it may be wax finished.

5.52   Greyish variegation on dull metal surface (satin surface only) Semi-matt

Copper nitrate	20 gm
Zinc sulphate	30 gm
Mercuric chloride	30 gm
Water	1 litre

Applied liquid (Several minutes)

The noted colour could only be obtained on a satin surface. On highly polished faces, only patchy stains on a dull metal surface could be produced. A more yellow-brown ground colour was produced on muntz brass (Cu 60:40 Zn) than on yellow brass (Cu 70:30 Zn).

▲ Mercuric chloride is very toxic and must be prevented from coming into contact with the skin and eyes. The dust must not be inhaled. If taken internally, very serious effects are produced.

▲ In view of the high risks involved, and the very limited colouring potential, this solution cannot be generally recommended.

The solution is wiped on to the surface of the object with a soft cloth, producing a slightly adherent grey-black layer. Continued rubbing with the cloth clears this layer and produces a surface which is evenly coated with mercury. This tends to be non-adherent when dry. The solution is then applied to the surface by dabbing with a soft cloth, and allowed to dry in air. The procedure may be repeated if necessary, omitting the initial wiping. When the colour has developed, the object is washed in cold water and allowed to dry in air. When dry it may be wax finished.

5.53   Yellowish-brown (slight petrol lustre) Semi-matt

Copper sulphate	25 gm
Lead acetate	25 gm
Water	1 litre

Hot immersion (Ten to fifteen minutes)

The article is immersed in the hot solution (80°C) producing green lustre colours after about one minute, which change gradually to produce an opaque orange after about six or seven minutes. This colour darkens with continued immersion. After ten or fifteen minutes the article is removed and washed thoroughly in hot water. After drying in sawdust, it is wax finished.

Small additions of ammonium chloride (at a rate of 2.5 gm/litre), introduced into the solution at a late stage in the immersion, produce a brightening of the colours. In this case however the colours thus produced darkened patchily on exposure to daylight, and the variation cannot be recommended.

▲ Lead acetate is highly toxic, and every precaution should be taken to prevent inhalation of the dust during weighing, and preparation of the solution. Colouring should be carried out in a well ventilated area, as the vapours evolved must not be inhaled.

5.54   Mid-brown Semi-gloss

Barium chlorate	52 gm
Copper sulphate	40 gm
Water	1 litre

Hot immersion (Fifteen minutes)

Immersion in the hot solution (70°C) produces an immediate golden lustre colour, which changes gradually to pink/green and then to a brownish lustre after about three or four minutes. Continued immersion produces a greater opacity of colour and a gradual development to a yellow-brown or orange brown colour. When the colour has fully developed, after fifteen minutes, the object is removed and washed in hot water. It is wax finished, after thorough drying in sawdust.

The lustre colours produced in the early stages of immersion are generally even, but are less bright than the equivalent colours produced by other methods.

▲ Barium compounds are poisonous and harmful if taken internally. Barium chlorate is a strong oxidising agent and should not be allowed to come into contact with combustible material. It should be prevented from coming into contact with the eyes, skin or clothing.

5.55   Orange-brown Gloss

Sodium hydroxide	100 gm
Sodium tartrate	60 gm
Copper sulphate	60 gm
Water	1 litre

Hot immersion (Forty minutes)

The object is immersed in the hot solution (60°C) producing dull lustre colours after about one minute. The surface gradually becomes more opaque and the colour slowly turns brown as immersion continues. After about forty minutes, when the colour has developed, the object is removed, washed in hot water and either dried in sawdust or allowed to dry in air. It may be wax finished when dry.

An even green lustre colour is produced after one minute in the hot solution. A mottled yellowish-brown surface with a greenish lustrous tinge is produced after about ten minutes. The final opaque orange-brown colour is even, and is not tinged with lustre.

▲ Sodium hydroxide is a powerful caustic alkali. The solid or solutions can cause severe burns. It must be prevented from coming into contact with the eyes or skin. When preparing the solution, small quantities of the sodium hydroxide should be added to large quantities of water and not vice-versa. The solution will become warm as the sodium hydroxide is dissolved. When it is completely dissolved, the other ingredients are added.

5.56   Pale yellowish-green bronzing Satin/semi-matt

Copper acetate	80 gm
Water	1 litre

Hot immersion and scratch-brushing (Thirty to forty minutes)

The object is immersed in the hot solution (50–60°C) and the surface colour allowed to develop through a series of lustrous colours, to a brownish colour. After thirty minutes it is removed and scratch-brushed with the hot solution, which initially removes some of the colour formed during immersion. Scratch-brushing is continued for several minutes until the surface colour has developed. The object is then washed in warm water and allowed to dry in air. When completely dry it is wax finished.

Tests carried out using the technique of direct scratch-brushing, without prior immersion, produced poor results with little colour development.

5.57   Light grey-brown Semi-matt

A	Potassium nitrate	100 gm
	Water	1 litre
B	Copper sulphate	25 gm
	Ferrous sulphate	20 gm
	Acetic acid (glacial)	1 cm³

Boiling immersion (Thirty minutes)

The object is immersed in solution A for about fifteen minutes, producing a slight brightening of the surface. The object is removed to a bath of hot water, while the copper sulphate, ferrous sulphate and acetic acid are separately added to solution A. The object is re-immersed in the boiling solution, which gradually produces a greyish-brown colour, for about fifteen minutes. It is washed in hot water, allowed to dry in air and when dry wax finished.

Even light grey-brown colours were obtained on yellow brass (Cu 70:30 Zn), in tests. A rather thin and tenuous colour layer tended to be produced on muntz brass (Cu 60:40 Zn).

▲ Potassium nitrate is a powerful oxidant and must be prevented from coming into contact with combustible materials.

▲ Glacial acetic acid is very corrosive and must be prevented from coming into contact with the eyes and skin.

5.58   Light greyish bronzing   Semi-matt

A	Copper sulphate	50 gm
	Water	1 litre
B	Potassium sulphide	12.5 gm
	Water	1 litre

Hot immersion and scratch-brushing (A) (A few minutes)
Cold immersion and scratch-brushing (B)

The article is immersed in the copper sulphate solution for about thirty seconds and transferred to the cold potassium sulphide solution, and immersed for about one minute. It is then removed and scratch-brushed with each solution in turn. The surface is finished with a light scratch-brushing in the potassium sulphide solution. The article is washed thoroughly in warm water, left to dry in air, and wax finished.

A dark brown colour tends to develop as the solution is applied, which is scratched by the brush back to the brass. The surface requires gentle working up until it begins to have a satin rather than a glossy appearance.

▲ Colouring should be carried out in a well ventilated area, as some hydrogen sulphide gas will be liberated.

5.59*   Brown (or variegated black on brown)   Semi-matt   Pl. X

Potassium permanganate	2.5 gm
Copper sulphate	50 gm
Ferric sulphate	5 gm
Water	1 litre

Boiling immersion (Fifteen minutes)

The article is immersed in the boiling solution, and the surface becomes coated with a superficial dark brown film, which should be removed by bristle-brushing in a bath of hot water. The article is re-immersed in the boiling solution and again becomes coated with dark brown.

If a plain brown surface is required, the brown film should be removed every few minutes as immersion continues. After about fifteen minutes the article is removed and washed thoroughly in hot water. After drying in air it may be wax finished.

If a variegated surface is required, then the brown film should be left on the surface as immersion continues. After fifteen minutes, the article is removed, and allowed to dry in air for a few minutes. The surface is then gently brushed with a dry bristle-brush, which will partially remove the brown layer, leaving a variegated surface. The article should be wax finished.

The solution should not be boiled for any length of time, prior to the immersion of the object, or the dark layer tends not to be deposited on the surface.

5.60   Reddish-ochre   Semi-matt

Copper sulphate	50 gm
Ferrous sulphate	5 gm
Zinc sulphate	5 gm
Potassium permanganate	2.5 gm
Water	1 litre

Boiling immersion (Thirty minutes)

The object is immersed in the gently boiling solution. A dark brown layer forms on the surface, which should not be disturbed. Immersion is continued for fifteen minutes, after which time the object is removed and the dark layer washed away in hot water using the bristle-brush. The object is then re-immersed for a further fifteen minutes and then removed and bristle-brushed gently with hot water. After thorough rinsing it is dried in sawdust, and wax finished when completely dry.

Lighter colours with a more green bias can be produced using this solution. See recipe 5.61.

5.61   Yellowish-green   Semi-gloss/semi-matt

Copper sulphate	50 gm
Ferrous sulphate	5 gm
Zinc sulphate	5 gm
Potassium permanganate	2.5 gm
Water	1 litre

Boiling immersion (Ten to fifteen minutes)

The object is immersed in the boiling solution, and becomes coated with a dark brown layer. After one or two minutes the object is removed and the dark layer brushed away under hot water, using a bristle-brush. It is then re-immersed for two minutes, and the bristle-brushing again carried out when the dark brown layer has re-formed. After two or three such immersions and bristle-brushings, the brown layer will tend not to be adherent when it re-forms, and can be removed by agitating the object while in the solution. Immersion is continued to ten or fifteen minutes, after which the object is removed and given a final bristle-brushing under hot water. It is thoroughly rinsed in warm water, dried in sawdust and wax finished when completely dry.

The yellowish-green was only obtained on yellow brass (Cu 70:30 Zn). On brass with a higher proportion of zinc, muntz brass (Cu 60:40 Zn), a reddish-brown colour was produced, which was prone to streaking.

Darker colours with a reddish bias can be produced using this solution, if the technique is modified. See recipe 5.60.

5.62   Dull mid-brown Semi-matt

Copper sulphate	25 gm
Nickel ammonium sulphate	25 gm
Potassium chlorate	25 gm
Water	1 litre

Boiling immersion (Twenty minutes)

▲ Nickel salts are a common cause of sensitive skin reactions.

▲ Potassium chlorate is a powerful oxidising agent and should not be allowed to come into contact with combustible material. The solid and solutions should not be allowed to come into contact with the eyes, skin or clothing.

Immersing the object in the boiling solution produces an immediate lustrous mottling of the surface. Continued immersion induces a change to a more even orange colour which darkens gradually. After twenty minutes the object is removed and washed in hot water. After thorough drying in sawdust it is wax finished.

5.63★   Mottled or marbled brown on orange-brown Semi-matt   *Pl. X*

Sodium chlorate	100 gm
Copper nitrate	25 gm
Water	1 litre

Boiling immersion (Forty minutes to one hour)

The mottling tends to increase and darken with longer immersion times. After one hour a grey-green colour begins to tinge the surface and develops slowly to a patchy grey-green patina.

▲ Sodium chlorate is a powerful oxidant and must not be allowed to come into contact with combustible material or acids. It should also be prevented from coming into contact with the eyes, skin or clothing.

The object is immersed in the boiling solution, and the surface is quickly coloured with a dark lustre, which develops to a thin yellowish-orange colour. Continued immersion produces a gradual change to a more opaque orange-red or orange-brown colour. When the colour has developed evenly and to the required depth, after about forty minutes to one hour, the object is removed and washed in hot water. It is dried in sawdust, and may be wax finished when dry.

5.64★   Ochre Gloss   *Pl. IX*

'Rokusho'	3.25 gm
Copper sulphate	3.25 gm
Potassium aluminium sulphate	2 gm
Water	1 litre

Boiling immersion (About one hour)

For description of 'Rokusho' and its method of preparation, see appendix 1.

The article is immersed in the boiling solution and after a few minutes the surface darkens and becomes more opaque as the colour develops. The surface develops gradually as immersion continues, and after about one hour the article is removed and washed in hot water. It is dried in sawdust and may be wax finished when completely dry.

5.65   Ochre Gloss   *Pl. IX*

'Rokusho'	5 gm
Copper sulphate	5 gm
Water	1 litre

Boiling immersion (About one hour)

For description of 'Rokusho' and its method of preparation, see appendix 1.

Increasing the copper sulphate content to 20–30 gm/litre produces darker and more tan brown colours. The sample shown in plate IX was prepared using this more concentrated solution.

The article is immersed in the boiling solution and after a few minutes the surface darkens and becomes more opaque as the colour develops. The surface develops gradually as immersion continues, and after about one hour the article is removed and washed in hot water. It is dried in sawdust and may be wax finished when completely dry.

5.66   Mid-brown (Semi-matt)

Potassium chlorate	50 gm
Copper sulphate	25 gm
Water	1 litre

Hot immersion (Fifteen or twenty minutes)

Tests showed that there is a tendency to some colour variation at the edges of sheet material, and around surface features such as punched marks.

▲ Potassium chlorate is a powerful oxidising agent and should not be allowed to come into contact with combustible material. Contact with the eyes, skin and clothing must be avoided.

▲ A nose and mouth face mask fitted with a fine dust filter should be worn when brushing away any dry powder from the surface.

The object is immersed in the hot solution (80°C) and is initially coloured with an uneven golden lustre finish which changes to a more opaque yellow-green. Continued immersion darkens the colour to orange and then finally to a more brown colour. After fifteen or twenty minutes, when the full colour has developed, the object is removed and washed in hot water. When it has been thoroughly dried in sawdust, it is wax finished. A green powdery film may develop when the sample is dried, which can be brushed away prior to waxing. The wax finish tends to make any residual green bloom invisible.

5.67   Mid-brown Semi-gloss

Copper sulphate	125 gm
Sodium acetate	12.5 gm
Water	1 litre

Boiling immersion (Ten to fifteen minutes)

After one minute of immersion in the boiling solution a slightly lustrous orange/pink colour develops on the surface of the object. This gradually darkens tending to become brown as immersion continues. When the colour has developed, after ten or fifteen minutes, the object is removed and washed in hot water. It is wax finished after thorough drying in sawdust.

An even mid-brown colour was produced on both yellow brass (Cu 70:30 Zn) and muntz brass (Cu 60:40 Zn). During the later stages of immersion the muntz brass samples tested tended to become more opaque and gradually acquired an even matt grey layer on the brown ground. This did not occur with yellow brass.

5.68   Brownish-yellow Semi-gloss/semi-matt

Copper sulphate	100 gm
Copper acetate	40 gm
Potassium nitrate	40 gm
Water	1 litre

Boiling immersion (Ten minutes)

The article is immersed in the boiling solution, which produces lustrous green colours after one minute. These gradually darken and become more opaque with continued immersion. The article is removed when the colour has developed, after about ten minutes, and washed in hot water. After thorough drying in sawdust, it is wax finished.

Similar rather drab colours were produced on both yellow brass and muntz brass. In tests, it was also found that both metals were prone to streaks of a slightly darker colour.

▲ Potassium nitrate is a powerful oxidising agent and must not be allowed to come into contact with combustible material.

5.69   Purplish-brown (gritty surface spots) Semi-matt/matt

Copper sulphate	6.25 gm
Copper acetate	1.25 gm
Potassium nitrate	1.25 gm
Water	1 litre

Boiling immersion (About one hour)

The article is immersed in the boiling solution and after a few minutes the surface darkens and becomes more opaque as the colour develops. The surface develops gradually as immersion continues, and after about one hour the article is removed and washed in hot water. It is dried in sawdust and may be wax finished when completely dry.

Coarse gritty spots tend to occur evenly on the surface as the colour develops.

▲ Potassium nitrate is a powerful oxidising agent and must not be allowed to come into contact with combustible materials.

5.70   Purplish-brown (gritty surface spots) Semi-matt/matt

Copper sulphate	8 gm
Copper acetate	3.25 gm
Potassium aluminium sulphate	2 gm
Water	1 litre
Acetic acid (6% solution)	20 cm³

Boiling immersion (About one hour)

The article is immersed in the boiling solution and after a few minutes the surface darkens and becomes more opaque as the colour develops. The surface develops gradually as immersion continues, and after about one hour the article is removed and washed in hot water. It is dried in sawdust and may be wax finished when completely dry.

Coarse gritty spots tend to occur evenly on the surface as the colour develops.

5.71*   Dark purplish-brown Semi-matt/semi-gloss

Copper nitrate	200 gm
Water	1 litre

Boiling immersion (Twenty minutes)

The article is immersed in the boiling solution. A greenish tinge appears slowly during the first few minutes, which gradually darkens to a brown colour as immersion continues. A drab grey layer subsequently forms. After twenty minutes the article is removed and washed thoroughly in hot water to clear the grey layer and reveal the brown colour beneath. After thorough washing and drying in sawdust the surface shows a characteristic grey bloom, which becomes almost invisible when the article is wax finished.

Brasses containing higher proportions of zinc (eg Cu 60:40 Zn, muntz) yield colours that are closer to a dull slate grey, and are subject to mottling and greater surface variation.

Tests carried out with various less concentrated solutions of copper nitrate, produced drab mid-tan colourations on all grades of brass. Similar results were obtained with concentrated hot solutions at temperatures below boiling.

Prolonged immersion in a concentrated cold solution produced distinctive results. See recipe 5.3.

5.72*   Mid-brown/dark brown Gloss

Copper nitrate	80 gm
Water	1 litre
Ammonia (.880 solution)	3 cm³

Boiling immersion (Thirty minutes)

The article is immersed in the boiling solution. After five minutes a pink tinge is apparent, which subsequently becomes more greenish-yellow and develops during the later stages of immersion to a brown colour. After thirty minutes the article is removed and washed thoroughly in hot water before drying in sawdust. The dry surface has a greyish bloom which becomes nearly invisible after wax finishing.

Tests produced similar results on both yellow brass (Cu 70:30 Zn) and muntz brass (Cu 60:40 Zn). There is a tendency for mottling and streaking to occur.

Increasing the ammonia content to 5–10 cm³ per litre produced paler and more translucent surfaces with a slight dull lustre and dark surface marks and irregular streaks.

5.73   Orange-brown Semi-gloss/gloss

Ammonium acetate	50 gm
Copper acetate	30 gm
Copper sulphate	4 gm
Ammonium chloride	0.5 gm
Water	1 litre

Boiling immersion (Seven to ten minutes)

After one or two minutes of immersion in the boiling solution, the object takes on an olive colouration, which gradually darkens to a more ochre colour as immersion continues. The full colour tends to develop in about seven minutes, although it may take a little longer. Prolonging the immersion tends to cause the colour to become more pallid, and does not improve surface consistency. The object is removed and washed in hot water. After thorough drying in sawdust, it is wax finished.

In tests, the colour produced on yellow brass (Cu 70:30 Zn) was darker than that produced on muntz brass (Cu 60:40 Zn), which tended to be more ochre in colour.

5.74*   Dark olive (green patina on dark brown ground) Semi-matt   *Pl. IX*

Potassium chlorate	50 gm
Copper sulphate	25 gm
Nickel sulphate	25 gm
Water	1 litre

Boiling immersion (Thirty to forty minutes)

After two or three minutes' immersion in the boiling solution an orange colour develops on the surface of the object, which gradually darkens with continued immersion. The full colour develops in about thirty minutes, together with a green tinge which begins to appear after about twenty minutes. If this green colour is to be kept to a minimum, the object should be removed after thirty minutes. Colour development and surface quality are generally better with a longer immersion, the green tending to enhance the surface. The object is removed after about forty minutes, washed in warm water, and either dried in sawdust or allowed to dry in air. When dry, it is wax finished.

The longer immersion time produces a surface in which the green patina is integrated with the brown ground, to produce a dark olive green colour. With shorter immersion times a slightly more glossy brown surface, tinged with green, is produced.

▲ Potassium chlorate is a powerful oxidising agent and must not be allowed to come into contact with combustible material. The solid or solutions must be prevented from coming into contact with the eyes, skin or clothing.

▲ Nickel compounds are a common cause of sensitive skin reactions. Precautions should be taken to prevent inhalation of the dust, during preparation of the solution.

5.75   Brown and dark brown (variegated) Semi-matt

Copper acetate	30 gm
Ammonium chloride	15 gm
Ferric oxide	30 gm
Water	1 litre

Hot immersion. Hot scratch-brushing (Ten to fifteen minutes)

The solution is brought to the boil and then removed from the source of heat. The object is then immersed, and becomes coated with a thick brown layer. After ten minutes' immersion the object is removed and gently heated on a hotplate, the surface being worked with the scratch-brush. More solution is applied and worked in with the scratch-brush, occasionally allowing the surface to dry off without scratch-brushing. When the colour has developed it is allowed to dry on the hotplate for a few minutes, removed and the dry residue removed with a stiff bristle-brush. When the surface is free of residue it is wax finished.

An alternative technique which is more suited to larger objects is to heat locally with a torch and apply the pasty liquid with the scratch-brush, without prior immersion.

▲ A nose and mouth face mask fitted with a fine dust filter should be worn when brushing off the dry residue.

5.76   Dark brown (variegated with lighter areas) Semi-matt

Copper sulphate	30 gm
Potassium permanganate	6 gm
Water	1 litre

Boiling immersion and scratch-brushing (Several minutes)

On immersion in the boiling solution, the object becomes coated with a brownish film, which develops to a dark purplish-brown after one minute. This brown film is loosely adherent and sloughs off gradually to reveal uneven underlying grey-brown colours. After three or four minutes the object is removed and gently scratch-brushed with the hot solution. Overworking will cut through to the underlying metal. If the surface is unsatisfactory then the process is repeated. When the desired colour has been obtained, the object is washed in hot water, dried either in sawdust or in air, and wax finished when dry.

The results of tests carried out with both yellow brass and muntz brass were variable, in both the depth of colour obtained and the quality of surface produced.

The development of the loose brown layer appears to be essential to successful colouring. If the solution is boiled for too long before the object is immersed, then a more dense precipitate is formed in the solution, and coating does not take place.

5.77   Dark orange-brown (slight petrol lustre) Semi-gloss

Copper sulphate	60 gm
Copper acetate	10 gm
Potassium aluminium sulphate	25 gm
Water	1 litre

Boiling immersion (Thirty minutes)

The object is immersed in the boiling solution, producing a gradual darkening of the surface to a drab ochre/brown colour after ten minutes. A slight lustre may also be apparent. Continued immersion produces further darkening, although colour development is slow. After thirty minutes the object is removed, washed in hot water and dried in sawdust. When dry, it is wax finished.

A darker purplish-brown was obtained on muntz brass (Cu 60:40 Zn), which also had a slight petrol lustre.

5.78   Reddish-brown Semi-matt

Copper sulphate	125 gm
Ferrous sulphate	100 gm
Acetic acid (glacial)	6.5 cm³
Water	1 litre

Boiling immersion (Thirty to forty minutes)

The object is immersed in the boiling solution. The surface darkens gradually during the course of immersion, initially tending to be drab and rather uneven, but later becoming more even and generally darker and richer in colour. When the colour has developed, after thirty to forty minutes, the object is removed and washed in hot water. It is dried in sawdust, and wax finished when dry.

A darker and more dull purplish-brown is produced on muntz brass (Cu 60:40 Zn).

There is a tendency for matt patches to occur, particularly with prolonged immersion.

▲ Acetic acid in the highly concentrated 'glacial' form is corrosive, and must not be allowed to come into contact with the skin, and more particularly the eyes. It liberates a strong vapour which is irritating to the respiratory system and eyes.

5.79   Brown (slightly variegated) Semi-matt

Copper sulphate	25 gm
Sodium chloride	100 gm
Potassium polysulphide	10 gm
Water	1 litre

Hot immersion (Fifteen minutes)

The object is immersed in the hot solution (60°C), producing a slight matting of the surface due to etching, and the gradual development of dull brown colours. After about fifteen minutes, when the brown has developed, the object is removed. Relieving may be carried out at this stage, using the hot solution with the scratch-brush. The object is then removed washed in warm water and allowed to dry in air. When dry, it is wax finished.

The resulting colour is rather dull and the surface tends to be uneven.

▲ The colouring should be carried out in a well ventilated area, as some hydrogen sulphide gas will be liberated.

5.80   Reddish-brown Semi-matt

Copper nitrate	80 gm
Nitric acid (10% solution)	100 cm³
Water	1 litre

Hot immersion (Five minutes)

The article is immersed in the hot solution (60–70°C), which causes etching of the surface and a gradual darkening. After five minutes it is removed and washed in warm water, and allowed to dry in air. When dry, it may be wax finished

(Some sources suggest using this as a ground for a greenish patina, produced by applying a solution of ammonium carbonate (100 gm/litre) by wiping with a soft cloth and allowing to dry. Tests for extended periods produced only the slightest traces of a green tinge.)

A dull brown colour was produced on yellow brass (Cu 70:30 Zn). Slightly lighter and partly non-adherent films tended to be obtained on muntz brass (Cu 60:40 Zn).

▲ Nitric acid is corrosive and may cause severe irritation or burns. It should be prevented from coming into contact with the eyes or skin.

5.81   Mid-brown Gloss

A	Ammonium carbonate	180 gm
	Copper sulphate	60 gm
	Acetic acid (6% solution)	1 litre
B	Oxalic acid	1 gm
	Ammonium chloride	0.25 gm
	Acetic acid (6% solution)	500 cm³

Hot immersion   (Fifteen minutes)

Solution A is boiled and allowed to evaporate gradually until its volume is reduced by half. Solution B is then added and the mixture heated to bring the temperature to 80°C. The article is immersed in the hot solution, the colour gradually darkening until it becomes brown after ten minutes. After fifteen minutes, when the colour has fully developed, the article is removed and rinsed carefully in warm water. The surface should not be disturbed at this stage. It is allowed to dry in air, and wax finished when completely dry.

A mid-brown with a slightly yellow bias was obtained with yellow brass (Cu 70:30 Zn). A mid-brown with a slightly red bias was produced on muntz brass (Cu 60:40 Zn).

▲ The ammonium carbonate should be added to the acetic acid in small quantities. The reaction is effervescent, and if large amounts are added the effervescence is very vigorous.

▲ Preparation of the solution, and colouring, should be carried out in a well ventilated area. Acetic acid vapour will be liberated when the solutions are boiled.

▲ Oxalic acid is harmful if taken internally, and irritating to the eyes and skin.

5.82*   Dark reddish-brown Gloss   Pl. X

Ammonium carbonate	60 gm
Copper sulphate	30 gm
Oxalic acid (crystals)	1 gm
Acetic acid (6% solution)	1 litre

Hot immersion   (Ten minutes)

The object is immersed in the hot solution (80°C) producing light brownish colours after two or three minutes, which may be uneven initially. With continued immersion the colours become even and darken gradually. When the colour has fully developed, after about ten minutes, the object is removed and washed in hot water. After thorough drying in sawdust, it may be wax finished.

Richer colours tended to be produced on muntz brass (Cu 60:40 Zn) than yellow brass (Cu 70:30 Zn). Plate X shows the effect on muntz brass.

▲ Oxalic acid is poisonous and may be harmful if taken internally.

▲ The ammonium carbonate should be added to the acetic acid in small quantities. The reaction is effervescent, and if large amounts are added the effervescence becomes very vigorous.

5.83*   Dark brown Gloss   Pl. IX

Ammonium carbonate	50 gm
Copper acetate	25 gm
Acetic acid (6% solution)	1 litre

Boiling immersion   (Several minutes)

The object is immersed in the boiling solution, and the surface gradually darkens to a reddish-brown colour. After several minutes, when the colour has fully developed, the object is removed and washed in warm water. After careful drying in sawdust, it may be wax finished.

In tests, a reddish brown was produced after three or four minutes, which became a dark brown after five minutes. After seven or eight minutes a slight purple bias to the dark brown is produced.

Slightly richer colours were produced on muntz brass (Cu 60:40 Zn) than on yellow brass (Cu 70:30 Zn), at the reddish-brown stage.

Tests involving the addition of 5 gm of ammonium chloride and 2 gm of oxalic acid per litre, to the solution when the colours have developed, produced poor results. The colour initially becomes lighter but is then subject to streaking and etching effects, producing very patchy matt yellow surfaces. The variation cannot be recommended. Very similar results were obtained with small additions of zinc chloride or copper chloride.

▲ The ammonium carbonate should be added to the acetic acid in small quantities. The reaction is effervescent, and if large amounts are added, the effervescence is very vigorous.

5.84   Variegated dark brown/black on dark brown ground   Semi-matt

Potassium polysulphide	4 gm
Water	1 litre

Torch technique

The metal is heated with a blow-torch and the solution applied sparingly with a scratch-brush using small circular motions and working across the surface until a matt black colour appears. The amounts of heat and solution used are critical. If too much solution is applied then the surface tends to strip, and if too little heat is used then the solution does not act on the metal. When the matt black colour has been obtained, it should be lightly brushed with a dry scratch-brush. This brings up the surface and highlights any faults, which can then be worked on with further heating and applications of solution. When treatment is complete, the object is allowed to cool and dry thoroughly, and may then be wax finished.

▲ The fumes evolved as the solution is applied to the heated metal should not be inhaled. Adequate ventilation must be provided.

▲ A face mask or goggles should be worn to protect the eyes from hot splashes as the solution is applied.

5.85   Brown/dark brown Gloss

Copper carbonate	200 gm
Ammonia (.880 solution)	250 cm³
Water	750 cm³

Cold immersion (Fifteen to thirty minutes)

The ammonia is added to the water, followed by the carbonate, which will partially dissolve. Some excess carbonate should remain in suspension. The article is suspended in the cold solution. The colour develops slowly throughout the period of immersion. When the desired depth of colour has been reached, the article is removed and washed in cold water. It should be dried carefully in fine sawdust, and wax finished when dry.

To obtain a black colour using this solution see recipe 5.97.

▲ This procedure should only be carried out where there is adequate ventilation. Ammonia vapour will be freely liberated when the solution is prepared, and in colouring, which can severely irritate the eyes and respiratory system. The skin, and particularly the eyes, must be protected against accidental splashes from this corrosive alkali.

5.86★   Dark brown Semi-matt

Nickel sulphate	35 gm
Copper sulphate	25 gm
Potassium permanganate	5 gm
Water	1 litre

Hot immersion (Five to thirty minutes)

The object is immersed in the hot solution (80–90°C) and the surface immediately becomes covered with a dark brown layer. After four or five minutes it is removed and gently bristle-brushed with hot water. The dark layer is removed revealing a brown 'bronzing' beneath. If darker colours are required then the process should be repeated.

If, on the other hand, a full very dark brown/black colour with a more matt finish is required, the object should be re-immersed, and the dark brown layer will then re-form and should be left on the surface as immersion is continued. After thirty minutes the object is removed, and gently bristle-brushed under hot water. After thorough rinsing, and drying in sawdust, it is wax finished.

▲ Nickel compounds are a common cause of sensitive skin reactions. Precautions should also be taken to avoid inhalation of the dust when preparing the solution.

5.87   Dark brown (slightly variegated) Semi-matt/matt

Sodium thiosulphate	60 gm
Copper sulphate	42 gm
Potassium hydrogen tartrate	22 gm
Water	1 litre

Hot immersion (Fifteen minutes)

After one minute of immersion in the solution (at 50–60°C), a yellow-orange colour forms on the surface of the article and gradually turns to bright green. The article is removed from the solution after fifteen minutes and allowed to dry in air, without washing in water. When the article is completely dry it is gently rubbed with a soft cloth, which removes the green layer, revealing the dark brown. The surface should not be overworked at this stage, or some of the brown layer may be removed, revealing patches of a reddish colour. The article is finally wax finished.

The green layer is superficial, and easily removed, but may leave a slight green tinge to the underlying colour.

5.88   Reddish-brown (greenish-grey bloom) Semi-matt

Copper sulphate	120 gm
Potassium chlorate	60 gm
Water	1 litre

Hot immersion (Ten or fifteen minutes)

The object is immersed in the hot solution (70–80°C). A rapid series of lustre colours is produced, becoming more opaque and darkening to an orange-buff within two or three minutes. Continued immersion produces a gradual darkening to a dark brown colour. After about ten minutes a greenish layer begins to be deposited on the object, as a dense precipitate is formed in the solution. If this is not required then the surface should be gently bristle-brushed in warm water, and the object re-immersed. After about fifteen minutes the object is removed and washed. It should be bristle-brushed under warm water if the greenish layer has re-formed. After thorough washing it is dried in sawdust and wax finished when dry. A greyish or greenish-grey bloom tends to remain.

A dense precipitate is formed as immersion progresses, causing 'bumping'. Glass vessels should not be used.

▲ Potassium chlorate is a powerful oxidising agent and must not be allowed to come into contact with combustible material. Contact with the eyes, skin or clothing must be avoided.

5.89   Brown Semi-matt/matt

Copper sulphate	125 gm
Sodium chlorate	50 gm
Water	1 litre

Boiling immersion (Five to ten minutes)

The article is immersed in the boiling solution, which produces pink/green lustre colours very rapidly. These darken and become more opaque with continued immersion. When the orange-brown colour has developed, after about five or ten minutes, the article is removed and washed thoroughly in hot water. After drying in sawdust, it is wax finished.

Prolonging the immersion will darken the colour, but will also cause the deposition of a grey-green and very uneven layer. This layer is difficult to make even, and hard to remove when immersion is complete.

▲ Sodium chlorate is a powerful oxidising agent and must not be allowed to come into contact with combustible materials or acids. Inhaling the dust must be avoided, and the eyes and skin should be protected from contact with either the solid or solutions.

5.90   Brownish-orange Semi-gloss/gloss

Copper sulphate	120 gm
Potassium chlorate	60 gm
Water	1 litre

Hot immersion (Five or six minutes)

The object is immersed in the hot solution (70–80°C), rapidly developing green lustre colours, which quickly become more opaque after about a minute. Continued immersion darkens the colour to a brownish-orange. After five or six minutes the object is removed and washed thoroughly in hot water. It may be wax finished after thorough drying in sawdust.

Prolonging the immersion produces further darkening to a deep brown colour. See recipe 5.88.

▲ Potassium chlorate is a powerful oxidising agent and must not be allowed to come into contact with combustible materials. Contact with the eyes, skin or clothing must be avoided.

5.91   Pink and yellowish-green Semi-matt

Copper nitrate	80 gm
Water	1 litre
Ammonia (.880 solution)	3 cm^3

Cold application (Ten days)

The solution is applied to the metal surface by dabbing with a soft cloth. It should be applied sparingly and may require vigorous dabbing in the initial stages to ensure adequate wetting. After each application the surface is allowed to dry completely in air. This procedure is repeated twice a day for ten days. The article should be allowed to dry for several days before wax finishing.

This treatment produces an irregularly mixed surface of pink and yellowish-green on yellow brass (Cu 70:30 Zn). On brasses with a higher proportion of zinc eg muntz brass (Cu 60:40 Zn) an even yellowish-green mottled surface is produced, which is similar to that obtained on gilding metal using this method.

5.92*   Reddish-brown/orange-brown (slightly variegated) Semi-matt

| Copper nitrate | 200 gm |
| Water | 1 litre |

Applied liquid (Twice a day for five days)

The solution is initially dabbed on vigorously with a soft cloth, to ensure that the surface is evenly wetted. Subsequent applications should be wiped on, and used very sparingly, to inhibit the formation of green patina. After each application the article is left to dry in air. This procedure is repeated twice a day for about five days. The article is then left to dry thoroughly in air, and wax finished when completely dry.

A reddish brown variegated with lighter browns and grey was produced in tests carried out on yellow brass (Cu 70:30 Zn). A more even mid-brown colour was produced in tests with muntz brass (Cu 60:40 Zn).

5.93   Dark-brown variegated with light grey-green Semi-gloss

Potassium chlorate	50 gm
Copper sulphate	25 gm
Ferrous sulphate	25 gm
Water	1 litre

Boiling immersion (Twenty minutes)

The object is immersed in the boiling solution. The colour develops gradually during immersion, becoming darker and tending to acquire a bloom of greenish-grey patina during the latter stages. If the bloom becomes excessive it should be removed by gently bristle-brushing the surface. When the required colour has developed, after about twenty minutes, the object is removed and washed in hot water. It is dried in sawdust, and may be wax finished when dry.

An orange colour develops first, and becomes brown after about ten minutes, this darkens and acquires patches of green. This effect is only produced on yellow brass (Cu 70:30 Zn); an even orange-brown colour is produced on muntz brass (Cu 60:40 Zn).

▲ Potassium chlorate is a powerful oxidant and must not be allowed to come into contact with combustible materials. It should also be prevented from coming into contact with the eyes, skin or clothing.

5.94   Green on reddish-brown Semi-matt

Copper sulphate	125 gm
Sodium chlorate	50 gm
Ferrous sulphate	25 gm
Water	1 litre

Boiling immersion (Thirty minutes)

The object is immersed in the boiling solution. Orange colours develop after a few minutes and gradually darken with continued immersion. After about fifteen minutes a greenish layer begins to be deposited on the now brown ground colour. Agitating the object within the solution tends to inhibit the deposition of the green layer. After thirty minutes the object is removed and washed thoroughly in hot water. After drying in sawdust, it is wax finished.

Tests produced very inconsistent results with yellow brass (Cu 70:30 Zn), which tended to yield patchy surfaces.

If the ground colour only is required, then the object should be intermittently bristle-brushed in hot water, during immersion. A residual greenish-grey bloom tends to remain.

▲ Sodium chlorate is a powerful oxidising agent and must not be allowed to come into contact with combustible material or acids. Inhalation of the dust must be avoided. The eyes and skin should be protected from contact with either the solid or solutions.

5.95   Dark grey and light olive green (variegated) Semi-matt

Copper acetate	40 gm
Copper sulphate	200 gm
Potassium nitrate	40 gm
Sodium chloride	65 gm
Acetic acid (6% solution)	1 litre

Cold immersion and scratch-brushing (Ten minutes)

The object is suspended in the cold solution, producing an immediate etching effect. After five minutes a pale cream or pink colour develops unevenly. After ten minutes the object is removed and scratch-brushed with the cold solution to even the surface. It should then be allowed to dry in air without washing. Any wet areas on the surface should be very gently brushed out with a soft brush, to prevent streaking or pooling. When dry, the object is exposed evenly to daylight, causing a darkening of the colours. When the colour has stabilised the object is wax finished.

The surface produced after exposure to daylight is very patchy and uneven. It does not appear to be possible to control either the degree of variegation or the surface quality.

▲ Potassium nitrate is a powerful oxidising agent and should not be allowed to come into contact with combustible material.

5.96   Variegated olive/orange-brown Semi-matt

Copper nitrate	20 gm
Zinc sulphate	30 gm
Mercurous chloride	30 gm
Water	1 litre

Applied liquid (Twice a day for ten days)

The solution is initially applied by vigorous dabbing with a soft cloth, to overcome resistance to wetting. Once some surface effect has been produced, the solution may be applied by wiping. The surface is allowed to dry completely after each application, which should be carried out twice a day for at least ten days. If the solution is used very sparingly then a patina-free ground colour tends to be produced. A more liberal use of the solution during the later stages of treatment will tend to encourage the formation of a blue-green patina. When treatment is complete, the article should be allowed to dry in air for several days before wax finishing.

Slightly streaky and uneven surfaces tend to be obtained. Some dull pink patches occurred on yellow brass (Cu 70:30 Zn).

In tests, the blue-green patina was found to be difficult to produce.

▲ Mercurous chloride may be harmful if taken internally. There are dangers of cumulative effects.

5.97*   Blue-black/black Gloss   *Pl. IX*

Copper carbonate	125 gm
Ammonia (.880 solution)	250 cm³
Water	750 cm³

Warm immersion (One hour)

The ammonia is added to the water, followed by the carbonate, which will partially dissolve. Some excess carbonate should remain in suspension. The solution is heated to about 50°C, and the object is immersed. The colour develops slowly during the period of immersion. After about one hour, when the colour has fully developed, the object is removed and washed thoroughly in warm water. It should be carefully dried in sawdust, and wax finished when dry.

▲ This procedure should only be carried out where there is adequate ventilation. Ammonia vapour will be freely liberated when the solution is prepared, and during colouring, which can severely irritate the eyes and respiratory system. The skin, and particularly the eyes, must be protected against accidental splashes of this corrosive alkali.

5.98*   Black/dark slate-grey   Gloss/semi-gloss   *Pl. X*

Sodium thiosulphate	250 gm
Water	1 litre
Copper sulphate	80 gm
Water	1 litre

Cold immersion   (Two hours)

The dark colours produced tend to have a slightly grey or cloudy appearance, although they are generally even.

The lustre colours produced during immersion tend to be less clear than those obtained with some other thiosulphate solutions.

The chemicals should be separately dissolved, and the two solutions then mixed together. The article is suspended in the cold solution, and after ten minutes' immersion shows signs of darkening. After twenty minutes a reddish-purple lustre appears and gradually changes to a blue-green lustre after thirty minutes. The sequence of lustre colours is repeated, becoming more opaque and producing a maroon colour after about one hour. With continued immersion the colour slowly darkens to black. After about two hours the article is removed, washed thoroughly and allowed to dry in air. When dry it may be wax finished.

5.99   Black/dark slate   Semi-gloss

Sodium thiosulphate	6.25 gm
Ferric nitrate	50 gm
Water	1 litre

Hot immersion   (One minute)

During preparation, the nitrate solution is darkened by the addition of the thiosulphate, but slowly clears to give a straw-coloured solution.

Higher temperatures should not be used. Tests carried out at 70–80°C caused stripping of the surface colour and strong etching effects.

A succession of lustre colours is rapidly produced when the article is immersed in the hot solution (50–60°C), changing to a more opaque purplish colour after about forty-five seconds. The colour darkens quickly and the article is removed after about one minute. After thorough washing it is allowed to dry in air, and then wax finished.

5.100   Black   Matt   *Pl. IX*

A	Copper sulphate	50 gm
	Water	1 litre
B	Sodium thiosulphate	50 gm
	Lead acetate	12.5 gm
	Water	1 litre

Hot immersion (A)   (One minute)
Hot immersion (B)   (Two minutes)

Tests carried out with yellow brass (Cu 70:30 Zn) produced an even overall black colour, with some evidence of more tenuous adherence at the edges. On muntz brass (Cu 60:40 Zn) the results tended to be only partly adherent, but producing a good variegated black and yellow brown surface. Plate IX shows this effect.

▲ Lead acetate is a highly toxic substance, and every precaution should be taken to prevent inhalation of the dust during preparation. The solution is very harmful if taken internally. Colouring should be carried out in a well ventilated area. The vapour evolved from the hot solution must not be inhaled.

The object is immersed in the hot copper sulphate solution (80°C) for one minute, producing a darkening of the surface. It is then transferred to solution B (80°C) for two minutes' immersion. These alternating immersions are repeated a number of times until an even black layer is produced on the surface. The object is then washed in warm water and allowed to dry in air. When dry, it is either rubbed gently with a soft cloth or bristle-brushed, and wax finished.

5.101   Black   Semi-gloss

Sodium thiosulphate	50 gm
Ferric nitrate	12.5 gm
Water	1 litre

Hot immersion   (Twenty minutes)

If the temperature is too high the colour layer will be patchy and more fragile when removed from solution, revealing a tan colour beneath the black.

The article is immersed in the hot solution (60–70°C) and after about one minute the surface is coloured with a purple-blue lustre, which gradually recedes. After five minutes a brown colour slowly appears which changes to a slate grey with continued immersion. After about twenty minutes the article is removed, washed in hot water and allowed to dry in air. The surface may be fragile at this stage and should be handled as little as possible. When completely dry the article is wax finished.

5.102 Grey patches on brown lustre Semi-gloss

Sodium thiosulphate	65 gm
Water	1 litre
Antimony trichloride	until cloudy

Hot immersion (Thirty minutes)

The sodium thiosulphate is dissolved in the water and the antimony trichloride added by the drop, to form a white precipitate. The solution begins to become orange and cloudy at the top, and then gradually throughout. The solution is heated to 50°C and when that temperature is reached, the object is immersed. The surface initially passes through the lustre cycle: golden after three minutes, purple after four minutes, pale metallic blue after ten minutes. After about fifteen minutes grey/white streaks appear on the surface, which darkens to produce an appearance like a black powdery layer. The article is removed after thirty minutes and washed gently in warm water, removing the powdery black deposit and leaving an uneven dark surface. After drying in air or sawdust, it may be wax finished.

A dull brownish lustrous surface, patched with cloudy grey matt areas, is produced on yellow brass (Cu 70:30 Zn). A similar poor surface quality is produced on muntz brass (Cu 60:40 Zn), but the underlying lustre is darker.

▲ Antimony trichloride is poisonous if taken by mouth. It is also irritating to the skin, eyes and respiratory system, and contact must be avoided. The vapour evolved from the hot solution must not be inhaled.

5.103 Dark slate Semi-gloss

A	Copper nitrate	100 gm
	Water	1 litre
B	Potassium sulphide	5 gm
	Water	1 litre
	Hydrochloric acid (35%)	5 cm³

Boiling immersion (A)   (Fifteen seconds)
Cold immersion (B)   (Thirty minutes)

The object is immersed in the boiling solution A for fifteen seconds and then transferred to solution B. The surface darkens initially to produce a series of uneven petrol blue/purple interference colours, and then gradually changes to a slate grey. After thirty minutes the object is removed and rinsed, relieved by gentle scratch-brushing or pumice if desired, and finally washed and dried in sawdust. When completely dry it is wax finished.

This recipe is suitable for producing antiquing and relieved effects only and cannot be generally recommended.

▲ Solution B should only be prepared and used in a fume cupboard or a very well ventilated area. When the acid is added to the potassium sulphide solution, quantities of hydrogen sulphide gas are evolved which is toxic and can be dangerous even in moderate concentrations.

5.104 Mottled blue-black/light grey (lustre) Semi-gloss

Selenous acid	6 cm³
Water	1 litre
Sodium hydroxide solution (250 gm/litre)	50 drops

Warm immersion (Five minutes)

The acid is dissolved in the water, and the sodium hydroxide added by the drop, while stirring. The solution is then heated to 25–30°C. The article to be coloured is immersed in the solution, producing a purplish-brown lustre after about fifteen seconds, which rapidly darkens to a brown lustre tinged with blue. Continued immersion produce a gradual development to a mottled slate grey after two or three minutes. After about five minutes, the article is removed and washed in cold water, allowed to dry in air and wax finished when dry.

The lustre colours produced at an early stage in the immersion tend to be uneven. Prolonging the immersion beyond about five minutes causes a gradual loss of the blue lustrous quality of the mottle, and tends to produce an uneven dull slate grey finish.

▲ Selenous acid is poisonous and very harmful if swallowed or inhaled. Colouring must be carried out in a well ventilated area. Contact with the skin or eyes must be avoided to prevent severe irritation.

▲ Sodium hydroxide is a powerful caustic alkali. Contact with the eyes and skin must be prevented.

5.105 Dark purplish-slate Semi-gloss

Ammonium sulphate	30 gm
Sodium hydroxide	40 gm
Water	1 litre

Boiling immersion (Thirty minutes)

The object is immersed in the boiling solution and the surface gradually darkens to produce a black colour, and acquires a slightly crystalline appearance. After thirty minutes the object is removed, washed in warm water and allowed to dry in air. Relieving is then carried out, and the object wax finished.

The dark coloured layer tends to be thin and only tenuously adherent before finishing. It can be worked with a bristle brush to reveal areas of the metal surface beneath. Tests suggest that although the solution can be used if relieved effects are required, it is not generally suitable for producing solid colour.

▲ Sodium hydroxide is a powerful caustic alkali, the solid and strong solutions causing severe burns to the eyes and skin. When preparing the solution, small quantities of the sodium hydroxide should be added to large quantities of water and not vice-versa. The reaction is exothermic and will warm the solution. When the sodium hydroxide is completely dissolved, the ammonium sulphate is added and the solution may be heated.

5.106 Dark slate on brown Semi-matt

Sodium thiosulphate	65 gm
Copper sulphate	12 gm
Copper acetate	10 gm
Arsenic trioxide	5 gm
Sodium chloride	5 gm
Water	1 litre

Hot immersion (Thirty minutes)

Immersion in the hot solution (50–60°C) produces patchy lustre colours during the first fifteen minutes. Continued immersion produces a progressively more opaque and darker colour. After about thirty minutes, the article is removed and washed in warm water. It is allowed to dry in air, and may be wax finished when dry.

The surface tends to be very uneven. The procedure is difficult to control and cannot generally be recommended.

▲ Arsenic trioxide is very toxic and will give rise to very harmful effects if taken internally. The dust must not be inhaled. The vapour evolved from the hot solution must not be inhaled.

5.107 Black dusting on dull pink Matt/semi-matt

Copper sulphate	50 gm
Potassium chlorate	20 gm
Potassium permanganate	5 gm
Water	1 litre

Boiling immersion (Fifteen minutes)

The object is immersed in the boiling solution, and rapidly becomes coated with a chocolate-brown layer. Immersion is continued for fifteen minutes, after which the object is removed and gently rinsed in a bath of warm water. It is then allowed to dry in air, leaving a dark powdery layer on the surface. When dry this is gently brushed with a soft brush to remove any dry residue. This treatment will remove some of the top layer, partly revealing the underlying colour, and should be continued until the desired surface effect is obtained. A stiffer bristle-brush may be required. After any residual dust has been removed with a soft brush, the object is wax finished.

When the loose dark layers are brushed away, a pink ground is revealed. Some black tends to be adherent, forming dark dusty lines or spotted areas.

▲ Potassium chlorate is a powerful oxidising agent, and must not be allowed to come into contact with combustible material. The solid or solutions should be prevented from coming into contact with the eyes, skin or clothing.

▲ The potassium chlorate and the potassium permanganate must not be mixed together in the solid state. They should each be separately dissolved in a portion of the water, and then mixed together.

5.108 Patchy slate-grey on dull metal Semi-matt

A	Barium sulphide	2 gm
	Potassium sulphide	2 gm
	Ammonium sulphide (16% solution)	2 cm³
	Water	1 litre
B	Copper nitrate	20 gm
	Ammonium chloride	20 gm
	Calcium hypochlorite	20 gm
	Water	1 litre

Cold immersion (A)   (Thirty seconds)
Cold immersion (B)   (A few minutes)

The article is immersed in the mixed sulphide solution, which immediately blackens the surface, and removed after about thirty seconds. If the surface is uneven then it can be lightly brushed with a scratch brush and re-immersed for a similar length of time. The article is washed and transferred to solution B, which changes the colour to a dark slate or grey-brown. After a few minutes the article is removed and allowed to dry in air. When dry it may be wax finished.

The secondary treatment in solution B tended to produce results that were only tenuously adherent, and cannot be generally recommended.

Similar poor results were also obtained with an after-treatment of a solution where the calcium hypochlorite is replaced by an equal quantity of calcium chloride.

▲ Ammonia vapour and hydrogen sulphide gas are liberated from the mixed sulphide solution. Colouring should be carried out in a fume cupboard or very well ventilated area. It should be prevented from coming into contact with the eyes and skin.

▲ Barium compounds are toxic, and harmful if taken internally.

▲ The fine dust from the calcium hypochlorite (bleaching powder) must not be inhaled.

5.109   Light grey on ochre   Matt

Bismuth nitrate	50 gm
Nitric acid (25% solution)	300 cm³
Water	700 cm³

Cold immersion   (Twenty minutes)

The article to be coloured is briefly pickled in sulphuric acid (50% solution), and rinsed in cold water. It is then immersed in the acidified bismuth nitrate solution, which initially darkens the surface and subsequently produces a grey colouration. When the grey colour has fully developed, after about twenty minutes, the object is removed and rinsed in a bath of cold water. It is allowed to dry in air, and may be wax finished when dry.

The grey colour is more adherent on a matted or satin surface than on a highly polished finish. A near-white colour tends to be retained in surface features such as punched marks or textured areas.

The grey colour is very fragile when wet and the surface should not be handled at this stage.

▲ Nitric acid and sulphuric acid are corrosive and must be prevented from coming into contact with the eyes and skin.

5.110   Dull metallic grey and black (mottled and grained) on brown   *Pl. X*
Semi-matt/matt

Nickel ammonium sulphate	50 gm
Sodium thiosulphate	50 gm
Water	1 litre

Hot immersion   (Fifteen minutes)

A purplish-grey colour quickly forms on the surface of the object on immersion, becoming patchy and metallic grey after one minute. After two minutes the object is removed from the solution and lightly scratch-brushed with water, to remove the non-adherent deposits that build up on the surface. The object is then re-immersed and the scratch-brushing repeated twice at intervals of five minutes. After fifteen minutes the object is again lightly scratch-brushed, briefly re-immersed and then washed in hot water. After thorough drying in sawdust, it may be wax finished.

Tests produced variable results. In all cases and with different grades of brass a linear 'grained' effect was produced. In some cases the brown ground was completely obscured, giving a greenish-grey colour overall.

In preparing the solution, the two chemicals should be separately dissolved in two portions of the water. The thiosulphate solution is warmed, and the nickel ammonium sulphate solution added gradually. The light green solution is used at a temperature of 60–80°C.

▲ Nickel salts are a common cause of sensitive skin reactions. The vapour evolved from the hot solution must not be inhaled.

5.111   Metallic grey on yellowish ground   Semi-gloss/semi-matt

Copper nitrate	300 gm
Water	100 cm³
Silver nitrate	2 gm
Water	10 cm³

Applied liquid   (Several minutes)

The copper nitrate and the silver nitrate are separately dissolved in the two portions of water, and then mixed together. The solution is warmed to about 40°C. The article to be coloured is immersed for a few minutes in a cold solution of sulphuric acid (50% solution), and then washed. The warm colouring solution is applied to the article by wiping with a soft cloth, producing a superficial grey layer which is washed away with warm water, revealing the underlying surface effect. When the article is completely dry, it may be wax finished.

A metallic grey partial chemical plating is produced. On yellow brass (Cu 70:30 Zn) this tended to take the form of areas of dark plating, nearly covering the surface. On muntz brass (Cu 60:40 Zn) a fine light 'veil' of plating tended to occur on a pale yellowish and pink ground.

▲ Silver nitrate should be prevented from coming into contact with the skin, and particularly the eyes which may be severely irritated.

▲ Sulphuric acid is very corrosive, causing severe burns, and must be prevented from coming into contact with the eyes and skin.

5.112*   Blue-green patina on grey/pink ground Matt   *Pl. IX*

Ammonium carbonate	24 gm
Potassium hydrogen tartrate	6 gm
Sodium chloride	6 gm
Copper sulphate	6 gm
Acetic acid (6% solution)	to form a paste

Applied paste (Several days)

The solid ingredients are made to a paste with a little of the acetic acid, using a pestle and mortar. The paste is applied with a soft brush. The object is then left to dry for about two days, and the application is then repeated. It is then left to dry thoroughly in air, generally taking several days to dry. When completely dry, the surface is brushed gently with a bristle-brush to remove any loose material, rubbed with a soft cloth to smooth the surface, and wax finished.

If the paste is made up with stronger solutions of acetic acid (15–30%) then there is a greater tendency for the blue-green patina to be non-adherent, partially revealing pink ground colours.

▲ A nose and mouth face mask fitted with a fine dust filter should be worn when brushing away the dry residue.

5.113*   Blue-green patina on light reddish-brown ground Semi-matt/matt

106

Ammonium chloride	35 gm
Copper acetate	20 gm
Water	1 litre

Applied liquid (Several days)

The ingredients are ground with a little of the water using a pestle and mortar, then added to the remaining water. The solution is applied to the object by dabbing and wiping, using a soft cloth. It should be applied sparingly, to leave an evenly moist surface. The object is then allowed to dry in air. This procedure is repeated once a day for several days, producing a gradual development of the ground colour and blue-green patina. When treatment is complete, the object is left to dry for several days, during which time there is further patina development. When completely dry, and there is no further surface change, the object is wax finished.

It is essential to ensure that all patina development is complete, and the surface is completely dry, before wax finishing is carried out. The final drying period may have to be extended to a matter of weeks in damp or humid conditions.

5.114*   Blue-green patina on pink ground Semi-matt   *Pl. IX*

Copper nitrate	110 gm
Water	110 cm³
Ammonia (.880 solution)	440 cm³
Acetic acid (6% solution)	440 cm³
Ammonium chloride	110 gm

Applied liquid (Five days)

The solution is applied to the surface by dabbing with a soft cloth, and the article then left to dry in air. The metal quickly darkens and a whiteish blue-green powdery patina forms gradually on the surface. When dry, this is rubbed with a soft dry cloth to remove any loose material and to smooth the surface, prior to the next application of solution. This procedure is repeated over a period of five days, during which time the patina darkens and becomes integral with the surface. When treatment is complete and the surface thoroughly dry, the article may be wax finished.

The surface produced is variegated and may include some yellow or yellow-green colouration.

▲ Colouring must be carried out in a well ventilated area. Ammonia vapour will be liberated from the strong solution, causing irritation of the eyes and respiratory system. The solution must be prevented from coming into contact with the eyes and skin, as it can cause burns or severe irritation.

5.115   Blue-green patina on brown/black ground Semi-matt   *Pl. X*

Copper nitrate	100 gm
Nitric acid (70% solution)	40 cm³
Water	1 litre

Torch technique

The object is heated with a blow-torch and the solution applied sparingly with a bristle-brush or paint brush. The liquid quickly turns dark brown, and becomes yellowish where it is not being directly heated. Continued heating makes the surface blacken and areas of blue-green patina begin to form. The blue-green patina tends to be superficial at this stage and the surface should be further heated and stippled with solution until a good dark brown or black ground has been established. The object is then gently heated and dabbed with an almost dry brush or cloth, barely damp with the solution, until an evenly distributed blue-green patina is obtained. When treatment is complete, the object is allowed to cool and dry thoroughly, and may then be wax finished.

▲ Nitric acid is very corrosive and must not be allowed to come into contact with the eyes, skin or clothing. A face shield or goggles should be worn to protect the eyes from hot splashes of solution as it is applied to the heated metal.

▲ The fumes evolved as the solution is applied to the heated metal should not be inhaled. Adequate ventilation must be provided, and a nose and mouth face mask fitted with the correct filter should be worn.

5.116* Blue-green patina (blue-green patina on black ground)

Copper nitrate	200 gm
Water	1 litre

Torch technique

▲ Colouring should be carried out in a well ventilated area, so that inhalation of the vapours is avoided.

▲ A face shield or goggles should be worn to protect the eyes from hot splashes of solution.

The metal surface is heated with a blow torch, and the solution applied with a soft brush until it is covered with an even blue-green patina. If the surface is then heated, without further application of the solution, it will become black. The solution may then be applied to this black ground, heating the surface as necessary, to form a blue-green patina on the ground colour. The surface quality obtained can be varied by the precise method of application—a stippled surface may be produced by stippling with a relatively dry brush, or a more mottled or marbled effect obtained using the same technique with a brush that is more 'loaded' with the solution. When the required finish has been achieved, the object is allowed to dry out, and may be wax finished when dry.

5.117* Olive green on black ground Matt   *Pl. X*

A	Copper nitrate	200 gm
	Water	1 litre
B	Potassium polysulphide	50 gm
	Water	1 litre

Torch technique (A)
Torch technique (B)

The concentrations of both solutions can be varied within quite wide limits.

▲ Colouring should be carried out in a well ventilated area, so that inhalation of the vapours and of the hydrogen sulphide gas that are evolved is avoided.

▲ A face shield or goggles should be worn to protect the eyes from hot splashes of solution.

The article is heated with a blow torch and the copper nitrate solution applied with a soft brush until an even thin layer of blue-green patina covers the surface. This is again heated, without further application of the solution, until it turns black. Any loose residue should be brushed off with a soft dry brush. The copper nitrate is then again applied, heating as necessary, until the blue-green colour re-appears and is evenly distributed. While the surface is still hot, the polysulphide solution is applied until the required depth of colour is achieved. When treatment is complete, the article is allowed to dry for some time, and may then be wax finished.

5.118* Blue-green patina on brown ground Semi-matt

Copper nitrate	80 gm
Water	1 litre
Ammonia (.880 solution)	3 cm³

Applied liquid (Three days)

The blue-green patina is difficult to induce on brass with this treatment. Tests on yellow brass (Cu 70:30 Zn) yielded a thin veil of patina on an even brown ground. Higher zinc brasses, eg muntz (Cu 60:40 Zn), yielded an irregular brown and black ground with only slight patina development.

Prolonging the treatment beyond three days does not appear to produce any further development. In tests extending to twenty days no notable changes could be detected after three days.

▲ Inhalation of the fine spray is harmful. It is essential to wear a suitable nose and mouth face mask provided with the correct filter, and to ensure adequate ventilation.

The solution is applied to the object by spraying with a moderately fine atomising spray to produce a misty coating. It is essential to apply only a fine misty coating and to avoid any pooling of the solution or runs on the surface. The object is then left to dry. This procedure is repeated twice a day for three days, after which the surface should be allowed to dry out thoroughly for several days before finishing with wax.

5.119* Blue-green patina on pale pink/ochre ground Matt

Ammonium carbonate	150 gm
Ammonium chloride	100 gm
Water	1 litre

Applied liquid (Several days)

Colour development is slow, producing a thin variegated layer of patina on a pale pink or ochre ground which is also variegated.

It is essential to ensure that the surface is completely dry and patina development complete, before wax finishing. The drying time may need to be extended in damp or humid atmospheric conditions.

The solution is applied to the object, by wiping it on to the surface with a soft cloth. The solution should be used sparingly, leaving the surface evenly moist. The object is then allowed to dry in air. This procedure is repeated once a day for several days, until the grey-green patina is sufficiently developed. When treatment is complete, the object should be allowed to dry out thoroughly for several days, and is wax finished when completely dry.

5.120*   Blue-green patina on pale ochre/pink variegated ground Semi-matt

Ammonium chloride	40 gm
Ammonium carbonate	120 gm
Sodium chloride	40 gm
Water	1 litre

Cold immersion and applied liquid (Several hours)

The object to be coloured is immersed in the cold solution for about one hour, producing a slight etching effect and a gradual development of the ground colour. It is then removed and allowed to dry thoroughly in air. Any excessive surface moisture should be gently brushed out with a soft brush, to prevent 'runs' or 'pooling' of the solution and to leave an evenly moistened surface. The patina develops slowly as the surface dries out. If further patina development is desired, then the solution should be applied sparingly with a soft cloth and the object again left to dry thoroughly. When patina development is complete and the surface dry, the object may be wax finished.

The ground colour obtained on yellow brass (Cu 70:30 Zn) is a pale variegated ochre and dull pink. On muntz brass (Cu 60:40 Zn) the pink colour tended to be absent. Patina development is slight in both cases, with a generally watery appearance.

It is essential to ensure that patina development is complete, and the surface dry, before wax finishing. In damp or humid conditions the drying time may need to be extended.

5.121   Green patina on brown ground Semi-matt

Copper nitrate	100 gm
Hydrochloric acid (35% solution)	10 cm³
Water	1 litre

Torch technique

The object is heated with a blow-torch and the solution sparingly applied with a bristle-brush or paint brush. A heavy deposit of yellowish-green patina quickly forms, but this is rather non-adherent at this stage. Continued heating and the application of small amounts of solution by brushing gradually darkens the surface. When the brown ground colour has been established, the green patina is built up either by stippling with a nearly dry brush, or by using a cloth which is barely damp with the solution, heat being frequently applied to the area being worked. When a good patina has been obtained, the surface can be burnished with a dry cloth. Wax finishing is carried out when the object is cool and thoroughly dry.

In some cases the patina may have a tendency to 'sweat' shortly after treatment. If this occurs then the object should be left in a warm dry place for some time, and the wax only applied when the surface is thoroughly dry.

▲ Hydrochloric acid is corrosive and should not be allowed to come into contact with the eyes and skin. A face shield or goggles should be worn to protect the eyes from hot splashes as the solution is applied.

▲ The fumes evolved as the solution is applied to the heated metal should not be inhaled. Adequate ventilation must be provided and a nose and mouth face mask fitted with the correct filter should be worn.

5.122*   Pale blue-green patina on light brown ground Semi-matt

Ammonium carbonate	150 gm
Sodium chloride	50 gm
Copper acetate	60 gm
Potassium hydrogen tartrate	50 gm
Water	1 litre

Applied liquid (Several days)

The solution is applied generously to the surface of the article using a soft cloth. It is then allowed to dry in air. This procedure is repeated twice a day for several days, during which time the colour gradually develops. The object is then left to dry in air, without further treatment, to allow the patina to develop. When patina development is complete and the surface is completely dry, the article is wax finished.

Excess moisture on the surface after application should be brushed out with a soft brush.

The pale patina tends to be variegated and marked with a fine stipple of a slightly darker tone. Very similar results were obtained with different grades of yellow brass.

5.123*   Blue patina on pink/reddish-brown ground Matt   *Pl. IX, Pl. X*

Copper nitrate	30 gm
Sodium chloride	30 gm
Ammonium chloride	15 gm
Potassium aluminium sulphate	7 gm
Acetic acid (6% solution)	1 litre

Boiling immersion (Ten minutes)

The article is immersed in the boiling solution. After two or three minutes a fine etching effect and pink colour develop on the surface. Immersion is continued to ten minutes, and the article is then removed. It is then dipped several times in the solution and removed. Excess solution is prevented from pooling or running on the surface by gently and rapidly brushing with a soft bristle-brush, and the article is left to dry thoroughly in air. When dry the surface is wax finished.

The results are difficult to control. The patches of brown tend to be irregular and the patina develops slowly and unevenly. Patina development can be encouraged by further sparing brushed applications of the solution when the object is dry and the solution has cooled to 50–60°C.

It is essential to ensure that the surface is completely dry before wax finishing. Drying may need to be prolonged in damp weather.

Plate IX shows the effect on yellow brass (Cu 70:30 Zn); Plate X shows the effect on muntz brass, (Cu 60:40 Zn).

5.124   Blue-green patina on red-brown/maroon ground Semi-matt

Copper nitrate	200 gm
Water	1 litre

Applied liquid (Five days)

The article is dipped in the cold solution for a few seconds, drained and allowed to dry in air. This procedure is repeated twice a day for about five days, during which time the surface darkens gradually to a red-brown. After about two days, patches of powdery blue-green patina also begin to appear on the surface. When treatment is complete, the article should be left to dry in air for a period of at least three days to allow the patina to develop fully and dry out. When dry, the article is wax finished.

Some sources suggest a scratch-brushed application of the solution. Tests carried out using this technique produced similar but slightly lighter results over the same period of time.

Tests carried out in which the metal was briefly 'pickled' in a 10% solution of nitric acid prior to the initial dip, as suggested by some sources, produced very similar results with somewhat duller surfaces.

The patina tends to develop in streaks and patches corresponding to the draining of the solution from the surface. This can be minimised by 'brushing out' with a soft brush after dipping, to prevent runs and pooling.

If the ground colour alone is required, then after two days of dipping and drying, any patina should be brushed off scrupulously and application continued by wiping-on very sparingly. A completely patina-free ground is difficult to achieve.

5.125   Slight blue-green patina on variegated pink/brown ground Matt/semi-matt

Ammonium carbonate	150 gm
Copper acetate	60 gm
Sodium chloride	50 gm
Potassium hydrogen tartrate	50 gm
Acetic acid (10% solution)	1 litre

Applied liquid (Twice a day for several days)

The solution is dabbed on to the object with a soft cloth, and then allowed to dry thoroughly in air. This process is repeated twice a day for several days until the desired colour is achieved. Any powder or other loose material that forms during drying periods should be gently brushed away with a soft dry cloth, prior to the next application of solution. When treatment is complete the surface should be allowed to dry in air for several days. When completely dry, it may be wax finished.

The surface tends to be fragile when the dry residue is rubbed away, particularly on yellow brass (Cu 70:30 Zn). Sparing use of the solution produces a 'glaze' of green/blue-green patina. A less sparing use in the latter stages of treatment encourages the development of a more variegated blue-green.

For results obtained using a more concentrated form of this solution, see recipe 5.139.

It is essential to ensure that patina development is complete, and the surface completely dry, before wax finishing. In damp or humid conditions, the drying time may need to be extended.

▲ The ammonium carbonate should be added to the acetic acid in small quantities. The reaction is effervescent and if large quantities are added, then the effervescence is very vigorous.

5.126   Variegated purple-pink and yellow-green Semi-matt

Copper nitrate	85 gm
Zinc nitrate	85 gm
Ferric chloride	3 gm
Hydrogen peroxide (3% solution)	1 litre

Applied liquid (Twice a day for five days)

The cold solution is dabbed on to the surface of the object with a soft cloth, and the object left to dry thoroughly in air. This procedure is repeated twice a day for a period of about five days. An initial lustrous colouration is gradually replaced by a more opaque colour. Traces of blue-green patina may appear during the later stages of treatment, if the application is not sufficiently sparing, and these should be brushed away when dry if not required. The object is wax finished when completely dry.

Darker and more coherent results are produced on muntz brass (Cu 60:40 Zn) than on yellow brass (Cu 70:30 Zn). Little patina development occurred with either metal.

▲ Hydrogen peroxide must be kept away from contact with combustible material. In the concentrated form it causes burns to the eyes and skin, and when diluted will at least cause severe irritation.

5.127   Variegated pink/brown (traces of green patina) Matt

Copper sulphate	20 gm
Zinc chloride	20 gm
Water	to form a paste

Applied paste (Twenty-four hours)

The ingredients are ground to a thin creamy paste with a little water, using a pestle and mortar. The paste is applied sparingly to the surface of the object with a soft brush. It is then allowed to dry completely, and the residual dry paste washed off with cold water. The object is then allowed to dry in air for about twelve hours, and the procedure repeated if required. When treatment is complete, the object should be allowed to dry for several days at least. Exposure to daylight causes a variegated darkening of the ground colours. When the surface is completely dry and colour change has ceased, the object may be wax finished.

Repeated applications tend to encourage the formation of a more dense green patina, but this tends to 'sweat' intermittently over long periods of time, and gradually breaks down to a powdery incrustation.

5.128   Variegated pink on dark orange-brown Semi-matt

Copper nitrate	30 gm
Zinc chloride	30 gm
Water	to form a paste

Applied paste (Two or three hours)

The ingredients are mixed by grinding together using a pestle and mortar. Water should be added by the drop, to form a creamy paste. Excess water should be avoided, as the paste tends to thin rather suddenly. The paste is applied to the object with a soft brush, and allowed to dry for two or three hours. The residual paste is gently washed away with cold water, and the object then allowed to dry in air for several days. Exposure to daylight will cause a variegated darkening of the ground colours. When the surface is completely dry, and colour change has ceased, it may be wax finished.

Some bluish-green patina may also form.

Repeated applications do not improve the results, and tend to require far longer final drying periods.

5.129   Light green patina on pink ground Semi-matt

Copper nitrate	35 gm
Copper chloride	35 gm
Calcium chloride	35 gm
Water	1 litre

Applied liquid to heated metal (Repeat several times)

The article is either heated on a hotplate or with a torch to a temperature such that the solution sizzles gently when it is applied. The solution is applied with a soft brush, and dries out to produce a superficial powdery green layer. The procedure is repeated several times, each fresh application tending to remove the green powdery layer, until an underlying colour has developed. Continued application produces an adherent pale green patina with a powdery appearance. When treatment is complete, the article should be left to cool and dry for some time before wax finishing.

5.130   Pale blue (or green) patina on light brown/ochre ground Semi-matt

Ammonium carbonate	180 gm
Copper sulphate	60 gm
Copper acetate	20 gm
Oxalic acid	1.5 gm
Ammonium chloride	0.5 gm
Acetic acid (10% solution)	1 litre

Torch technique

The ingredients are dissolved and the solution boiled for about ten minutes. It is then left to stand for several days. The mixture should be shaken immediately prior to use. The surface of the object to be coloured is heated with a blow-torch, and the solution applied with a soft brush. This process is continued until a pale blue-patina is produced, on a light brown or ochre ground. If the torch is then gently played across the surface, the pale blue will change to green. Varying effects can be obtained by 'stippling' or 'dragging' the brush on the surface. When the required finish is obtained, the object is left to cool and dry for some time and may then be wax finished.

Continuous application of liberal quantities of the solution will tend to inhibit patina formation, but encourages the development of the ground colour.

▲ The ammonium carbonate should be added to the acetic acid in small quantities. The reaction is effervescent, and if large amounts are added, this may be very vigorous.

▲ Colouring should be carried out in a well ventilated area, so that inhalation of the vapours is avoided.

▲ A face shield or goggles should be worn to protect the eyes from hot splashes of solution.

5.131   Dark green on mid-brown ground Matt

Ammonium sulphate	105 gm
Copper sulphate	3.5 gm
Ammonia (.880 solution)	2 cm^3
Water	1 litre

Torch technique

The metal surface is heated with a blow-torch, and the solution applied with a soft brush, until a dark green colour on a mid-brown ground is obtained. If the metal is too hot when the solution is applied then impermanent light grey or pinkish-red colours are produced. Continued application of the solution as the metal cools will remove these colours and encourage the brown ground and the dark green patina. If the green patina is heated once it has developed, by playing the torch across the surface, then it will tend to become black. When the desired surface finish has been achieved, the object is left to cool and dry for some time, and may then be wax finished.

A wide variety of surface effects are obtained, which depend on the temperature of the metal, and the manner in which the solution is applied (eg liberally or sparingly; by 'stippling' or by 'painting' etc).

▲ Colouring should be carried out in a well ventilated area, to avoid inhalation of the vapours.

▲ A face shield or goggles should be worn to protect the eyes from hot splashes of solution.

5.132   Greyish-blue patina and brown spotting on pink ground   Semi-matt/matt

Copper nitrate	200 gm
Sodium chloride	200 gm
Water	1 litre

Applied liquid   (Twice a day for five days)

The solution is dabbed on to the object with a soft cloth and left to dry in air. This process is repeated twice a day for five days. The object is then allowed to dry in air, without treatment, for a further five days. The ground colour develops fairly rapidly and is followed by the gradual appearance of brown spotting or mottling. The grey-blue patina develops very slowly as the surface dries out. Finishing should only be carried out when it is certain that the surface is dry and patina development complete.

Scratch-brushing may be used to overcome initial resistance to 'wetting', and unevenness.

A less sparing application should be used in the later stages to induce greater patina development.

In damp weather conditions the surface may take several weeks to dry completely, and may 'sweat' intermittently. The drying process should not be rushed and wax finishing only carried out when it is certain that chemical action has ceased.

5.133   Dark green and pale blue 'watery' surface   Semi-matt

Ammonium carbonate	180 gm
Copper sulphate	60 gm
Copper acetate	20 gm
Oxalic acid	1.5 gm
Ammonium chloride	0.5 gm
Acetic acid (10% solution)	1 litre

Boiling immersion   (Twenty-five minutes)

The object is immersed in the boiling solution, which produces some etching effects and light colouring after a few minutes. As the immersion continues streaky greenish films gradually develop, beginning to become dark after about ten minutes. The object is removed after twenty-five minutes and the hot solution applied with a soft brush, to leave the surface evenly moist. The object should then be allowed to dry thoroughly in air for several hours, without washing. When dry, it is wax finished.

The surface dries to a dark green initially. Pale blue patches gradually develop, to give a surface with a 'watery' or glazed appearance. The surface tends to be fragile until completely dry.

▲ The ammonium carbonate should be added to the acetic acid in small quantities. The reaction is effervescent, and this may be very vigorous if large amounts are added.

▲ Oxalic acid is harmful if taken internally and irritating to the eyes and skin.

5.134   Brown and yellow-green (variegated)   Semi-matt   *Pl. X*

Copper sulphate	10 gm
Lead acetate	10 gm
Ammonium chloride	5 gm
Water	to form a paste

Applied paste   (Several days)

The ingredients are ground to a creamy paste with a little water, using a pestle and mortar. The paste is applied to the object with a soft brush, and is left to dry out completely. When completely dry, after several hours, the residue is brushed away with a bristle-brush. This process is repeated three or four times. When treatment has finished, the object should be allowed to stand in a dry place for several days to ensure thorough drying. After a final brushing with the bristle-brush, it is wax finished.

The variegated surfaces produced include some darker 'figure'. On yellow brass (Cu 70:30 Zn) the brown tends to predominate, while on muntz brass (Cu 60:40 Zn) the surface is predominantly yellow-green. Plate X shows the effect on muntz brass.

▲ Lead acetate is highly toxic, and every precaution should be taken to prevent the inhalation of the dust during weighing, and preparation of the paste. A mask fitted with a fine dust filter must be worn when brushing away the dry residue. A clean working method is essential throughout.

5.135   Blue-green patina on brown ground   Semi-matt/matt

Copper acetate	30 gm
Copper carbonate	15 gm
Ammonium chloride	30 gm
Hydrochloric acid (18% solution)	2 cm³
Water	to form a paste

Applied paste   (Two or three days)

The ingredients are ground to a paste with a little water and the hydrochloric acid, using a pestle and mortar. The paste is applied to the object with a soft brush and then left for several hours to dry out. The dry residue is removed with a stiff bristle-brush. The procedure is repeated until a brown ground has developed with a thin bloom of patina. The paste is then thinned out to a liquid and applied using the same procedure, but with a soft brush to remove residue. When the surface is sufficiently developed and well dried out, the object may be wax finished.

A thick incrustation tends to form if the paste is made too thick. This is difficult to remove by brushing, when dry, but tends to flake eventually.

▲ A nose and mouth face mask should be worn when brushing away the dry residue.

5.136   Blue-green patina on yellowish ground Matt

Ammonium carbonate	300 gm
Acetic acid (10% solution)	35 cm³
Water	1 litre

Warm scratch-brushing (Two or three minutes)

The ammonium carbonate is added to the water, while stirring, and the acetic acid then added very gradually. The solution is then warmed to 30–40°C. The solution is applied to the object liberally with a soft brush, and then immediately scratch-brushed. The surface is worked for a short time, while applying more solution, and then allowed to dry in air. Any excess moisture on the object should be brushed out with a soft bristle-brush, to prevent 'pooling' or 'running' and to leave an evenly moist surface. When the surface is completely dry, which may take several days, it is wax finished.

In tests on yellow brass (Cu 70:30 Zn) a yellowish ground marred by streaks was produced. Patina development is also streaky. Little patina or ground colour development could be produced on muntz brass (Cu 60:40 Zn), other than a slightly stained etched surface.

The results obtained in tests with this solution were poor. Although the colours obtained are interesting, the surfaces were very prone to streaking and patchiness, and little control over this could be gained. The technique cannot therefore be generally recommended.

5.137   Dark brown and ochre patches Semi-matt

Ammonium sulphate	105 gm
Copper sulphate	3.5 gm
Ammonia (.880 solution)	2 cm³
Water	1 litre

Applied liquid (Several days)

The solution is applied to the surface of the object with a soft cloth. It is then allowed to dry in air. This procedure is repeated twice a day for several days until the ground colour has developed and traces of patina begin to appear. The solution should then be applied more sparingly by dabbing with a soft cloth. The object is allowed to dry slowly in a damp atmosphere to encourage the development of patina. It is finally allowed to dry in a warm dry atmosphere and wax finished when completely dry.

The surface produced cannot be described as variegated, but rather consists of distinct patches of brown and of an ochre-coloured etched surface. Patina development tends to be slight.

This recipe, which has been used commercially to induce patination on copper roofing, requires weathering in the open air to achieve the best results. In tests carried out in workshop conditions, only slight patination and patchy ground colours could be obtained.

5.138   Slight blue-green patina on brown ground

A	Ammonium sulphate	80 gm
	Water	1 litre
B	Copper sulphate	100 gm
	Sodium hydroxide	10 gm
	Water	1 litre

Applied liquid (A)   (Five to ten days)
Applied liquid (B)   (Two or three days)

The ammonium sulphate solution is applied to the surface of the object by dabbing vigorously with a soft cloth, leaving it sparingly moist. It is then allowed to dry completely. This procedure is repeated twice a day for up to ten days. Solution B is then applied in the same way for two or three days. When treatment is complete the object is allowed to dry in air for several days. When completely dry, it may be wax finished.

In tests, the solution tended to retract into pools and failed to 'wet' the surface. The addition of wetting agents produced no improvement, and the surface developed in patches rather than evenly.

▲ Sodium hydroxide is a powerful caustic alkali, and must be prevented from coming into contact with the eyes and skin.

5.139   Slight blue-green patina on dull olive ground Matt

Ammonium carbonate	600 gm
Sodium chloride	200 gm
Copper acetate	200 gm
Potassium hydrogen tartrate	200 gm
Acetic acid (10% solution)	1 litre

Applied liquid   (Twice a day for several days)

The solution is applied to the article to be coloured, by wiping sparingly with a soft cloth. It should then be allowed to dry thoroughly in air. This procedure should be repeated twice a day for several days, until the ground colour has developed and the patina has formed. The article should then be left to dry in air for several days, during which time the patina will continue to develop. When patina development has ceased and the surface is completely dry, the article may be wax finished.

An uneven dull olive/yellow-brown ground tends to be produced, on which patina development is slight. The results are difficult to control and the procedure cannot be generally recommended.

For results obtained with a more dilute form of this solution, see recipe 5.125.

It is essential to ensure that the surface is dry and patina development complete before wax finishing. In damp or humid conditions the drying time may need to be extended.

▲ The ammonium carbonate should be added to the acetic acid in small quantities. The reaction is effervescent and if large amounts are added the effervescence is very vigorous.

5.140   Thin blue-green patina on dark pink ground Semi-matt

Ammonium chloride	10 gm
Sodium chloride	6.5 gm
Water	1 litre

Torch technique

The object is heated with a blow torch and the solution dabbed on with a cloth. The surface gradually changes to a darkish pink. When this ground colour is established the dabbing should be continued to bring up the light blue-green patina. When treatment is complete, the object is allowed to cool and dry before wax finishing.

▲ The fumes evolved as the solution is applied to the heated metal should not be inhaled. Adequate ventilation should be provided.

▲ A face shield or goggles should be worn to protect the eyes from hot splashes as the solution is applied.

5.141   Very slight blue-green patina on dull metal Semi-matt

Ammonium chloride	10 gm
Ammonium acetate	10 gm
Water	1 litre

Applied liquid   (Twice a day for ten days)

The solution is applied to the surface by wiping with a soft cloth or by brushing with a soft brush. It is then allowed to dry in air. This procedure is repeated twice a day for about ten days. The article is then allowed to dry in air, without treatment, for a further five days. When completely dry, and when patina development has ceased, it may be wax finished.

Colour development is generally slow, the ground developing unevenly at first, and the patina appearing later during the periods of air-drying. Patina development is more pronounced in surface features such as punched marks or textured areas. Test results were poor, and the procedure cannot be generally recommended.

5.142* Purplish-brown lustre Gloss

Copper nitrate	115 gm
Tartaric acid (crystals)	50 gm
Sodium hydroxide	65 gm
Ammonia (.880 solution)	15 cm³
Water	1 litre

Boiling immersion (Fifteen minutes)

Immersion in the boiling solution produces a series of lustre colours, which begins with a blue after about one minute, and continues to an even green lustre after about two minutes. If the green colour is required, the object should be immediately removed, washed in hot water, allowed to dry in air, then wax finished. Continued immersion causes a gradual pink tinge to appear which darkens to produce the final colour after about fifteen minutes. The object is removed and finished as above.

Although a number of colour changes occur, the two colours noted are the only ones in this lustre series which are both rich and even.

▲ Sodium hydroxide is a powerful corrosive alkali, and the solid or solutions must be prevented from coming into contact with the eyes and skin. It should be added to the solution in small quantities.

▲ Preparation and colouring should be carried out in a well ventilated area. Ammonia vapour will be liberated, which if inhaled causes irritation of the respiratory system. A combination of caustic vapours will be liberated from the solution as it boils.

▲ A dense precipitate is formed which may cause severe bumping. Glass vessels should not be used.

5.143* Golden lustre Gloss   *Pl. X*

Potassium permanganate	10 gm
Water	1 litre

Hot immersion (Three to five minutes)

The article is immersed in the hot solution (90°C). A golden lustre colour develops within one minute, gradually becoming more intense. When the lustre colour is fully developed, which may take from three to five minutes, the article is removed and washed in hot water which is gradually cooled during washing. The article is finally washed in cold water before being carefully dried either in sawdust, or by gently blotting the surface with absorbent tissue paper to remove excess moisture and allowing it to dry in air. When dry, it is wax finished.

The effect is only obtained on polished, or directionally grained satin surfaces. On rough, as-cast and matt surfaces a dull orange or orange-brown is produced.

Extending the immersion time produces a slight darkening initially to a more pink colour. Which rapidly fades to a pale silvery grey. Colours produced at longer immersion times tend to be more fragile when wet, and should not be handled at that stage.

5.144 Bluish-green lustre Gloss   *Pl. IX*

Sodium hydroxide	25 gm
Copper carbonate	75 gm
Water	1 litre

Hot immersion (Two or three minutes)

The sodium hydroxide is dissolved in the water, and the copper carbonate then added. The solution is heated to 70–80°C. The article is immersed in the hot solution, and the surface gradually acquires lustre colours. When the green colour has developed, after two or three minutes, the article is removed and washed in warm water. It is allowed to dry in air, and may be wax finished when dry.

In tests, better colours were obtained on muntz brass (Cu 60:40 Zn) than on yellow brass (Cu 70:30 Zn). Prolonging the immersion tends to cloud the surface and eventually produces an opaque brown colour. Plate IX shows a muntz brass sample after two to three minutes' immersion.

▲ Sodium hydroxide is corrosive and must be prevented from coming into contact with the eyes and skin. It should be added to the solution in small quantities.

5.145* Golden brown lustre (pink tinge) Gloss   *Pl. IX*

Sodium hydroxide	25 gm
Potassium persulphate	10 gm
Water	1 litre

Hot immersion (Thirty minutes)

The potassium persulphate is dissolved in the water, and the sodium hydroxide then added in small quantities while stirring. The solution is then brought up to the correct temperature (70–80°C) and the article immersed. A golden lustre colour is produced, which darkens very gradually to a golden-brown colour which is tinged with pink. When the desired colour is reached, the article is removed and washed in cold water, and allowed to dry in air. When dry, it may be wax finished.

▲ Potassium persulphate is a powerful oxidant and should not be allowed to come into contact with combustible material. The sodium hydroxide and the persulphate must not be allowed to come into contact in the solid state.

▲ Sodium hydroxide is a powerful caustic alkali which can cause severe burns. Both the solid and solutions should be prevented from coming into contact with the eyes and skin.

5.146* Mottled reddish-brown lustre Gloss

Ammonium persulphate	10 gm
Sodium hydroxide	30 gm
Water	1 litre

Boiling immersion (Thirty minutes)

The ammonium persulphate is dissolved in half the water, and the sodium hydroxide separately dissolved in the remaining water. The two solutions are then mixed together and the mixture heated to boiling. The article to be coloured is immersed in the boiling solution, which produces golden and then brown lustre colours. Continued immersion darkens these colours to an opaque but glossy finish. When the colour has developed, after about thirty minutes, the article is removed, washed in hot water and dried in sawdust. When dry, it may be wax finished.

▲ Ammonium persulphate (peroxodisulphate) is a powerful oxidant, and must be kept out of contact with all combustible materials. It should be prevented from coming into contact with the eyes, skin or clothing.

▲ Sodium hydroxide is a powerful caustic alkali and must be prevented from coming into contact with the eyes and skin.

5.147   Variegated pink/green lustre on greenish-brown Gloss

Potassium hydroxide	100 gm
Copper sulphate	30 gm
Sodium tartrate	30 gm
Water	1 litre

Warm immersion (About ten minutes)

Pl. XV

Similar colours are produced on muntz brass (Cu 60:40 Zn). An interim golden lustre colour is produced on both yellow brass and muntz brass, after about five minutes.

▲ Potassium hydroxide is a powerful corrosive alkali and contact of the solid or solutions with the eyes or skin must be prevented.

The potassium hydroxide is added gradually to the water and stirred. Heat is evolved as it dissolves, raising the temperature to about 35°C. The other ingredients are added and allowed to dissolve. There is no need to heat the solution. The object is immersed in the warm solution, and immersion continued until the lustre colour develops. The object is then removed and washed in warm water. It is allowed to dry in air, after any excess moisture has been removed by dabbing with absorbent tissue. When dry, it is wax finished.

5.148   Green lustre (pink tinge) Gloss

Potassium sulphide	10 gm
Water	1 litre
Ammonia (.880 solution)	1 cm³

Hot immersion (About ten minutes)

The green lustre tends to be unevenly tinged with pink.

As an alternative to immersion, the solution can be applied liberally using a soft brush, without 'working' the surface. The results tend to be more variegated.

▲ Some hydrogen sulphide gas will be liberated from the solution. Colouring should be carried out in a well ventilated area.

The object is immersed in the warm solution (40°C) and the surface gradually takes on a lustrous quality, which develops to a green lustre after about ten minutes. The object is then removed, washed in cold water, and allowed to dry in air. When dry it may be wax finished.

5.149   Green and pink lustre Gloss

Copper sulphate	120 gm
Potassium chlorate	60 gm
Water	1 litre

Hot immersion (Thirty seconds)

Beyond the green lustre phase, the surface rapidly becomes orange-buff coloured. See also recipe 5.90.

▲ Potassium chlorate is a powerful oxidising agent and must not be allowed to come into contact with combustible material. Contact with the eyes, skin or clothing must be avoided.

The article is immersed in the hot solution (70–80°C) and quickly removed after thirty seconds' immersion, when the green lustre has developed. It is immediately washed thoroughly in hot water. After drying in sawdust, it is wax finished. The colour appears to change on drying, and again on waxing, due to the change in the interference colour caused by the presence of the layer of water, and then the layer of wax on the surface. The final colour is a green lustre variegated with pink.

5.150   Purple-pink lustre on brown Gloss

Sodium thiosulphate	20 gm
Antimony potassium tartrate	10 gm
Water	1 litre

Hot immersion (Forty-five minutes)

On Muntz brass (Cu 60:40 Zn) an uneven pale blue lustre on a brown ground is produced. The surfaces produced on sheet brass in general were prone to marks. Handling should be avoided.

Timings for the immersion may be variable. If the desired colour is produced in a shorter time, the object should be removed at that stage. The immersion should not be continued longer than necessary.

▲ Antimony potassium tartrate is poisonous and harmful if taken internally. The vapour evolved from the hot solution should not be inhaled.

The article is immersed in the hot solution (60°C) for ten minutes, producing a slight darkening of the surface and an orange tinge. The temperature of the solution is then gradually raised to 80°C and immersion continued. After forty-five minutes the article is removed, washed in warm water and allowed to dry in air. When dry, it is wax finished.

5.151   Dark reddish-brown (green lustre) Gloss

Potassium permanganate	10 gm
Sodium hydroxide	25 gm
Water	1 litre

Boiling immersion (Thirty minutes)

A light purple colour with a reddish sheen is produced on a directionally grained satin surface.

▲ Sodium hydroxide is a powerful caustic alkali which can cause severe burns. Both the solid and solutions must be prevented from coming into contact with the eyes and skin. When preparing the solution, small quantities of sodium hydroxide should be added to large amounts of water and *not* vice-versa. After it has dissolved, the potassium permanganate is added.

The object is immersed in the boiling solution, and after two or three minutes the surface shows signs of darkening. After ten minutes this develops to an uneven dark lustrous surface. Continued immersion causes some further darkening. After thirty minutes the object is removed, washed in warm water and allowed to dry in air. When dry, it may be wax finished.

5.152   **Lustre series Gloss**

Sodium thiosulphate	240 gm
Copper acetate	25 gm
Water	1 litre
Citric acid (crystals)	30 gm

Cold immersion (Various)

The sodium thiosulphate and the copper acetate are dissolved in the water, and the citric acid added immediately prior to use. The article is immersed in the cold solution and a series of lustre colours is produced in the following sequence: golden-yellow/orange; brown; purple; blue; pale grey. This sequence takes about thirty minutes to complete, the brown changing to purple after about fifteen minutes. It is followed by a second series of colours, in the same sequence, which take a further thirty minutes to complete. The latter colours in the second series tend to be slightly more opaque. A third series of colours follows, if immersion is continued, but these tend to be indistinct and murky. (If the temperature of the solution is raised then the sequences are produced more rapidly. However, it was found in tests that the results tended to be less even. Temperatures in excess of about 35–40°C tend to cause the solution to decompose, with a concomitant loss of effect). When the required colour has been produced, the article is removed and washed in cold water. It is allowed to dry in air, any excess moisture being removed by gentle blotting with an absorbent tissue. When dry it may be wax finished or coated with a fine lacquer.

5.153   **Lustre series Gloss**

Sodium thiosulphate	125 gm
Lead acetate	35 gm
Water	1 litre

Hot immersion (Various)

More dilute solutions produce the lustre colour sequence, but take a correspondingly longer time to produce.

▲ Lead acetate is very toxic and will give rise to serious conditions if taken internally. The dust, and the vapour evolved from the hot solution, must not be inhaled.

The article is immersed in the hot solution (50–60°C), which produces a series of lustre colours in the following sequence: golden-yellow/orange; purple; blue; pale blue; pale grey. The timing tends to be variable, but it takes roughly five minutes for the purple stage to be reached, and about fifteen minutes for pale blue. (Tests failed to produce a further series of colours, the surface tending to remain pale grey after twenty minutes.) When the desired colour has been produced, the article is removed and immediately washed in cold water. It is allowed to dry in air, any excess moisture being removed with an absorbent tissue. When dry it may be wax finished or coated with a fine lacquer.

5.154   **Lustre series Gloss**

Sodium thiosulphate	280 gm
Lead acetate	25 gm
Water	1 litre
Potassium hydrogen tartrate	30 gm

Warm immersion (Various)

Citric acid may be used instead of potassium hydrogen tartrate, and in the same quantity. Tests carried out with both these alternatives produced identical results.

▲ Lead acetate is very toxic and will give rise to serious conditions if taken internally. The dust, and the vapour evolved from the hot solution, must not be inhaled.

The sodium thiosulphate and the lead acetate are dissolved in the water, and the temperature is adjusted to 30–40°C. The potassium hydrogen tartrate is added immediately prior to use. The article to be coloured is immersed in the warm solution and a series of lustre colours is produced in the following sequence: golden-yellow/orange; purple; blue; pale blue/pale grey. The colour sequence takes about ten minutes to complete at 40°C. (Hotter solutions are faster, cooler solutions slower.) A second series of colours follows in the same sequence, but these tend to be less distinct. When the desired colour has been reached, the article is removed and immediately washed in cold water. It is allowed to dry in air, any excess moisture being removed by dabbing with an absorbent tissue. When dry it may be wax finished, or coated with a fine lacquer.

5.155   **Dark straw/brown lustre**   **Gloss**

Barium sulphide	10 gm
Water	1 litre

Hot immersion   (Two or three minutes)

The object is immersed in the hot solution (50°C), and the surface quickly takes on a golden colour, which darkens to a straw-coloured or brown lustrous finish. The object should be quickly removed and rinsed thoroughly in warm water. It is then allowed to dry in air, and wax finished when completely dry.

If immersion is continued for too long, the colour will quickly become very pale and may acquire a blue tinge.

Cold solutions will produce similar results, given very extended immersion times. In tests it was found that prolonged immersion in a cold solution tends to produce less even results.

▲ Barium sulphide is poisonous and may be very harmful if taken internally. Some hydrogen sulphide gas will be liberated as the solution is heated. Colouring should be carried out in a well ventilated area.

5.156   **Pink/green lustre on orange ground**   **Gloss**

Copper sulphate	25 gm
Sodium hydroxide	5 gm
Ferric oxide	25 gm
Water	1 litre

Hot immersion and applied liquid and bristle-brushing   (A few minutes)

The copper sulphate is dissolved in the water and the sodium hydroxide added. The ferric oxide is added, and the mixture heated and boiled for about ten minutes. When it has cooled to about 70°C, the object is immersed in it for about ten seconds and becomes coated with a brown layer. The object is transferred to a hotplate and gently heated until the brown layer has dried. The dry residue is brushed off with a soft bristle-brush. The turbid solution is then applied to the metal with a brush, as the object continues to be heated on the hotplate. The residue is brushed off, when dry. The application is repeated two or three times, until a green/pink lustre is produced. After a final brushing with the bristle-brush, the object is wax finished.

The results tend to be variegated rather than even, and may appear patchy when viewed in some lights.

▲ Contact with the sodium hydroxide in the solid state must be avoided when preparing the solution. It is very corrosive to the skin and in particular to the eyes.

5.157   **Pink/green lustre on mottled brown ground**   **Gloss**   *Pl. X*

Copper sulphate	25 gm
Nickel ammonium sulphate	25 gm
Potassium chlorate	25 gm
Water	1 litre

Boiling immersion   (One to two minutes)

The object is immersed in the boiling solution and quickly removed when the mottled surface and lustre colour have developed. It should be washed immediately in hot water. After drying in sawdust, it is wax finished.

▲ Nickel salts are a common cause of sensitive skin reactions.

▲ Potassium chlorate is a powerful oxidising agent and should not be allowed to come into contact with combustible material. The solid, and solutions, should be prevented from coming into contact with the eyes, skin or clothing.

5.158   **Uneven pink/green on golden lustre**   **Semi-gloss**

A	Copper sulphate	50 gm
	Water	1 litre
B	Potassium sulphide	12.5 gm
	Water	1 litre

Hot immersion (A)   (Thirty seconds)
Cold immersion (B)   (One minute)

The article is immersed in the hot copper sulphate solution (80°C) for about thirty seconds to produce a green/pink lustre colour. It is then transferred to the cold potassium sulphide solution and immersed for about one minute, producing a clouding of the surface and a greenish-pink colour. This tends to be uneven and can be improved by applying the solutions alternately using a bristle-brush. When treatment is complete, the article is washed in warm water and allowed to dry in air. When dry, it is wax finished.

If the results produced in solution A are streaky, then the streaks will tend to be visible, in some lights, after treatment with solution B. Application with the bristle-brush darkens and improves the results. Producing an adequate finish evenly is difficult, and the recipe cannot be generally recommended.

For results obtained using a scratch-brushed technique see recipe 5.58.

▲ Colouring should be carried out in a well ventilated area as some hydrogen sulphide gas will be liberated.

5.159　Dull pink lustre　Gloss

Antimony trichloride	50 gm
Olive oil	to form a paste

Applied paste　(Five minutes)

The antimony trichloride is ground to a creamy paste with a little olive coil using a pestle and mortar. The paste is brushed on to the surface of the object to be coloured. A purple lustre colour develops after about two minutes, changing to a pink lustre after about five minutes. A steel blue appears if the paste is left for longer. When the desired colour has developed. The residual paste is washed from the surface with pure distilled turpentine. The object is finally rubbed with a soft cloth, and wax finished.

Tests tended to result in surfaces which were slightly uneven.

Localised effects can be produced by the selective application of brush strokes.

If the paste is left to dry on the surface, it becomes difficult to remove without damaging the colour finish.

Tests carried out using aqueous solutions produced similar results, but these tended to be patchy.

▲ Antimony trichloride is highly toxic and every precaution must be taken to avoid contact with the eyes and skin. It is irritating to the respiratory system and eyes, and can cause skin irritation and dermatitis.

5.160　Lustre sequence　Gloss

Antimony trisulphide	50 gm
Ammonium sulphide (16% solution)	100 cm³
Water	1 litre

Cold immersion　(Two to ten minutes)

The article is immersed in the cold solution, producing a rapid sequence of lustre colours initially. After about two minutes the colour obtained is predominantly blue with some local red or golden surface variation. This changes gradually to a greenish lustre after four minutes, which slowly fades. The final colours will have developed after about six to eight minutes, and immersion beyond ten minutes produces no detectable change. When the desired colour has been reached, at any stage in the immersion, the article is removed, washed in cold water and allowed to dry in air. When dry, it is wax finished.

The colours produced at an early stage in the immersion are the most intense, but they are also subject to unpredictable variegated effects.

The best results obtained in tests occurred after about two or three minutes, when a dark golden brown lustre appears. This was the most even colour produced. Later colours tended to be pale and patchy.

▲ The preparation and colouring should be carried out in a well ventilated area. Ammonium sulphide solution liberates a mixed vapour of ammonia and hydrogen sulphide, which must not be inhaled, and which will irritate the eyes. The solution must be prevented from coming into contact with the eyes and skin, as it will cause burns or severe irritation.

5.161　Pale blue lustre on grey-brown metallic ground　Gloss

Sodium thiosulphate	300 gm
Antimony potassium tartrate	15 gm
Ammonium chloride	20 gm
Water	1 litre

Hot immersion　(Twenty minutes)

Immersion in the hot solution (60°C) rapidly produces a series of lustre colours on the surface of the object. These are pale and pass from purple to brown and blue, becoming a very pale blue grey after about two minutes. This is followed by a gradual darkening of the surface, producing dark greys and browns. After about twenty minutes the object is removed, washed in warm water and dried in sawdust. It may be wax finished when dry.

A mottled and uneven pale blue lustre on a glossy metallic grey-brown ground is produced. Very similar results were produced with different grades of brass.

▲ Antimony potassium tartrate is poisonous and may be harmful if taken internally. The vapour evolved from the hot solution must not be inhaled.

5.162　Pale brown (lustre)　Gloss

Potassium sulphide	2 gm
Ammonium sulphate	0.5 gm
Water	1 litre

Cold immersion　(One or two minutes)

The article is immersed in the cold solution. A golden lustre colour develops fairly quickly, and in roughly one to two minutes will begin to change to a mustard colour. As soon as this occurs the object should be transferred quickly to a container of cold water and thoroughly rinsed. It should then be washed in clean cold water and dried in sawdust. When completely dry it is wax finished.

The technique is difficult to control. The colour produced tends to be variegated rather than even.

The lustre colour is only produced on highly polished surfaces. Matt or other unpolished surfaces yield only patchy dull brownish colours.

5.163　Uneven pale lustre　Semi-gloss

Butyric acid	20 cm³
Sodium chloride	17 gm
Sodium hydroxide	7 gm
Sodium sulphide	5 gm
Water	1 litre

Boiling immersion　(One hour)

The solid ingredients are added to the water and allowed to dissolve, and the butyric acid then added, and the solution heated to boiling. The object is immersed in the boiling solution, which rapidly produces a series of patchy lustre colours. Continued immersion gradually produces a more opaque dark purplish-green colour, which tends to be uneven. After about one hour the object is removed and washed in hot water. It is allowed to dry in air, and may be wax finished when dry.

A variegated blue lustre on a brownish ground was obtained after ten minutes' immersion.

The results produced tend to be uneven or patchy, and the procedure cannot be generally recommended.

▲ Butyric acid should be prevented from coming into contact with the eyes and skin, as it can cause severe irritation. It also has a foul odour which tends to cling to the hair and clothing. Preparation and colouring should be carried out in a well ventilated area.

▲ Both sodium hydroxide and sodium sulphide are corrosive, and must be prevented from coming into contact with the eyes and skin.

5.164* Dark green and buff stipple Semi-gloss   *Pl. IX*

Ammonium chloride	16 gm
Sodium chloride	16 gm
Ammonia (.880 solution)	30 cm³
Water	1 litre

Sawdust technique (Twenty-four hours)

Some development of blue-green patina may occur, particularly in any surface features such as punched marks.

The surface is locally etched by the solution where the granular sawdust is in close contact with the metal. These points of contact and surrounding areas tend to be coloured a light buff. The remaining areas tend to be coloured green.

▲ Ammonia solution causes burns or severe irritation, and must be prevented from coming into contact with the eyes or skin. The vapour should not be inhaled as it irritates the respiratory system. Colouring should be carried out in a well ventilated area.

The object to be coloured is laid-up in sawdust which has been evenly moistened with the solution. It is then left for about twenty-four hours. The object should not be disturbed while colouring is in progress, as re-packing tends to cause a loss of surface quality and definition. Progress should be monitored by including a small sample of the object metal, when laying-up, which can be examined as necessary. When treatment is complete, the object is washed thoroughly in cold water and allowed to dry in air. It may be wax finished, when completely dry.

5.165* Buff/green (stipple) Semi-matt   *Pl. X*

Ammonium chloride	10 gm
Copper sulphate	4 gm
Potassium binoxalate	20 gm
Water	1 litre

Sawdust technique (Several days)

The surface is locally etched where the granular particles of sawdust are in close contact with the metal, producing an even stipple of 'bitten' areas which take on a buff and pink colouration. The surrounding unetched areas acquire a slightly lustrous green or olive colour.

▲ Potassium binoxalate is harmful if taken internally. It can also cause irritation, and should be prevented from coming into contact with the eyes or skin.

The object to be coloured is laid-up in sawdust which has been moistened with the solution, and left for several days. Colour development is very slow, and should be monitored using a sample so that the object remains undisturbed. When the colour has developed satisfactorily, the object is removed, washed in cold water and allowed to dry in air for several hours. It may be wax finished when dry.

5.166* Yellowish-green/pink mottle Semi-matt   *Pl. IX*

Ammonium chloride	15 gm
Potassium aluminium sulphate	7 gm
Copper nitrate	30 gm
Sodium chloride	30 gm
Acetic acid (6% solution)	1 litre

Sawdust technique (Twenty to thirty hours)

Some development of a light blue-green stipple of patina may also take place. In tests, this occurred more readily on muntz brass (Cu 60:40 Zn) than yellow brass (Cu 70:30 Zn).

The object to be coloured is laid-up in sawdust which has been evenly moistened with the solution, and left for about twenty or thirty hours. The object is then removed and allowed to dry in air for several hours, without washing. When dry, the surface is brushed with a stiff bristle-brush to remove any particles of sawdust or loose material. The object is then left for a further period of several hours, after which it may be wax finished.

5.167 Pink etched stipple on brass Semi-matt   *Pl. X*

Copper nitrate	200 gm
Sodium chloride	100 gm
Water	1 litre

Sawdust technique (Twenty or thirty hours)

The surface is selectively etched by the solution, where the granular sawdust is in close contact with the metal, causing an even stipple of etched areas that are coloured pink. The remainder of the surface is little affected.

▲ No patina development occurred in tests carried out using this solution with yellow brass.

The article to be coloured is laid-up in sawdust which has been moistened evenly with the solution. It should then be left for a period of twenty or thirty hours. After twenty hours the object should be examined, and re-packed for shorter periods of time, as necessary. The sawdust should be kept moist throughout the period of treatment. When treatment is complete, the article should be washed thoroughly in cold water and allowed to dry in air. When completely dry, after several hours, it may be wax finished.

5.168 Black with blue patina on grey and ochre ground (stipple) Semi-matt

Ammonium carbonate	150 gm
Copper acetate	60 gm
Sodium chloride	50 gm
Potassium hydrogen tartrate	50 gm
Water	1 litre

Sawdust technique (Ten to twenty hours)

The surface is blackened by the solution and selectively etched where the granular sawdust is in close contact with the metal, producing a stipple of etched areas, which become coloured light grey, greyish-mauve and ochre. The blackened, unetched areas tend to acquire some blue patina.

The object to be coloured is laid-up in sawdust which has been evenly moistened with the solution. It is left for a period of from ten to twenty hours, colour development being monitored with a sample, so that the object remains undisturbed. When treatment is complete, the object is removed, washed in cold water and allowed to dry in air for several hours. When dry, it may be wax finished.

5.169* Light metallic grey stipple on variegated golden-brown/purple-brown Gloss

Butyric acid	20 cm³
Sodium chloride	17 gm
Sodium hydroxide	7 gm
Sodium sulphide	5 gm
Water	1 litre

Sawdust technique (Twenty hours)

The object to be coloured is laid-up in sawdust which has been evenly moistened with the solution, and left for a period of up to twenty hours. The timing of colour development tends to be unpredictable, and it is essential to include a sample of the same material as the object, when laying-up, so that progress can be monitored without disturbing the object. When treatment is complete, the object is removed and washed thoroughly with cold water. After several hours' air-drying, it may be wax finished.

▲ Butyric acid causes burns, and contact with the skin, eyes and clothing must be prevented. It also has a foul pungent odour, which tends to cling to the clothing and hair, and is difficult to eliminate.

▲ Sodium hydroxide is a powerful caustic alkali causing burns, and must be prevented from coming into contact with the eyes or skin.

▲ Sodium sulphide is corrosive and causes burns. It must be prevented from coming into contact with the eyes or skin. When preparing the solution it must not be mixed directly with the acid, or toxic hydrogen sulphide gas will be evolved. The acid should be added to the water and then neutralised with the sodium hydroxide, before the remaining ingredients are added.

5.170* Light brown/orange stipple on dark lustre Semi-matt

Ammonium sulphate	85 gm
Copper nitrate	85 gm
Ammonia (.880 solution)	3 cm³
Water	1 litre

Sawdust technique (Thirty hours)

The article to be coloured is laid-up in sawdust which has been moistened evenly with the solution. It should then be left for a period of twenty or thirty hours. The sawdust should be kept moist throughout the period of treatment. After thirty hours the object should be examined, and re-packed for shorter periods of time as necessary. When treatment is complete, the object should be washed thoroughly in cold water and allowed to dry in air. When dry it may be wax finished.

The surface is selectively etched where the granular sawdust is in close contact with the metal, producing a stipple of etched areas that take on a light brown or orange colour. The remainder of the surface is unetched and becomes dark and lustrous.

Etching is more pronounced and extensive if a moister medium is used. The small etched areas are then more yellow in colour and have a frosted appearance.

5.171 Reddish-purple and dark greenish-grey (stipple) Semi-matt

Copper nitrate	280 gm
Water	850 cm³
Silver nitrate	15 gm
Water	150 cm³

Sawdust technique (Twenty hours)

The copper nitrate and the silver nitrate are dissolved in separate portions of water, and the two solutions then mixed. The article to be coloured is laid-up in sawdust which has been evenly moistened with this mixed solution. It is left for a period of about twenty hours, ensuring that the sawdust remains moist throughout this time. When treatment is complete, the article is removed and washed in cold water. After allowing it to dry thoroughly in air for some time, it may be wax finished.

The surface is coloured reddish-purple by the treatment, and selectively etched where the granular sawdust is in close contact with the metal, producing a stipple of etched spots. These tend to be coloured a reddish-pink and acquire a fringe of dark grey or greenish-grey.

▲ Silver nitrate must be prevented from coming into contact with the skin, and more particularly the eyes, as it may cause burns or severe irritation.

5.172 Pale blue-green/purple stipple on brass Semi-matt

Copper nitrate	100 gm
Water	200 cm³
Ammonia (.880 solution)	400 cm³
Acetic acid (6% solution)	400 cm³
Ammonium chloride	100 gm

Sawdust technique (Twenty to thirty hours)

The copper nitrate is dissolved in the water and the ammonia gradually added. The acetic acid is then added, followed by the ammonium chloride. The article to be coloured is laid-up in sawdust which has been evenly moistened with this solution, and is then left for a period of twenty to thirty hours. When treatment is complete, the article is washed in cold water and allowed to dry in air. When completely dry, after several hours, it may be wax finished.

The surface is selectively etched where the granular sawdust is in close contact with the metal, producing a stipple of etched spots which acquire a pale blue-green patina. These spots are fringed with purple. The remainder of the surface is a slightly matt brass colour. A wetter medium tends to result in more extensive etching over the whole surface and a greater development of the blue-green and pale purple colours.

Colour development should be monitored by using a sample.

▲ Preparation and colouring should be carried out in a well ventilated area. Ammonia vapour and some acetic acid vapour will be liberated, which cause irritation of the eyes and respiratory system.

▲ The strong ammonia solution must be prevented from coming into contact with the eyes and skin, as it will cause burns or severe irritation.

5.173* Reddish-brown mottle Semi-matt   *Pl. IX*

Copper nitrate	200 gm
Water	1 litre

Sawdust technique (Two days)

The article to be coloured is laid-up in sawdust which has been evenly moistened with the solution, and left for about two days. Colour development may be monitored using a sample, but no deleterious effects were produced, in tests, by unpacking and re-packing the object itself. When the colour has developed satisfactorily, the article is removed and washed in cold water. After several hours drying in air, it may be wax finished.

The red/reddish-brown ground develops first, and may be evened and intensified by removing the object before the moist medium becomes too dry, and re-packing in freshly moistened sawdust. If the medium is too wet, or unevenly moistened, then localised crystalline encrustations of copper tend to develop as the medium dries out. Some green patina also tends to develop as the medium becomes drier.

Richer ground colours tend to be produced on muntz brass (Cu 60:40 Zn) than on yellow brass (Cu 70:30 Zn).

5.174* Grey-green patina (stipple) on dark ground Matt

Ammonium carbonate	120 gm
Ammonium chloride	40 gm
Sodium chloride	40 gm
Water	1 litre

Sawdust technique (Twenty to thirty hours)

The article to be coloured is laid-up in sawdust which has been evenly moistened with the solution, and is then left for a period of about twenty or thirty hours. The sawdust should be kept moist throughout the period of treatment. When treatment is complete, the article should be washed in cold water and allowed to dry in air. When it is completely dry, it may be wax finished.

The surface is etched by the solution, producing a stippled texture which is coloured dark brown or black. This is overlaid with a fine incrustation of predominantly grey-green patina. The overall appearance is of a mixed stippled surface of pale green and grey-blue.

If the medium is too moist initially, or is re-moistened during treatment, dull brown etched areas tend to be exposed.

5.175 Blue-green patina (stipple) on pale purple/buff ground Matt

Ammonium chloride	100 gm
Ammonium carbonate	150 gm
Water	1 litre

Sawdust technique (To twenty or thirty hours)

The article to be coloured is laid-up in sawdust which has been moistened evenly with the solution. It may then be left for periods of up to twenty or thirty hours. When treatment is complete, the object is washed in cold water and allowed to dry in air. When it is completely dry, it may be wax finished.

The surface is etched by this treatment, producing a textured surface which takes on a greyish/pale purple colour, mixed with the light buff colour of the etched metal. The blue-green patina develops in the form of spots of incrustation on this surface, giving a stippled appearance.

5.176 Pale blue/lilac and grey (stipple) on yellow-brown ground Matt

Ammonium chloride	350 gm
Copper acetate	200 gm
Water	1 litre

Sawdust technique (Ten to twenty hours)

The object to be coloured is laid-up in sawdust which has been moistened evenly with the solution, and left for a period of ten to twenty hours. After ten hours it should be examined and re-packed if further treatment is required. It should subsequently be examined every few hours until the required surface finish is obtained. The sawdust should be kept moist during the full period of treatment. The object should then be removed, washed in cold water and allowed to dry in air. When completely dry, it may be wax finished.

The metal is selectively etched by this treatment, producing fine 'pits' in the surface that take on a variously coloured pale patina. The remainder of the surface takes on a slightly frosted yellow-brown finish. Some darker grey spots also tend to occur. Colour development tends to be unpredictable, and is best monitored using a sample.

5.177* Dark brown/black and yellowish-green stipple Gloss

Potassium hydroxide	100 gm
Copper sulphate	30 gm
Sodium tartrate	30 gm
Water	1 litre

Sawdust technique (Twenty hours)

The object is laid-up in sawdust which has been evenly moistened with the solution, and is then left undisturbed for a period of about twenty hours. The sawdust should remain moist throughout this period. When treatment is complete, the object is removed and washed thoroughly in cold water. After it has been allowed to dry in air, it may be wax finished.

The greenish-yellow stipple tends to be lustrous.

▲ Potassium hydroxide is a strong caustic alkali, and precautions should be taken to prevent it from coming into contact with the eyes and skin.

▲ When preparing the solution, the potassium hydroxide should be added to the water in small quantities, and *not* vice-versa. When it has dissolved completely, the other ingredients may be added.

5.178   Pale yellow-brown stipple on greenish ground Gloss

'Rokusho'	30 gm
Copper sulphate	5 gm
Water	1 litre

Sawdust technique (Several days)

The surface produced tends to have a pale and translucent quality.

For description of 'Rokusho' and its method of preparation, see appendix 1.

The 'Rokusho' and the copper sulphate are added to the water, and the mixture boiled for a short time and then allowed to cool to room temperature. The object to be coloured is laid-up in sawdust which has been moistened with the solution. It is then left for several days. Colour development is very slow, and should be monitored using a sample so that the object remains undisturbed. When the colour has satisfactorily developed, the object is removed, washed in cold water and allowed to dry in air. It may be wax finished when dry.

5.179   Metallic stipple (dark lustre) on yellowish etched ground Matt

Ammonium chloride	10 gm
Ammonia (.880 solution)	20 cm³
Acetic acid (30% solution)	80 cm³
Water	1 litre

Sawdust technique (One or two days)

The surface is selectively etched by the solution, producing a yellowish etched ground, and leaving an even stipple of unetched metal with a lustrous finish. The colour of the lustrous finish varies from reddish-brown to pale brown with a blue tint, depending on the length of treatment. Longer treatment times give rise to a predominance of the etched ground.

▲ Preparation and colouring should be carried out in a well ventilated area. Ammonia vapour and some acetic acid vapour will be liberated, which can irritate the eyes and respiratory system. These solutions should be prevented from coming into contact with the eyes and skin, or severe irritation or burns may result.

The object to be coloured is laid-up in sawdust which has been evenly moistened with the solution, and left for one or two days. Colour development should be monitored using a sample, so that the object itself remains undisturbed. When the colour has developed satisfactorily, the object is removed, washed in cold water and allowed to dry in air for several hours. When dry, it may be wax finished.

5.180   Slight buff and greyish stipple Semi-matt

Di-ammonium hydrogen orthophosphate	100 gm
Water	1 litre

Sawdust technique (Thirty hours)

The surface is selectively etched by the solution to produce a stipple of dull 'bitten' areas that take on a buff or greyish colour, and leaving a slightly dull quality to the unetched areas of the surface. A slight greyish bloom of patina may also develop.

Tests carried out with this solution, using various other techniques, including cold and hot direct application, and hot and cold immersion, produced little or no effect.

The object to be coloured is laid-up in sawdust which has been evenly moistened with the solution. It is left for about thirty hours and then removed and washed in cold water. If further colour development is required the treatment should be repeated for additional shorter periods. When treatment is complete, the object is finally washed and allowed to dry in air, and wax finished when completely dry.

5.181   Greenish-brown and blue patina (stipple) on dull brass Semi-matt

Copper carbonate	300 gm
Ammonia (.880 solution)	200 cm³
Water	1 litre

Sawdust technique (To twenty or thirty hours)

▲ The strong ammonia solution will liberate ammonia vapour, which irritates the eyes and respiratory system. Preparation and colouring must therefore be carried out in a well ventilated area.

▲ Ammonia solution causes burns and severe irritation, and must be prevented from coming into contact with the eyes and skin.

The article to be coloured is laid-up in sawdust which has been evenly moistened with the above solution, and left for a period of up to twenty or thirty hours. The sawdust should remain moist throughout the period of treatment. When treatment is complete, the article is removed and washed in cold water. It is then allowed to dry in air for some time before wax finishing.

5.182   Yellow-brown stipple on etched pink ground Semi-matt

Copper sulphate	100 gm
Lead acetate	100 gm
Ammonium chloride	10 gm
Water	1 litre

Sawdust technique (Ten to twenty hours)

The surface is coloured a slightly lustrous variegated yellow-brown, and selectively etched, producing a stipple of pink ground. The etched pink ground tends to become more extensive as treatment is prolonged.

▲ Lead acetate is a toxic substance, and every precaution should be taken to prevent inhalation of the dust, when preparing the solution. It is very harmful if taken internally. A clean working method is essential throughout.

The object to be coloured is laid-up in sawdust which has been evenly moistened with the solution. It is then left to develop for a period of up to about twenty hours. When treatment is complete, the object is removed, washed in cold water and allowed to dry in air. It may be wax finished when completely dry.

5.183   Grey/black and brown stipple on frosted yellow ground Matt

Ammonium carbonate	300 gm
Acetic acid (10% solution)	35 cm³
Water	1 litre

Sawdust technique (To twenty hours)

The metal is strongly etched, producing a stippled grey and brown textured surface. This is broken by areas of frosted yellow, that become more extensive as treatment proceeds.

The ammonium carbonate is added to the water and the acetic acid then added gradually, while stirring. The object to be coloured is laid-up in sawdust which has been evenly moistened with this solution. It is then left for a period of up to twenty hours, ensuring that the sawdust remains moist throughout the period of treatment. After about ten hours, and at intervals of several hours subsequently, the object should be examined to check its progress. When the desired surface finish has been obtained, the object is removed, washed in cold water and allowed to dry in air. When thoroughly dry, it may be wax finished.

5.184   Orange/dark brown/black stipple on frosted yellow etch Semi-matt

Copper acetate	60 gm
Copper carbonate	30 gm
Ammonium chloride	60 gm
Water	1 litre
Hydrochloric acid (15% solution)	5 cm³

Sawdust technique (Ten to twenty hours)

A stippled orange and dark brown glossy surface is produced initially, but this is rapidly etched back to a frosted yellow matt finish.

▲ Hydrochloric acid can cause burns or severe irritation, and must be prevented from coming into contact with the eyes and skin.

The object to be coloured is laid-up in sawdust which has been evenly moistened with the solution. It is left for a period of about ten hours and then examined. If further treatment is required, it should be re-packed and examined at intervals of a few hours. When treatment is complete, the object is removed, washed thoroughly in cold water and allowed to dry in air. When completely dry, it may be wax finished.

5.185   Light buff stipple on dull metal ground Semi-matt

Copper nitrate	20 gm
Ammonium chloride	20 gm
Calcium hypochlorite	20 gm
Water	1 litre

Sawdust technique (Twenty or thirty hours)

▲ The corrosive fine dust of the calcium hypochlorite (bleaching powder) must not be inhaled, as it can severely irritate the respiratory system. The solution must be prevented from coming into contact with the eyes, which would be severely irritated.

The object to be coloured is laid-up in sawdust which has been moistened evenly with the solution, and left for a period of twenty or thirty hours. The sawdust should be kept moist throughout this time. After treatment, the object is washed in cold water and allowed to dry in air. When thoroughly dry, it may be wax finished.

5.186*   Variegated reddish-brown Matt   *Pl. X*

Copper nitrate	200 gm
Sodium chloride	200 gm
Water	1 litre

Cotton-wool technique (Twenty hours)

The surface is etched to a matt finish and coloured reddish-brown by the solution. The colouring tends to be uneven, and patches of a copper-enriched frosted surface can occur, where the solution tends to pool.

Cotton-wool, moistened with the solution, is applied to the surface of the object to be coloured. It is then left for a period of about twenty hours. It is essential to ensure that the cotton-wool is in full contact with the surface and is kept moist throughout the period of treatment. Breaks in the colour and patches of patina may result if the cotton-wool is allowed to dry out or lift from the surface. When treatment is complete, the object should be washed thoroughly in cold water, and allowed to dry in air. When dry, it may be wax finished.

5.187   Variegated blue-green/brown/frosted yellow Semi-matt

Ammonium carbonate	150 gm
Copper acetate	60 gm
Sodium chloride	50 gm
Potassium hydrogen tartrate	50 gm
Water	1 litre

Cotton-wool technique (Ten to twenty hours)

The surface is etched back by the solution. The least etched areas are coloured brown and tend to acquire some blue-green patina. Areas that are more positively etched become yellow in colour and have a frosted appearance. The result is a very variegated finish.

Cotton-wool, moistened with the solution, is applied to the surface of the article. It is important to ensure that the cotton-wool is in full contact with the surface. It is left for ten to twenty hours, during which time the cotton-wool should remain moist. Colour development should be monitored with a sample, so that the article remains undisturbed. When treatment is complete, the article is removed and washed in cold water. After drying in air for several hours, it may be wax finished.

5.188   Variegated lustre Gloss

Ammonium sulphate	85 gm
Copper nitrate	85 gm
Ammonia (.880 solution)	3 cm³
Water	1 litre

Cotton-wool technique (Thirty hours)

The technique is only suitable for use on polished surfaces. A golden lustre colour, variegated or mottled with pink and green, is produced.

Cotton-wool, moistened with the solution, is applied to the surface of the object to be coloured. It is then left for a period of about thirty hours. It is essential to ensure that the moist cotton-wool is in full contact with the surface of the object, or breaks in the colour will occur. The cotton-wool should be kept moist throughout the period of treatment. When treatment is complete, the object should be washed thoroughly in cold water and allowed to dry in air. When dry, it is wax finished.

5.189*   Red and dark brown mottle on a yellow etched ground Semi-matt

Copper nitrate	280 gm
Water	850 cm³
Silver nitrate	15 gm
Water	150 cm³

Cotton-wool technique (Twenty hours)

A mottled red and dark brown colour develops on a slightly frosted, etched yellow ground. The etching effect will be increased if a more moist medium is used. Colour development should be monitored using a sample.

▲ Silver nitrate must be prevented from coming into contact with the skin, and more particularly the eyes, as it may cause burns or severe irritation.

The copper nitrate and the silver nitrate are dissolved in separate portions of water, and the two solutions then mixed. Cotton-wool, moistened with this mixed solution, is applied to the surface of the object to be coloured. It is important to ensure that the cotton-wool is in full contact with the surface, and that it remains moist during the period of treatment. The object is removed after about twenty hours and washed in cold water. After thorough drying in air, it may be wax finished.

5.190*   Pale blue/grey-green/brown (variegated) Semi-matt

Ammonium carbonate	120 gm
Ammonium chloride	40 gm
Sodium chloride	40 gm
Water	1 litre

Cotton-wool technique (Twenty to thirty hours)

A very variegated surface with a smooth finish is produced, consisting of thin deposited layers of blue and grey-green on a brown ground.

Cotton-wool, moistened with the solution, is applied to the surface of the object to be coloured. It is important to ensure that the moist cotton-wool is in full contact with the surface of the object. It is then left for a period of about twenty to thirty hours, ensuring that the cotton-wool remains moist throughout the period of treatment. When treatment is complete, the object should be washed in cold water, and allowed to dry in air. When completely dry, it is wax finished.

5.191   Pale blue/greenish-blue on light buff (variegated) Semi-matt

Copper nitrate	100 gm
Water	200 cm³
Ammonia (.880 solution)	400 cm³
Acetic acid (6% solution)	400 cm³
Ammonium chloride	100 gm

Cotton-wool technique (Twenty to thirty hours)

A very variegated surface is produced, with a finish that is smooth to the touch. The surface consists of a smooth and even incrustation, which forms on, and covers, an etched surface.

▲ Preparation and colouring should be carried out in a well ventilated area. Ammonia vapour and some acetic acid vapour will be liberated, which cause irritation of the eyes and respiratory system.

▲ The strong ammonia solution must be prevented from coming into contact with the eyes and skin, as it will cause burns or severe irritation.

The copper nitrate is dissolved in the water and the ammonia gradually added. The acetic acid is then added, followed by the ammonium chloride. Cotton-wool is wetted with the solution and applied to the surface of the object to be coloured. It is then left for a period of about twenty to thirty hours. The cotton-wool should be kept moist throughout the period of treatment. When treatment is complete, the object is washed in cold water and allowed to dry in air. When dry, it is wax finished.

5.192   Pale greenish-blue patina on brown ground (variegated) Matt/semi-matt

Ammonium chloride	100 gm
Ammonium carbonate	150 gm
Water	1 litre

Cotton-wool technique (To twenty or thirty hours)

The metal is etched by the solution, producing a relatively smooth brown ground. The patina develops as a thin deposited layer, giving a variegated appearance. In tests, some localised defects in the patina layer tended to occur.

Cotton-wool, moistened with the solution, is applied to the surface of the object to be coloured. It is then left for periods of up to twenty or thirty hours. The cotton-wool should be kept moist throughout the period of treatment. After about ten or fifteen hours the surface should be examined, by carefully exposing a small portion, to determine the extent of colour development. Subsequently a check should be made every few hours until the desired surface finish is obtained. When treatment is complete, the object is washed in cold water and allowed to dry in air. When completely dry, it may be wax finished.

5.193   Yellow-brown (variegated and slightly 'frosted')   Matt

Ammonium chloride	350 gm
Copper acetate	200 gm
Water	1 litre

Cotton-wool technique   (Ten to twenty hours)

Cotton-wool, moistened with the solution, is applied to the surface of the object. This is then left for a period of ten or twenty hours, ensuring that the cotton-wool remains moist. After the first few hours, the surface should be periodically examined by exposing a small portion, to check the progress. When the desired surface finish has been reached, the object is removed, washed in cold water and allowed to dry in air. When completely dry, it may be wax finished.

The surface is evenly etched to a 'frosted' dull yellow-brown finish, which is variegated with slightly darker patches. Some areas may occur with an incrustation of blue-green patina.

5.194   Dark brown with lustrous patches   Gloss

Potassium hydroxide	100 gm
Copper sulphate	30 gm
Sodium tartrate	30 gm
Water	1 litre

Cotton-wool technique   (Twenty to thirty hours)

Cotton-wool, moistened with the solution, is applied to the surface of the object, which is then left for a period of twenty to thirty hours. It is important to ensure that the moist cotton-wool is in full contact with the surface, and that it remains moist throughout the period of treatment. When treatment is complete, the object is washed in cold water and allowed to dry in air. When dry, it may be wax finished.

The surface passes through a series of lustre colour changes, but tends to a dark brown colour. Some patches of lustre tend to occur on the dark brown surface.

▲ Potassium hydroxide is a strong caustic alkali, and should be prevented from coming into contact with the eyes or skin.

▲ When preparing the solution, the potassium hydroxide should be added to the water in small quantities, and *not* vice-versa.

5.195*   Yellow-brown/dark brown (mottled)   Gloss

Copper carbonate	300 gm
Ammonia (.880 solution)	200 cm³
Water	1 litre

Cotton-wool technique   (Twenty hours)

Cotton-wool, moistened with the solution, is applied to the surface of the object to be coloured, which is then left for a period of twenty hours. It is important to ensure that the moist cotton-wool is in full contact with the surface, and that it remains moist throughout the period of treatment. When treatment is complete, the article is removed and washed in cold water. It is then allowed to dry in air for some time before wax finishing.

▲ The strong ammonia solution will liberate ammonia vapour, which irritates the eyes and respiratory system. Preparation and colouring must therefore be carried out in a well ventilated area.

▲ Ammonia solution causes burns and severe irritation, and must be prevented from coming into contact with the eyes and skin.

5.196   Variegated pink (slight grey patina)   Matt

Copper sulphate	100 gm
Lead acetate	100 gm
Ammonium chloride	10 gm
Water	1 litre

Cotton-wool technique   (Ten to twenty hours)

Cotton-wool, moistened with the solution, is applied to the surface of the object to be coloured, and is left in place for a period of up to twenty hours. The cotton-wool should remain moist throughout the period of treatment. When treatment is complete, the object is removed and washed in cold water. It is then dried in air, and gradually darkens on exposure to daylight. When colour change has ceased, the object may be wax finished.

The metal is etched by the solution to produce a patchy or variegated pink surface. Some patches of light grey patina tend to develop on the pink ground.

▲ Lead acetate is a toxic substance, and every precaution should be taken to prevent inhalation of the dust when preparing the solution. It is very harmful if taken internally. A clean working method is essential throughout.

5.197   Cloudy black/brown and bluish patina (variegated)   Matt

Ammonium carbonate	300 gm
Acetic acid (10% solution)	35 cm³
Water	1 litre

Cotton-wool technique   (Twenty hours)

The ammonium carbonate is added to the water and the acetic acid then added gradually, while stirring. Cotton-wool, moistened with this solution, is applied to the surface of the object, and left for a period of about twenty hours. It is important to ensure that the cotton-wool is in full contact with the surface, and that it remains moist throughout. When treatment is complete, the object is washed in cold water and allowed to dry in air. When completely dry, it may be wax finished.

A smooth cloudy black surface with a bluish patina is produced. This tends to be unevenly thin, giving a brownish appearance in places. Localised breaches of this surface tended to occur, in tests, revealing spots of frosted yellow.

5.198   Dull pink/brown mottle   Semi-matt

Copper acetate	60 gm
Copper carbonate	30 gm
Ammonium chloride	60 gm
Water	1 litre
Hydrochloric acid (15% solution)	5 cm³

Cotton-wool technique   (Ten to twenty hours)

Cotton-wool, moistened with the above solution, is applied to the surface of the article, and left for a period of about ten hours. The surface is then examined, by carefully lifting a small portion, to determine the stage it has reached. When the desired surface finish has been obtained, the article is washed thoroughly in cold water and allowed to dry in air. When completely dry, it may be wax finished.

Some traces of greenish patina also occurred, in tests on yellow brass (Cu 70:30 Zn).

▲ Hydrochloric acid can cause burns or severe irritation, and must be prevented from coming into contact with the eyes and skin.

5.199   Buff and 'frosted' yellow (mottled)   Matt

Copper nitrate	20 gm
Ammonium chloride	20 gm
Calcium hypochlorite	20 gm
Water	1 litre

Cotton-wool technique   (Twenty hours)

Cotton-wool, moistened with the solution, is applied to the surface of the object to be coloured, and left for a period of about twenty hours. It is important to ensure that the cotton-wool is in close contact with the surface, and that it remains moist throughout the period of treatment. When treatment is complete, the object is removed and washed thoroughly in cold water. After a period of drying in air, it may be wax finished.

A mottled buff colour on a slightly frosted yellow ground is produced. The surface is smooth to the touch. Although the effect obtained in tests is interesting, it is difficult to control. Surface development should be monitored by using a sample.

▲ The corrosive fine dust of the calcium hypochlorite (bleaching powder) must not be inhaled, as it can severely irritate the respiratory system. The solution must be prevented from coming into contact with the eyes, which would be severely irritated.

5.200   Blue-green patina on variegated brown   Semi-matt

Ammonium sulphate	90 gm
Copper nitrate	90 gm
Ammonia (.880 solution)	1 cm³
Water	1 litre

Cloth technique   (Twenty hours)

Soft cotton cloth which has been soaked with the solution is applied to the surface of the object, and stippled into place with stiff bristle-brush. The object is then left for a period of about twenty hours. The cloth should be removed when it is very nearly dry, and the object then left to dry in air without washing. The blue-green patina tends to develop during the drying period. (If a patina-free ground colour is required, the cloth should be removed at an earlier stage when it is still wet, and the object thoroughly washed. The cloth application may be repeated, for a more developed ground colour.) When treatment is complete and the surface thoroughly dried out, it may be wax finished.

The ground colour tends to develop in patches, which vary in colour from light brown to near black. The blue-green patina tends to take on the texture of the applied cloth.

Tests carried out involving direct application of this solution were not successful. The surface stubbornly resists 'wetting', even with additions of wetting agents to the solution.

5.201   Blue-green patina on dark grey/black ground   Semi-matt

Ammonium carbonate	20 gm
Oxalic acid (crystals)	10 gm
Acetic acid (6% solution)	1 litre

Cloth technique   (Twenty-four hours)

Strips of soft cotton cloth, or other absorbent material, are soaked with the solution. These are applied to the surface of the object and stippled into place using the end of a stiff brush which has been moistened with the solution. They are left on the surface until very nearly dry, which may take up to twenty-four hours, and then removed. If patina development is to be encouraged, the surface is then moistened by wiping the solution on using a soft cloth. If the ground colour and a minimum of patina is required, then the surface is allowed to dry without applying any more solution. When the surface is completely dry, it may be wax finished.

A dark grey ground is produced which tends to take on the texture of the applied cloth. Cloudy areas of blue-green patina develop on this ground.

If the cloth is allowed to dry completely on the surface, it becomes very adherent and tends to damage the surface when removed.

A variety of tests were carried out using this solution. Immersion at temperatures from cold to boiling produced no effect, even after prolonged immersion. Direct application of the cold or hot solution by dabbing and wiping also produced no effect, as the surface resists wetting.

▲ The ammonium carbonate should be added to the acetic acid in small quantities to avoid violent effervescence.

▲ Oxalic acid is harmful if taken internally. Contact with the eyes and skin should be avoided.

6 Silver (rolled sheet) and silver-plate

The silver sheet used in tests, and to which the recipes and results refer, is standard silver (925 standard). The main colour heading for each recipe refers to the results obtained on polished, fire-stain free surfaces, unless otherwise stated, and includes a reference to both the colour and surface quality. The results obtained on satin-finished surfaces, where these are significantly different, are given in the marginal notes.

Tests on silver-plate were carried out on gilding metal which was flash-plated with copper prior to silver plating. The results obtained are given in the marginal notes accompanying the silver recipes. Any bi-metallic colouring effects produced are also noted.

	Recipe numbers
Brown Brownish-purple/grey-brown/dark brown	6.1 – 6.14
Grey Grey/dark grey/black	6.15 – 6.29
Green patina	6.30 – 6.34
Lustre colours	6.35 – 6.37
Special techniques	6.38 – 6.57
Partial-plate colours Silver-plate/copper plate on gilding metal	6.58 – 6.69

6.1*   **Brownish-purple Gloss**   *Pl. XI*

Copper chloride	50 gm
Water	1 litre

Hot immersion (A few minutes)

The article is immersed in the hot solution (50–60°C) for a few minutes, during which time the colour develops gradually. When the colour has developed, the article is removed and washed in warm water. It is then allowed to dry in air, and may be wax finished when completely dry.

An even brownish-purple colour is produced on standard silver. Continued immersion tends to produce a dark brown colour with a slightly less glossy finish. The brown colour can also be produced using a more dilute solution for a longer period of immersion.

A variegated pink and grey lustrous surface is produced on silver plate. This is better if a more dilute solution (10–15 gm/litre) is used.

A bi-metallic effect can also be produced, see recipe 6.69.

6.2*   **Dark brown (*or* dark purple-brown/black) Semi-gloss**   *Pl. XI*

Barium sulphide	5 gm
Water	1 litre

Cold immersion or cold scratch-brushing (Ten or fifteen minutes)

The article is immersed in the cold solution for a period of about ten or fifteen minutes, producing a slightly uneven dark brown colour. When the colour has developed, the article is removed, washed in cold water and allowed to dry in air. It may be wax finished when dry. (A similar but darker colour is produced by direct scratch-brushing with the cold solution. Direct scratch-brushing with a hot solution (60°C) produces a lighter reddish-brown, that tends to be tonally variegated, and is difficult to obtain evenly.)

A mid-brown bronzing colour with a gloss or semi-gloss finish is produced on silver-plated surfaces. Plate XI shows this effect. Variations in the procedure tend to produce the same colour.

▲ Barium compounds are poisonous and may be very harmful if taken internally.

6.3*   **Dark brown Semi-matt/semi-gloss**

Copper nitrate	20 gm
Sodium chloride	16 gm
Ammonium chloride	4 gm
Potassium hydrogen tartrate	12 gm
Water	to form a paste

Applied paste (Four hours)

The ingredients are ground together with a little water, using a pestle and mortar, to form a creamy paste. This is applied to the surface of the object to be coloured, using a soft brush, and allowed to dry for about four hours. The dry residue is brushed away, and the object then washed in cold water and allowed to dry in air. When completely dry, it may be wax finished.

The surface produced tends to have a slightly mottled quality.

The effect of this paste on silver-plated surfaces is not known.

6.4*   **Mid-brown Gloss**

Copper nitrate	20 gm
Ammonium chloride	20 gm
Calcium hypochlorite	20 gm
Water	1 litre

Cold immersion (Fifteen minutes)

The article is immersed in the cold solution, which produces a brown colour that develops gradually during the period of immersion. After about fifteen minutes, the article is removed and washed in cold water. It is dried in sawdust and may be wax finished when dry.

The brown colour is only produced on standard silver. Silver-plated surfaces are unaffected at first, but gradually acquire a slight grey tarnish. A slight bi-metallic colouring effect is produced on silver-plated gilding metal, the underlying metal taking on an ochre colour. This effect is difficult to obtain without tarnishing the silver-plate.

▲ A fine dust may be raised from the calcium hypochlorite (bleaching powder) which is corrosive and must not be inhaled. The powder and strong solutions should be prevented from coming into contact with the eyes and skin.

6.5*   **Mid grey-brown Semi-gloss/gloss**

Copper sulphate	125 gm
Sodium chlorate	50 gm
Ferrous sulphate	25 gm
Water	1 litre

Boiling immersion (Ten minutes)

The article is immersed in the boiling solution and gradually darkens, becoming slate grey within five minutes. After about ten minutes the warmer grey-brown colour develops, and the article is removed, washed in hot water and allowed to dry in air. When dry it may be wax finished.

The mid grey-brown colour is only produced on standard silver. Silver-plated surfaces are coloured slate grey, but are rapidly attacked by the solution, revealing the underlying metal and finally stripping the plate completely.

Bi-metallic effects can be produced with brief immersions, see recipe 6.63.

▲ Sodium chlorate is a powerful oxidant and should be kept out of contact with combustible material or acids. It should be prevented from coming into contact with the eyes, skin and clothing.

6.6   Greyish-brown Semi-gloss/gloss

Ammonium carbonate	180 gm
Copper sulphate	60 gm
Copper acetate	20 gm
Oxalic acid (crystals)	1.5 gm
Ammonium chloride	1.5 gm
Acetic acid (10% solution)	1 litre

Boiling immersion (Twenty minutes)

The article is immersed in the boiling solution, producing a light pink-buff colour after about two or three minutes. This gradually darkens with continued immersion to a grey-brown colour which develops within twenty minutes. When the colour has developed, the article is removed and washed in warm water and dried in sawdust. When dry it may be wax finished.

The grey-brown colour is only produced on standard silver. Silver-plated surfaces are unaffected initially, but are later attacked by the solution, which tends to remove the plating in spots. A bi-metallic effect can be obtained, if the silver-plate is cut through to the underlying copper, which becomes dark grey. However this is difficult to obtain without general damage to the remainder of the plate surface.

▲ The ammonium carbonate should be added to the acid in small quantities. The reaction is effervescent, and if large amounts are added, this effervescence becomes very vigorous.

6.7   Variegated greys and light browns Semi-gloss

Copper nitrate	15 gm
Zinc nitrate	15 gm
Ferric chloride	0.5 gm
Hydrogen peroxide (100 vols.)	to form a paste

Applied paste (Four hours)

The ingredients are ground using a mortar and pestle, and the hydrogen peroxide added in small quantities to form a thin creamy paste. The paste is applied to the object with a soft brush, and left to dry in air. After four hours, the residue is washed away with cold water and the object is allowed to dry in air for several hours. When completely dry, it may be wax finished.

The surface of the object should be handled as little as possible until dry, and should not be brushed during washing.

▲ Hydrogen peroxide should not be allowed to come into contact with combustible material. It should be prevented from coming into contact with the eyes and skin, as it may cause severe irritation. It is very harmful if taken internally.

6.8*   Mid-brown Satin/semi-matt

Sodium sulphide	50 gm
Antimony trisulphide	50 gm
Water	1 litre
Acetic acid (10% solution)	100 cm³

Hot scratch-brushing (A few minutes)

The solid ingredients are dissolved in the water, and the solution brought up to temperature (50°C). The acetic acid is added immediately prior to use. The surface is directly scratch-brushed with the hot solution, alternating with hot water, until the mid-brown colour is produced. The article is then washed in warm water and allowed to dry in air. When dry it may be wax finished.

The procedure produces an even mid-brown colour on standard silver, but is not suitable for use with silver-plated surfaces.

▲ Preparation and colouring must be carried out in a fume cupboard or a well ventilated area, as hydrogen sulphide gas is readily evolved from the acidified solution.

▲ Sodium sulphide is corrosive to the eyes and skin, particularly in the solid state.

6.9   Dark greyish-brown Satin/semi-matt

Potassium sulphide	1 gm
Ammonium chloride	4 gm
Water	1 litre

Cold scratch-brushing (A few minutes)

The surface is directly scratch-brushed with the cold solution, producing a reddish-brown colouration which tends to be uneven. Continued working with the cold solution produces a more even surface which is darker in colour. When the desired surface quality has been obtained, the article is washed in cold water and allowed to dry in air. It may be wax finished when dry.

The surface produced by this treatment tends to be tonally variegated. It is difficult to produce an even finish at the reddish-brown stage, and continued working causes a gradual loss of the reddish tinge.

▲ Colouring should be carried out in a well ventilated area, as some hydrogen sulphide gas is evolved.

6.10   Light bronzing Semi-gloss

Antimony trisulphide	12.5 gm
Sodium hydroxide	40 gm
Water	1 litre

Applied liquid and hot scratch-brushing (A few minutes)

The hot solution (70°C) is applied to the surface liberally with a soft brush, to allow the colour to build up, and is then lightly worked with a scratch-brush, which is intermittently wetted with the solution. A light warm grey-brown tone is produced, which is finished by light burnishing under hot water. The article is allowed to dry in air, and may be wax finished when dry.

The light brownish tone is only produced on standard silver. The procedure is not suitable for use with silver-plated surfaces, which rapidly become badly corroded.

▲ Sodium hydroxide is a powerful caustic alkali and must be prevented from coming into contact with the eyes and skin. When preparing the solution, small quantities of the alkali should be added to large quantities of water and *not* vice-versa.

6.11   Mid-brown mottled with dark brown Semi-matt

Ammonium carbonate	48 gm
Potassium tartrate	12 gm
Sodium chloride	12 gm
Copper sulphate	12 gm
Acetic acid (30% solution)	5 cm³
Water	to form a paste

Applied paste (Two or three days)

A mid-brown, mottled with dark brown, is produced on standard silver. Some blue-green patina may also form.

An uneven grey ground is produced on silver-plated surfaces, which tend to acquire a more substantial blue-green patina taking the form of a bright incrustation. It is difficult to determine from test results the extent to which this is the result of a breakdown of the plated surface, accompanied by colouring of the underlying metal.

The ingredients are ground to a creamy paste with a little water, using a pestle and mortar. The paste is applied to the object with a soft brush and left to dry out completely for about two days. The residue is washed off in cold water and the object left to dry in air for several days. When completely dry it may be wax finished.

6.12   Reddish-brown Semi-gloss

Potassium permanganate	5 gm
Potassium chlorate	20 gm
Copper sulphate	50 gm
Water	1 litre

Boiling immersion (Twenty minutes)

A non-adherent patchy grey is produced on silver-plated surfaces.

The brown colour produced on standard silver is very fragile when wet. It also tends to be fragile when dry, and can be brushed back to produce a variegated or 'relieved' brown on silver finish. The residual brown can be burnished and becomes more adherent. An adherent brown colour was produced on a satinised finish rather than a high polish. The results tend to be variable, and the procedure cannot be generally recommended.

▲ Potassium chlorate is a powerful oxidant and should not be allowed to come into contact with combustible material. It should not be mixed with the permanganate in the solid state. It should also be prevented from coming into contact with the eyes, skin and clothing.

The article is immersed in the boiling solution, which produces a brown colour within about five minutes. Continued immersion darkens the colour very gradually. After about twenty minutes the article is removed and rinsed in warm water and allowed to dry in air. When dry it may be wax finished.

6.13   Salmon pink (variegated) Semi-matt

A	Copper sulphate	125 gm
	Ferrous sulphate	100 gm
	Water	1 litre
B	Ammonium chloride	1 gm

Boiling immersion (A) (Fifteen minutes)
Boiling immersion (A+B) (Ten minutes)

A salmon pink bloom is produced on standard silver. This develops gradually but tends to remain variegated, rather than covering the entire surface.

Silver-plate is not coloured, but is gradually attacked by the solution, particularly where this is thin or flawed.

Bi-metallic colouring effects can be obtained, see recipe 6.62.

The object is immersed in the boiling solution of copper sulphate and ferrous sulphate for a period of ten to fifteen minutes. It is then removed to a bath of hot water, while the ammonium chloride is added to the colouring solution. The object is re-immersed and immersion continued for about ten minutes. When the colour has developed, it is removed and washed in hot water, allowed to dry in air, and may be wax finished when dry.

6.14   Dark brown Gloss

Antimony trisulphide	20 gm
Ferric oxide	20 gm
Potassium polysulphide	2 gm
Ammonia (.880 solution)	2 cm³
Water	to form a paste (approx 4 cm³)

Applied paste (sulphide paste) (Several minutes)

A dark brown gloss is produced on standard silver, and a dark grey semi-gloss on silver-plated surfaces. Some mottling or slight tonal variation may occur.

A variegated surface can be obtained by brushing the paste onto the whole surface and allowing it to dry for about three hours, until it is superficially dry and cracking, and then burnishing it away with a nylon brush or a brass scratch-brush. This may be repeated as necessary to darken the colour.

▲ A nose and mouth face mask should be worn when the residues are brushed away.

The potassium polysulphide is ground using a pestle and mortar and the other solid ingredients added and mixed. These are made into a thick paste by the gradual addition of the ammonia and a small quantity of water. The paste is applied to the whole surface of the object to be coloured, using a bristle-brush. A stiff nylon brush is then used to work the surface gradually in small areas. The nylon brush should be intermittently dipped in a mixture of the dry ingredients of the paste (in the same proportion as above), and the surface worked in this way until the paste becomes dry and falls from the surface. When the whole surface is dry and completely free of any residue or dust, it may be wax finished.

6.15   Warm grey   Semi-gloss

Sodium chlorate	100 gm
Copper nitrate	25 gm
Water	1 litre

Boiling immersion (Forty minutes to one hour)

The object is immersed in the boiling solution, and the surface gradually acquires a warm grey tone. When the colour has developed sufficiently, the object is removed and washed in warm water. It is dried in sawdust, and may be wax finished when dry.

A warm grey is produced on standard silver. Silver-plated surfaces are initially dulled slightly, and subsequently acquire patches of a greyish tarnish.

A bi-metallic colouring effect can be obtained on silver-plated gilding metal, see recipe 6.58.

▲ Sodium chlorate is a powerful oxidant and must not be allowed to come into contact with combustible material or acids. It should also be prevented from coming into contact with the eyes, skin or clothing.

6.16   Very pale grey   Semi-gloss

Copper carbonate	125 gm
Ammonia (.880 solution)	250 cm³
Water	750 cm³

Hot immersion (Ten to twenty minutes)

The article is immersed in the hot solution (50°C) for a period of about ten to twenty minutes, producing a very pale grey colour. It is then removed, washed in cold water and allowed to dry in air. It may be wax finished when dry.

The grey colouration is very pale, and is only produced on standard silver. Silver-plated surfaces are slightly tarnished after ten minutes immersion, and tend to be breached where the plating is thin or inconsistent, revealing the underlying metal.

The procedure cannot be recommended for colouring silver or silver plate, but good bi-metallic colouring effects are obtained with silver-plated gilding metal, see recipe 6.67.

▲ Ammonia vapour will be freely liberated from the solution, which irritates the eyes and respiratory system. Preparation and colouring should be carried out in a well ventilated area. The skin and particularly the eyes, should be protected from accidental splashes of this corrosive alkali.

6.17   Light grey   Semi-matt

Ammonium persulphate	10 gm
Sodium hydroxide	30 gm
Water	1 litre

Boiling immersion (Ten to fifteen minutes)

The ammonium persulphate is dissolved in half the water, and the sodium hydroxide separately dissolved in the remaining water. The two solutions are then mixed together and the mixture heated to boiling. The article to be coloured is immersed in the boiling solution until the colour has developed. After about ten or fifteen minutes, it is removed and washed in hot water. After thorough drying in sawdust it may be wax finished.

A light grey is produced on standard silver.

A very variegated metallic grey is produced on silver-plated surfaces. Prolonged immersion tends to remove the plate. A bi-metallic colouring effect can be produced, see recipe 6.66.

▲ Ammonium persulphate (peroxodisulphate) is a powerful oxidant, and must be kept out of contact with all combustible materials. It should be prevented from coming into contact with the eyes, skin or clothing.

▲ Sodium hydroxide is a powerful caustic alkali and must be prevented from coming into contact with the eyes and skin.

6.18   Light grey (dusty appearance)   Semi-matt

Copper sulphate	120 gm
Potassium chlorate	60 gm
Water	1 litre

Hot immersion (To fifteen minutes)

The article is immersed in the hot solution (80°C), producing a buff colour after a few minutes, which becomes slightly more grey with continued immersion. During the later stages of immersion a dusty light green tends to be deposited on the surface. After about fifteen minutes, the article is removed, washed in warm water and dried in sawdust. When dry it may be wax finished.

The light grey colour produced on standard silver has a dusty appearance.

Silver-plated surfaces are unaffected initially, but any local defects or thin areas in the plating tend to be attacked as immersion continues. A bi-metallic colouring effect can be produced with short immersion times, see recipe 6.60.

▲ Potassium chlorate is a powerful oxidant and must not be allowed to come into contact with combustible material. It should be prevented from coming into contact with the eyes and skin.

6.19   Light warm grey (brown patches)   Semi-matt

Sodium hydroxide	25 gm
Potassium persulphate	10 gm
Water	1 litre

Hot immersion (Several minutes)

The potassium persulphate is dissolved in the water, and the sodium hydroxide then added in small quantities while stirring. The solution is then brought up to the correct temperature (70–80°C) and the article immersed. A light warm grey is gradually produced on standard silver, taking several minutes to develop. If immersion is continued for any length of time, irregular brown patches tend to appear. When the desired colour has been produced, the article is removed and washed in warm water. It is dried in sawdust, and may be wax finished when dry.

Silver-plated surfaces are initially dulled by the solution, producing a light metallic grey colouration, which is slightly variegated. Continued immersion tends gradually to remove the plating, to expose and colour the underlying copper plate. A bi-metallic effect can be obtained, see recipe 6.65.

▲ Potassium persulphate is a powerful oxidant and should not be allowed to come into contact with combustible material. The sodium hydroxide and the persulphate must not be allowed to come into contact in the solid state.

▲ Sodium hydroxide is a powerful caustic alkali which can cause severe burns. Both the solid and solutions should be prevented from coming into contact with the eyes and skin.

6.20   Dull grey patches on plate (silver-plate only) Semi-matt

Potassium permanganate	10 gm
Water	1 litre

Hot immersion (Fifteen minutes)

The article is immersed in the hot solution (80°C) for a period of about fifteen minutes. Dull grey patches gradually develop on the plate, producing a variegated surface. The article is removed and washed in hot water and allowed to dry in air. It may be wax finished when dry.

A dark grey layer is produced on standard silver, but this is totally non-adherent when dry.

On silver-plated surfaces, a more even grey colour was obtained on a matted or satin surface. Some bi-metallic effects are also obtained where the silver-plate is deliberately cut through to expose the underlying copper plate and gilding metal, which are coloured dark grey and brown respectively.

The results in all cases tend to be unpredictable and the procedure cannot be generally recommended.

6.21★   Mid-grey Matt

Potassium sulphide	10 gm
Water	1 litre
Ammonia (.880 solution)	Five drops

Warm immersion (Fifteen minutes)

The article is immersed in the solution (40–50°C), which initially produces a series of uneven lustre colours, that gradually change to an opaque grey after about ten minutes. After fifteen minutes, the article is removed and allowed to dry in air. When dry, it may be wax finished.

A darker grey is produced on silver-plated surfaces, which tends to be uneven. The grey colour is best produced on silver-plate by a brief immersion, followed by light scratch-brushing with the solution and hot water.

The solution may also be applied to standard silver by scratch-brushing, which produces a satin surface of a similar colour.

▲ Preparation and colouring should be carried out in a well ventilated area, as some hydrogen sulphide gas will be evolved.

6.22★   Grey (to dark purplish-grey) Semi-gloss   *Pl. XI*

Sodium dichromate	150 gm
Nitric acid (10% solution)	20 cm³
Hydrochloric acid (15% solution)	6 cm³
Ethanol	1 cm³
Water	1 litre

Cold immersion (Five to twenty minutes)

The sodium dichromate is dissolved in the water, and the other ingredients are then added separately while stirring the solution. The article is immersed in the cold solution which produces a grey colour after four or five minutes, that then darkens gradually to a dark purplish-grey after about twenty minutes. When the desired colour has been obtained, the article is removed and washed in cold water. It is allowed to dry in air, and may be wax finished when dry.

Silver-plated surfaces are darkened by the solution initially, but are then gradually corroded, producing colouring effects that are attributable to the underlying metal. No strong bi-metallic effect was obtained.

▲ Sodium dichromate is a powerful oxidant, and must not be allowed to come into contact with combustible material. The dust must not be inhaled. The solid and solutions should be prevented from coming into contact with the eyes and skin, or clothing. Frequent exposure can cause skin ulceration, and more serious effects through absorption.

▲ Both hydrochloric and nitric acids are corrosive and should be prevented from coming into contact with the eyes, skin or clothing.

6.23   Dark grey Satin/semi-matt

Potassium sulphide	25 gm
Ammonium chloride	38 gm
Water	1 litre

Cold immersion (Several minutes)

The article is immersed in the cold solution, which produces a grey colouration within two or three minutes. This tends to be uneven at first but becomes more even if immersion is continued for a further two or three minutes. The article is removed and washed in cold water and allowed to dry in air. When dry it may be wax finished.

A dark grey colour is also produced on silver-plated surfaces, which tend to retain a glossy finish. Immersion should not be prolonged, as any weaknesses or thin areas in the plating tend to be badly affected.

An after-treatment suggested by some sources, consisting of a brief immersion in a cold dilute solution of barium sulphide (2 gm/litre), did not appear to alter the results.

▲ Preparation and colouring should be carried out ·in a well ventilated area, as some hydrogen sulphide gas will be evolved.

6.24★   Dark grey Matt

Potassium sulphide	3 gm
Ammonium carbonate	6 gm
Water	1 litre

Hot immersion (Five to ten minutes)

The article is immersed in the hot solution (50°C), which rapidly colours the surface grey. Immersion is continued for several minutes, and the article is then removed and washed in warm water. It is allowed to dry in air, and may be wax finished when dry.

The even dark grey colour is only produced on standard silver. Silver-plated surfaces initially acquire a series of lustre colours, and then gradually become grey as the plating begins to break down.

▲ Preparation and colouring should be carried out in a well ventilated area, as some hydrogen sulphide gas is evolved from the hot solution.

6.25*   Dark grey Matt

Pl. XI

Ammonium sulphide (16% solution)	150 cm³
Water	850 cm³

Hot immersion (To ten minutes)

The article is immersed in the hot solution (50°C), which gradually produces a grey colouration within ten minutes. When the colour has developed, the article is removed, washed in cold water and allowed to dry in air. When dry it may be wax finished.

A similar colour is also produced on silver-plated surfaces, but immersion should not be prolonged as the solution tends to attack the plating.

▲ Ammonium sulphide solution liberates both ammonia and hydrogen sulphide gas. Preparation and colouring must be carried out in a fume cupboard or well ventilated area. The solution is also corrosive, and must be prevented from coming into contact with the eyes and skin.

6.26   Dark purplish-grey Satin/semi-gloss

Sodium sulphide	30 gm
Sulphur (flowers of sulphur)	4 gm
Water	1 litre

Hot scratch-brushing (A few minutes)

The surface is directly scratch-brushed with the hot solution (60°C), which initially produces a reddish-brown colouration. Continued working darkens the colour to a purplish-grey. When the desired colour has been obtained, the article is washed in warm water and allowed to dry in air. When dry it may be wax finished.

The procedure produces an even dark purplish-grey colour on standard silver. The interim reddish-brown colour is difficult to obtain evenly.

The procedure is not suitable for silver-plated surfaces, which tend to be destructively attacked by the treatment.

▲ Sodium sulphide is corrosive and can cause severe burns to the eyes and skin, particularly in the solid form.

▲ Preparation and colouring should be carried out in a well ventilated area, as some hydrogen sulphide gas is evolved.

6.27*   Dark grey/black Satin/semi-matt

Pl. XI

Ammonium sulphide (16% solution)	

Applied liquid (A few minutes)

The solution is applied directly to the surface of the object to be coloured, using a soft cloth, and the dark colour quickly develops. The surface is then scratch-brushed lightly, additional solution being applied if necessary, until an even colour is produced. (Silver-plated surfaces should not be scratch-brushed, but washed in cold water after the colour has developed.) When treatment is complete, the object is washed in cold water and allowed to dry in air. When dry, it may be wax finished.

Treatment should not be prolonged in the case of silver-plated surfaces. The colour produced is similar to that for standard silver, but the surface tends to be glossy and has a slightly cloudy quality and some lustre colour.

▲ Ammonium sulphide solution freely liberates both ammonia and hydrogen sulphide gas. It is essential to carry out the colouring in a well ventilated area. The solution is also corrosive, and must be prevented from coming into contact with the eyes and skin.

6.28   Black Satin

Antimony trisulphide	30 gm
Ferric oxide	10 gm
Ammonium sulphide (16% solution) to form a paste	

Applied paste (sulphide paste) (Several minutes)

The ingredients are mixed to a thick paste using a pestle and mortar, by the very gradual addition of a small quantity of ammonium sulphide solution. The paste is applied to the whole surface of the article to be coloured, using a bristle-brush. A stiff nylon brush is then used to work the surface gradually in small areas. The nylon brush should be intermittently dipped in a mixture of the dry ingredients of the paste (in the same proportion as above), and the surface worked in this way until the paste becomes dry and falls from the surface. When the whole surface is dry and completely free of any residue or dust, it may be wax finished.

A dark reddish-brown satin finish is produced on silver-plated surfaces. Some tonal variation tends to occur.

A variegated surface can be obtained by brushing the paste onto the whole surface and allowing it to dry for about three hours, until it is superficially dry and cracking, and then burnishing it away with a nylon brush or a brass scratch-brush. This may be repeated, as necessary, to darken the colour.

▲ Ammonium sulphide solution is very corrosive and precautions should be taken to protect the eyes and skin while the paste is being prepared. It also liberates harmful fumes of ammonia and of hydrogen sulphide, and the preparation should therefore be carried out in a fume cupboard or well ventilated area. The paste once prepared is relatively safe to use but should be kept away from the skin and eyes.

▲ A nose and mouth face mask fitted with a fine dust filter should be worn when the residues are brushed away. The dust produced in this case contains antimony compounds, which are toxic.

6.29   Dark grey   Matt

Potassium polysulphide	1.5 gm
Water	1 litre

Hot immersion   (Five minutes)

In tests with standard silver, an adherent grey colour could only be obtained on a matted or grained satin surface. The colour tended to be non-adherent on highly polished surfaces.

An even dark grey with a metallic gloss finish was obtained on silver-plated surfaces. Prolonging the immersion tends to produce some slightly dull areas.

The article is immersed in the hot solution (50°C), which produces a rapid series of lustre colours within the first minute. The surface then acquires a grey colour. After about five minutes, the article is removed, washed in warm water and allowed to dry in air. It may be wax finished when dry.

6.30*   Green patina on dark grey ground   Matt   *Pl. XI*

Ammonium chloride	350 gm
Copper acetate	200 gm
Water	1 litre

Applied liquid   (Twice a day for five days)

A similar green patina is produced on a silver-plated surface, but the ground colour tends to be brown.

The solution is applied to the object by wiping with a soft cloth, to leave a sparingly moist surface. This is then left to dry in air, before the next application. The procedure is repeated twice a day for about five days. The patina develops during the periods of drying. When treatment is complete, the object is allowed to dry in air for a further five days. When the surface is completely dry and patina development has ceased, it may be wax finished.

6.31*   Pale blue-green patina on brown ground   Matt

Copper nitrate	30 gm
Zinc chloride	30 gm
Water	to form a paste

Applied paste   (Several hours)

Similar results are produced on silver-plated surfaces.

The ingredients are ground to a paste with a little water, using a pestle and mortar. The paste is applied to the object to be coloured with a soft brush, and is then left to dry out for several hours. The residual paste is then removed by washing the object in cold water. It is then left to dry in air for three or four days. When completely dry, it may be wax finished.

6.32   Dark brown (blue-green patina)   Semi-matt

Copper sulphate	20 gm
Zinc chloride	20 gm
Water	to form a paste

Applied paste   (Several hours)

A dark brown surface with a thin covering of blue-green patina is produced on standard silver. The patina may be removed during treatment if a plain brown colour is required.

A greenish patina on a dark grey variegated ground is produced on silver-plated surfaces. In the case of silver-plated gilding metal, a bright blue-green patina tended to form where the plate had been deliberately breached.

The ingredients are ground to a thin creamy paste with a little water, using a pestle and mortar. The paste is applied sparingly to the object using a soft cloth or brush. The surface darkens as the paste is applied. The residual paste is washed away after about an hour, and the object allowed to dry in air. A thin even coating of blue-green patina tends to form, which may either be left, or removed by bristle-brushing under cold water and further drying. The procedure may be repeated if necessary. When treatment is complete and the dry surface shows no signs of further patina development, it may be wax finished.

6.33   Mid-brown (blue-green patina)   Semi-matt   *Pl. XI*

Copper nitrate	100 gm
Water	200 cm³
Ammonia (.880 solution)	400 cm³
Acetic acid (6% solution)	400 cm³
Ammonium chloride	100 gm

Applied liquid   (Five or six days)

An even brown colour with a slightly grey tone is produced on standard silver. Blue-green patina will tend to form in any surface features such as punched marks or textured areas.

A blue-green patina on a dark brown ground is produced on silver-plated surfaces, as in plate XI. Any weak areas in the plating will tend to be corroded, and treatment should not be prolonged.

▲ Ammonia vapour will be liberated from the solution, which irritates the eyes and respiratory system. Preparation and colouring should be carried out in a well ventilated area. The skin and eyes should be protected from accidental splashes of the strong solution.

The copper nitrate is dissolved in the water and the ammonia added, followed by the acetic acid and finally the ammonium chloride. The solution is applied to the surface by dabbing with a soft cloth, and the article then allowed to dry. The surface should be sparingly moist. This procedure is repeated several times on the first day, and then twice a day for about five days. The article should be left for a further five days without treatment, and may be wax finished when completely dry.

6.34* Greenish-grey Satin

Pl. XI

Copper sulphate	125 gm
Sodium chlorate	50 gm
Water	1 litre

Boiling immersion (Fifteen minutes)

The article is immersed in the boiling solution, which produces a creamy surface layer after about five minutes. This gradually darkens to a greenish-grey, within about fifteen minutes. When the desired colour has developed, the article is removed, washed in warm water, and allowed to dry in air. When dry, it may be wax finished.

In tests with standard silver, a more green colour was obtained on a grained satin surface.

A greenish-grey colour was also obtained on silver-plated surfaces, but this tended to be uneven and cloudy in appearance.

The green colouration is less apparent when the article is in the colouring bath, and appears to develop on drying and exposure to daylight.

▲ Sodium chlorate is a powerful oxidant and must not be allowed to come into contact with combustible materials. The dust must not be inhaled. It should be prevented from coming into contact with the eyes, skin and clothing.

6.35* Lustre series Gloss

Pl. XI

Sodium thiosulphate	240 gm
Copper acetate	25 gm
Water	1 litre
Citric acid (crystals)	30 gm

Cold immersion (Various)

The sodium thiosuphate and the copper acetate are dissolved in the water, and the citric acid added immediately prior to use. The article is immersed in the cold solution and a series of lustre colours are produced in the following sequence: golden-yellow/orange; brown; purple; blue; pale grey. This sequence takes about thirty minutes to complete, the brown changing to purple after about fifteen minutes. It is followed by a second series of colours, in the same sequence, which take a further thirty minutes to complete. The latter colours in the second series tend to be slightly more opaque. A third series of colours follows, if immersion is continued, but these tend to be indistinct and murky. (If the temperature of the solution is raised then the sequences are produced more rapidly. However, it was found in tests that the results tended to be less even. Temperatures in excess of about 35–40°C tend to cause the solution to decompose, with a concomittant loss of effect.) When the required colour has been produced, the article is removed and washed in cold water. It is allowed to dry in air, any excess moisture being removed by gentle blotting with a tissue. When dry it is wax finished or coated with a fine lacquer.

The orange colour in the second series tends to be more red than in the first. At a temperature of 30–35°C this was reached in from seven to ten minutes. The blue colours may also tend to be more turquoise.

Plate XI shows a silver-plated sample after 50 minutes' immersion.

6.36 Lustre series Gloss

Pl. XI

Sodium thiosulphate	280 gm
Lead acetate	25 gm
Water	1 litre
Potassium hydrogen tartrate	30 gm

Warm immersion (Various)

The sodium thiosulphate and the lead acetate are dissolved in the water, and the temperature is adjusted to 30–40°C. The potassium hydrogen tartrate is added immediately prior to use. The article to be coloured is immersed in the warm solution and a series of lustre colours are produced in the following sequence: golden-yellow/orange; purple; blue; pale blue/pale grey. The colour sequence takes about ten minutes to complete at 40°C. (Hotter solutions are faster, cooler solutions slower.) A second series of colours follows in the same sequence, but these tend to be less distinct. When the desired colour has been reached, the article is removed and immediately washed in cold water. It is allowed to dry in air, any excess moisture being removed by dabbing with an absorbent tissue. When dry it may be wax finished, or coated with a fine lacquer.

The silver-plated sample shown in plate XI illustrates the sequence of colours obtained. It was prepared by gradually lifting the sample from the solution as colour changes occurred.

Citric acid may be used instead of potassium hydrogen tartrate, and in the same quantity. Tests carried out with both these alternatives produced identical results.

▲ Lead acetate is very toxic and will give rise to serious conditions if taken internally. The dust, and the vapour evolved from the hot solution, must not be inhaled.

6.37 Reddish-brown lustre with blue tinge Gloss

Selenous acid	6 cm³
Water	1 litre
Sodium hydroxide solution (250 gm/litre)	50 drops

Warm immersion (Forty seconds)

The acid is dissolved in the water, and the sodium hydroxide added by the drop, while stirring. The solution is then heated to 25–30°C. The article to be coloured is immersed in the solution, producing a purplish-brown lustre after about fifteen seconds. The colour develops rapidly, and the article should be removed after about forty seconds, and washed thoroughly in cold water. It is allowed to dry in air, and wax finished when completely dry.

Prolonging the immersion causes a gradual fading of the colour. If immersion times are extended to from thirty to fifty minutes, some pale mottled grey colours are produced.

Silver-plated surfaces were unaffected for short immersion times. Prolonged immersion for periods from thirty to fifty minutes caused some irregular markings to appear, apparently through damage to the plated surface.

▲ Selenous acid is poisonous and very harmful if swallowed or inhaled. Colouring must be carried out in a well ventilated area. Contact with the skin or eyes must be avoided to prevent severe irritation.

▲ Sodium hydroxide is a powerful caustic alkali. Contact with the eyes and skin must be prevented.

6.38*   Mid-brown mottle on silver   Semi-matt

Copper nitrate	100 gm
Water	200 cm³
Ammonia (.880 solution)	400 cm³
Acetic acid (6% solution)	400 cm³
Ammonium chloride	100 gm

Sawdust technique (Twenty to thirty hours)

The copper nitrate is dissolved in the water and the ammonia gradually added. The acetic acid is then added, followed by the ammonium chloride. The article to be coloured is laid-up in sawdust which has been evenly moistened with this solution, and is then left for a period of twenty to thirty hours. When treatment is complete, the article is washed in cold water and allowed to dry in air. When completely dry, after several hours, it may be wax finished.

A mid-brown mottle on a slightly dulled metal surface is produced on standard silver. The surface tends to be flecked with fine dark spots.

Silver-plated gilding metal surfaces were strongly etched by this solution, ultimately producing a blue, green and brown stippled surface. If the treatment is halted at an earlier stage, this patination is interspersed with spots or areas of bright silver-plate. The multicoloured patinated surface has a matt finish.

▲ Preparation and colouring should be carried out in a well ventilated area. Ammonia vapour and some acetic acid vapour will be liberated, which cause irritation of the eyes and respiratory system.

▲ The strong ammonia solution must be prevented from coming into contact with the eyes and skin, as it will cause burns or severe irritation.

6.39*   Mid- and dark purplish-brown stipple   Semi-matt   *Pl. XI*

Ammonium chloride	100 gm
Ammonium carbonate	150 gm
Water	1 litre

Sawdust technique (To twenty or thirty hours)

The article to be coloured is laid-up in sawdust which has been moistened evenly with the solution. It may then be left for periods of up to twenty or thirty hours. When treatment is complete, the object is washed in cold water and allowed to dry in air. When it is completely dry, it may be wax finished.

A mottled mid-purplish-brown surface, stippled with flecks of a darker purplish-brown, is produced on standard silver.

Silver-plated surfaces were selectively etched by this treatment, producing a scattering of fine pits. A pale green patina developed in these pits, in the case of silver-plated gilding metal. The unetched areas of the surface became grey.

6.40*   Grey-brown mottle on buff-tinged silver   Semi-gloss

Copper nitrate	200 gm
Sodium chloride	100 gm
Water	1 litre

Sawdust technique (Twenty or thirty hours)

The article to be coloured is laid-up in sawdust which has been moistened evenly with the solution. It should then be left for a period of twenty or thirty hours. After twenty hours the object should be examined, and re-packed for shorter periods of time, as necessary. The sawdust should be kept moist throughout the period of treatment. When treatment is complete, the article should be washed thoroughly in cold water and allowed to dry in air. When completely dry, after several hours, it may be wax finished.

A grey-brown mottle on a buff-tinted ground is produced on standard silver.

In tests carried out with silver-plated gilding metal, the surfaces were selectively etched by the solution, where the granular sawdust was in close contact with the plate, cutting through the plate to an etched ochre. Some blue-green patina developed in the region of the etched areas. The remainder of the plated surface acquired a very slight grey tint. The etched spots were evenly distributed on the surface.

6.41*   Dark brown stipple   Semi-gloss

Copper acetate	60 gm
Copper carbonate	30 gm
Ammonium chloride	60 gm
Water	1 litre
Hydrochloric acid (15% solution)	5 cm³

Sawdust technique (Ten to twenty hours)

The object to be coloured is laid-up in sawdust which has been evenly moistened with the solution. It is left for a period of about ten hours and then examined. If further treatment is required, it should be re-packed and examined at intervals of a few hours. When treatment is complete, the object is removed, washed thoroughly in cold water and allowed to dry in air. When completely dry, it may be wax finished.

A dark brown stipple on a bright metal surface is produced on standard silver. As treatment progresses, the surface tends to become completely covered with a mid-brown, stippled with dark brown.

A dark grey stipple is produced initially on silver-plated surfaces, but the plating tends to become pitted. In tests on silver-plated gilding metal a greenish patina developed in the pits.

▲ Hydrochloric acid can cause burns or severe irritation, and must be prevented from coming into contact with the eyes and skin.

6.42*   Brown and purple-brown mottle   Semi-matt

Ammonium carbonate	120 gm
Ammonium chloride	40 gm
Sodium chloride	40 gm
Water	1 litre

Sawdust technique (Twenty to thirty hours)

The article to be coloured is laid-up in sawdust which has been evenly moistened with the solution, and is then left for a period of about twenty or thirty hours. The sawdust should be kept moist throughout the period of treatment. When treatment is complete, the article should be washed in cold water and allowed to dry in air. When it is completely dry, it may be wax finished.

An even buff-brown and purple-brown mottle (or stipple) is produced on standard silver. If the medium used is fairly fine, then the whole surface will be coloured. If the medium is coarse, then spots of uncoloured silver tend to remain.

Silver-plated gilding metal was selectively attacked by the solution, producing a scattering of fine pits in the plated surface, in which a pale green patina developed. The remainder of the surface was coloured a pale grey.

6.43*   Mottled mid-brown Semi-matt

Ammonium chloride	350 gm
Copper acetate	200 gm
Water	1 litre

Sawdust technique (Ten to twenty hours)

An even mid-brown colour with a slightly mottled appearance is obtained with standard silver.

Silver-plated surfaces are darkened initially and acquire a dark brown stipple, if relatively dry media are used. If the medium is moist or wet, the plate tends to be heavily corroded, particularly with longer treatment times, exposing the underlying metal.

The object to be coloured is laid-up in sawdust which has been moistened evenly with the solution, and left for a period of ten to twenty hours. After ten hours it should be examined and re-packed if further treatment is required. It should subsequently be examined every few hours until the required surface finish is obtained. The sawdust should be kept moist during the full period of treatment. The object should then be removed, washed in cold water and allowed to dry in air. When completely dry, it may be wax finished.

6.44   Purplish-brown stipple Semi-gloss

Copper nitrate	20 gm
Ammonium chloride	20 gm
Calcium hypochlorite	20 gm
Water	1 litre

Sawdust technique (Twenty or thirty hours)

A purplish-brown stipple is produced on standard silver. Prolonged treatment tends to produce a slightly yellow tint to the uncoloured areas of silver.

Silver-plated surfaces were slightly tarnished by this treatment. Some localised pitting occurred with more extended periods.

▲ The corrosive fine dust of the calcium hypochlorite (bleaching powder) must not be inhaled, as it can severely irritate the respiratory system. The solution must be prevented from coming into contact with the eyes, which would be severely irritated.

The object to be coloured is laid-up in sawdust which has been moistened evenly with the solution, and left for a period of twenty or thirty hours. The sawdust should be kept moist throughout this time. After treatment, the object is washed in cold water and allowed to dry in air. When thoroughly dry, it may be wax finished.

6.45*   Dark brown mottle on dull silver Semi-gloss

Copper sulphate	100 gm
Lead acetate	100 gm
Ammonium chloride	10 gm
Water	1 litre

Sawdust technique (Ten to twenty hours)

A dark brown stipple is produced on a slightly dulled metal surface, on standard silver, which tends to spread to an extensive mottle as treatment proceeds.

A similar effect is produced on silver-plated surfaces, but the colour tends to be grey rather than brown.

▲ Lead acetate is a toxic substance, and every precaution should be taken to prevent inhalation of the dust, when preparing the solution. It is very harmful if taken internally. A clean working method is essential throughout.

The object to be coloured is laid-up in sawdust which has been evenly moistened with the solution. It is then left to develop for a period of up to about twenty hours. When treatment is complete, the object is removed, washed in cold water and allowed to dry in air. It may be wax finished when completely dry.

6.46   Mixed light grey and greyish-brown stipple Semi-matt

Ammonium carbonate	150 gm
Copper acetate	60 gm
Sodium chloride	50 gm
Potassium hydrogen tartrate	50 gm
Water	1 litre

Sawdust technique (Ten to twenty hours)

A mixed stipple of light grey and greyish-brown colours is produced on standard silver.

Little effect was produced on a silver-plated surface other than a slight tarnish where the medium was excessively moist. Where the silver-plated gilding metal had been deliberately cut through by abrasion, some blue-green patina and ochre colouring was produced on the underlying metal.

The object to be coloured is laid-up in sawdust which has been evenly moistened with the solution. It is left for a period of from ten to twenty hours, colour development being monitored with a sample, so that the object remains undisturbed. When treatment is complete, the object is removed, washed in cold water and allowed to dry in air for several hours. When dry, it may be wax finished.

6.47   Light grey stipple Semi-gloss

Copper nitrate	280 gm
Water	850 cm³
Silver nitrate	15 gm
Water	150 cm³

Sawdust technique (Twenty hours)

A whiteish-grey stipple, with some darker stippling, is produced on standard silver. The non-stippled areas of metal are slightly dulled by the treatment.

Similar effects are produced initially on silver-plated surfaces, but these tend to be attacked progressively as treatment continues, and become pitted, revealing the underlying metal.

▲ Silver nitrate must be prevented from coming into contact with the skin, and more particularly the eyes, as it may cause burns or severe irritation.

The copper nitrate and the silver nitrate are dissolved in separate portions of water, and the two solutions then mixed. The article to be coloured is laid-up in sawdust which has been evenly moistened with this mixed solution. It is left for a period of about twenty hours, ensuring that the sawdust remains moist throughout this time. When treatment is complete, the article is removed and washed in cold water. After allowing it to dry thoroughly in air for some time, it may be wax finished.

6.48*   **Mid-brown Semi-matt**

Ammonium chloride	100 gm
Ammonium carbonate	150 gm
Water	1 litre

Pl. XI

Cotton-wool technique (To twenty or thirty hours)

An even mid-brown was obtained on standard silver.

Tests on silver-plated gilding metal produced a gradual 'pitting' of the plated surface. Pale green patina developed in the pitted areas. The remainder of the surface was coloured grey.

Cotton-wool, moistened with the solution, is applied to the surface of the object to be coloured. It is then left for periods of up to twenty or thirty hours. The cotton-wool should be kept moist throughout the period of treatment. After about ten or fifteen hours the surface should be examined, by carefully exposing a small portion, to determine the extent of colour development. Subsequently a check should be made every few hours until the desired surface finish is obtained. When treatment is complete, the object is washed in cold water and allowed to dry in air. When completely dry, it may be wax finished.

6.49   **Mid-brown Semi-matt**

Copper acetate	60 gm
Copper carbonate	30 gm
Ammonium chloride	60 gm
Water	1 litre
Hydrochloric acid (15% solution)	5 cm³

Cotton-wool technique (Ten to twenty hours)

An even mid-brown colour is produced on standard silver.

Silver-plated surfaces acquire a dark tarnish initially, but are rapidly attacked by this treatment, revealing the underlying metal. In the case of silver-plated gilding metal, a greenish patina was formed. This was much more even than could be obtained directly on a gilding metal surface.

▲ Hydrochloric acid can cause burns or severe irritation, and must be prevented from coming into contact with the eyes and skin.

Cotton wool, moistened with the above solution, is applied to the surface of the article, and left for a period of about ten hours. The surface is then examined, by carefully lifting a small portion, to determine the stage it has reached. When the desired surface finish has been obtained, the article is washed thoroughly in cold water and allowed to dry in air. When completely dry, it may be wax finished.

6.50   **Mid-brown Semi-matt**

Ammonium carbonate	120 gm
Ammonium chloride	40 gm
Sodium chloride	40 gm
Water	1 litre

Cotton-wool technique (Twenty to thirty hours)

A mid-brown colour, with some variation in tone over the surface, was produced on standard silver.

Tests with silver-plated gilding metal produced a patchy pale green surface where the plating had been gradually corroded by the solution.

Cotton-wool, moistened with the solution, is applied to the surface of the object to be coloured. It is important to ensure that the moist cotton-wool is in full contact with the surface of the object. It is then left for a period of about twenty to thirty hours, ensuring that the cotton-wool remains moist throughout the period of treatment. When treatment is complete, the object should be washed in cold water, and allowed to dry in air. When completely dry, it is wax finished.

6.51   **Mid-brown Semi-matt**

Ammonium chloride	350 gm
Copper acetate	200 gm
Water	1 litre

Cotton-wool technique (Ten to twenty hours)

A mid-brown, with some tonal variation, is produced on standard silver.

Silver-plated surfaces are initially darkened to an uneven grey colour, and are then progressively attacked by the solution, with continued treatment. The underlying metal is exposed and coloured by the solution. Some of the interim stages in this development show interesting surface effects, but these are difficult to control.

Cotton-wool, moistened with the solution, is applied to the surface of the object. This is then left for a period of ten or twenty hours, ensuring that the cotton-wool remains moist. After the first few hours, the surface should be periodically examined by exposing a small portion, to check the progress. When the desired surface finish has been reached, the object is removed, washed in cold water and allowed to dry in air. When completely dry, it may be wax finished.

6.52*   **Mid-brown Semi-matt**

Copper nitrate	20 gm
Ammonium chloride	20 gm
Calcium hypochlorite	20 gm
Water	1 litre

Cotton-wool technique (Twenty hours)

A variegated mid-brown colour is produced on standard silver.

Silver-plated surfaces were coloured a dull metallic grey initially, but were subsequently corroded locally, revealing the underlying metal.

▲ The corrosive fine dust of the calcium hypochlorite (bleaching powder) must not be inhaled, as it can severely irritate the respiratory system. The solution must be prevented from coming into contact with the eyes, which would be severely irritated.

Cotton-wool, moistened with the solution, is applied to the surface of the object to be coloured, and left for a period of about twenty hours. It is important to ensure that the cotton-wool is in close contact with the surface, and that it remains moist throughout the period of treatment. When treatment is complete, the object is removed and washed thoroughly in cold water. After a period of drying in air, it may be wax finished.

6.53   Mid-brown Semi-matt

Copper nitrate	100 gm
Water	200 cm³
Ammonia (.880 solution)	400 cm³
Acetic acid (6% solution)	400 cm³
Ammonium chloride	100 gm

Cotton-wool technique (Twenty to thirty hours)

The copper nitrate is dissolved in the water and the ammonia gradually added. The acetic acid is then added, followed by the ammonium chloride. Cotton-wool is wetted with the solution and applied to the surface of the object to be coloured. It is then left for a period of about twenty to thirty hours. The cotton-wool should be kept moist throughout the period of treatment. When treatment is complete, the object is washed in cold water and allowed to dry in air. When dry, it is wax finished.

An even mid-brown colour is produced on standard silver.

Silver-plated surfaces were strongly attacked by this treatment. The silver-plate is rapidly removed and an incrustation tends to form, which appears to be associated with the underlying copper-plating, and is bright green in colour. These areas are unevenly etched through to the underlying brass which takes on a frosted and corroded appearance. After thirty hours only slight traces of the silver-plate are left. Results suggest that a controlled use of this technique might produce interesting effects.

▲ Preparation and colouring should be carried out in a well ventilated area. Ammonia vapour and some acetic acid vapour will be liberated, which cause irritation of the eyes and respiratory system.

▲ The strong ammonia solution must be prevented from coming into contact with the eyes and skin, as it will cause burns or severe irritation.

6.54   Mid-brown Semi-gloss/Semi-matt

Copper nitrate	200 gm
Sodium chloride	200 gm
Water	1 litre

Cotton-wool technique (Twenty hours)

Cotton-wool, moistened with the solution, is applied to the surface of the object to be coloured. It is then left for a period of about twenty hours. It is essential to ensure that the cotton-wool is in full contact with the surface and is kept moist throughout the period of treatment. Breaks in the colour and patches of patina may result if the cotton-wool is allowed to dry out or lift from the surface. When treatment is complete, the object should be washed thoroughly in cold water, and allowed to dry in air. When dry, it may be wax finished.

An even mid-brown colour is produced on standard silver.

Silver-plated surfaces are completely stripped, revealing the etched surface of the underlying metal.

6.55★   Dark grey-brown Semi-gloss

Copper sulphate	100 gm
Lead acetate	100 gm
Ammonium chloride	10 gm
Water	1 litre

Cotton-wool technique (Ten to twenty hours)

Cotton-wool, moistened with the solution, is applied to the surface of the object to be coloured, and is left in place for a period of up to twenty hours. The cotton-wool should remain moist throughout the period of treatment. When treatment is complete, the object is removed and washed in cold water. It is then dried in air, and gradually darkens on exposure to daylight. When colour change has ceased, the object may be wax finished.

An even dark grey-brown is produced on standard silver.

A mottled pale grey-green and grey-brown is produced on silver-plated surfaces.

▲ Lead acetate is a toxic substance, and every precaution should be taken to prevent inhalation of the dust when preparing the solution. It is very harmful if taken internally. A clean working method is essential throughout.

6.56   Mid-brown Semi-matt

Ammonium carbonate	150 gm
Copper acetate	60 gm
Sodium chloride	50 gm
Potassium hydrogen tartrate	50 gm
Water	1 litre

Cotton-wool technique (Ten to twenty hours)

Cotton-wool which has been moistened with the solution is applied to the surface of the article to be coloured. It is important to ensure that the cotton-wool is in full contact with the surface. It is left for a period of from ten to twenty hours, during which time the cotton-wool should remain moist. Colour development should be monitored with a sample, so that the article remains undisturbed. When treatment is complete, the article is removed and washed in cold water. After being allowed to dry in air for several hours, it may be wax finished.

An even mid-brown colour was produced on standard silver.

A light greyish-purple colour occurred on silver-plated surfaces. Some blue-green patina was produced where the plated gilding metal had been deliberately breached by abrasion prior to colouring.

6.57   Light grey mottle Semi-matt

Copper nitrate	280 gm
Water	850 cm³
Silver nitrate	15 gm
Water	150 cm³

Cotton-wool technique (Twenty hours)

The copper nitrate and the silver nitrate are dissolved in separate portions of water, and the two solutions then mixed. Cotton-wool, moistened with this mixed solution, is applied to the surface of the object to be coloured. It is important to ensure that the cotton-wool is in full contact with the surface, and that it remains moist during the period of treatment. The object is removed after about twenty hours and washed in cold water. After thorough drying in air, it may be wax finished.

A slight, uneven, light grey mottle is produced on standard silver. The results obtained in tests were generally poor, and the procedure cannot be recommended. Similar poor results were obtained with silver-plated surfaces. Prolonged treatment tended to expose the underlying metal.

▲ Silver nitrate must be prevented from coming into contact with the skin, and more particularly the eyes, as it may cause burns or severe irritation.

6.58*   Silver-plate cut through to red Semi-gloss

Sodium chlorate	100 gm
Copper nitrate	25 gm
Water	1 litre

Boiling immersion (Forty minutes to one hour)

The object is immersed in the boiling solution, after the surface has been suitably prepared, and immersion continued until the red colour has developed on the underlying gilding metal. The immersion time should be kept as short as possible, to avoid grey patches on the silver-plate. When the surface has developed satisfactorily, the object is removed and washed in warm water. It should be dried in sawdust, and may be wax finished when dry.

The silver-plate is dulled by the solution, and tends to acquire patches of grey tarnish if immersion is prolonged. The exposed copper-plate and gilding metal are coloured red.

▲ Sodium chlorate is a powerful oxidant and must not be allowed to come into contact with combustible material or acids. It should also be prevented from coming into contact with the eyes, skin or clothing.

6.59*   Silver-plate cut through to dark red and brown Semi-matt   *Pl. XI*

Copper acetate	70 gm
Copper sulphate	40 gm
Potassium aluminium sulphate	6 gm
Acetic acid (10% solution)	10 cm³
Water	1 litre

Boiling immersion (Thirty minutes)

Immersion in the boiling solution produces no effect on the silver-plate, but only on the copper-plate and gilding metal exposed by surface preparation. Red and brown colours develop during the course of immersion on these metals. When the colours have developed sufficiently, the article is removed, washed in hot water and dried in sawdust. When dry it may be wax finished.

No effect is produced on standard silver by immersion in this solution.

The lighter colours exhibited by sample 6.59ii in plate XI are produced by a very brief immersion in the solution.

6.60   Silver-plate cut through to orange-brown Semi-gloss   *Pl. XI*

Copper sulphate	120 gm
Potassium chlorate	60 gm
Water	1 litre

Hot immersion (A few minutes)

The article is immersed in the hot solution (80°C), after suitable surface preparation to expose the underlying copper-plate and gilding metal. The silver-plate remains unaffected, but the underlying metals are coloured different tones of orange-brown. When the colour has developed, the article should be removed immediately, and immersed in hot water. (Immersion should not be continued any longer than is necessary, or damage may occur to the silver plate.) The article is dried in sawdust and may be wax finished when dry.

▲ Potassium chlorate is a powerful oxidant and should not be allowed to come into contact with combustible material. It should be prevented from coming into contact with the eyes and skin.

6.61*   Slate grey cut through to orange-brown and dark brown Semi-gloss   *Pl. XV*

Copper sulphate	125 gm
Potassium chlorate	50 gm
Ferrous sulphate	25 gm
Water	1 litre

Boiling immersion (Thirty minutes)

The article is immersed in the boiling solution, after suitable surface preparation to expose the underlying copper-plate and gilding metal. A slate grey colour gradually develops on the silver-plate, and orange colours on the exposed underlying metals. These colours darken as immersion continues. After about thirty minutes, the article is washed in hot water. It is dried in sawdust, and may be wax finished when dry.

A very pale greyish tone is produced on standard silver with this solution. An even slate grey is produced on silver-plated surfaces. The colour is produced readily and evenly on polished surfaces, but tends to be slight on a plated satinised surface, producing a texture of grey lines.

Immersion should not be prolonged more than is necessary, as some blistering of the surface tends to occur if extended times are used.

The orange-brown colours produced on the underlying metals tend to darken slightly on exposure to daylight.

▲ Potassium chlorate is a powerful oxidant and must not be allowed to come into contact with combustible material. It should be prevented from coming into contact with the eyes and skin.

6.62★   Silver-plate cut through to orange and pinkish-red   *Pl. XVI*
Semi-matt/semi-gloss

A	Copper sulphate	125 gm
	Ferrous sulphate	100 gm
	Water	1 litre

B	Ammonium chloride	1 gm

Boiling immersion (A) (Five minutes)
Boiling immersion (A+B) (Two minutes)

The silver-plate remains uncoloured, the differential colouring being produced on the copper-plate and the gilding metal respectively. The dual colour is most developed if the surface is prepared by gradual polishing, so that the underlying copper-plate is preserved.

It is essential to monitor progress very carefully, as the colouring action, after the addition of the ammonium chloride is very rapid. Although the silver-plate remains uncoloured, it may recede rapidly at this stage by erosion at the edge of plated areas.

The object is immersed in the boiling solution of copper sulphate and ferrous sulphate, after suitable surface preparation, for a period of about five minutes. It is then removed to a bath of hot water while the ammonium chloride is added to the colouring solution. The object is re-immersed for about two minutes, rapidly colouring the underlying copper-plate and gilding metal, orange and pinkish-red. When the colour has developed, the object is immediately removed and washed in hot water. It is dried in sawdust, and may be wax finished when dry.

6.63★   Variegated orange-brown and ochre with grey patches   *Pl. XI*
Semi-gloss

Copper sulphate	125 gm
Sodium chlorate	50 gm
Ferrous sulphate	25 gm
Water	1 litre

Boiling immersion (Two to ten minutes)

The particular effect shown in plate XI is produced on a silver-plated, directionally satinised surface.

▲ Sodium chlorate is a powerful oxidant and should be kept out of contact with combustible material and acids. It should be prevented from coming into contact with the eyes, skin or clothing.

The article is immersed in the boiling solution, after suitable surface preparation to expose the underlying copper-plate and gilding metal. The silver-plated areas are rapidly coloured grey, while the copper-plate is coloured a variegated orange brown, and the gilding metal a light ochre. As immersion continues, the colours darken, and the silver-plate and copper-plate are progressively attacked and removed. Timing is critical, and the article should be removed as soon as the desired effect is obtained, and plunged into warm water to stop the reaction. It may then be dried in sawdust. Exposure to daylight will darken the colours slightly. When colour change has ceased, the article may be wax finished.

6.64   Light grey cut through to brown Semi-matt

Ammonium carbonate	50 gm
Copper acetate	25 gm
Acetic acid (6% solution)	1 litre
Zinc chloride	5 gm

Boiling immersion (Fifteen minutes)

Standard silver is very little affected by this treatment, acquiring a slight grey clouding.

Silver-plated surfaces tend to be gradually attacked by the solution; immersion should not be continued any longer than is necessary.

▲ The ammonium carbonate should be added to the acid in small quantities. The reaction is effervescent, and if large amounts are added this effervescence becomes very vigorous.

The article is immersed in the boiling solution, after suitable surface preparation to reveal the underlying copper-plate plate and gilding metal. The silver-plate is gradually dulled to a cloudy light grey by the solution, which colours the underlying metals brown. After about fifteen minutes the article is removed, washed in warm water and dried in sawdust. When it is completely dry it may be wax finished.

6.65★   Dull plate/metallic grey cut through to black/dark brown Semi-matt

Sodium hydroxide	25 gm
Potassium persulphate	10 gm
Water	1 litre

Hot immersion (Several minutes)

▲ Potassium persulphate is a powerful oxidant and should not be allowed to come into contact with combustible material. The sodium hydroxide and the persulphate must not be allowed to come into contact in the solid state.

▲ Sodium hydroxide is a powerful caustic alkali which can cause severe burns. Both the solid and solutions should be prevented from coming into contact with the eyes and skin.

The potassium persulphate is dissolved in the water, and the sodium hydroxide then added in small quantities while stirring. The solution is then brought up to the correct temperature (70–80°C) and the article immersed, after suitable surface preparation. The exposed gilding metal is rapidly coloured a dark reddish-brown, which is fringed with a coppery tone and black. The immersion should not be prolonged any more than is necessary, or the silver-plate tends to be affected. When the colour has developed satisfactorily, the article is removed and washed in warm water. It is dried in sawdust, and may be wax finished when dry.

6.66*   Variegated metallic grey cut through to black Semi-matt/semi-gloss

Ammonium persulphate	10 gm
Sodium hydroxide	30 gm
Water	1 litre

Boiling immersion (About ten minutes)

The ammonium persulphate is dissolved in half the water, and the sodium hydroxide separately dissolved in the remaining water. The two solutions are then mixed together and the mixture heated to boiling. The article to be coloured is immersed in the boiling solution, after suitable surface preparation, and removed after about ten minutes. It is washed in hot water, dried thoroughly in sawdust, and wax finished when dry.

The silver-plate is coloured grey and the underlying gilding metal, black. If the surface is worn through by a polishing method, the black tends to be fringed with a coppery colour.

▲ Ammonium persulphate (peroxodisulphate) is a powerful oxidant, and must be kept out of contact with all combustible materials. It should be prevented from coming into contact with the eyes, skin or clothing.

▲ Sodium hydroxide is a powerful caustic alkali and must be prevented from coming into contact with the eyes and skin.

6.67   Silver-plate cut through to dark green and black Semi-gloss   *Pl. XI*

Copper carbonate	125 gm
Ammonia (.880 solution)	250 cm³
Water	750 cm³

Hot immersion (Ten minutes)

The article is immersed in the hot solution (50°C), after suitable surface preparation to expose the underlying copper-plate and gilding metal. The silver-plated areas remain unaffected, while the copper plate is coloured a dark grey-green and the gilding metal black. When the desired colour has been obtained, the article should be removed, washed in hot water and dried in sawdust. (The immersion should not be prolonged, as this tends to tarnish the silver-plate.) The article may be wax finished.

A simple contrast between the uncoloured silver-plate and a black ground can be obtained, if the copper-plating is omitted. The solution should then be used cold.

▲ Ammonia vapour, which irritates the eyes and respiratory system, will be freely liberated by the solution. Preparation and colouring should be carried out in a well ventilated area. The skin and particularly the eyes should be protected from accidental splashes of this corrosive alkali.

6.68*   Silver plate cut through to dark green/black Semi-gloss   *Pl. XI*

Ammonium carbonate	60 gm
Copper sulphate	30 gm
Oxalic acid	1 gm
Acetic acid (10% solution)	1 litre

Hot immersion (Thirty minutes)

Immersion in the hot solution (80°C) produces no effect on the silver plate, but only on the copper plate and gilding metal exposed by surface preparation. A dark green colour gradually develops on these metals during the course of immersion. When the colour has developed sufficiently, the article is removed, washed in hot water and dried in sawdust. When dry it may be wax finished.

No effect is produced on standard silver by immersion in this solution.

▲ The ammonium carbonate should be added to the acid in small quantities. The reaction is effervescent and if large quantities are added the effervescence becomes very vigorous.

▲ Oxalic acid is poisonous and harmful if taken internally.

6.69   Variegated pink/grey lustre cut through to light brown Semi-matt

Copper chloride	50 gm
Water	1 litre

Hot immersion (A few minutes)

The article is immersed in the hot solution (50–60°C) for a few minutes, during which time the colour develops gradually. When the colour has developed, the article is removed and washed in warm water. It is then allowed to dry in air, and may be wax finished when completely dry.

A variegated pink/grey lustre is produced on the plated surface. Where this has been cut through to the underlying copper and gilding metal, a light orange-brown colour is produced.

A more dilute solution may be used to produce a less grey plated surface and a lighter more yellow colour on the underlying metal.

SAFETY

Safety in metal colouring

There are two aspects of safety which need consideration. Firstly there are the inherent properties of chemicals and their interactions, which may give rise to dangers because the chemicals or their products are toxic, corrosive or harmful in various ways. Information relating to the use of a chemical in a specific recipe is included in the notes that accompany that recipe. The more general properties and interactions of chemicals that are potentially hazardous are given in the list of chemicals and hazards.[1] The other aspect of safety is more general, but equally important, and concerns the preparation of chemicals and their use in the workshop in the context of the different colouring techniques.

1. See page 339. This list only gives information about chemicals that occur in the recipes included in this book, and their interactions.

Food, drink, smoking

Food and drink should never be prepared or consumed in an area where chemicals are weighed, prepared or used. The danger of accidental contamination is perfectly obvious. This stricture also applies to smoking, where the danger from contamination may be worsened by the fact that the chemical may be inhaled through a cigarette, for example. In addition, some of the chemicals used, notably degreasing solutions and oxidising agents, present a significant fire hazard in the presence of naked flames. A strict no-smoking rule should be applied.

General considerations

The principal general safeguard in metal colouring is to adopt a clean and systematic working method. The workshop layout should be carefully planned so as to provide adequate space for carrying out the various activities involved. As far as possible the different parts of the process should be separated, and in particular a separate area provided for the storage and preparation of chemicals. Adequate ventilation must be provided, preferably in the form of localised extraction, which serves the area where colouring is carried out and that where concentrated liquids are diluted. Personal safety equipment should be available for use, including nose and mouth face masks with appropriate filters, face shields and goggles, gloves, plastic or rubber aprons and protective footwear. A standard first-aid kit should be available, containing supplies of sterile dressings for burns. An emergency eyewash bottle should also be available, and an accessible water supply provided. Fire extinguishers should be provided for dealing with fires that may involve oxidising agents, and flammable liquids such as the organic solvents used in degreasing.

Chemicals should never be used casually, or used experimentally without expert advice. Recipes should not be modified on an *ad hoc* basis. Any proposed additions or alterations should be considered in consultation with a qualified chemist.

Degreasing and cleaning

The cleaning cycle will in most cases include the use of an organic polishing compound remover or degreasing agent. It is essential when using these chemicals to ensure that adequate ventilation is provided, preferably in the form of localised extraction, to minimise the risk of a build-up in concentration of the vapours. Inhaling them in any quantity can have serious effects, initially giving rise to feelings of light-headedness, headaches and nausea, and leading to loss of consciousness in extreme cases.[2] High concentrations of many of these vapours also constitute a serious explosive

2. If exposure to concentrated vapours occurs, or if exposure is continuous or repeated, then more serious effects may result. See 'Chemicals and their hazards', final section.

hazard and fire risk. It is important therefore to ensure that there are no naked flames in the vicinity, and that the area is adequately ventilated.

In any manual cleaning with these chemicals, and cleaning agents in general, it is essential to wear gloves. The cumulative effect of the degreasing action of cleaners can lead to serious skin conditions. It was found that both household and industrial rubber and synthetic rubber gloves were attacked by cleaning agents, causing them to swell and disintegrate. Similar though less pronounced effects occurred with PVC gloves. Nitrile and high grade PVC gloves were the least affected, and are probably the most suitable for use with organic degreasing agents.

Preparation of solids

The weighing and preparation of solids, powders and crystalline substances should always be carried out carefully. Many chemicals in the dry state will yield fine airborne dusts if carelessly handled, and care should always be taken to avoid inhaling these, particularly in the case of toxic or corrosive chemicals or those classed as oxidising agents. A nose and mouth face mask fitted with a filter suitable for the exclusion of these fine dusts may be worn to provide added protection, but it is important to note that these do not provide total protection.[3] All chemicals should be prevented from coming into contact with the skin, and gloves may be worn to protect the hands. These should be fine rather than heavy duty to avoid the loss of tactile sensitivity that is required when handling containers, and well fitting surgical gloves were found to be suitable in practice. It is clearly essential to ensure that the inside of gloves and more particularly masks do not become contaminated with chemicals.

Some chemicals are supplied in a coarse form, as solid lumps or large crystals, and may require grinding with a pestle and mortar during preparation. It should be noted that chemicals listed as oxidising agents should not be dry-ground. They can become highly reactive when finely divided, and sensitive to small amounts of contamination, producing potentially explosive or flammable conditions in some cases.

Spatulas, scoops and containers used for weighing and transferring chemicals should always be clean and dry prior to use and washed immediately after use, so that the risk of contaminating one chemical with another is minimised. All spillages of chemicals should be cleared promptly. The best general safeguard is a clean and systematic working method in which only one chemical is dealt with at a time, and containers are closed as soon as a chemical has been dispensed.

Solids should not be mixed unnecessarily. When preparing solutions with a number of ingredients, it is preferable to dissolve each chemical separately and then mix them in the form of solutions, rather than to mix them as solids.

Preparation of solutions

Solutions are prepared either by dissolving solids or by the dilution of concentrated liquids. As a general rule it can be stated that solids or liquids should be added in small amounts to relatively large quantities of water, and not vice-versa, when preparing solutions. This is essential when diluting or dissolving concentrated acids and alkalis.

The dilution of concentrated liquids is clearly potentially hazardous, as the chemical is present in its most mobile and powerfully acting form. Adequate precautions must be taken against spillage and accidental splashes. A full face shield or goggles, gloves of a sufficient length to cover all exposed areas of the hands and arms, a full length well-fitting plastic or rubber apron, and chemically resistant footwear, should all be worn.[4] Decanting the concentrated liquid should be carried out in plastic trays, so that any accidental spillage is contained. A supply of running water should be available, and an emergency eye-wash bottle kept immediately to hand.[5]

Concentrated liquids such as acids, ammonia and ammonium sulphide solution, for example, are also liable to liberate strong vapours. These must not be inhaled, and decanting or dilution should therefore be carried out in an area which is provided with good localised ventilation.[6]

Liquids can be measured using measuring jugs, measuring cylinders or graduated beakers made of glass or a plastic such as polypropylene. Where small quantities of solutions are involved, a pipette can be used. This must not be mouth-aspirated. A variety of automatic pipettes are available which range in complexity from a valved rubber bulb which is attached to a simple

3. Masks are widely used to reduce the inhalation of potentially harmful dusts, vapours, fumes, mists and gases. Reliance on masks is only justified in the case of short exposures (minutes) and low frequency (10–20 times daily). It is essential that filters for masks are correctly chosen and maintained. If not they may simply engender a false sense of security. Filters are generally one of three main types: A, chemical absorbers (specific to certain chemicals, ranges or molecular sizes); B, mechanical filters, blocking particles of a known size; C, mechanical/chemical, combining some specific properties of both types. Consultation with the manufacturer is recommended when selecting a filter for a particular application. Masks should generally be considered as an additional safeguard, rather than as an isolated precaution.

4. Masks may be worn as an added safeguard, see note 3 above.

5. The water supply should be readily accessible. It should be possible to drench any part of the body quickly and easily. This may necessitate the provision of a suitable hose from taps, for example.

6. Ideally, good ventilation implies the provision of localised extraction that draws vapours away from the operator.

glass pipette, pump-type glass pipettes, to fully automatic trigger-operated devices. These obviate the need to draw up the solution by mouth suction, which is clearly highly dangerous, particularly in the case of toxic or corrosive chemicals.

A systematic and clean working method is essential throughout. Only one chemical should be dealt with at a time, and containers should be closed as soon as the chemical has been dispensed. All apparatus should be clean and dry prior to use and washed immediately after use. Any contamination of the skin should be washed away, and any spillages cleared immediately[7].

Storage of chemicals

Chemicals as supplied from the manufacturers should be carefully stored so as to minimise the risk of accidental breakage. Concentrated liquids which are supplied in glass 'Winchester' bottles are often delivered in an outer plastic container, and can be stored in these in the workshop. Chemicals that are potentially interactive should be stored separately. The chemical list can be consulted to determine which chemicals should be stored apart.[8] Acids, for example, should be stored away from sulphides, since in the event of an accident their interaction will yield toxic hydrogen sulphide gas. Similarly, oxidising agents should be stored apart from flammable liquids, combustible materials and acids. If large quantities of concentrated corrosive liquids or flammable liquids are to be stored they may require specialised storage; the local safety authority should be consulted to determine the allowable storage levels; and special facilities required, for particular chemicals. Toxic substances and poisons should be stored in a separate, clearly labelled, lockable cabinet.

Storage of prepared solutions.

Prepared colouring solutions can be stored in polypropylene containers, which are available in a range of sizes. These are preferable to glass, which is prone to breakage, and more resistant to heat and abrasion than the alternative plastics such as PVC, which is an advantage when they are cleaned. All stored solutions must be clearly and completely labelled, and marked with the appropriate hazard warning. Self-adhesive tapes marked with hazard warning symbols are available, and are convenient for this purpose.[9]

Immersion colouring

Potential hazards that occur during the preparation of chemicals will also be present during immersion colouring. Although in most cases the chemicals will be in a more dilute form by this stage, and therefore less hazardous, this is offset by the fact that many of them are used at high temperatures or at boiling point. Precautions should therefore be taken to prevent chemicals from reaching the eyes, skin or clothing, by wearing the appropriate protective clothing. Risks will be considerably reduced if attention is paid to workshop layout and tank positioning. There should be sufficient clear space around the tank to allow free access. In addition it is an advantage if the tank is set at the correct height. In practice it was found that the top of the tank should be roughly at table height. If the tank is too low then there is a tendency to stoop over it to examine and remove objects, while if it is too high then visual monitoring and removing the object becomes difficult and therefore hazardous.

The major additional hazard in immersion colouring arises from the vapours that are liberated when solutions are heated. These hazards are potentially greatest in the case of solutions of toxic or corrosive chemicals, but all vapours should be regarded as a potential danger and should not be inhaled. Adequate ventilation must be provided, preferably in the form of localised extraction, ensuring that the vapours are drawn away from the operator and are not allowed to accumulate in the air even in low concentrations.[10] Continuous or repeated exposure to vapours, even at low levels, will produce harmful effects which may be serious in the case of toxic or corrosive chemicals and oxidising agents.

Application techniques

Applying chemicals by dabbing, wiping or brushing is generally unproblematic so long as contact with the eyes and skin is avoided. Protective clothing should always be worn, and special attention given in the case of solutions of toxic or corrosive chemicals.

7. Spillages of chemicals classed as corrosive, toxic, or oxidising agents should be cleared with copious amounts of water. The use of plastic trays when preparing chemicals offers an effective means of containing spillages, and is to be recommended.

8. The list relates to the chemicals required for use in recipes included in this book. Other chemicals, cyanides for example, may require more specialised provisions.

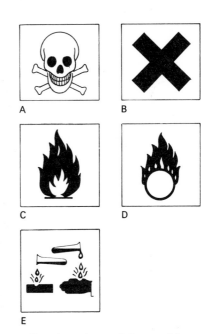

9. Hazard warning symbols as used for proprietary chemicals and available in the form of adhesive tapes.
A Toxic, poisonous.
B Harmful.
C Flammable.
D Oxidising agent.
E Corrosive.

10. Masks fitted with the correct filter may be worn for additional protection, but reliance on masks is not recommended as they do not offer adequate protection in isolation. See note 3 above.

Application of chemicals in the form of a fine spray is potentially much more hazardous, since the airborne mists produced may be easily inhaled and can give rise to serious harmful effects. This technique should not be used in the case of chemicals that are listed as being significantly harmful if taken internally, including toxic and corrosive substances and oxidising agents. When used with chemicals that are less inherently hazardous, adequate ventilation must be provided, preferably in the form of localised extraction. Masks fitted with suitable filters should be worn for additional protection, but should not be relied on in isolation.

Application techniques normally include stages in which any loose material produced is brushed or rubbed away from the dry surface. This will give rise to fine dusts which must not be inhaled. Masks fitted with filters suitable for the removal of fine dusts should be worn.

Bristle-brushing and scratch-brushing

The general safety considerations associated with the application of solutions also apply in the case of application by bristle-brushing or scratch-brushing. In addition it should be noted that the use of these techniques will tend to cause a spray of chemical as the surface is brushed.

Applied pastes

As in the case of other application techniques, contact with the eyes and skin must be avoided. The technique may also involve brushing away dry residues, which will cause fine dusts. These must not be inhaled, particularly where the paste contains toxic or corrosive substances, or oxidising agents. A mask, fitted with a filter suitable for the removal of fine dusts, should be worn.

Torch technique

The general safety considerations associated with the application of solutions also apply in this case. Protective clothing and face shields should be worn as the solution may tend to spit when it is applied to the hot metal. Fumes and vapours will inevitably be produced as the solutions are applied. These must not be inhaled. Adequate ventilation must be provided, preferably in the form of localised extraction, and a nose and mouth face mask worn as an added precaution. This should be fitted with the appropriate filter.[11]

11. Masks should be worn as an added safeguard. See note 3 above. In the majority of cases solutions used in torch techniques yield acid vapours.

Vapour technique

Where potentially harmful vapours are involved, the technique requires careful planning. Details of a suitable procedure are given in the relevant section of the chapter devoted to metal colouring techniques. The more general considerations relating to the preparation and use of solutions also apply in this case.

Particle and appressed techniques

Where sawdusts, wood shavings, cotton wool or cloth are used as media for the retention of solutions in particle and appressed techniques, certain chemicals must be avoided. These materials should not be moistened with oxidising agents, as potentially explosive or flammable conditions may result as they dry out. Acids and alkalis in a concentrated form should also be avoided.

Moisture-retaining materials used for these techniques should be thoroughly washed and dried prior to disposal as normal refuse, particularly in the case of materials which have been impregnated with toxic or corrosive chemicals.

Finishing

If fine sawdust is used for drying objects after colouring, a mask should be worn to prevent the dust from being inhaled.

Where lacquers are used for finishing, the safety precautions indicated by the manufacturer should be strictly adhered to. Many lacquers give rise to vapours that must not be inhaled, and many are flammable.

Disposal of wastes

In general nearly all solutions and light precipitates may be run to waste provided that they are diluted with very large quantities of water. Each solution should be disposed of separately, followed by sufficient running water to ensure that the chemical has cleared the system before any further chemicals are run to waste.[12] Heavy precipitates should be allowed to dry out, and should generally be mixed with sand before disposing of them in containers, in small quantities, as normal refuse.

In the case of solutions and solid residues of the toxic substances, guidance should be sought from the local authority. Although in some cases small quantities of the solutions of lead, mercury, arsenic and antimony can be run to waste, these substances can affect the efficiency of water filtration and purification plants. The local authority will recommend acceptable levels or suitable disposal procedures, in the light of local conditions. They will also advise on the disposal of the solid residues of the toxic poisons, and oxidising agents, which should not be included in normal refuse.[13]

Symptoms and first aid

The most common occurrence is likely to be the result of skin contact with chemicals of one sort or another. All chemicals that contaminate the skin must be washed away promptly with copious amounts of water. This is particularly important in the case of the more corrosive and toxic chemicals.[14] If the eyes are contaminated then they must be immediately irrigated with water using an eyewash bottle or a gently running stream of water. Care should be taken to ensure that the water being used to clear a contaminant from one eye is not allowed to run into the other eye. In some cases a reflex action may occur which keeps the eyelids tightly shut. They should be gently prised apart and washing carried out promptly. All injuries to the eye require prompt medical attention.

In addition to the immediate effects of corrosive chemicals whose action is likely to be obvious, skin reactions can occur more slowly as a result of repeated or continuous exposure to chemicals. This generally becomes apparent in the form of itching or irritation, reddening of the skin, rashes, or slight blistering.[15] In all cases where this occurs, exposure to the chemical should cease until the symptoms subside, and medical attention be sought if the effects are more than trivial. If the symptoms recur when work is resumed, then medical advice should be taken. Although the hands are most commonly affected, the face, neck and arms are also prone to these effects by the action of airborne dusts or vapours. More rarely, the skin may be sensitised by a chemical, producing no immediate effects for weeks or even months. On subsequent exposure, however, serious inflammation and damage can occur, and healing may be prolonged and difficult. If contact is renewed, then the skin often becomes so responsive that it reacts violently in the presence of the chemical, and the sufferer may not even be able to enter the same room as the chemical without developing a recurrence which may affect the whole surface of the skin.[16] This is clearly a very serious condition and all contact with the chemical must be avoided. If it is suspected that this type of reaction is taking place, then medical advice should be sought immediately.

Inhalation is the most dangerous and rapidly acting route by which toxic chemicals can enter the body. Most toxic gases are acutely irritating to the respiratory tract or possess a warning odour which is detectable at concentrations well below the danger level.[17] In all cases where there is a build-up of gases or vapours the essential first measure is to remove the sufferer to fresh air. (In extreme cases, where breathing has stopped, artificial respiration should also be applied. If the victim is unconscious then he should be placed face down in order to maintain a clear airway. Immediate medical attention should be sought.) It is more likely that the effects of inhaling gases, vapours from heated solutions, and dusts, will occur through low levels of continuous or repeated exposure. Symptoms such as irritation of the respiratory system, headaches, nausea, light-headedness and a tendency to feelings of lethargy or debility, may all be indications of the effects of inhalation. The most effective way of dealing with this hazard is to ensure that adequate ventilation is provided and to seek medical advice promptly if symptoms occur.

The risk of taking in chemicals by mouth should be negligible if a clean working method is used. The greatest danger is from the mouth-aspirated pipette, which should *never* be used. If ingestion of chemicals does occur then the treatment will depend on the type of chemical involved. Corrosive substances will result in burns of the mouth, and should be treated by repeated mouth washes with water. The mouth wash must not be swallowed. A similar treatment should be given initially in the case of

12. Interactions between chemicals will occur as readily in the waste system as in the beaker. The list of chemicals and hazards (page 339) may be consulted to determine which interactions to avoid, and consequently which solutions in particular should not be disposed of at the same time.

13. The dry residues of oxidising agents can present a significant fire risk in some circumstances.

14. It should be noted that the presence of cuts or grazes or other skin blemishes can significantly increase the danger from toxic substances.

15. The developed form of these symptoms is a condition known as contact dermatitis.

16. This condition, which is known as sensitization dermatitis, depends largely on individual susceptibility, and a list of chemicals to avoid cannot be simply stated. Of the chemicals included in this book, nickel compounds in particular and metal compounds in general have produced this effect in some people. It has also been caused by chromates, dichromates, hypochlorites, and degreasing solutions.

17. Although the sense of smell provides a useful early warning that a gas or vapour is present, it cannot be relied upon to assess whether the level of chemical is increasing. With continued exposure, one tends to become less and less sensitive to odours.

poisonous substances, followed by drinking quantities of water or milk to dilute the chemical, once the mouth has been cleared. It is sometimes suggested that vomiting should be induced, but this can worsen the situation if the chemical involved is corrosive and is not without its own dangers, and is therefore not recommended. Medical attention should be promptly sought.

Chemical burns from corrosive chemicals should be treated by washing with copious amounts of water, as noted above. Dry burns from flames and heat sources should be covered with loosely applied sterile dressings. If blisters occur, care should be taken not to burst them. To reduce the danger of infection, handling should be reduced to a minimum and any temptation to clean the surface must be resisted. All burns or scalds of more than a trivial nature should be referred to medical attention.

There are no specific antidotes to the chemicals involved in the colouring recipes included in this book. In the event of an accident the simple measures noted above should be promptly applied, and medical attention sought. It is important to record full details of any chemical involved in an accident and any treatments that have been applied, and to ensure that this information accompanies a patient to hospital.

Chemicals and their hazards

The following list has been compiled from the available safety literature, and from recommendations made by the manufacturers and suppliers of the chemicals. The main sources of reference used were—Muir (Bib. 269), Bretherick (Bib. 72), Steere (Bib. 360), Sax (Bib. 317), McCann (Bib. 258). The list only deals with chemicals that are required for the recipes included in this book.

Chemical	Formula	Hazards	Additional notes
Acetic acid	$CH_3.COOH$	Concentrated solutions, especially the highly concentrated form known as glacial acetic acid or ethanoic acid, are flammable. They can also cause severe burns. The vapour irritates the respiratory system and eyes. The liquid burns the eyes severely and causes ulceration and burns to the skin. Violent reactions may occur with nitric acid, chromium trioxide, potassium permanganate, peroxides, and other oxidising agents.	Glacial acetic acid freezes to a crystalline solid in cool weather. The dilute solutions generally used in colouring are relatively safe. They are best prepared by dilution of a stock solution of about 35% which is commonly supplied, rather than from more concentrated forms.
Ammonia	NH_3	Ammonia solutions are corrosive and can cause severe burns. The vapour is very irritating to the eyes and all parts of the respiratory system. The solution burns the eyes severely, burns the skin, and will cause severe internal damage if swallowed.	Commonly supplied as a 35% solution with a specific gravity of .880, often referred to as 880 ammonia. Pressure develops in the bottle in warm weather, and the cap should be released with caution.
Ammonium acetate	$CH_3.COO.NH_4$	Harmful if taken internally.	
Ammonium carbonate	NH_4HCO_3+ $NH_2.COO.NH_4$	Strongly effervescent when mixed with acid solutions. Although this is not generally hazardous, careless addition of large amounts to acid solutions will cause them to overflow.	The solid yields a strong smell of ammonia. When strong solutions are used, especially in the case of particle techniques, the vapour tends to irritate the eyes.
Ammonium chloride	NH_4Cl	Harmful if taken internally.	
di-Ammonium hydrogen orthophosphate	$(NH_4)_2HPO_4$	Harmful if taken internally.	
Ammonium molybdate	$(NH_4)_6Mo_7O_{24}$	Harmful if taken internally.	
Ammonium nitrate	NH_4NO_3	A powerful oxidising agent that may cause fire if mixed with combustible material. Mixtures with combustible materials are readily ignited, and may be explosive if these are finely divided. Fires should be extinguished with a water spray. Spillages should be washed down with large amounts of water. If organic materials such as wood, paper or fabrics are contaminated, they should be thoroughly washed, or there may be a risk of fire when the chemical dries out. Violent reactions may occur with acids, chlorates, potassium permanganate, sulphur.	This chemical should not be used for particle techniques involving organic materials such as sawdust or wood shavings. It should not be used with appressed techniques involving cotton-wool, cloth or paper.

Ammonium persulphate (Ammonium peroxodisulphate)	$(NH_4)_2S_2O_8$	A powerful oxidising agent that can cause fire if mixed with combustible materials. It should be kept out of contact with all combustible materials. Contact with the eyes, skin and clothing must be avoided. Fires should be extinguished with a water spray. The vapours from heated solutions should not be inhaled.	This chemical should not be used for particle techniques involving organic materials such as sawdust or wood shavings. It should not be used for appressed techniques involving cotton-wool, cloth, paper etc.
Ammonium sulphate	$(NH_4)_2SO_4$	May be harmful if taken internally.	
Ammonium sulphide	$(NH_4)_2S$	The liquid is corrosive and will severely irritate and may burn the eyes and skin. It will cause severe internal damage if swallowed. The vapour can cause unconsciousness if inhaled in high concentrations, and can cause headaches and lethargy at lower concentrations. If mixed with acids, toxic hydrogen sulphide gas is evolved.	Commonly supplied as a 16% solution, it is a yellow liquid with a strong offensive smell. Pressure may develop in the container, which should therefore be opened with caution.
Antimony potassium tartrate	$KSbO.C_4H_4O_6$	Harmful if taken by mouth, causing nausea and vomiting, and burning of the mouth and throat. It is irritating to the eyes, skin and respiratory system. In certain conditions the action of acidic reducing agents on antimony compounds may yield stibine, an extremely poisonous gas that can cause blood destruction, and damage to the liver and kidneys.	Although hazardous reactions can occur under certain circumstances, as indicated, it is commonly used as a small addition to some solutions and is relatively safe in these applications.
Antimony trichloride	$SbCl_3$	Antimony trichloride is very toxic and poisonous if taken internally. It is also corrosive, causing burns to the eyes and skin. In case of contact with the eyes, it should be washed out immediately with copious amounts of water and medical advice sought. In certain conditions the action of acidic reducing agents on antimony compounds may yield stibine, an extremely poisonous gas that can cause blood destruction, and damage to the liver and kidneys.	It has only a limited application in colouring, and in view of its toxicity is not generally recommended for workshop use.
Antimony trisulphide	Sb_2S_3	Harmful if taken internally. Addition of acids will liberate toxic hydrogen sulphide gas.	The dust produced when dry residues of pastes containing this chemical are removed, are toxic and must not be inhaled.
Arsenic trioxide (arsenious oxide)	As_2O_3	Arsenic trioxide is very toxic and presents a serious risk of poisoning by inhalation or if swallowed. There is also an added danger from cumulative effects as inhalation of very small concentrations of dust or fumes over a long period will cause poisoning. It is irritating to the eyes, skin and respiratory system. If contact occurs, then wash with plenty of water, and seek medical advice if any symptoms occur. The extremely poisonous gas arsine is liberated under certain conditions, by the action of acid reducing agents on arsenic compounds. This gas causes damage to the liver and kidneys, and can be fatal from a few inhalations.	This chemical cannot be recommended for use in the average workshop as it constitutes such a high risk. Although there are extant processes for producing grey colourations with this chemical, the conditions must be strictly controlled as the processes are potentially very dangerous.
Barium chlorate	$Ba(ClO_3)_2.H_2O$	A strong oxidising agent which can cause fire or risk of explosion if mixed with combustible materials. Clothing contaminated with it, should be removed and thoroughly washed. It is also poisonous and may be very harmful if inhaled or swallowed. Violent reactions may occur if it is mixed with ammonium salts, acids, sulphur.	This chemical should not be used with particle techniques involving organic materials such as sawdust or wood shavings. It should not be used for appressed techniques involving cotton-wool, cloth, paper, etc.

Barium sulphide	BaS	Harmful by inhalation or if swallowed. If it comes into contact with the skin, it should be washed off with plenty of water. Addition of acids will liberate toxic hydrogen sulphide gas.	
Bismuth nitrate	$Bi(NO_3)_3.5H_2O$	Harmful if taken internally.	
Butyric acid (normal)	$C_3H_7.COOH$	Butyric acid is corrosive, causing irritation or burns to the eyes and skin. It will also be very harmful if taken internally.	It is a colourless and oily liquid with an extremely strong and pungent smell. The vapour tends to cling to the hair and clothing for some time, and the smell, which resembles the smell of goats, can be difficult to eliminate.
Calcium chloride	$CaCl_2.6H_2O$	Calcium chloride is an irritant, and contact with the eyes and skin should be avoided. It is also harmful if taken internally.	
Calcium hypochlorite	$Ca(ClO)_2$	Causes irritation of the eyes and skin, and is harmful if taken internally, causing irritation of the alimentary system. Care should be taken to avoid inhaling the dust, which irritates the respiratory system. The addition of acids will liberate chlorine gas, which must not be inhaled.	More commonly referred to as bleaching powder. It is a white powder from which light dusts are easily raised. Pressure may build up in the container during storage.
Chromium trioxide (chromic acid)	CrO_3	A very powerful oxidant that can cause fire if allowed to come into contact with combustible material. The solid and its solutions cause severe eye burns and will burn the skin. The dust irritates all parts of the respiratory system and must not be inhaled. If swallowed it would cause severe internal injury. Frequent exposure to this chemical can result in skin ulceration. Violent reactions may occur with acetic acid, ethanol, turpentine, butyric acid, sulphur. Suspected carcinogen.	Chromium trioxide, which is supplied in the form of dark red crystalline flakes, dissolves readily in water forming chromic acid. When preparing solutions, small quantities of the solid should be added to large quantities of water.
Citric acid	$C_6H_8O_7$	Contact with the eyes and skin should be avoided. May be harmful if taken internally.	
Copper acetate	$(CH_3.COO)_2Cu$ H_2O	Copper acetate is harmful if swallowed. Contact with the eyes and skin should be avoided. The dust should not be inhaled as it will irritate the respiratory system.	Light dusts are easily raised if this chemical is handled carelessly.
Copper carbonate	$CuCO_3.Cu(OH)_2$ H_2O	Harmful if taken internally. Contact with the eyes and skin should be avoided. The dust should not be inhaled, as it will irritate the respiratory system.	Light dusts are easily raised if this chemical is handled carelessly.
Copper chloride	$CuCl_2.2H_2O$	Harmful if taken internally. Contact with the eyes and skin should be avoided.	
Copper nitrate	$Cu(NO_3)_2.3H_2O$	Harmful if taken internally. Contact with the eyes and skin should be avoided.	
Copper sulphate	$CuSO_4.5 H_2O$	Harmful if taken internally. Contact with the eyes and skin should be avoided.	

Ethanol (ethyl alcohol)	C_2H_5OH	Ethanol is used, either in the form of 'absolute' alcohol or in the denatured forms known as methylated spirits, in very small quantities as a wetting agent with some solutions. It is highly flammable. The absolute form is extremely intoxicating, while the denatured forms contain additives that will result in serious risks to health if taken internally. The vapour, in high concentrations, should not be inhaled. Reacts with varying degrees of violence with a wide range of oxidants. May be explosive in contact with silver nitrate.	It can be safely used as a small additive to solutions, as a wetting agent where indicated. If any other use is contemplated, the safety literature should be consulted. It should not be stored in any quantities without special provisions.
Ferric chloride	$FeCl_3.6H_2O$	Ferric chloride is corrosive when wet and contact with the eyes and skin should be avoided. If the solid is in contact with the skin it will tend to absorb moisture and become corrosive. It is also very harmful if taken internally.	
Ferric nitrate	$Fe(NO_3)_3.9H_2O$	Ferric nitrate is harmful if swallowed. Contact with the eyes and skin should be avoided.	
Ferric oxide	Fe_2O_3	No significant hazards.	
Ferric sulphate	$Fe(SO_4)_3+aq.$	Harmful if taken internally.	
Ferrous sulphate	$FeSO_4.7H_2O$	Harmful if swallowed. Contact with the eyes and skin should be avoided.	
Hydrochloric acid	HCl	A corrosive acid which can severely burn the eyes and skin. If swallowed it would cause severe internal damage. The vapour irritates all parts of the respiratory system and must not be inhaled. The vapour severely irritates the eyes.	Commonly available in strong concentrations of 32% and 35%. The liquid fumes when the container is opened to the air. When preparing solutions, small amounts of strong solutions should be added to large quantities of water and not vice-versa. Pressure may develop in the container, which should therefore be opened with caution.
Hydrogen peroxide	H_2O_2	A powerful oxidising agent which assists fire. Contact with combustible material may cause fire. In higher concentrations it is corrosive to the eyes and skin, causing burns. If swallowed it would cause severe internal injury. Violent reactions may occur if mixed with flammable liquids, combustible materials, metal salts, copper, iron, under certain conditions.	This chemical should not be used for particle techniques involving organic materials such as sawdust or wood shavings. It should not be used for appressed techniques involving cotton-wool, cloth, paper etc. It is commonly supplied as a concentrated solution, 28%, which is referred to as 100 vols.
Iron filings (cast iron)	Fe	No significant hazards.	
Lead acetate	$(CH_3.COO)_2Pb.3H_2O$	Lead salts are highly toxic, and if swallowed may cause vomiting, internal injury and collapse. Similar effects may be caused by inhaling the dust, or through skin contact with the damp solid or solutions. There is also a danger of cumulative effects through slight exposure over a long period. Suspected carcinogen.	This chemical cannot be recommended for use in the average workshop, as it constitutes such a high risk.

Lead dioxide (lead peroxide)	PbO_2	Lead dioxide is toxic, and harmful when inhaled and swallowed, and may cause internal injury. There is also a danger of cumulative effects through slight exposure over a long period.	
Mercuric chloride	$HgCl_2$	This chemical is very toxic, and presents a serious risk of poisoning by swallowing, inhalation or skin contact. There is a serious danger of cumulative effects through slight exposure over a long period. It is also irritating to the skin and eyes, and may cause severe damage to the eyes.	This chemical cannot be recommended for use in the average workshop as it constitutes such a high risk. It has only a very limited use in metal colouring.
Mercurous chloride	Hg_2Cl_2	Harmful cumulative effects may occur if this chemical is taken internally.	
Nickel ammonium sulphate (ammonium nickel sulphate)	$(NH_4)_2SO_4.NiSO_4$ $6H_2O$	Nickel salts may be irritating to the eyes and skin. They are noted as a common cause of sensitive skin reactions, and may cause dermatitis. The dust should not be inhaled. It may be very harmful if taken internally.	
Nickel sulphate	$NiSO_4+aq.$	Nickel salts may be irritating to the eyes and skin. They are noted as a common cause of sensitive skin reactions, and may cause dermatitis. The dust should not be inhaled. It may be very harmful if taken internally.	
Nitric acid	HNO_3	A powerful corrosive acid which causes severe burns to the eyes and skin. If swallowed it would cause severe internal injury. It is also a powerful oxidising agent and may cause fires if mixed with combustible materials. The vapour irritates all parts of the respiratory system, and must not be inhaled. The vapour also severely irritates the eyes and skin. Violent reactions can occur with a wide range of chemicals; acetic acid, ethanol, chromium trioxide, flammable liquids and gases. If strong solutions of nitric acid are added to copper or brass, then brown fumes of nitrogen dioxide are evolved, which are toxic and must not be inhaled.	Commonly supplied as a 95% solution (fuming nitric acid), and as a 70% solution. Nitric acid should not be used in conjunction with organic materials such as sawdust or wood shavings in particle techniques, or with materials such as cotton-wool, cloth, paper in appressed techniques. Dilute solutions can be conveniently prepared from a 70% stock solution. Fuming nitric acid should not be used, as it is not necessary, and the risk from vapour is greatly increased. When diluting the acid, small amounts of the concentrated acid should be added to large quantities of water and not vice versa. Pressure may develop in the container, which should therefore be opened with caution.
Oleic acid (octadec-9-enoic acid)	$C_{17}H_{33}COOH$	Oleic acid is irritating to the eyes, skin and respiratory system. It is harmful if taken internally.	
Orthophosphoric acid (phosphoric acid)	H_3PO_4	A strong corrosive acid that causes severe irritation or burns to the eyes and skin. If taken internally, it would lead to severe internal irritation and injury.	When diluting the acid, small amounts of the concentrated acid should be added to large amounts of water, and not vice-versa.
Oxalic acid	$(COOH)_2.2H_2O$	Causes severe internal pain and collapse, if swallowed. The dust and solutions irritate the eyes and skin. The dust irritates the respiratory system.	

Potassium aluminium sulphate (aluminium potassium sulphate)	$Al.K(SO_4)_2.12H_2O$	Harmful if taken internally.	
Potassium binoxalate (potassium hydrogen oxalate)	KHC_2O_4	Causes severe internal pain and collapse, if swallowed. Contact with the eyes and skin should be avoided.	
Potassium chlorate	$KClO_3$	A powerful oxidising agent which may be explosive or cause fires when mixed with combustible materials. It is very harmful if taken internally, by inhalation or swallowing, causing severe irritation of the respiratory system, intestinal tract and kidneys. If organic materials such as wood, fabrics, paper or clothing become contaminated by the solid or solutions, they should be thoroughly washed with large amounts of water or they may become dangerously flammable when dry. Violent reactions can occur with acids, sulphur, metal sulphides, ammonium salts, sulphuric acid.	This chemical must not be used in conjunction with organic materials such as sawdust or wood shavings in particle techniques, or with materials such as cotton-wool, cloth or paper in appressed techniques.
Potassium dichromate	$K_2Cr_2O_7$	A powerful oxidant that can cause fire if mixed with combustible materials. If organic materials such as wood, fabrics, paper or clothing become contaminated by the solid or solutions, they should be thoroughly washed with large amounts of water or they may become dangerously combustible when dry. It is very harmful if taken internally, by inhalation or swallowing, causing irritation and damage internally. Contact with the eyes and skin must be avoided. Frequent exposure to the dust, or long term absorption via the skin can cause skin ulceration, and liver and kidney disease. Suspected carcinogen.	This chemical must not be used in conjunction with organic materials such as sawdust or wood shavings in particle techniques, or with materials such as cotton-wool, cloth, paper in appressed techniques.
Potassium hydrogen tartrate (potassium bi-tartrate)	$COOH.CH(OH).CH(OH).COOK$	May be harmful if taken internally.	
Potassium hydroxide	KOH	A corrosive caustic alkali that causes severe burns to the eyes and skin. If swallowed it would cause severe internal irritation and injury.	When preparing solutions, the solid, which is commonly supplied in the form of flakes, should be added in small amounts to a large quantity of water and not vice versa. Heat will be evolved as the solid dissolves.
Potassium nitrate	KNO_3	A powerful oxidant that is explosive if mixed with combustible materials. If organic materials such as wood, fabrics, paper or clothing become contaminated by the solid or solutions, they should be thoroughly washed with large amounts of water or they may become dangerously combustible when dry. Explosive reactions may occur if it is mixed with antimony trisulphide, barium sulphide, sulphur. Mixtures with combustible materials are readily ignited, or explosive if they are finely divided.	This chemical must not be used in conjunction with organic materials such as sawdust or wood shavings in particle techniques, or with materials such as cotton-wool, cloth, or paper in appressed techniques.

Potassium permanganate	$KMnO_4$	An oxidising agent that can cause fire if mixed with combustible materials. It is harmful if taken internally, and contact with the eyes and skin should be avoided. Violent reactions may occur if it is mixed with acetic acid, hydrogen peroxide, concentrated hydrochloric acid, concentrated sulphuric acid.	This chemical should not be used in conjunction with organic materials such as sawdust or wood shavings in particle techniques or with materials such as cotton-wool, cloth, or paper in appressed techniques.
Potassium persulphate (potassium peroxodisulphate)	$K_2S_2O_8$	An oxidising agent that can cause fire if mixed with combustible materials. It is harmful if swallowed, and if allowed to come into contact with the eyes or skin. The vapour from heated solutions should not be inhaled. A violent reaction may occur if it is mixed with potassium hydroxide.	This chemical should not be used in conjunction with organic materials such as sawdust or wood shavings in particle techniques or with materials such as cotton-wool, cloth, or paper in appressed techniques.
Potassium sulphate	K_2SO_4	May be harmful if taken internally.	
Potassium sulphide (potassium polysulphide)	K_xS_x	The solid and strong solutions are corrosive and may cause burns to the eyes or skin. If swallowed it would cause severe internal injury. The dust should not be inhaled. Addition of acids will liberate toxic hydrogen sulphide gas.	
Selenous acid	H_2SeO_3	Selenous acid is a toxic substance that presents a serious risk of poisoning by inhalation or swallowing. There is also a danger from cumulative effects through slight exposure over a long period. It is also corrosive and may cause burns. It is irritating to the eyes, skin and respiratory system and contact must be avoided.	
Silver nitrate	$AgNO_3$	Silver nitrate is harmful if swallowed, or if allowed to come into contact with the eyes or skin. It can cause severe burns to the eyes.	
Sodium acetate (hydrated)	$CH_3.COONa.$ $3H_2O$	Sodium acetate is harmful if swallowed, or if allowed to come into contact with the eyes or skin.	
Sodium chlorate	$NaClO_3$	Sodium chlorate is a powerful oxidant which may become explosive if mixed with combustible materials. If organic materials such as wood, fabrics, paper or clothing become contaminated by the solid or solutions, they should be thoroughly washed with large amounts of water or they may become dangerously combustible when dry. It is very harmful if taken internally, by inhalation or swallowing, causing severe irritation, nausea and vomiting and even kidney damage. The dust or strong solutions irritate the eyes and skin. Violent reactions may occur if it is mixed with fibrous or absorbent organic materials, ammonium salts, sulphur, sulphides, concentrated sulphuric acid, acids generally.	This chemical must on no account be used in conjunction with organic materials such as sawdust or wood shavings in particle techniques, or with materials such as cotton-wool, cloth, or paper in appressed techniques. Explosive conditions may result.
Sodium chloride	$NaCl$	No significant hazards.	

Sodium dichromate	$NaCr_2O_7.2H_2O$	A powerful oxidant that can cause fire if mixed with combustible material. If organic materials such as wood, fabrics, paper or clothing become contaminated with the solid or solutions, they should be thoroughly washed with large amounts of water or they may become dangerously combustible when dry. It is very harmful if taken internally, by inhalation or swallowing, causing irritation and damage internally. Contact with the eyes and skin must be avoided. Frequent exposure to the dust, or long term absorption via the skin can cause skin ulceration, and liver and kidney disease. Suspected carcinogen.	This chemical must not be used in conjunction with organic materials such as sawdust or wood shavings in particle techniques or with materials such as cotton-wool, cloth, or paper in appressed techniques.
Sodium hydroxide	$NaOH$	A corrosive caustic alkali that causes severe burns to the eyes and skin. If swallowed it would cause severe internal irritation and injury. A violent reaction can occur if the solid is mixed with a small amount of water, or particularly if a small amount of water is added to quantities of the solid.	When preparing solutions, the solid which is commonly supplied in the form of small pellets, should be added in small amounts to a large quantity of water and not vice versa. Heat will be evolved as the solid dissolves.
Sodium metabisulphite	$Na_2S_2O_5$	Harmful if taken internally.	
Sodium sulphide	$Na_2S.2H_2O$	The solid and strong solutions cause severe burns to the eyes, and burns to the skin. It is very irritant and poisonous if swallowed. Addition of acids will liberate toxic hydrogen sulphide gas.	
Sodium tartrate	$(CHOH.COONa)_2$ $.2H_2O$	Harmful if taken internally.	
Sodium tetraborate (borax)	$Na_2B_4O_7.10H_2O$	Borax is irritating to the eyes and skin, and harmful if taken internally. The dust should not be inhaled.	
Sodium thiosulphate	$Na_2S_2O_3.5H_2O$	Sodium thiosulphate irritates the eyes and may be harmful if taken internally. The vapours produced from hot solutions should not be inhaled. If the solid is heated or treated with acids, sulphur dioxide is liberated, which must not be inhaled.	
Stannic chloride (hydrated)	$SnCl_4.5H_2O$	Stannic chloride is corrosive, causing irritation or burns to the eyes and skin. It is irritating to the respiratory system and harmful if swallowed.	
Sulphur (flowers of sulphur)	S	Explosive mixtures can result if sulphur is added to oxidising agents.	
Sulphuric acid	H_2SO_4	A powerful corrosive acid which causes severe burns to the eyes and skin. If swallowed it would give rise to severe irritation and internal injury. The vapour must not be inhaled. Violent reactions can occur if it is mixed with chlorates, permanganates. The addition of sulphuric acid to nitrates will liberate nitrogen dioxide gas, which is toxic and must not be inhaled.	When diluting the concentrated acid, small amounts of acid should be added to large amounts of water and not vice versa.

Tartaric acid	(CHOH.COOH)$_2$	Harmful if taken internally.
Zinc chloride	ZnCl$_2$	Zinc chloride causes burns and should be prevented from coming into contact with the eyes or skin. It is harmful if taken internally. The fumes are irritating to the respiratory system, and should not be inhaled.
Zinc nitrate	Zn(NO$_3$)$_2$.4H$_2$O	May be harmful if taken internally.
Zinc sulphate	ZnSO$_4$.7H$_2$O	May be harmful if taken internally.

Degreasing agents

A wide range of organic solvents, and proprietary organic solvents are available for use as degreasing agents. Most of these are volatile and hazardous to some degree, and many which are used in the metal-finishing trades are only suitable for use with specialised equipment. It is essential, when obtaining degreasing agents, to ensure that they are suitable for use with manual cleaning techniques. The manufacturer's warnings should always be heeded. It is absolutely essential to ensure that there is adequate ventilation, preferably localised extraction, when using these chemicals, so that even a slight build-up of vapour is avoided. Since some organic solvents may be absorbed via the skin, contact should be avoided. They must not be ingested. Most organic solvents are volatile and flammable to some degree, and should not be used in the presence of naked flames.

Aromatic hydrocarbons in general should be avoided, and benzene (benzol) in particular should not be used.[18] Chlorinated hydrocarbons should also be avoided, including tetrachloroethane (acetylene tetrachloride), carbon tetrachloride, chloroform, trichloroethylene, ethylene dichloride(1,2-dichloroethane), methylene chloride (methylene dichloride, dichloromethane), perchloroethylene (tetrachloroethylene), ortho-dichlorobenzene, monochlorotoluene (chlorotoluene). Of the chlorinated hydrocarbons, methyl chloroform (1,1,1-trichloroethane) is relatively less harmful.[19] Petroleum distillates such as VM&A naphtha are probably the most suitable for use, but those with a high flammability, such as petroleum ether, should be avoided.

18. Benzene is listed as a known human carcinogen. The following chemicals are suspected of being carcinogenic: tetrachloroethane, carbon tetrachloride, chloroform, trichloroethylene, ethylene dichloride, perchloroethylene.

19. Aromatic and chlorinated hydrocarbons are generally highly toxic, with the exception of methyl chloroform which is moderately toxic by inhalation and only slightly toxic by contact.

Excluded colouring procedures

A small number of colouring procedures were excluded from the research programme on the grounds that they constitute a serious threat to life, unless carried out in extremely well controlled conditions. They are entirely unsuitable for use in the average workshop, requiring specialised safety provisions and a high degree of control.

Cyanides

Some colouring techniques noted in the literature involve the use of cyanide-containing compounds. The addition of acids to cyanides liberates the lethal gas hydrogen cyanide.[20] When cyanide complexes are used in colouring techniques, it is very difficult to determine whether a particular use of them can be considered to be safe. The risks to life are so great, in this case, that the storage and use of cyanide-containing compounds in the workshop must be totally discouraged.

20. Death can result from one or two inhalations of the gas. Although cyanide solutions are used in the electroplating and metal finishing industries, their use is subject to stringent control, and the provision of specialised antidotes and safety measures.

Arsenic and antimony compounds

There are a number of recipes for producing grey colourations on the cuprous metals, which involve the use of arsenic or antimony compounds dissolved in acids. Under these conditions, the highly toxic gases arsine and stibine may be produced.[21] These present a serious risk to health or even life.

21. A few inhalations of these gases can cause blood destruction, kidney and liver damage, and death in extreme cases.

Although these processes have been used industrially, and are relatively well known, the high risks involved suggest that they are entirely unsuitable for use in the average workshop.

Molten salts

One colouring technique that is mentioned in a number of sources involves immersing the metal in molten salts, often a mixture of oxidising agents, at high temperatures.[22] This is potentially very dangerous since minute amounts of grease or water contaminating the metal surface can cause the molten salts to erupt violently. These techniques can only be carried out in specialised conditions and require extensive safety equipment similar to that used in experimental foundry work, and cannot be recommended for general use in metal colouring.

Acidification of sulphides

Many of the older books on metal colouring suggest the addition of acids when metals are coloured with sulphides. Adding acids to sulphides liberates the toxic gas hydrogen sulphide, which can be fatal in moderate to high concentrations. Although the use of sulphides will in any case generally liberate some of this gas, if adequate ventilation is provided, the technique is generally relatively safe and is extensively used. However, the addition of acids dramatically increases the levels of hydrogen sulphide gas produced and presents a serious risk to health.[23] The addition of acids is generally unnecessary in sulphide colouring, and cannot be recommended. If acidified sulphides are used, then preparation and colouring must be carried out in a fume cupboard so that the gas produced is totally controlled.

22. The technique is reported to have been used in the colouring of small parts in the German clockmaking industries, at one time.

23. Some restricted tests were carried out during the course of research. The results were generally poor in comparison with the use of unacidified sulphides. A few recipes have nevertheless been included for information, which require the use of a fume cupboard.

APPENDICES

Appendix 1: Japanese alloys and colouring

The integration of colouring with other metalworking techniques is probably more developed in Japanese metalwork than in any other tradition. Although most traditions have used colouring or patination, Japanese metalwork is distinguished by the use of a number of specialised alloys, some of which have been specifically developed for the colours that they produce after patination. These are used in conjunction with copper, silver, gold, brass and bronze. Although this book is specifically concerned with the colouring of a range of commonly available cuprous metals and silver, and the Japanese alloys fall outside this range, the difference in approach is of interest and requires some discussion. Also various versions or adaptations of Japanese colouring solutions have found their way into western metal colouring, and are included in the recipe section.[1] One of the important colouring agents used in Japan is referred to as *rokusho,* and a small number of recipes involving the use of this chemical have also been included for their effects on some of the common cuprous metals.

Japanese alloys

A wide range of alloys are used, of which the following are noted as being of importance:[2]

Karakane A leaded tin bronze with a variable composition which generally falls within the following range, copper 71–89%, tin 2–8%, lead 5–15%. Additions of arsenic and antimony are common, often in the form of *shirome,* a by product of the de-silverisation of copper with lead, particularly where a grey patina is produced.

Shintyuu Brass containing from 15–20% zinc and 2–5% lead.

Sentoku A yellow bronze, intermediate between bronze and brass which commonly has the composition copper 72%, tin 8%, zinc 13% and lead 6%. This appears to have been used more frequently than *shintyuu.*

In addition, the following alloys are often used in conjunction with copper, silver and gold in a variety of metalworking techniques, and are noted for the range of colours that they produce after patination.

Shakudo An alloy with the composition copper 95–98%, gold 2–5%.

Shibuichi An alloy with a variable composition usually falling within the range copper 60–75%, silver 25–40%.

Kuromi-do An alloy consisting of copper with an addition of 1% arsenic.

Shiro-shibuichi An alloy with the approximate composition silver 60%, copper 40%.

Kuro-shibuichi A composite alloy with a variable composition usually falling within the range *shakudo* 60–85%, *shibuichi* 15–40%.

These metals and alloys are used in various combinations by inlaying or laminating techniques of different kinds,[3] and then coloured using a patinating agent which yields different colours on each metal and alloy. Variation in colour combinations can be very subtle, and is usually achieved by the careful selection of particular alloys rather than by the use of a wide range of colouring agents.

1. These commonly involve the use of solutions containing copper sulphate and copper acetate, with various additions of alum, acetic acid or ammonia.

2. The information relating to Japanese alloys and colouring contained in this section has been compiled from the following sources: Roberts-Austen (Bib. 301 and 302), Gowland (Bib. 156 and 157), Pijanowski (Bib. 281), Niiyama (Bib. 273), Savage and Smith (Bib. 315), and from correspondence with Eiro Niiyama and Hara Kinjiro. A large Japanese literature relating to these techniques exists, but very little information is currently available in English. The range of alloys is much broader and more subtle than is represented here. In addition, more highly specialised alloys also exist such as *sawari* (copper 67%, tin 33%) which is melted to fill engraved lines.

3. Including the technique known as *mokume-gane* in which a range of metal plates are laminated, rolled out, cut and further laminated to build a sheet which has a large number of fine layers. These are exposed by carving to different depths. The sheet is then again rolled out flat, stretching the exposed laminations, which become differentially coloured after patination. The technique can be controlled to produce various effects and patterns.

Rokusho

A colouring agent which is particularly important in Japanese metalworking is known as *rokusho*. It is commercially available in Japan, but as far as is known is not available elsewhere. Used in conjunction with copper sulphate, and sometimes also with additions of alum or plum vinegar, it produces different colours on the various metals and alloys. Gold is made slightly more yellow by treatment with *rokusho*, while silver is whitened very slightly from its polished state. Copper and gilding metal are coloured from orange to orange-brown or reddish-brown, while cast brass yields an orange-brown grain-enhanced surface. *Shakudo* is coloured from aubergine to black, depending on the gold content of the alloy. *Shibuichi* is coloured grey, while *shiro-shibuichi* is coloured light grey and *kuro-shibuichi* dark grey. *Kuromi-do* takes on a colour which varies from dark brown to black. The colour that is obtained on the leaded bronzes is not known, but tin bronzes such as the LG3 and LG4 used in tests were little affected even with prolonged treatment. Variations in the composition of the various alloys is said to produce subtle differences in the colours obtained.

Colouring with *rokusho* and copper sulphate, which is usually carried out by immersing the article in the boiling solution, is preceded by careful surface preparation and cleaning. In addition to degreasing and cleaning, a further treatment is often employed, in which a paste is applied to the surface of the metal immediately prior to immersion. The paste used is prepared by grinding a particular species of long white radish known as *daikon*,[4] in a pestle and mortar with about five parts of water. The paste is applied and the object immersed for a short period, agitating it constantly while it is in the solution. This process is then repeated a number of times. The precise purpose of the radish paste is not known, but it is often credited either with protecting the surface from tarnish or uneven colouring in the initial stages, or with activating the surface of the metal.[5]

In the absence of commercially produced *rokusho*, the following preparation (referred to in recipes as 'Rokusho') can be used: copper acetate (6.5 gms), sodium hydroxide (2.25 gms) and calcium carbonate (2.25 gms) are added to between 150–175 cm³ of distilled water and thoroughly mixed. The mixture is then allowed to stand and separate for at least a week. The supernatant is carefully drained, yielding the equivalent of about five grams of commercial *rokusho*.[6] This is typically used with about five grams of copper sulphate in a litre of water for immersion colouring, sometimes with the addition of small quantities of alum or dilute acetic acid. The technique used is as described above.

4. The long white radish known as *mooli* which is more familiar as a West Indian vegetable is apparently the same species as *daikon*. We are grateful to Nick Stanbury for drawing this to our attention.

5. *Daikon* contains, in addition to the starch, cellulose, water content associated with species of this type, substantial quantities of a plant glucoside. This has been identified as 4-methylthio-3-butenylthioglucoside. When the radish is ground in the presence of water, the hydrolytic enzyme present is released, encouraging the breakdown of the glucoside into glucose, potassium bi-sulphate, and a thiocyanate complex. The latter chemicals may have the effect of acting as a mild surface activating agent, while the body of the paste may encourage a more even colour by ensuring that the initial colouring action is gradual. It is of interest to note that a weak solution of potassium bi-tartrate is sometimes used commercially in metal finishing as a 'holding' solution in which objects are kept immersed after cleaning and prior to colouring, to prevent tarnish.

6. It is thought that some commercially produced Japanese *rokusho* sometimes includes small quantities of additional ingredients. The nature of these additions is not known.

Appendix 2: Lustre colours

Some of the finishes recorded in the recipe section are described as 'lustre' colours. These differ from the majority of colours produced in that they have, as the name implies, a lustrous and transparent quality. The observed colour in these cases is not due to the inherent colour of a layer formed at the surface of the metal, as is usual, but rather is the result of a phenomenon known as 'optical interference'. The colours produced are therefore sometimes referred to as interference colours. Optical interference colours occur where thin transparent or near-transparent layers are deposited on the surface of the metal.[1] When light strikes the surface of the layer some is reflected, while the remainder passes into the layer and is reflected at the surface of the metal. The observed colour is caused by the interference of the two portions of reflected light as they pass back to the eye of the observer.[2] The phenomenon is familiar as the 'rainbow colours' seen in a thin film of oil on water. If the thickness of the film is even then a single uniform colour is seen, rather than a multi-coloured effect. The visible colour depends on the thickness of the deposited layer. In ideal conditions violet is seen first, followed by indigo, blue, green, yellow, orange and red. As the layer becomes thicker, this sequence is repeated a second or even a third time.

In practice the distinct steps described are not always seen. Some colours, notably orange, blue and purple, tend to be produced more readily than others in the sequence.[3] In many cases the sequence will appear to begin with an orange lustre colour, as the earlier colours can be faint and may pass quite rapidly. In addition, second and third series colours may be of a slightly different hue than those of the first series, and often lack their clarity. In many cases the later series take an impracticably long time to develop, tending to become pallid and grey rather than appearing as full lustre colours.

The colour quality obtained will also depend on other factors. Since the layers produced are transparent, the colour of the underlying metal will contribute to the observed colour. Silver and silver-plated surfaces produce results that are closest to spectral colours, as they are generally more reflective and have less inherent colour than copper or its alloys. The clearest and most lustrous results are produced on highly polished surfaces. Although good results can also be obtained on directionally satinised finishes, matt surfaces in general tend to scatter the light more randomly and produce only dull and pallid colouring.

Precise colour development can be difficult to monitor and control, as the colours may appear and recede fairly quickly. Rapid transfer to a bath of water is essential to halt the action of the chemical once the required colour is obtained. Monitoring may also be complicated by the fact that the colour of the object when dry can differ from its apparent colour when wet or immersed. Practice may be required to determine the point at which an object should be removed from the solution, in order to achieve a particular final colour when dry. Wax finishing and lacquering also tend to have the effect of altering the observed colour slightly.

Since the observed colour depends on the thickness of the film, the colour may change as the surface is viewed from more oblique angles. This change in observed colour will also occur on uneven or textured surfaces, which can considerably enhance the overall effect. On plain surfaces, the colour may be totally lost when the object is viewed from certain angles, revealing the colour of the metal surface.

1. Thiosulphate solutions, for example, produce a thin layer of metal sulphide.

2. A concise description of the mechanism of optical interference is given by Mr. P. Gainsbury in an appendix to 'The colouring and working of the refractory metals', J. Brent Ward, Project Report No. 34/1, Worshipful Company of Goldsmiths, London, 1978.

3. Some recipes will only produce one colour, rather than a sequence. Indications are given in the notes accompanying individual recipes.

In all cases where a lustre colour is to be produced, the surface of the metal must be thoroughly prepared, and absolutely clean and grease-free. Even the slightest traces of grease or fingermarks will tend to mar the finish.

Slight petrol lustre.

In some cases a slight lustre effect may occur as a superficial effect on an otherwise opaque colour. These effects, which are generally very slight, take the form of a very pale multicoloured sheen which tends to be uneven. Where this occurred in tests, it is recorded in the notes to individual recipes, where it is referred to as a 'slight petrol lustre'.

Appendix 3: Unsuccessful recipes

The following recipes produced little or no colouring effect in tests. It should be noted that the recipes which failed to colour silver or silver plate might form the basis for bi-metallic or partial plate colouring.

---

1   No effect on silver or silver plate

Copper nitrate	200 gm
Water	1 litre

Boiling immersion (Thirty minutes)

Tests carried out with the boiling solution produced no effect for a period of immersion of thirty minutes.

---

2   No effect on silver or silver plate

Copper nitrate	80 gm
Water	1 litre
Ammonia (.880 solution)	10 cm³

Boiling immersion (Thirty minutes)

Immersion in the boiling solution produced no effect during a period of thirty minutes.

---

3   Non-adherent bloom on silver or silver plate

Copper sulphate	80 gm
Water	1 litre

Boiling immersion (Thirty minutes)

Immersion in the boiling solution for a period of thirty minutes produced no effect other than a slight powdery greenish bloom, which was found to be totally non-adherent.

---

4   No effect on silver or silver plate

Copper sulphate	25 gm
Water	1 litre
Ammonia (.880 solution)	3 cm³

Boiling immersion (Thirty minutes)

Immersion in the boiling solution for a period of thirty minutes produced no effect other than a very slight non-adherent bloom.

---

5   Slight tarnish on silver and silver plate

Potassium permanganate	30 gm
Copper sulphate	6 gm
Water	1 litre

Boiling immersion (Fifteen minutes)

Immersion in the boiling solution for a period of fifteen minutes produced no effect on either standard silver or silver plated surfaces. Scratch-brushing with the hot solution produced a slight yellow stain on standard silver and slight grey stains on silver plate.

---

6   Non-adherent bloom on silver or silver plate

Potassium chlorate	50 gm
Copper sulphate	25 gm
Nickel sulphate	25 gm
Water	1 litre

Boiling immersion (Thirty minutes)

▲ Potassium chlorate is a powerful oxidant and should not be allowed to come into contact with combustible materials. It should be prevented from coming into contact with the eyes, skin and clothing.

▲ Nickel salts are a common cause of sensitive skin reactions.

Immersion in the boiling solution for a period of thirty minutes produced a slight bloom on both standard silver and silver plated surfaces. This was found to be non-adherent, when the samples were washed.

---

7   No effect on silver or silver plate

Potassium hydroxide	100 gm
Copper sulphate	30 gm
Sodium tartrate	30 gm
Water	1 litre

Immersion   (Twenty minutes)

▲ Potassium hydroxide is a powerful caustic alkali and both the solid and solutions should be prevented from coming into contact with the eyes and skin. When preparing the solution, small quantities of the potassium hydroxide should be added to large quantities of water, and *not* vice-versa.

Immersion in the solution at temperatures from warm to boiling produced no effect on either standard silver or silver plate during a period of twenty minutes.

8   No effect on silver or silver plate

Sodium hydroxide	100 gm
Sodium tartrate	60 gm
Copper sulphate	60 gm
Water	1 litre

Hot immersion   (Thirty minutes)

▲ Sodium hydroxide is a powerful caustic alkali. The solid or solutions can cause severe burns. It must be presented from coming into contact with the eyes or skin. When preparing the solution, small quantities of the sodium hydroxide should be added to large quantities of water and *not* vice-versa. The solution will become warm as the sodium hydroxide is dissolved. When it is completely dissolved, the other ingredients are added.

Immersion in the hot solution produced no effect on either standard silver, or silver plated surfaces, during a period of thirty minutes.

9   No effect on copper or copper alloys

Potassium nitrate	80 gm
Potassium aluminium sulphate	40 gm
Hydrochloric acid (10% solution)	80 cm³
Water	1 litre

Immersion   (Various)

▲ Potassium nitrate is a powerful oxidant which must be kept out of contact with all combustible materials. It should be completely dissolved in water before acid is added.

▲ Hydrochloric acid is corrosive and must be prevented from coming into contact with the eyes and skin.

Immersion in the solution at temperatures ranging from warm to boiling, for various lengths of time up to one hour, produced no effect. Varying the concentration of the ingredients produced no improvement. Additions of sodium chloride, and sodium chloride with ammonia, as suggested by some sources produced patchy non-adherent films only.

10   No effect on copper, copper alloys or silver

Ammonium sulphate	15 gm
Copper nitrate	15 gm
Ammonia (.880 solution)	to form a paste

Applied paste

As the consistency of the paste alters to a watery film, some separation of the crystalline ingredients occurs.

The solid ingredients were ground using a pestle and mortar, and the ammonia added by the drop to form a creamy paste. The paste was applied to the object using a soft brush. After a short time the consistency of the paste changed becoming watery, and producing no colouring effect.

11   No effect on copper, copper alloys or silver

Zinc chloride	50 gm
Olive oil	to form a paste

Applied paste   (Several days)

The test was repeated with a number of variations, in paste consistency and timing of application, but no adherent results could be obtained. No significant results were produced by removing the paste before it had completely dried, and increasing the frequency of application.

The ingredients were ground to a paste using a pestle and mortar. The paste was applied to the test surface with a soft brush, and allowed to dry thoroughly in air. The residue was brushed away with a bristle-brush. This procedure was repeated twice a day for several days. Although colours were produced in the pastel green and grey range, these were found to be superficial and totally non-adherent.

12   Slight tarnish on copper and copper alloys

Potassium hydrogen tartrate	6 gm
Stannic chloride (hydrated)	3 gm
Sodium metabisulphite	19 gm
Water	1 litre

Hot immersion   (Thirty minutes)

▲ Stannic chloride is corrosive and must be prevented from coming into contact with the eyes and skin.

Immersion in the hot solution (80°C) for thirty minutes produced no effect other than a very slight tarnish. (Additions of sodium tartrate and tartaric acid at a rate of 10 gm/litre were also ineffective. An addition of copper acetate at a rate of 10 gm/litre produced patchy brown finishes, but the technique cannot be recommended.)

13   Slight staining on copper and copper alloys

A Ammonia (10% solution)
B Acetic acid (glacial)
C Oleic acid (undiluted)

Cold immersion (A)
Application (B)
Application (C)

▲ The procedures should be carried out in a well ventilated area, as Ammonia vapour and acid vapours are liberated, which irritate the eyes and respiratory system.

▲ Glacial acetic acid is very corrosive and can cause severe burns. It must be prevented from coming into contact with the eyes and skin. Both the ammonia solution and the oleic acid should be prevented from coming into contact with the eyes and skin.

Initially, tests were carried out in which the metal was steeped in the ammonia solution for about five minutes, and the acetic acid brushed onto the surface, followed by the oleic acid which was also applied by brushing. No colouring effect could be obtained. Variations in the procedure, over a wide range, also produced no effect other than some very slight stains.

14   Slight bloom on copper and copper alloys

Copper sulphate	25 gm
Ferric oxide	32 gm
Sodium hydroxide	20 gm
Water	1 litre

Hot immersion (Thirty minutes)

▲ Sodium hydroxide is a powerful caustic alkali and must be prevented from coming into contact with the eyes and skin. It should be added to the solution in small quantities

No colouring action was produced in tests carried out with this solution. Immersion in the hot solution (80°C) produced a 'bright dip' effect, and slight traces of bloom. A similar result was produced by applying the solution to the heated metal.

15   Non-adherent golden films on copper and copper alloys

Chromium trioxide	100 gm
Water	1 litre

Immersion (To thirty minutes)

▲ Chromium trioxide is a very powerful oxidant, and must be prevented from coming into contact with combustible material. It is also highly corrosive, causing severe burns, and must be prevented from coming into contact with the eyes and skin. The dust must not be inhaled.

The cold solution produced no effect. Hot and boiling solutions gave rise to even golden films within ten minutes, but these proved to be totally non-adherent. Changes in concentration, and period of immersion produced the same result.

16   Non-adherent golden films on copper or copper alloys

Sodium dichromate	150 gm
Water	1 litre
Orthophosphoric acid (90% solution)	10 cm³
Ethanol	3 cm³

Immersion (To thirty minutes)

▲ Sodium dichromate is very corrosive and must be prevented from coming into contact with the eyes and skin. The dust irritates the eyes, and must not be inhaled. It must not be allowed to come into contact with combustible material.

▲ Phosphoric acid is very corrosive and must not be allowed to come into contact with the eyes or skin. The sodium dichromate should be completely dissolved in the water before the addition of the phosphoric acid.

Immersion in the cold solution produced golden films within ten minutes, but these were found to be totally non-adherent. The underlying metal surface appeared to have been chemically polished. Variations in timing, temperature and concentration produced similar results. The golden films tended not to form when the test samples were agitated while in the solution.

17   Non-adherent black layer on copper and copper alloys

A	Copper acetate	25 gm
	Ammonia (.880 solution)	20 cm³
	Water	500 cm³
B	Potassium sulphide	50 gm
	Ammonium chloride	75 gm
	Water	500 gm

Various

▲ Preparation and colouring should be carried out in a well ventilated area, as ammonia vapour and some hydrogen sulphide gas will be liberated.

▲ Potassium sulphide in the solid form, or as a concentrated solution, is corrosive and should be prevented from coming into contact with the eyes and skin.

▲ Concentrated ammonia solution should be prevented from coming into contact with the eyes and skin.

The two solutions are prepared separately and then mixed together. Immersion in the solution at temperatures from warm to boiling, produced a black surface colouring, which proved to have little resistance to washing when wet, and little adherence if allowed to dry without washing. Scratch-brushing with the solution produced little improvement.

Appendix 4: Glossary of chemical names

The older books listed in the bibliography often refer to chemicals by archaic names, and other books use common names for some chemicals. This glossary identifies the substances in question by systematic chemical names.

Acetic ether	Ethyl acetate
Acid of sugar	Oxalic acid
Ackey	Nitric acid
Ackey	Also used to refer to a mixture of sulphuric and nitric acids with free chlorine ions
Alum or Alum flower or Alum meal	Potassium aluminium sulphate
Alumino-ferric	A mixture of aluminium sulphate and sodium sulphate
Amalgam	An alloy of a metal with mercury
Ammonium sulfhydrate	Ammonium hydrosulphide
Antichlor	Sodium thiosulphate
Antimony black	Antimony trisulphide
Antimony bloom	Antimony trioxide
Antimony glance	Antimony trisulphide
Antimony red or Antimony vermillion	Antimonous oxysulphide
Antimony white	Antimous oxide
Antimony yellow	Basic lead antimonate
Aqua fortis	Nitric acid
Aqua regia	A mixture of nitric and hydrochloric acids
Argol	Crude form of potassium hydrogen tartrate
Aromatic spirits of ammonia	A 10% solution of ammonia in ethyl alcohol (ethanol)
Arsenic glass	Arsenic trioxide
Azurite	Mineral form of basic copper carbonate
Baking soda	Sodium bicarbonate
Barium sulphuret	Barium sulphide
Barium white	Barium sulphate
Baryta	Barium oxide
Barytes	Natural mineral form of barium sulphate
Bicarbonate of soda	Sodium hydrogen carbonate (sodium bicarbonate)
Bichrome	Potassium dichromate
Bitter salt	Magnesium sulphate
Black ash	Crude form of sodium carbonate
Black lead	Graphite
Blanc-fixe	Barium sulphate
Bleaching powder	Calcium hypochlorite
Blende	Natural mineral form of zinc sulphide
Blue copperas	Copper sulphate (crystals)
Blue dip	A solution of mercury salts
Blue lead	Lead sulphate
Blue salts	Nickel sulphate
Blue stone	Copper sulphate (crystals)
Blue verditer	Basic copper carbonate
Blue vitriol	Copper sulphate (crystals)
Bone ash	Crude calcium phosphate
Bone black	Crude animal charcoal
Boracic acid	Boric acid
Borax	Sodium tetraborate
Bremen blue	Basic copper carbonate
Brimstone	Sulphur
Brunswick green	Copper oxychloride
Burnt alum	Anhydrous potassium aluminium sulphate
Burnt lime	Calcium oxide
Burnt ochre or burnt ore	Ferric oxide
'Butter of' a metal	The chloride of the metal concerned eg. 'Butter of antimony' is antimony trichloride
Cadmium yellow	Cadmium sulphide
Caliche	Impure form of sodium nitrate
Calomel	Mercurous chloride
Carbolic acid	Phenol
Carbonic acid	A solution of carbon dioxide gas in water
Carbonic anhydride	Carbon dioxide
Carnallite	Mineral form of magnesium potassium chloride
Caustic	Usually used to refer to the hydroxide, e.g. caustic soda is sodium hydroxide. There are variants however, e.g. lunar caustic is silver nitrate
Caustic lime	Calcium hydroxide
Caustic potash	Potassium hydroxide
Caustic soda	Sodium hydroxide
Ceruse	Basic lead carbonate
Chalk	Calcium carbonate
Chevreul salt	Cupro-cupric sulphate
Chili nitre or chili saltpetre	Sodium nitrate
Chinese red	Basic lead chromate

Chinese white	Zinc oxide
Chloride of lime	Calcium hypochlorite
Chloride of soda	Sodium hypochlorite
Chlorinated lime	Calcium hypochlorite
Chrome alum	Potassium chromium sulphate
Chrome green	Chromium oxide
Chrome red	Basic lead chromate
Chrome yellow	Lead chromate
Chromic acid	Chromium trioxide
Cinnabar	Mercuric sulphide
Cobalt black	Cobalt oxide
Cobalt green	Cobalt zincate
Common salt	Sodium chloride
Copperas (blue)	Copper sulphate (crystals)
Copperas (green)	Ferrous sulphate
Copperas (white)	Zinc sulphate
Corrosive sublimate	Mercuric chloride
Corundum	Chiefly aluminium oxide
Cream of tartar	Potassium hydrogen tartrate
Crocus	Ferric oxide
Crystal ammonia	Ammonium carbonate
Crystal carbonate	Sodium carbonate
Derby red	Basic lead chromate
Derinatol	Basic bismuth gallate
Double nickel salts	Ammonium nickel sulphate
Eau de Javelle	Potassium hypochlorite (soln.)
Eau de Labarraque	Sodium hypochlorite (soln.)
Emerald green	Copper aceto-arsenite
Epsom salts	Magnesium sulphate
Everitts salt	Potassium ferrous ferrocyanide
Farina	Starch
Ferro-prussiate	Potassium ferrocyanide
Ferrum	Iron
Fixed white	Barium sulphate
Flowers of sulphur	Sulphur
'Flowers' of a metal	The oxide of the metal e.g. flowers of zinc is zinc oxide
Freezing salt	Crude form of sodium chloride
French chalk	Hydrated magnesium silicate
French verdigris	Basic copper acetate
Fustic	A yellow wood used as a dye-stuff
Galena	Natural form of lead sulphide
Glauber's salt	Sodium sulphate
Green verditer	Basic copper carbonate
Green vitriol	Ferrous sulphate (crystals)
Gypsum	Calcium sulphate
Hartshorn salt	Ammonium carbonate carbamate
Heavy spar	Barium sulphate
Hepar calcis	Calcium sulphide
Hepar sulphuris	Potassium sulphide
Horn silver	Silver chloride
Hypo	Sodium thiosulphate
Indian red	Ferric oxide
Iron black	Precipitated antimony

Iron mordant	Ferric sulphate
Jewellers' rouge	Ferric oxide
Isinglass	Agar-agar, gelatine
Kainit	Double salt of potassium magnesium sulphate and magnesium chloride
Kieserite	Mineral magnesium sulphate
Kings green	Copper aceto-arsenite
Kings yellow	Arsenious sulphide
Lampblack	Crude form of carbon
Lead peroxide	Lead dioxide
Lead protoxide	Lead oxide
Lemon chrome	Barium chromate
Lime (slaked)	Calcium hydroxide
Lime	Calcium oxide
Liquor ammonia	Ammonium hydroxide (soln.)
Litharge	Lead monoxide
Liver of sulphur	Potassium sulphide or potassium polysulphide
Lunar caustic	Silver nitrate
Lye (soda lye)	Sodium hydroxide
Lysol	Cresol soap solution
Magnesia	Magnesium oxide
Magnesite	Mineral magnesium carbonate
Malachite	Mineral form of basic copper carbonate
Manganese black	Manganese dioxide
Marble	Calcium carbonate
Massicot	Lead monoxide
Milk of barium	Barium hydroxide
Milk of lime	Calcium hydroxide
Milk of magnesium	Magnesium hydroxide
Milk of sulphur	Precipitated sulphur
Minium	Lead tetroxide
Mohrs salt	Ferrous ammonium sulphate
'Muriate' of a metal	The metal chloride
Muriatic acid	Hydrochloric acid
Natron	Sodium carbonate
Nitre	Potassium nitrate
Nordhausen acid	Fuming sulphuric acid
Oil of vitriol	Conc. sulphuric acid
Oleic acid	Octadec-9-enoic acid
Oleum	Fuming sulphuric acid
Orpiment	Arsenic trisulphide
Paris blue	Ferric ferrocyanide
Paris green	Copper aceto-arsenite
Pearl ash	Anhydrous potassium carbonate
Permanent white	Barium sulphate
Plaster of Paris	Calcium sulphate
Plumbago	Graphite
Precipitated chalk	Calcium carbonate
Prussian blue	Ferric ferrocyanide
Prussic acid	Hydrocyanic acid
Pyrites	Natural mineral form of ferrous di-sulphide

Pyroligneous acid	Crude form of acetic acid
Quicklime	Calcium oxide
Quicksilver	Mercury
Realgar	Arsenic disulphide
Red antimony	Antimony oxysulphide
Red lead	Lead tetroxide
Red liquor	Aluminium acetate (soln.)
Red oil	Oleic acid (see above)
Red precipitate	Oxide of mercury
Red prussiate of potash	Potassium ferricyanide
Red prussiate of soda	Sodium ferricyanide
Rochelle salt	Potassium sodium tartrate
Rock salt	Sodium chloride
Rouge	Ferric oxide
Sal acetosella	Potassium binoxalate
Sal ammoniac	Ammonium chloride
Salt	Sodium chloride
Salt cake	Crude sodium sulphate
Salt of lemon or salt of sorrel	Potassium hydrogen oxalate
Salt of tartar or salt of wormwood	Potassium carbonate
Saltpetre	Potassium nitrate
Satin white	Calcium sulphate
Scheele's green	Copper hydrogen arsenite
Schlippes salt	Sodium thioantimonate
Single nickel salts	Nickel sulphate
Slaked lime	Calcium hydroxide
Soda or washing soda (crystals)	Sodium carbonate (crystals)
Soda ash	Anhydrous sodium carbonate
Soda lime	Mixture of calcium oxide and sodium hydroxide
Soda monohydrate	Sodium carbonate (crystals)
Soda nitre	Sodium nitrate
Sodium hyposulphite	Sodium thiosulphate
Soluable tartar	Potassium tartrate
Sour water	Dilute sulphuric acid
Spirit of hartshorn	Ammonia solution
Spirit of salt	Hydrochloric acid
Spirit of wine	Ethanol (ethyl alcohol)
Stibnite	Natural form of antimony sulphide
Sugar of lead	Lead acetate
Sulphuret of ammonia	Ammonium sulphide
Talc	Hydrated magnesium silicate
Tartar	Crude form of potassium hydrogen tartrate (potassium bi-tartrate)
Tartaric acid	Racemic acid
Tartar emetic	Antimony potassium tartrate
Tin crystals	Stannous chloride
Tin white	Stannic hydroxide
Tri or trike	Trichloroethylene
Trona	Natural form of sodium carbonate
Turnbull's blue	Ferrous ferricyanide
Ultramarine yellow	Barium chromate
Unslaked lime	Calcium oxide
Venetian red	Ferric oxide
Verdigris	Basic copper acetate
Vermillion	Red mercuric sulphide
Vienna lime	Calcium carbonate
Vinegar	Crude form of acetic acid
Vitriol	Sulphuric acid
'Vitriolate of' x	Sulphate of x
Washing soda	Sodium carbonate (crystals)
Waterglass	Sodium silicate (soln.)
White acid	A mixture of hydrofluoric acid and ammonium fluoride
White arsenic	Arsenic trioxide
White caustic	Sodium hydroxide
White lead	Basic lead carbonate
White vitriol	Zinc sulphate (crystal)
Whiting	Calcium carbonate
Yellow prussiate of potash	Potassium ferrocyanide
Yellow prussiate of soda	Sodium ferrocyanide
Zinc vitriol	Zinc sulphate
Zinc white	Zinc oxide

Appendix 5: Conversion tables

Solid measure

Metric weight

Kilogram		Gram
1Kg	=	1000
		1gm

Avoirdupois (Imperial) weight

Pound		Ounces	Drachms
1lb	=	16	256
		1oz =	16
			1dr.av

Apothecaries' weight

Pound		Ounces	Drachms	Scruples	Grains
1lb	=	12	96	288	5760
		1ℨ =	8	24	480
			1ℨ =	3	60
				1℈ =	20
					1Gr

Troy weight

Pound		Ounces		Pennyweights	Grains
1lb	=	12		240	5760
		1oz	=	20	480
			1dwt	=	24
					1Gr

Metric	**Avoirdupois**		**Apothecary**		**Troy**
1Kg (Kilogram) =	2·205lb	=	2·68lb	=	2·68lb
1gm (Gram)	0·0353oz		0·0321oz*		0·0321oz*
			*(15·43Gr)		*(15·43Gr)

Avoirdupois	**Metric**		**Apothecary**		**Troy**
1lb =	0·45Kg*	=	1·215lb	=	1·215lb
1oz	28·35gms		0·911oz		0·911oz
	*(453·6gms)				

Apothecary		**Troy**		**Metric**		**Avoirdupois**
1lb (Pound)	=	1lb	=	373gms	=	0·823lb (13·17oz)
1ℨ (Ounce)		1oz		31·1gms		1·097oz
1ℨ (Drachm)				3·88gms		0·137oz
1℈ (Scruple)				1·296gms		0·046oz
		1dwt		1·555gms		0·055oz
1Gr (Grain)		1Gr		0·065gms		0·0023oz

Liquid measure

Metric

Litre		Cubic Centimetres (Millilitres)
1l	=	1000
		1cm³ (ml)

Imperial

Gallon		Quarts		Pints		Gills		Fluid Ounces		Fluid Drachms
1gal	=	4		8		32		160		1280
		1qt	=	2		8		40		320
				1pt	=	4		20		160
						1gill	=	5		40
								1fl.oz.	=	8
										1fl.dr.

U.S. measure

Gallon		Quarts		Pints		Gills		Fluid Ounces		Fluid Drachms
1gal	=	4		8		32		128		1024
		1qt	=	2		8		32		256
				1pt	=	4		16		128
						1gill	=	4		32
								1fl.oz	=	8
										1fl.dr.

Metric		**Imperial**		**U.S. measure**
1litre	=	1·76pts	=	2·11pts

Imperial		**Metric**		**U.S. measure**
1gal	=	4·546 litres	=	1·201gals
1pt		0·568 litres (568·25cm³)		1·201pts

U.S. measure		**Imperial**		**Metric**
1gal	=	0·833gals	=	3·785 litres
1pt		0·833pts		0·473 litres (473·17cm³)

To convert	**into**	**multiply by**
fl.oz/Imperial gal	cm³/litre	6·25
fl.oz/U.S. gal	cm³/litre	7·5
cm³/litre	fl.oz/Imperial gal	0·16
cm³/litre	fl.oz/U.S. gal	0·13
oz/Imperial gal	gms/litre	6·24
oz/U.S. gal	gms/litre	7·5
gms/litre	oz/Imperial gal	0·16
gms/litre	oz/U.S. gal	0·134
dwt/Imperial gal.	gms/litre	0·34
dwt/U.S. gal	gms/litre	0·41
gms/litre	dwt/Imperial gal	2·92
gms/litre	dwt/U.S. gal	2·44

Temperature

Centigrade and Fahrenheit

To convert degrees Centigrade to Fahrenheit, multiply by nine, divide the result by five, and add thirty-two.

To convert degrees Fahrenheit to Centigrade, subtract thirty-two, multiply the result by five and divide by nine.

Degrees Centigrade °C	Degrees Fahrenheit °F	Degrees Centigrade °C	Degrees Fahrenheit °F
0	32		
5	41	55	131
10	50	60	140
15	59	65	149
20	68	70	158
25	77	75	167
30	86	80	176
35	95	85	185
40	104	90	194
45	113	95	203
50	122	100	212

BIBLIOGRAPHY

1   Abbey, S.   The Goldsmiths and Silversmiths Manual.
                                        The Technical Press Ltd. 1952.

2   Achard, K.F.   Recherches sur les Proprietes des Alliages Metalliques.
                                        Berlin. 1788.

3   Addicks, L. (Ed.)   Silver in Industry.
                                        Reinhold Publishing Corporation. USA. 1940.

4   Ainsworth, J.   Mechanical Colouring of White Brass.
                                        Metal Finishing. 75(6). 1977.

5   Allied Chemical   Brightening and Passivating of Copper.
     Corporation. USA.   Belgian Patent. 657,099. 1965.

6   American Brass Company.   Corrosion Ratings of Copper and Copper Alloys.
                                        Materials in Design Engineering. 52(12). 1960. pp 129–131

7   Anderson, L.J.   Japanese Armour.
                                        Arms and Armour Press. London. 1968.

8   Anon.   The Application of Coloured Patina Films on Metals.
                                        Trattimenti e Finiture. 6(2). 1966. pp 43–46. (Italian)

9   —   Application of Colour Microphotography to Microscopic Investigations of Oxidised Copper.
                                        Rudy Metale Niezelazne. 13(6). 1968. pp 274–276. (Polish)

10   —   Arsenical Colouring Baths for Copper.
                                        Electroplating and Metal Finishing. 11(12). 1958. p 451.

11   —   Artificial Patination.
                                        Copper. 27. 1966. pp 3–7.

12   —   Barbedienne Finish.
                                        Electroplating and Metal Finishing. 13(12). 1960. p 467.

13   —   Blackening Copper: Ebonol C.
                                        Chemical and Metallurgical Engineering. 49(6). 1942. p 142.

14   —   Black Finishes for Ferrous and Non-Ferrous Parts: Oxidine Finishes and the Jetal Process.
                                        Iron Age. 151(24/6). 1943. p 58.

15   —   Blue Gold.
                                        Metal Finishing. 62(12). 1964. p 79.

16   —   Chemical Colour Finishes for Metals.
                                        Chemical Age. 4/12/1937. supplement. pp 34–35.

17   —   Chemical Colouring of Copper and its Alloys.
                                        Galvano Tecnica. 27(6). 1976. pp 103–111. (Italian)

18   —   Chemical and Electrochemical Colouring of Copper Articles.
                                        Galvano. 39(398). 1970. pp. 207–209. (French)

19   —   Chemical Etchants for Dislocations in Alpha Brass.
                                        AIME.Met.Soc. Transactions. 239(5). 1967. pp 762–763.

20   —   Cleaning and Finishing of Copper and Copper Alloys.
                                        Metals Handbook. Vol. 2. American Society of Metals. 1964.

21   —   Colour Controlled Copper Alloys for Decorative Use.
                                        Metal Progress. 88(5). 1965. pp 54–58.

22   —   Coloured Finishes for Copper and Copper Alloys.
                                        Plating. 55(12). 1968. pp 1255, 1257–1260.

23   —   Colouring of Metals. Part 1.
                                        Product Finishing. 13(5). 1960. pp 71–73.
                                        (also, Galvano. 29. 1960. (French))

24   —   Colouring of Metals. Part 2.
                                        Product Finishing. 13(6). 1960. pp 68–71.
                                        (also, Galvano. 29. 1960. (French))

25   —   Colouring of Metals. Part 3.
                                        Product Finishing. 13(8). 1960. pp 88–90.
                                        (also, Galvano. 29. 1960. (French))

26   —   Colouring of Metals. Part 4.
                                        Product Finishing. 13(12). 1960. pp 76–78.
                                        (also, Galvano. 29. 1960. (French))

27   —   Colour to Choice.
                                        Copper. 16. 1962. pp 1–5.

28 —   Contribution to the Phenomenology of Corrosion and Patina Formation of Antique Copper
                                          Alloys.
                                          Praktische Metallographie. 4(1). 1967. pp 3–15. (German-English)

29 —   How to Colour Copper and its Alloys Part 1.
                                          Cuivres Laitons Alliages. (110). 1969. pp 3–7. (French)

30 —   How to Colour Copper and its Alloys. Part 2.
                                          Cuivres Laitons Alliages. (111). 1969. pp 3–9 (French)

31 —   Incralac: New Laquer for Outdoor Use.
                                          Copper. 20(10). 1964.
                                          (also, Metal Industry. 104(21). 1964. p 698.)
                                          (also, Tyd. Oppernlaktechnica. 7(11). 1964. pp 331–332.)

32 —   Incralac Progress.
                                          Copper. 28. 1966. pp 19–21.

33 —   Metal Colouring Processes for Copper.
                                          Metal Finishing. 54(4). 1956. p 74.

34 —   Metallographic Reagents based on Sulphide Films.
                                          Praktische Metallographie. 7(5). 1970. pp 242–248. (German-English)

35 —   Most Non-Ferrous Metals and Alloys Weather Well.
                                          Materials Engineering. 79(4). 1974. pp 64–66.

36 —   New Surface Protective Coatings for Copper Alloys.
                                          Precious Metal Moulding. 23(10). 1965. pp 57, 60, 62.

37 —   Observations on the Natural Patination of Copper.
                                          Journal of the Institute of Metals. August 1970. pp 98, 238.

38 —   Obtaining Oxidised Finish.
                                          Foundry. 87(11). 1959. p 102.

39 —   Occurrence of Zinc and Brass in Antiquity and the Middle Ages.
                                          Erzmetall. 23(6). 1970. pp 259–269. (German)

40 —   Patinas and Metal Colouring.
                                          Galvano. 34(346). 1965. pp 781–784. (French)

41 —   Pre-Patination of Unique Copper Tube Sculpture.
                                          Copper. 3(2). March 1969. pp 2–4.

42 —   Problems in the Preservation of Corroded Metal Antiquities.
                                          Schweiz.Arch. 37(5). May 1971. pp 160–167. (German)

43 —   Some Patinas and Metal Colourations.
                                          Galvano Organo. 43(445). June 1974. pp 559–560. (French)

44 —   Spectrochemical Analysis of Antique Copper Alloys.
                                          Z.Analyt.Chem. 250(1). 1970. pp 17–23 (German)

45 —   Tiffany Green. Colouring of Brass.
                                          Electroplating and Metal Finishing. 13(10). 1960. p 387.

46 —   Ultrasonic Degreasing.
                                          Galvano Tecnica. 9(11). 1958. pp 273–276. (Italian)

47 —   Weather Resistance of Copper and Copper Alloys.
                                          Metall. 25(11). 1972. pp 1299–1303. (German)

48 Aoyama, Y.   Copper Corrosion Products on Electrical Traction Wires.
                                          Werkstoffe und Korrosion. 12(3). 1961. pp 148–150. (German)

49 Aoyama, Y.   Green Corrosion Product on Copper.
                                          Naturwissenschaften. 47. 1960. p 202. (German)

50 Arrivaut, G.   On the Formation of a Violet Alloy of Copper, Cu_2Sb.
                                          Journal of the Institute of Metals. 44. 1930. p 493.

51 Artioli, O. and Pezza, N.   Incralac for Architectural Applications.
                                          Il Rame. 11(8). 1964. pp 29–35. (Italian)

52 Asaro, R.J.   Tarnish Films and Stress Corrosion Cracking of Alpha Brass.
                                          Phil.Mag. 26(2). August 1972. pp 425–442.

53 Bacquais, G.   A New Step in the Corrosion Protection of Metals and Common Alloys.
                                          Galvano Organo. (489) October 1978. pp 655–657. (French)

54 Bailey, K.C.   The Elder Pliny's Chapters on Chemical Subjects.
                                          London. 1929.

55 Ballczo, H. and   Non-Destructive Ultramicroanalysis of Archaeological Objects.
    Mauterer, R.   Part 1. Complete Analysis of Streak Samples of Antique Metal Artefacts.
                                          Fresenius Z. Anal. Chem. 295(1). 1979. pp 36–44. (German)

56 Barnard, N.   Bronze Casting and Bronze Alloys in Ancient China.
                                          Monumenta Serica Monograph. XIV. Canberra and Nagoya. 1961.

57 Bearzi, B.   Technical Considerations regarding the S. Ludovico and the Guiditta of Donatello.
                                          Bollettino d'Arte. Vol. XXXVI. Serie IV. 1951. pp 119–123. (Italian)

58 Bearzi, B.   Considerations regarding the Formation of Patinas and Corrosion on Ancient Bronzes.
                                          Studi Etrusci. Vol. XXI. Serie II. 1950–1951. pp 261–266. (Italian)

59 Bearzi, B.   Technological and Metallurgical Examinations of the Statue of San Pietro.
                                          Commentari. Vol. XI(1). 1960. pp 30–32. (Italian)

60 Bearzi, B.   Technical Notes on the Restoration of Bronzes.
                                          Bollettino d'Arte. Vol. XLV. Serie IV. (I–II). 1960. pp. 42–44. (Italian)

61 Benckiser, J.A. (GmbH)   Chemical Polishing of Copper.
                                          French Patent. 1,329,250. 18/7/1962.
                                          German Patent. 1,175,523. 6/8/1964.

62 Bennett, H.   The Chemical Formulary.
                                          New York. 1935.

63	Bennett, H.E., Peck, R.L., Burge, D.K. and Bennett, J.M.	Formation and Growth of Tarnish on Evaporated Silver Films. Journal of Applied Physics. 40(8). 1969.
64	Berthelot, M.	The Colouring of Metals. In-Les Origines de l'Alchimie. Paris. 1885. (French)
65	Beutel, H.	Bewahrte Arbeitsweisen der Metallfärbung. 2nd. Edition. Vienna. 1925. (1st. Edition, Leipzig. 1913.) (German)
66	Bhanot, D. and Chawla, S.L.	Corrosion Resistance of Pure Copper, its Alloys and Stainless Steel in Chlorinated Lime Slurry and Calcium Hypochlorite Solution. Trans. Soc. Adv. Electrochem. Sci. Technol. 13(3). 1978. pp 227–237.
67	Birley, S.S.	Mechanism of Cu_2O Tarnish Formation on Cu/30Zn. J. Electrochem.Soc. 118(4). 1971. pp 636–637.
68	Birley, S.S. and Tromans, D.	Corrosion Fatigue of Copper and Alpha Brass. J. Electrochem. Soc. 119(10). 1972. pp 1278–1285.
69	Boizard, J.	Traité de Monoyes. Paris. 1696. (French)
70	Booker, C.J.L. and Salim, M.	Electrometric Characterisation of Tarnish Films on Cu/Zn Alloys. Nature Phys. Sci. 239(91), September 1972. pp 62–63.
71	Braunt, D.	The Colouring of Alloys. Appendix to—The Metallic Alloys. Philadelphia. 1922.
72	Bretherick, L.	Handbook of Reactive Chemical Hazards. 2nd. Edition. Butterworth. London and Boston, Mass. 1979.
73	Brinkley, F.	Japan, Its History, Art and Literature. (Eight volumes.) Volume 3. Metallurgy and Metalworking. London. 1903–1904.
74	Brown, B.F. et al (Eds.)	Corrosion and Metal Artefacts. A Dialogue between Conservators and Archaeologists and Corrosion Scientists. U.S. Department of Commerce. N.B.S. Special Publication 479. July 1977.
75	Brown, W.N.	Japanning and Enamelling. Scott Greenwood and Co. London. 1901.
76	Buchner, G.	Die Metallfärbung. Berlin. 1907 and 1922.
77	Burkart, W., Silman, H. and Draper, C.R.	Mechanical Polishing. Robert Draper Ltd. Teddington. 1960.
78	Burns, R.M. and Bradley, W.W.	Protective Coatings for Metals. 3rd Edition. Reinhold Publishing Corporation. USA.
79	Butts, A. (Ed.)	Silver; Economics, Metallurgy and Use. Van Nostrand Co. Inc. Princeton, New Jersey. 1967.
80	Caley, E.R.	The Leyden Papyrus X: An English Translation with Brief Notes. Journal of Chemical Education. 3. 1926. pp 1149–1166.
81	Canning, W. and Co.	The Canning Practical Handbook on Electroplating, Polishing, Bronzing, Lacquering and Enamelling. 11th Edition. Birmingham and London. 1932. (21st Edition. Birmingham and London. 1970.) (22nd Edition. Birmingham and London. 1978.)
82	Capp, R.S.	Finishing Handbook and Directory. Product Finishing. 30th issue. Sawell publications. London. 1980.
83	Carpenter, A.M.	Techniques for the Colouring of Metals. Worshipful Company of Goldsmiths. Technical Advisory Committee Special Report No. 36. London. 1978.
84	Cartwright, P.A.	Metal Finishing Handbook. 1st Edition. Blackie and Son Ltd. London. 1950.
85	Cellini, B.	The Life of Benvenuto Cellini Written by Himself. (Translation by A. McDonnell.) J.M. Dent. London. 1907.
86	Cellini, B.	The Treatises of Benvenuto Cellini on Goldsmithing and Sculpture. (Translation by C.R. Ashbee.) Edward Arnold. London. 1888. Re-published—Dover. New York. 1967.
87	Chaplet, A.	Les Patines Multicolorés des Métaux. Le Chimiste. Paris. 1911. (Journal Monograph)
88	Chase, W.T.	What is the Smooth Lustrous Black Surface on Ancient Bronze Mirrors? In—B.F. Brown et al. (Eds.) Corrosion and Metal Artefacts. NBS Special Publication 479. U.S. Department of Commerce. Washington. 1977.
89	Chase, W.T. and Franklin, U.M.	Early Chinese Black Mirrors and Pattern Etched Weapons. Ars Orientalis. 11. 1979. pp 216–258.
90	Cialdea, U.	Restoration of Antique Bronzes. Mouseion. XVI. 1931.
91	Clarke, S.G and Andrew, J.F.	Surface Etching Pre-Treatment. Transactions of the Institute of Metal Finishing. 32(262). 1955.
92	Clarke, S.G. and Andrew, J.F.	Passivation of Copper. British Patent 974,800. 11/11/1964.
93	Collins, W.F.	The Corrosion of Early Chinese Bronzes. Journal of the Institute of Metals. 45. 1931. pp 23–55.
94	Collins. W.F.	The Mirror-Black and Quicksilver Patinas of certain Chinese Bronzes. Journal of the Royal Anthropological Society of Great Britain and Ireland. 64. 1934. pp 69–79.
95	Compton, W.A. et al. (Eds)	Nippon-to: The Art Sword of Japan. New York. 1976.

96	Cope, L.H.	Silver Surfaced Ancient Coins. In—E.T. Hall and D.M. Metcalf, Methods of Chemical and Metallurgical Investigation of Ancient Coinage. London. 1972.
97	Copper and Brass Research Association.	Colouring Copper and Brass. C.A.B.R.A., New York. (nd)
98	Copper and Brass Research Association.	Copper Handicraft. C.A.B.R.A., New York. 1958.
99	Copper and Brass Research Association.	The Maintenance, Cleaning, Finishing and Colouring of Copper, Brass and Bronze. C.A.B.R.A., Publication No. 4. New York. 1953.
100	Copper Development Association.	Copper and Brass Sheet Metalwork. C.D.A., Publication No. 66. 1965.
101	Copper Development Association.	Dinanderie. Copper. Summer 1959. pp 23–25.
102	Copper Development Association.	Surface Treatment for Copper and Copper Alloys. C.D.A. Publication No. 67. 1967.
103	Copper Development Association. (New York)	Coloured Finishes for Copper and Copper Alloys. Technical Report 121/8. C.D.A. (New York). 1968.
104	Copper Development Association. (New York)	Copper, Brass and Bronze in Architecture. C.D.A. (New York). 1965.
105	Cotton, J.B.	The Nature of Colours in Metals. The Metallurgist and Materials Technologist. 7(7). 1975. pp 350–352, 354.
106	Curti-Jung, T.	How is Brass Coloured? Metall-Rein.U.Vorbehandlung. 9(11). 1960. pp 191–192. (German)
107	Daney, L.J,	The Role of Oxide and Replacement Coatings in Metal Finishing. Part 2. Non-Ferrous Alloys. Metal Progress. 82(1). 1962. pp 73–77.
108	Debonliez, G. and Malpeyre, F.	Bronzage des Métaux et du Plâtre. Roret. Paris. 1890.
109	De Hart, H.G.	Pre-Treatment for Copper. (International Protected Metals Inc. NJ/USA.) U.S. Patent 3,198,672. 3/8/1965.
110	Desai, M.N. and Shah, Y.C.	The Effect of some Colloids on the Corrosion of Brass in Acetic Acid. Werkstoffe und Korrosion. 14(9). 1963. (German)
111	Desai, M.N. and Trivedi, M.	Influence of Hydrogen Peroxide on the Corrosion of Brass by Acids. Indian Journal of Applied Chemistry. 21(4). 1958. pp 137–141.
112	Deutsches Kupfer Institut.	Chemische Farbungen Von Kupfer Und Kupferlegierungen. Deutsches Kupfer Institut. Berlin. 1936.
113	Diana, S., Fiorentino, P. Marabelli, M. and Santini, M.	Techniques to protect the Horses of San Marco. Cleaning Tests. In—The Royal Academy (London). The Horses of San Marco. Venice. 1979. (Trans. by John and Valerie Wilton-Ely from the Italian edition. Procuratoria di San Marco/Olivetti. 1977.)
114	Dobrosavljevic, J.S. and Antonijevic, V.G.	Spectrochemical Analysis of Ancient Bronze. Doc. Chem. Yugosl. 43(9). 1978. pp 613–619. (Serbo-Croatian)
115	Dutta, K.P., Roy, S.K. and Bose, S.K.	Kinetics of Copper Tarnishing in Iodine Atmosphere at Low Temperatures. Indian J. Technol. 11(11). 1973. pp 609–612.
116	Dutta, K.P., Roy, S.K., Bose, S.K. and Sircar, S.C.	Kinetics of Copper Tarnishing in Iodine Atmosphere at Low Temperatures. International Symposium on Defect Interactions in Solids. Indian Institute of Science. Bangalore. 1974. pp 362–365.
117	Eberhardt, R.D., Arneson, E.W. and Imhoff, D.	Composition of Corrosion Products Formed on Metals in Steam Condensate. Corrosion. 17(4). 1961. pp 171–172.
118	Ebermeyer, T.	Metallfaerbung. Zeitschrift Fur Die Chemische Industrie. Berlin. 1887.
119	Engel, E.	Language and Application of Colours to Metals. Metal Finishing. 41(8). 1943. pp 488–490.
120	Eppensteiner, F.W. and Jenkins, M.R.	Chromate Conversion Coatings. Metal Finishing Guidebook Directory. 41st Annual Edition. (Ed.) N. Hall. Metals and Plastics Publications. 1973.
121	Evans, U.R.	Corrosion of Metals. 2nd Edition. London. 1926.
122	Farnsworth, M.	The Use of Sodium Metaphosphate in Cleaning Bronzes. Technical Studies in the Field of the Fine Arts. Vol. IX. Fogg Art Museum. 1940.
123	Ferguson, J.C.	Examination of Chinese Bronzes. In—Smithsonian Institution Report. Washington. 1915.
124	Field, S. and Bonney, S.R.	The Chemical Colouring of Metals and Allied Processes. Chapman and Hall. London. 1925.
125	Fink, C.G. et al.	Restoration of Ancient Bronzes and Other Alloys. Metropolitan Museum of Art. New York. 1925.
126	Fink, C.G. and Polushkin, E.P.	Microscopic Study of Ancient Bronze. Transactions of the American Institute of Mining Engineers. CXXII. 1936.
127	Fink, C.G. and Polushkin, E.P.	Microscopic Study of Ancient Copper and Bronze. Metals Technology. III. 1936.
128	Fiorentino, P. and Marabelli, M.	Techniques to protect the Horses of San Marco. Evaluation of the Examinations and Tests. In—The Royal Academy (London) The Horses of San Marco. Venice. 1979. (Trans. John and Valerie Wilton-Ely from the Italian edition. Procuratoria di San Marco/Olivetti. 1977.)
129	Fishlock, D.	The Art and Science of Colouring Metals. Pincturas Y Acabados. 7(38). 1965. pp 39–42. (Spanish)

130 Fishlock, D.   The Art and Science of Colouring Metals.
Trattimenti e Finiture. 4(6). 1964. pp 237–239. (Italian)

131 Fishlock, D.   The Art and Science of Colouring Metals.
Galvano. 32(320). 1963. (French)

132 Fishlock, D.   Black Finishes for Metals.
Product Finishing. 17(3). 1964. pp 56–63.

133 Fishlock, D.   Chromate Conversion Treatments.
Product Finishing. 12(12). 1959. pp 87–93.

134 Fishlock, D.   Metal Colouring.
Robert Draper Ltd. Teddington. 1962.

135 Fitzgerald, L.D.   Protection of Cuprous Metals from Atmospheric Corrosion.
In—Atmospheric Factors Affecting the Corrosion of Engineering Metals.
ASTM. STP. No. 46. Philadelphia. 1978. pp 152–159.

136 Forbes, R.J.   Studies in Ancient Technology.
E.J. Brill. Leiden, Netherlands. 1964.

137 Forbes, R.J.   Metallurgy in Antiquity.
E.J. Brill. Leiden, Netherlands. 1950.

138 Fortnum, C. Drury   Bronzes.
London. 1877.

139 Freeman, J.R. and
Kirby, P.   Green Patina on Copper with Ammonim Sulphate.
Metals and Alloys. 190(9). 1932.

140 Freeman, J.R. and
Kirby, P.   Green Patina on Copper with Ammonium Sulphate: Spray Process.
Metals and Alloys. 67(4). 1934.

141 Fujimori Kogyo Co. Ltd.
(Japan)   Cracked Oxide Coating.
Japanese Patent 4528. 11/3/1965.

142 Furukawa Electric Co. Ltd.
(Japan)   Chemical Brightening of Copper.
Japanese Patents 6053; 18301; 18302. 1964.

143 Gabel, H., Beavers, J.A.,
Woodhouse, J. and Pugh, E.   Structure and Composition of Thick Tarnish Films on Alpha Phase Copper Alloys.
Corrosion. 32(6). 1976. pp 253–257.

144 Gaillard, L.   Patinage des Bijoux.
Revue de la Bijouterie. Paris. 1900.

145 Gainsbury, P.E.   Project No. 6. The Chemical Colouring of Metals.
In—Technical Work at the 1971 Goldsmiths Company Summer Seminar.
Worshipful Company of Goldsmiths. Technical Advisory Committee, Special Report 15.

146 Gardam, G.E.   The Colours of Some Metals and Alloys.
Transactions of the Institute of Metal Finishing. 44(5). 1966. pp 186–188.

147 Gee, G.   The Goldsmiths Handbook.
Crosby, Lockwood and Son. London. 1903.

148 Geerlings, G.K.   Metal Crafts in Architecture.
New York. 1957.

149 Gettens, R.J.   Mineral Alteration Products on Ancient Metal Objects.
In—G. Thomson (Ed.), Recent Advances in Conservation. Butterworths. 1963.

150 Gettens, R.J.   Some Observations Concerning the Lustrous Surface of Ancient Eastern Bronze Mirrors.
Technical Studies in the Field of the Fine Arts. 3. 1934. pp 29–37.

151 Gettens, R.J.   The Corrosion Products of Metal Antiquities.
Smithsonian Institution. Washington. 1964.

152 Gettens, R.J.   Tin Oxide Patina of Ancient High-Tin Bronze.
Bulletin of the Fogg Museum of Fine Art. 11(1). 1949. pp 16–26.

153 Ghersi, M.   Dorure, Argenture, Colouration des Métaux.
Paris. 1910. .

154 Girard, J.   La Coloration des Métaux.
Paris. 1898.

155 Girard, J.   Recherches sur la Coloration des Métaux.
Comptes Rendus du Congrès de la Société des Gens de Science. Paris. 1900.

156 Gowland, W.   Metals and Metalworking in Old Japan.
Transactions of the Japan Society. 30. 1914–1915. pp 20–100.

157 Gowland, W.   The Art of Casting Bronze in Japan.
Journal of The Society of Arts. 43. 1895. p 522.

158 Graham, A.K. and
Pinkerton, H.   Electroplating Engineering Handbook.
2nd. Edition. Reinhold Publishing Corporation. New York. 1962.

159 Groebler, H.   Metal Colouring 150 Years Ago.
Metalloberflache. 18(11). 1964. pp 337–340. (German)

160 Gross, G.   New Knowledge in the Lustrous Hot Colouring of Metals.
Z.Metallkunde. 27. 1935. pp 230–241. (German)

161 Gross, G.   Researches into the Grey Colouring of Copper Alloys using Arsenic Mordants.
Oberflachentechnik (Coburg). 23. 1933. (German)

162 Gross, G.   Persulphate Coatings on Copper.
Chemiker Zeitung. 19. March, 1935. (German)

163 Hacker, W.   Metal Colouring.
Metalloberflache. 13(5). 1959. pp 136–138. (German)

164 Hacker, W.   Patina on Copper.
Galvano Tecnica. 29(284). 1960. pp 513–516. (Italian)

165 Hall, E.T. and
Metcalf, M. (Eds.)   Methods of Chemical and Metallurgical Analysis of Ancient Coinage.
Royal Numismatic Society, Special Publication No. 8. London. 1972.

166   Hall, N.
Colouring of Metals.
In—N. Hall (Ed.), Metal Finishing Guidebook Directory. 41st Annual Edition.
Metals and Plastics Publications Inc. 1973.

167   Hall, N.
Immersion Plating.
In—N. Hall (Ed.), Metal Finishing Guidebook Directory. 41st Annual Edition.
Metals and Plastics Publications Inc. 1973.

168   Hall, N. (Ed.)
Directory of Metal Finishing Chemicals.
Metals and Plastics Publications Inc. Westwood. New Jersey. 1963.

169   Hampel, K.R.
Finishing Pointers. A Chromic Acid Bright Dip for Copper and its Alloys.
Metal Finishing. 63(2). 1965. p 68.

170   Hanamura, N.
Colouring Copper. Japanese Patent 74,027,730. 10/6/1970.

171   Hanson, M.
Constitution of Binary Alloys.
McGraw-Hill. New York. 1958.

172   Harris, C.
Colouring Miscellaneous Metals: Silver, Tin, Gold, Zinc, Cadmium, Nickel.
Metal Industry. 78. 1951. pp 71–73.

173   Harris, C.
Modern Metal Colouring: Recent Developments in the Science.
Metal Industry. 54. 1939. pp 613–616.

174   Harris, C.
Modern Metal Colouring.
Journal of the Electrodepositors Technical Society. 15. 1938–1939. pp 97–108.

175   Harry, A.J.
The Tarnishing of Silver.
The Worshipful Company of Goldsmiths. Technical Advisory Committee Project Report
No.5a/4. London. 1976.

176   Hawley, W.M.
Japanese Swordsmiths.
Hollywood, California. 1967.

177   Healey, J.F.
Mining and Metallurgy in the Greek and Roman World.
Thames and Hudson. London. 1978.

178   Hedfors. H.
Compositiones ud Tingenda Musiva. (Codex Lucensis 490.)
Uppsala. 1932.

179   Heim, A.
Coloured Finishes on Copper.
Product Engineering. 135(12). 1956.

180   Hickman, B.
Japanese Crafts; Materials and their Applications.
Fine Books Oriental. London. 1977.

181   Hilpke, H.
The 'Precious' Patina or Cold Patina.
Formenbau und Fertigungstechnik. 85(9). 1960. p 204. (German)

182   Hiorns, A.H.
Metal Colouring and Bronzing.
MacMillan and Co. London and New York. 1892.

183   Hiorns, A.H.
Mixed Metals or Metallic Alloys.
3rd Edition. MacMillan, London. 1912. (1st Edition. 1890.)

184   Hiscox, G.D. (Ed.)
Henley's Twentieth Century Book of Formulas. Processes and Trade Secrets.
Norman W. Henley Publishing Company. New York. 1927. (Avenel Books 1979.)
Originally published as Henley's Twentieth Century Formulas, Recipes and Processes.
1907.

185   Hitchin, J.N.
Protective and Decorative Finishes for Copper and Copper Alloys.
Metal Finishing J. 19(216). 1973. pp 19–20, 22.

186   Hoffman, R.A. and
Hull, R.O.
Black Nickel Coating.
Proceedings of the American Electroplaters Society. 25(45). 1937.

187   Holt, M.C.N. and
Ward, A.M.
The Chemical Colouring of Metals: Some Reactions Involving the Slow Liberation of
Sulphur, Selenium and Tellurium.
Journal of the Electrodepositors Technical Society. 24.1949. pp 33–39.

188   Holt, M.C.N.
and Ward, A.M.
Chemical Colouring: Reactions Involving Liberation of Sulphur, Selenium and Tellurium.
Metal Industry. 74. 1949. pp 186–189.

189   Hubicka, B. et al.
Chemical Brightening of Brass in Solutions Containing $Fe_3{}^+$.
Rudy Metale Niezelazne. 20(6). 1975. pp 311–316. (Polish)

190   Hudson, J.C.
Third Report to the Atmospheric Corrosion Research Committee. (British Non-Ferrous
Metals Research Association)
Transactions of the Faraday Society. 25(177). 1929.

191   Imperial Chemical Industries.
(Metals Division)
Patination of Copper. British Patent 697,294. (n.d.)

192   Imperial Chemical Industries.
(Metals Division)
Inhibition of Copper Corrosion by Surface Treatment with a Solution containing
Benzotriazole. British Patent. 967,086. 19/8/1964.

193   Inami, H.
The Japanese Sword.
Tokyo. 1948.

194   Indira, K., Gowri, S.
and Shenai, B.
Chemical Polishing.
Metal Finishing. 64(11). 1966.

195   Institute of Metals.
Observations on the Natural Patination of Copper.
In—Copper and its Alloys. Institute of Metals Monograph. 34. 1970. pp 353–354.

196   International Copper Research
Association (Inc)
Incralac Laquer Formulation.
Precious Metal Moulding. 23(10). 1965.

197   International Copper Research
Association (Inc)
Interference Colours with Selenous Acid.
I.C.R.A. Report. Research Project No. 51. New York. 1965.

198   Irion, C.E. and Craig, G.L.
Sodium Bicarbonate Production of Green Patina on Copper.
U.S. Patent 1,974,140. 1933.

199   Jackson, H.
Lost Wax Bronze Casting.
Von Nostrand—Reinhold. New York. 1979.

200   Janjua, M., Toguri, J.
and Cooper, W.
Coloured Finishes for Copper and Copper Alloys using Selenous Acid.
Metal Finishing. 65(5). 1967. pp 54–57.

201 Janjua, M. and Toguri, J.
The Nature of Coloured Films Produced on Copper by Selenous Acid.
Can. Met. Quart. 7(3). 1968. pp 133–137.

202 Kagan, R.
Study of Tartrate Solutions for Chemical Copper Plating.
Zaschita Metallov. 8(1). 1972. pp 90–92. (Russian)

203 Kang, T.
Study of Corrosion Films on Bronze Objects from the Koryo-Lee Dynasties.
J. Korean Institute of Metals. 7(1). 1969. pp 15–20. (Korean)

204 Karlbeck, O.
Notes on Some Early Chinese Mirrors.
The China Journal of Science and Arts. 4(1). 1926. pp 3–9.

205 Kermani, M. and Scully, J.
Stress Corrosion in Alpha Brass in Ammoniacal Solutions.
Corrosion Science. 18(10). 1978. pp 883–905.
Corrosion Science. 19(2). 1979. pp 111–122.

206 Kluge, K.
Die Antiken Grossbronzen.
Berlin and Leipzig. 1927.

207 Kramer, O.P.
Metallfärbung.
Eugen G. Leuze Verlag. Saulgau, Wurtenburg. 1954.

208 Krause, H.
Colouring Metals.
Metallwirtschaft. 14. 1935. pp 1015–1017. (German)

209 Krause, H.
Chemical Colouring (Patination) of Metals.
Metalloberflache. 1951. pp B1–B22. (German)

210 Krause, H.
Metallfärbung.
3rd Edition. Munich. 1951.

211 Krause, H.
Metal Colouring and Finishing.
Authorised English Translation of Metallfärbung, 2nd Revised Edition.
Chemical Publishing Co. New York. and E. and F.H. Spon. London. 1938.

212 Krause, H.
Research into the Potassium Permanganate-Copper Sulphate Mordant.
Oberflachentechnik (Coburg). 9. 1933. (German)

213 Kuni-Ichi, T.
Nikon-To No Kagakuteki Kenkyu. (Scientific Study of Japanese Swords) Tokyo. 1953.
(Summary by the author in—Kikai Gakkai Shi. 54. 1918. pp 1–39. (Japanese). English translation of summary article—Bulletin of the Japanese Sword Society of America. 6. 1966. pp 1–20.)

214 Kushner, J.B.
Metal Colouring Theory and Practice.
Metals and Alloys. 12. August 1940. pp 191–192.

215 Kushner, J.B.
Metal Colouring Theory and Practice. Part 1.
Products Finishing. 4(4). 1940 pp 9–11, 14–15, 17.

216 Kushner, J.B.
Metal Colouring Theory and Practice. Part 2.
Products Finishing. 4(5). 1940 pp 32–34, 36, 38.

217 Kushner, J.B.
Metal Colouring Theory and Practice. Part 3.
Products Finishing. 4(6). 1940 pp 46–48, 50.

218 Kushner, J.B.
Metal Colouring Theory and Practice. Part 4.
Products Finishing. 4(7). 1940 pp 34–36, 38–39, 42.

219 Kushner, J.B.
Metal Colouring Theory and Practice. Part 5.
Products Finishing. 4(8). 1940 pp 38–39, 42–47.

220 Kushner, J.B.
Metal Colouring Theory and Practice. Part 6.
Products Finishing. 4(9). 1940.

221 Kushner, J.B.
Metal Colouring Theory and Practice. Part 7.
Products Finishing. 4(10). 1940 pp 14–16, 18, 20, 22.

222 Kushner, J.B.
Metal Colouring Theory and Practice. Part 8.
Products Finishing. 4(11). 1940 pp 26–28, 30, 32.

223 Lacal, R.
Chemical Polishing of Copper.
French Patent. 1,375,201. 2/9/1963.

224 Lacombe, M.S.
Nouveau Manuel Complet de Bronzage des Métaux et du Plâtre.
(Debonliez et Malpeyre, Nouvelle Edition. Revue, Clasée et Augmentée) Encyclopedie Roret. Paris. 1910.

225 Laffineur, R.
L'Incrustation a l'Époque Mycenienne.
L'Antique Classique. 43. 1974. pp 5–37. (French)

226 Lavorko, P.K.
Oxide Coatings of Metals.
(Translation of articles from Oksidnyye Pokrytiya Metallov.)
State Scientific, Technical, Machine, Building Publishing House. Moscow. 1963.

227 Lechat, H.
Sur les Patines des Bronzes Antiques.
Revue Archaeologique, Paris. 1896.

228 Lechtman, H.
Gilding of Metals in Pre-Columbian Peru.
In—W.J. Young (Ed.), Application of Science in the Examination of Works of Art.
Boston. 1973.

229 Lee, W.T.
A Survey of Non-Electrolytic Chromate Treatments and Coatings.
Metal Finishing Journal. 7. May 1961. pp 169–175, 193.

230 Leoni, M.
Copper Alloys—Their Structure and Corrosive Phenomena.
In—The Royal Academy (London), The Horses of San Marco, Venice. 1979.
(Trans. John and Valerie Wilton-Ely from the Italian Edition. Procuratoria di San Marco/Olivetti. 1977.)

231 Leoni, M.
Metallographic Examination of the Horses of San Marco.
In—The Royal Academy (London), The Horses of San Marco, Venice. 1979.
(Trans. John and Valerie Wilton-Ely from the Italian Edition. Procuratoria di San Marco/Olivetti. 1977.)

232 Levey, M.
Chemistry and Chemical Technology of Ancient Mesopotamia.
Philadelphia. 1959.

233 Li Hsun, and Luo Chi-Hsun.
The oxidation of Copper at 200–900°C.
Acta Metallurgica Sinica. 8(3). 1965. pp 311–318.

234 Lilenfield, S. and
White, C.
A Study of the Reaction between Hydrogen Sulphide and Silver.
Journal of the American Chemical Society. 52. 1930. pp 885–892.

235 Limet, H.
Le Travail du Métal au Pays de Sumer au temps de la Troisième Dynastie d'Ur.
Bibliothèque de la Faculté de Philosophie et Lettres de l'Université de Liège.
Paris. 1960.

236 Lindsay, J.
The Origins of Alchemy in Graeco-Roman Egypt.
Muller. London. 1970.

237 Lins, P.A. and
Oddy, W.A.
The Origins of Mercury Gilding.
Journal of Archaeological Science. 2. 1975. pp 365–373.

238 Llopis, J., Gamboa, J.M.
and Arizmendi, L.
Sulphuration of Copper with Polysulphide Solutions, Studied with Radioactive Tracers.
Electrochimica Acta. I(I). 1959. pp 39–57.

239 Lucas, A. and
Harris, J.R.
Ancient Egyptian Materials and Industries.
4th Edition. Edward Arnold. London. 1962.

240 Machine Moderne (Paris)
Comment Colorer les Métaux.
Brochure Publiée de Machine Moderne. Paris. 1923.

241 Machinery's Publishing Co.
(New York)
Metal Colouring and Finishing.
Monograph. New York. 1914.

242 Mackay, J.
The Animaliers. The Animal Sculptors of the 19th and 20th Centuries.
Ward Lock. London. 1973.

243 Maguire, S.
Techniques of Metal Colouring.
Gems. 4(2). 1972. pp 20–22.

244 Maher, M.F. and
Macstoker, F.
Metal Colouring: Brass.
Metal Finishing. 44(44)4. 1946. pp 153–154.

245 Maher, M.F. and
Macstoker, F.
Metal Colouring: Bronze.
Metal Finishing. 44(8). 1946. pp 331–333.

246 Malherbe, P.
Coloration des Métaux.
In—Revue de Physique et de Chimie. Paris. 1901.

247 Marchesini, L. and
Badan, B.
Corrosion Phenomena on the Horses of San Marco.
In—The Royal Academy (London), The Horses of San Marco, Venice. 1979.
(Trans. John and Valerie Wilton Ely from the Italian Edition. Procuratoria di San
Marco/Olivetti. 1977.)

248 Margival, F.
Les Patins de Bronze.
In—Revue de Chimie Industrielle. Paris. 1912.

249 Markovic, T. and
Atlic, E.
Anodic and Cathodic Behaviour of Copper in Aqueous Solutions.
Werkstoffe Und Korrosion. 16(11). 1965. pp 958–962. (German)

250 Marre, F.
Coloration Interferentielle des Métaux.
In—Arts, Sciences, Nature. Paris. 1905.

251 Maryon, H.
Metalwork and Enamelling.
Chapman and Hall. London. 1954.

252 Mason, W.A.
Product for Effecting a Cold Chemical Oxidation of Copper and its Alloys.
U.S. Patent 2878149. 1959.

253 Matsumo, T.
Constituents of Ancient Bronze and the Constitutional Relations between the Original
Alloy and the Patina.
Journal of the Chemical Industry of Japan. XXIV. 1921.

254 Mattiello, J.J. (Ed.)
Protective and Decorative Coatings.
John Wiley and Sons Inc. 1944.

255 Mattson, E.
Properties and Applications of Pre-Patinated Copper Sheet.
Sheet Metal Industry. 45(492). 1968. pp 270–274.

256 Mattson, E. and
Holm, R.
Green Patina on Copper Roofs.
Metallen. 20(1). 1964. pp 17–25. (German)

257 Mattson, E. and
Holm, R.
Green Patina on Copper and Alloys with Alkaline Copper Salts in the Form of a Paste.
U.S. Patent 3,152,927. 13/10/1964.

258 McCann, M.
Artist Beware. The Hazards and Precautions in Working with Art and Craft Materials.
Watson-Guptill. New York. 1979.

259 McKerrell, H. and
Tylecote, R.F.
The Working of Copper-Arsenic Alloys.
Proceedings of the Prehistoric Society. 38. 1972. pp 209–218.

260 McMullen, A.L.
Architectural Metalwork in Copper and its Alloys.
Copper Development Association Publication No. 63. London. 1963.

261 McTigue, P.T.
Phototarnishing of Silver and Copper.
Australian Journal of Chemistry. 18(11). 1965. pp 1851–1853.

262 Merriman, A.D.
A Dictionary of Metallurgy.
MacDonald and Evans. London. 1958.

263 Meyer, A.
Colouring Processes for Copper and Copper Alloys.
Pro-Metal. 71. 1959. pp 203–217. (French and German)

264 Meyer, W.R.
Cleaning and Preparation of Metals for Plating.
Transactions of the Institute of Metal Finishing. 31. 1955. pp 290–292.

265 Meyer, W.R.
Metal Colouring.
Metals and Alloys. 21(1). 1944. p 132.

266 Michel, J.
Coloration des Métaux.
(1st Edition. 1912) Revised and Enlarged Edition. Librairie Des Forges. Paris. 1931.

267 Mills, J.W. and
Gillespie, M.
Studio Bronze Casting.
New York and London. 1969.

268 Moran, S.F.
The Gilding of Ancient Bronze Statues in Japan.
Artibus Asiae. 31. 1969. pp 55–65.

269 Muir, G.D. Hazards in the Chemical Laboratory. 2nd. Edition.
The Chemical Society. London. 1977.

270 Nagago, T. Asiya-Kei No Kama. (Tea Kettles of the Asiya District)
Kyoto. 1957.

271 Nelson, G.A. Corrosion Data Survey.
3rd Revised Edition. Shell Development Company. Eneryville, California. 1960.

272 Nichols, H.W. Restoration of Ancient Bronzes and the Cure of Malignant Patina.
Field Museum of Natural History. Chicago. Museum Techniques Series No. 3. 1930.

273 Niiyama, E. Chokin Tankin No Giho I.
Nihon Kinko Sattka Kyokai. Tokyo. 1969.

274 Nippon Soda Co. Ltd. Black Coating on Zinc and Zinc Alloys.
Japanese Patent 1096/65. 21/1/1965.

275 Nowak, J., Lambertin, M. Brass Sulphidation by Hydrogen Sulphide and Sulphur Vapour at Low
and Colson, J. Pressure—Morphological and Kinetic Aspects.
Corrosion Science. 17(7). 1977. pp 603–613.

276 Oddy, W.A. et al. The Gilding of Bronze Statues in the Greek and Roman World.
In—The Royal Academy (London), The Horses of San Marco, Venice. 1979.
(Trans. John and Valerie Wilton-Ely from the Italian Edition. Procuratoria di San
Marco/Olivetti. 1977.)

277 Ostrander, C.W. Chromate Treatments.
Plating. 38(10). 1951. pp 1033–1035.

278 Pearson, G. Observations on Ancient Metallic Arms.
Philosophical Transactions. London. Read 6th June 1796.

279 Perring, J.W. Chemical Colouring of Metals.
Journal of the Electrodepositors Technical Society. 11. 1936. pp 75–86.

280 Phillip, T. Mappae Clavicula.
Archaologica. 32. 1847. pp 183–244.

281 Pijanowski, H.S. and Lamination of Non-Ferrous Metals by Diffusion: Adaptations of the Traditional Japanese
Pijanowski, E. Technique of Mokume-Gane.
Society of North America Goldsmiths. Goldsmiths Journal. 3(4). 1977.

282 Pijanowski, H.S. and Update: Mokume-Gane.
Pijanowski, E. Society of North American Goldsmiths. Goldsmiths Journal. 1979.

283 Pinnel, M.R., Tompkins, H.G. Oxidation of Copper in Controlled Clean Air and Standard Laboratory Air at 50–150°C.
and Heath, D.E. Appl. Surf. Sci. 2(4). 1979. pp 558–577.

284 Planiscig, I. Piccoli Bronzi Italiani del Renascimento.
Treves. Milan. 1930.

285 Plenderleith, H.J. Conservation of Antiquities and Works of Art.
Oxford University Press. 1966.

286 Plenderleith, H.J. Metals and Metal Technique.
In—C.L. Wooley (Ed), The Ur Excavations. Vol. II. London. 1934.

287 Plenderleith, H.J. Technical Notes on Chinese Bronzes with Special Reference to Patina and Incrustation.
Transactions of the Oriental Ceramic Society. 6. 1938–1939. pp 33–55.

288 Plinius Secundus. Natural History.
(Pliny the elder) Trans. H. Rackham. Harvard University Press. 1967.

289 Pollock. G.F. Chemical Blacking.
Products Finishing (London). 31(7). 1978. pp 16–17.

290 Pope, D. and Moss, R.L. The Effect of Sulphide Inclusions on the Sulphidising of Silver Foil.
Corrosion Science. 5(11). 1965. pp 773–777.

291 della Porta, G.B. Magiae Naturalis Libri Viginti.
Naples. 1589. (English Translation—London. 1658.)

292 Proctor, R.P.M. The Formation of Cuprous Oxide Film on Alpha-Brass Stress-Corrosion Fracture Surfaces.
Corros. Sci. Aus. 15(6–7). 1975. pp 349–359.

293 Proctor, R.P.M. and Fracture of Tarnish Films On Alpha-Brass.
Islam, M. Corrosion. 32(7). 1976. pp 267–270.

294 Pumpelly, R. Notes On Japanese Alloys.
American Journal of Science. 42. 1866. pp 43–45.

295 Rabate, H. Laquering Copper and its Alloys.
Trav. Peint. 20(5). 1965. pp 151–153. (French)

296 Rathgen, F. Die Konservirung Von Altertunsfunden.
Berlin. 1898.
(English translation—The Preservation of Antiquities. Cambridge University Press 1905)

297 Rein, J.J. The Industries of Japan.
New York. 1889.

298 Rich, J.C. The Materials and Methods of Sculpture.
Oxford University Press. New York. 1947.

299 Roast, H.J. Cast Bronze.
The American Society for Metals. Cleveland, Ohio. 1953.

300 Roberts, E. et al. Some Observations on the Films Formed on Alpha Brasses Tarnished in an Ammoniacal
Solution of Copper Sulphate.
Corrosion Science. 14(5). 1974. pp 307–320.

301 Roberts-Austen, W.A. Cantor Lectures on Alloys: Colours of Metals and Alloys Considered in Relation to their
Application to Art.
Journal of the Society of Arts. 36. 1888. pp 1137–1146.

302 Roberts-Austen, W.A. Cantor Lectures on Alloys: Application of Alloys in Metalwork.
Journal of the Society of Arts. 41. 1893. pp 1022–1043.

303   Roberts-Austen, W.A.   On the Liquation, Fusibility and Density of Certain Alloys of Copper and Silver.
                                       Proceedings of the Royal Society. 23. 1875. pp 481–495.

304   Robinson, B.W.   The Arts of the Japanese Sword.
                                       London. 1970.

305   Robinson, B.W.   Arms and Armour of Old Japan.
                                       HMSO. London. 1951.

306   Robinson, H.R.   A Short History of Japanese Armour.
                                       HMSO. London. 1965.

307   Ronnquist, A. and   The Oxidation of Copper: A review of Published Data.
       Fischmeister, H.   Journal of The Institute of Metals. 89(2). 1960. pp 65–76.

308   Rossetti, V.A. and   Examinations of the Patina and Incrustations. (Horses of San Marco)
       Marabelli, M.   In—The Royal Academy (London), The Horses of San Marco, Venice. 1979.
                                       (Trans. John and Valerie Wilton-Ely from the Italian Edition. Procuratoria di San
                                       Marco/Olivetti. 1977.)

309   Roy, S.K. and   Tarnishing Behaviour of Copper in Humid Atmospheres.
       Sircar, S.C.   Br. Corros. J. 13(4). 1978. pp 191–192.

310   Royal Academy, London.   The Horses of San Marco. Venice.
                                       Royal Academy. London. 1979. (Trans. John and Valerie Wilton Ely from the Italian
                                       Edition. Procuratoria di San Marco/Olivetti. 1977.)

311   Saeger, K.E. and   The Colour of Gold and its Alloys.
       Rodies, J.   Gold Bulletin. 10(1). 1977. pp 10–14.

312   Safranek, W.H.   Coloured Finishes for Copper and Copper Alloys.
                                       Plating. 55(12). 1968. pp 1257–1260.

313   Sakurai, Y.   Colouring Copper Plate.
                                       Japanese Patent. 4443/66. 12/3/1966.

314   Sanderson, M.D.   The Initial Stages of Oxide Film Formation on Copper and Some Copper Alloys.
                                       Corrosion. 25(7). 1969. pp 291–299.

315   Savage, E. and   The Techniques of the Japanese Tsuba Maker.
       Smith, C.S.   Ars Orientalis. 11. 1979. pp 291–328.

316   Savage, G.   Concise History of Bronzes.
                                       Thames and Hudson. London. 1968.

317   Sax, N.I.   Dangerous Properties of Industrial Materials.
                                       5th Edition. Van Nostrand Reinhold. New York. 1979.

318   Schikorr. and   Tarnishing of Silver, Copper and Copper-Zinc Alloys in the Presence of Cardboard and
       Volz.   Paper.
                                       Metall. 14(11). 1960. pp 1073–1076. (German)

319   Schmerling, G.   Emulsion Cleaning of Metals.
                                       Journal of the Electrodepositors Technical Society. 28. 1952. pp 179–186.

320   Scientific American.   The Colouring of Metals.
                                       Scientific American Cyclopedia of Formulas. pp 433–449. New York. 1939.

321   Sease, C.   Benzotriazole: A Review for Conservators.
                                       Studies in Conservation. 23. 1978. pp 76–85.

322   Sharma, S.P.   Atmospheric Corrosion of Silver, Copper and Nickel—Environmental Test.
                                       J. Electrochem. Soc. 125(12). 1978. pp 2005–2011.

323   Shaw, R.E.   Surface Preparation and Surface Coating.
                                       In—Institution of Metallurgists, Metal Finishing Review Course. Series 2. No. 8. 1972. pp
                                       8–16.

324   Shearman, W.M.   Metal Alloys and Patinas for Castings; for Metalsmiths, Jewellers and Sculptors.
                                       Kent State University Press. 1976.

325   Sheffield City Libraries.   Select Bibliography on the Colouring of Metals.
                                       INFRA. Research Bibliographies. New Series. No 28. Sheffield City Libraries.
                                       2nd Edition. 1948.

326   Shenoi, B.A. et al.   Ultrasonics in Metal Finishing.
                                       Metal Finishing. 68(8). 1970. pp 57–60. and Metal Finishing. 68(9). 1970. pp 56–59.

327   Shreir, L.L, (Ed)   Corrosion.
                                       George Newnes Ltd. London. 1963.

328   Silman, H.   Chemical and Electroplated Finishes.
                                       1st Edition. Chapman and Hall. London. 1948.

329   Silman, H., Isserlis, G.   Protective and Decorative Coatings for Metals.
       and Averill, A.F.   Finishing Publications Ltd. Teddington. UK. 1978.

330   Simonds, H.R. and   Colouring of Metals. 1–4.
       Young, C.   Iron Age. 26th March. 16th April. 18th June, 3rd September. 1936.

331   Smith, B.U.   Black Oxide Finish Fights Corrosion, Reduces Friction on Most Metals.
                                       Western Metalworking. 17(2). 1959. pp 46–47.

332   Smith, C.S.   Biringuccio's Pirotechnia—A Neglected Italian Metallurgical Classic.
                                       Mining and Metallurgy. 21. 1940. pp 189–192.

333   Smith, C.S.   Corrosion of an Ancient Bronze and an Ancient Silver Alloy—A Discussion.
                                       In—T.N. Rhodin (Ed), Physical Metallurgy of Stress Corrosion Fracture. pp 293–294.
                                       Interscience. New York. 1959.

334   Smith, C.S.   Decorative Etching and the Science of Metals.
                                       Endeavour. 16. 1957. pp 199–208.

335   Smith, C.S.   The Equilibrium Diagram of the Copper-rich, Copper-Silver Alloys.
                                       Transactions of the American Institute of Mining and Metallurgical Engineers. 99. 1932. pp
                                       101–114.

336   Smith, C.S.   An Examination of the Arsenic-rich Coating on a Bronze Bull.
                                       In—W.J. Young. Application of Science in the Examination of Works of Art. Boston. 1973.

BIBLIOGRAPHY

337 Smith, C.S.
Historical Notes on the Colouring of Metals.
In—A. Bishay (Ed) Recent Advances in the Science and Technology of Materials. Vol. III.
pp 157–167. Plenum. New York. 1974.

338 Smith, C.S.
A History of Metallography.
University of Chicago Press. 1960.

339 Smith, C.S.
Metallographic Study of Early Artefacts made from Native Copper.
Actes du XI Congrès International d'Histoire des Sciences. Vol 6. pp 237–252. Warsaw.
1968.

340 Smith, C.S.
Metallurgy of the Rennaissance.
Technology Review. 43. 1941. pp 155–157, 174–177.

341 Smith, C.S.
Notes on the Crystal Structure of Copper-Gold Alloys.
Mining and Metallurgy. 9. 1927. pp 458–459.

342 Smith, C.S.
A Search for Structure.
MIT. Press. Cambridge, Mass. London, England. 1981.

343 Smith, C.S.
Sectioned Textures in the Decorative Arts.
In—Stereology. Proceedings of the Second International Congress for Stereology, Chicago.
Springer-Verlag. New York. 1967.

344 Smith, C.S.
Seventeenth Century Brassmaking.
Metal Industry. 56. 1940. pp 285–287.

345 Smith, C.S.
Speculations on the Corrosion of Metals.
Archaemetry. 18. 1976. pp 114–116.

346 Smith, C.S. and
Gnudi, M.T.
The Pirotechnia of Vannoccio Biringuccio.
American Institute of Mining and Metallurgical Engineers and Yale University Press.
New York. 1942. New edition. MIT Press. 1966.

347 Smith, C.S. and
Hawthorne, J.
Mappae Clavicula: A Little Key to Medieval Techniques.
American Philosophical Society. Philadelphia. 1974.

348 Smith, C.S. and
Hawthorne, J.
Notes on a Japanese Magic Mirror.
Archives of the Chinese Art Society of America. 17. 1963. pp 23–25.

349 Smith, C.S. and
Hawthorne, J.
On Divers Arts: The Treatise of Theophilus.
University of Chicago Press. 1963.

350 Smith, C.S. and
Sisco, A.G.
Lazarus Etcher's Treatise on Ores and Assaying.
Trans. from the German Edition of 1580. University of Chicago Press. 1951.

351 Smith, G.
Laboratory and School of Arts.
London. 1740.

352 Smith, M. and
Stead, D.
Corrosion Problems Associated with the Use of Copper and Copper Alloys.
Australasian Corrosion Engineering. 8(9). 1964. pp 9–14.

353 Sozin, Y, and
Gorbachuck, G.
The Mechanism of Oxide Film Formation on the Surface of Electroplated Copper.
Zhurnal Fizicheskoi Khimii. 37(4). 1963. (Russian)

354 Spahn, H. and
Wells, F.
Chemical Polishing of Copper.
Canadian Patent 659,626. 19/3/1963.

355 Spon, E. and Spon, F.H.
Workshop Receipts.
Fifth Series. London. 1892. (Metallurgy). 2nd revised edition, London 1906. (Vol. 3
contains recipes for metal colouring.)

356 Stambalov. T.
The Corrosion and Conservation of Metallic Antiquities and Works of Art.
A Preliminary Study.
Central Research Laboratory for Objects of Art and Science. Amsterdam. (nd)

357 Starace, G.
Decorative Patinas on Copper and its Alloys.
Trattimenti e Finiture. 4(3). 1964. pp 100–102. (Italian)

358 Steel, M.C.
Factors in the Corrosion of Brasses.
Australasian Corrosion Engineering. 7(1). 1963.

359 Stefanskii, I. et al.
Bath for Etching the Contact Surfaces of Copper and Bronze Bi-Metals.
Metall. Gordnorudn. Prom-St. 3. 1978. pp 78–79. (Russian)

360 Steere, N.V.
Handbook of Laboratory Safety. 2nd Edition.
Chemical Rubber Company. Cleveland, Ohio. 1971.

361 Stone, G.C.
A Glossary of the Construction, Decoration and Use of Arms and Armour in all Countries
and at All Times, together with some Closely Related Subjects.
Portland, Maine. 1934.

362 Tanimura, H.
Development of the Japanese Sword.
J. Metals. 32(2). 1980. pp 63–73.

363 Tavernor-Parry, J.
Dinanderie.
London. 1910.

364 Theophilus
De Diversis Artibus.
Text with Trans. C.R. Dodwell. London. 1961.
(Translation with Technical Notes. J.H. Hawthorne and C.S. Smith. Chicago. 1963.)

365 Thews, E.R.
The Production of Coloured Gold Finishes.
Metal Finishing. 48(9). 1951. pp 80–85.

366 Toguri, J.M., Janjua, M.B.
and Cooper, W.C.
Development of Superior Coloured Finishes for Copper and Copper Alloys.
Final Report. International Copper Research Association. 1965.

367 Toguri, J.M., Janjua, M.B.
and Cooper, W.C.
Superior Coloured Finishes for Copper and Copper Alloys.
Electrodeposition and Surface Treatment. 1(1). 1972. pp 77–102.

368 Tong, H.
Stress Corrosion Cracking of an Alpha Phase Copper Alloy in Ammoniacal Solutions.
University of Illinois. Diss. Abstract. Int. 38(10). 1978. p 171.

369 Turner, G.L.
A Magic Mirror of Buddhist Significance.
Oriental Art. 12. 1966. pp 1–5.

370 Tylecote, R.F.
A History of Metallurgy.
The Metals Society. London. 1976.

371   Uno, D.   Japanese Alloys and Colouring.
Korrosion. J. Metallschutz. 6. 1929. (German)

372   Untracht, O.   Metal Techniques for Craftsmen.
Robert Hale. London. 1969.

373   Upton, N. and Walton, C.   Ultrasonic Cleaning Techniques for the Precious Metal Manufacturing Industry.
Worshipful Company of Goldsmiths. Technical Advisory Committee. Special Report 31. London. 1977.

374   Vaders, E.   Corrosion of Copper-Zinc Alloys by Sea-water and Chemical Solutions.
Metall. 16(12). 1962. (German)

375   Vasari.   Vasari on Technique.
Trans. L.S. Maclehose. G.B. Brown (Ed). Dent, London. 1907. Reprinted, New York, 1960.

376   Vernon, W.H.J.   First Report to the Atmospheric Corrosion Research Committee. British Non-Ferrous Metals Research Association.
Transactions of the Faraday Society. 19. 1924.

377   Vernon, W.H.J.   Second Report to the Atmospheric Corrosion Research Committee. British Non-Ferrous Metals Research Association.
Transactions of the Faraday Society. 23. 1927.

378   Vernon, W.H.J. and Whitby, L.   The Open Air Corrosion of Copper 1. A chemical Study of the surface Patina
Journal of the Institute of Metals. 42(2). 1929. pp 181–202.

379   Vernon, W.H.J. and Whitby, L.   The Open Air Corrosion of Copper. 2. The Mineralogical Relationships of Corrosion Products.
Journal of the Institute of Metals. 44(2). 1930. pp 389–408.

380   Vernon, W.H.J. and Whitby, L.   The Open Air Corrosion of Copper. 3. Artificial Production of Green Patina.
Journal of the Institute of Metals. 49(2). 1932. pp 153–167.

381   Vevers, H. and Gardam, G.   Lacquering to Obtain Brilliance and Metallic Lustre.
Journal of the Electrodepositors Technical Society. 28. 1952. pp 179–186.

382   Villenoisy, M.   La Patine du Bronze Antique.
Revue Archaeologique. Paris. 1896.

383   Villon, M.   Bronzage.
In—Dictionaire de Chimie Industrielle. Paris. 1898.

384   Villon, M.   Metallochromie.
In—Villon. Dorure et Argenture. Paris. 1903.

385   Vittori, O.   Pliny the Elder on Gilding: A New Interpretation of His Comments.
Gold Bulletin. 12. 1979. pp 35–39.

386   Wakayama, H.   To-Sogu Kodogu Koza. (Japanese Sword Furniture: Terminology)
Tokyo. 1974.

387   Watson, W.   Ancient Chinese Bronzes.
Faber. London. 1962.

388   Watanabe, M.   The Japanese Magic Mirror.
Archives of the Chinese Art Society of America. 19. 1965. pp 45–51.

389   Werner, E.   Electrolytic and Chemical Metallic Deposits and Metal Colouring.
Werkstattskniffe. Vol. 4–5. 5th Edition. Carl Langer Verlag. Munich. 1959.

390   Wesley, W.A. and Knapp, B.B.   Chloride Solution for Blackening Copper.
U.S. Patent 2,844,530. 22/7/1958.

391   Weymuller, C. and Siegrist, F.L.   Metal Colouring.
Metal Progress. 55. 1965.

392   Wiederholt, W.   The Atmospheric Corrosion of Copper and Copper Alloys.
Werkstoffe und Korrosion. 15(8). 1964. pp 633–644. (German)

393   Wiederholt, W.   The Chemical Surface Treatment of Metals.
Robert Draper Ltd. Teddington, UK. 1965.

394   Wiederholt, W.   Survey of the Resistance of Copper to Attack by Chemicals in Practical Use.
Kupfer Mitteilungen. 9. 1960. pp 309–321. (German)

395   Williams, C.R.   Gold and Silver Jewellery and Related Objects.
New York Historical Society. New York. 1924.

396   Wilson, H.   Silverwork and Jewellery.
Revised and Enlarged Edition. Pitman. London. 1948.

397   Wolff, M.   Patinage des Métaux.
In—Revue de Chimie Industrielle. Paris. 1897.

398   Wolynec, S. and Gabe, D.R.   Film Dissolution during Sulphide Tarnishing of Copper in Aqueous Solutions of Thiourea and Alkaline Polysulphide.
Corrosion Sci. 12(5). 1972. pp 437–450.

399   Wyatt, M. Digby   Metalwork and its Artistic Design.
London. 1852.

400   Yates, E.L.   The Change in Colour of a Silver-Gold Alloy.
Australian Journal of Physics. 16(1). 1963. pp 40–46.

401   Yetts, W.P.   Problems of Chinese Bronzes.
Journal of the Royal Central Asian Society. 18(3). 1931. pp 399–402.

402   Yoon, S.R.   Investigation of the Patination of Copper in Acidic Copper Sulphate Solution.
J. Metal Finishing Society Korea. 5(3). 1972. pp 77–85. (Korean)

403   Young, C.B.F.   Metal Colouring.
Transactions of the American Society of Metals. 26(12). 1938. pp 1019–1034.

404   Zheugelis, M.   Études des Bronzes Antiques.
Chimie et Industrie. Paris. 1930.